HVAC

Design Criteria, Options, Selection

William H. Rowe III, AIA, PE

HVAC

Design Criteria, Options, Selection

William H. Rowe III, AIA, PE

Illustrated by Carl W. Linde

R.S. MEANS COMPANY, INC.
CONSTRUCTION CONSULTANTS & PUBLISHERS
100 Construction Plaza
P.O. Box 800
Kingston, Massachusetts 02364-0800
(617) 585-7880

This book was edited by Dianne Harrison, John Moylan, and Julia Willard. Typesetting was supervised by Helen Marcella. The book and jacket were designed by Norman R. Forgit. Illustrations by Carl Linde.

Printed in the United States of America

10 9 8 7 6 5 4

Library of Congress Catalog Card Number 88-194511

ISBN 0-87629-102-7

To Mary Blade
Professor of Mechanical Engineering
The Cooper Union for the Advancement of Science and Art

Thanks for teaching the arts of science.

TABLE OF CONTENTS

PART II — EQUIPMENT SELECTION

TABLE OF FIGURES

Figure

FOREWORD

This book is designed to be used as a guide in the selection and analysis of HVAC systems. A simple step-by-step process is used to determine the most appropriate overall type of system for a project and to estimate the sizes (loads) of the various system components. This process includes choosing the appropriate generation equipment, distribution piping and/or ductwork, necessary accessories, controls, and terminal units.

The book is divided into three sections. Part I is a review of basic HVAC systems and principles, such as heating and cooling loads, codes, and selection criteria. Part II provides descriptions of the system components that supply the heating and cooling for a building. Equipment data sheets at the end of each chapter in this section summarize the important characteristics of each piece of equipment in an HVAC system. In Part III, the selection of appropriate HVAC systems for two complete projects— multi-family housing and a commercial office building— are analyzed in detail.

This book is the first in a series of titles from R.S. Means Company, Inc. adressing specific major components of building construction. Each title will explore a building "system" and will include discussions of the materials available, the methods of installation, and provide information for the selection of the most appropriate type of components. These books are useful to the facilities manager and the contractor, as well as to the architect, designer, and engineer.

Author's Note

As both a practitioner and an educator in architecture and engineering, I have tried to combine the two fields. In design, the architect is more than a general synthesizer of concepts; and the engineer's experience in many building types goes beyond specific systems design. In the end, a design must function well on both a large and a small scale. A building's environment is composed of both its image and its reality — and the challenge of designing is to never stop trying to meet and match both parts. In those cases where the owner also understands and encourages a dedicated design team, the formula for success is made. This book tries to make clear what the choices are and how to make them, so that the final design is as complete as possible.

Acknowledgments

For use of valuable copyright information

- The American Society of Heating, Refrigerating and Air Conditioning Engineers, Inc.
- The Carrier Corporation, McGraw-Hill Book Company
- ITT—Bell and Gossett
- R. S. Means Co., Inc.

For taking time to help and teach

- John Moylan, Dianne Harrison, Julia Willard, and Ted Wetherill — editors, R. S. Means Co., Inc.
- Rich Scogland, Bob Smith, and the membership of the Boston A.S.H.R.A.E. chapter
- John and Paul Kennedy — contractors, P.J. Kennedy and Son
- Joe Walsh and Kevin Cotter — officers, Plumbers and Gasfitters Union, Local No. 12
- Michael O'Connell — The Gutierrez Corporation
- Taichi Ohtaki—Director of Asset Management Asahi Life Insurance Co.

For encouragement when it was needed most

- Irene and Earl Morse
- Bonnie Wilson
- June Rowe

PART ONE

INITIAL DESIGN CONSIDERATIONS

In order to design, price, and lay out an HVAC system, it is necessary to understand heating, ventilating, and air conditioning systems, the calculation of heating and cooling loads, and the codes and regulations to be observed. Part 1 addresses these initial design considerations. In this section of the book, an overview of heating and cooling systems and how they operate, and the basic components that comprise a system, from generation through termination, are explained for heating, cooling, and ventilating systems. The procedures for selection and quantitative analyses provide the basis for the rest of the book.

CHAPTER ONE

BASIC HVAC SYSTEMS

Nature does not produce an environment that is always ideally comfortable to man. Excessive wind, rain, heat, humidity, and cold are undesirable; and mechanical systems are used to create more acceptable climates by heating, ventilating, and/or air conditioning (HVAC). An understanding of the fundamentals of these functions is necessary in order to select appropriate HVAC systems.

Heating is required in a building when the ambient temperatures are low enough to demand additional warmth for comfort. Boilers or furnaces *generate* the heat for a building; pipes or ducts *distribute* the heat; and convectors/radiators or diffusers are the terminal units that *deliver* the heat. A typical hydronic (hot water or steam) heating system is shown in Figure 1.1.

Cooling systems utilize a refrigeration cycle or other heat rejection method to supply cool air to occupied spaces. Chilled water, cool air, or refrigerant is distributed by pipes and ducts throughout the building to terminal units (diffusers or fan coils). These end units deliver the cooling to the desired spaces.

While cooling is rarely required by code, it is almost universally expected in commercial environments. Cooling systems may be independent of heating systems (such as a simple window air conditioner) or integrated with the heating system (such as a rooftop unit). Examples of cooling systems are shown in Figure 1.2.

Ventilating systems operate to provide fresh outdoor air to minimize odors and to reduce unhealthy dust or fumes. In many spaces, simple operable windows satisfy ventilation requirements. On the other hand, ventilation may be provided to a building by exhaust fans or fresh air intakes. Some of these methods are illustrated in Figure 1.3.

Air conditioning usually combines all of the features of heating, cooling, and ventilating systems, and may also provide additional "conditioning" of the overall environment, such as noise control, air cleaning (filtration), humidity control, and energy-efficient controls (free-cooling options). A basic rooftop HVAC unit, illustrated in Figure 1.4, is a common example of air conditioning.

The process from generation to distribution to terminal units, as shown in Figure 1.5, is common to all systems. **Generation** equipment produces heat (heating) or removes heat (cooling) to or from the building. Boilers, furnaces, or supplied steam add heat; cooling towers, chillers, or heat pumps reject heat. The equipment for generation systems is the most expensive component of the HVAC system and is generally located in the mechanical equipment room. (See Chapters 5 and 6 for more information on generation and generation systems.)

The warm air or cold water that the various pieces of generation equipment produce is then *distributed* throughout the building. A distribution system basically consists of pipes (water) or ducts (air) that take the heated or cooled medium from the equipment that generated it, through the building to the terminal unit. In

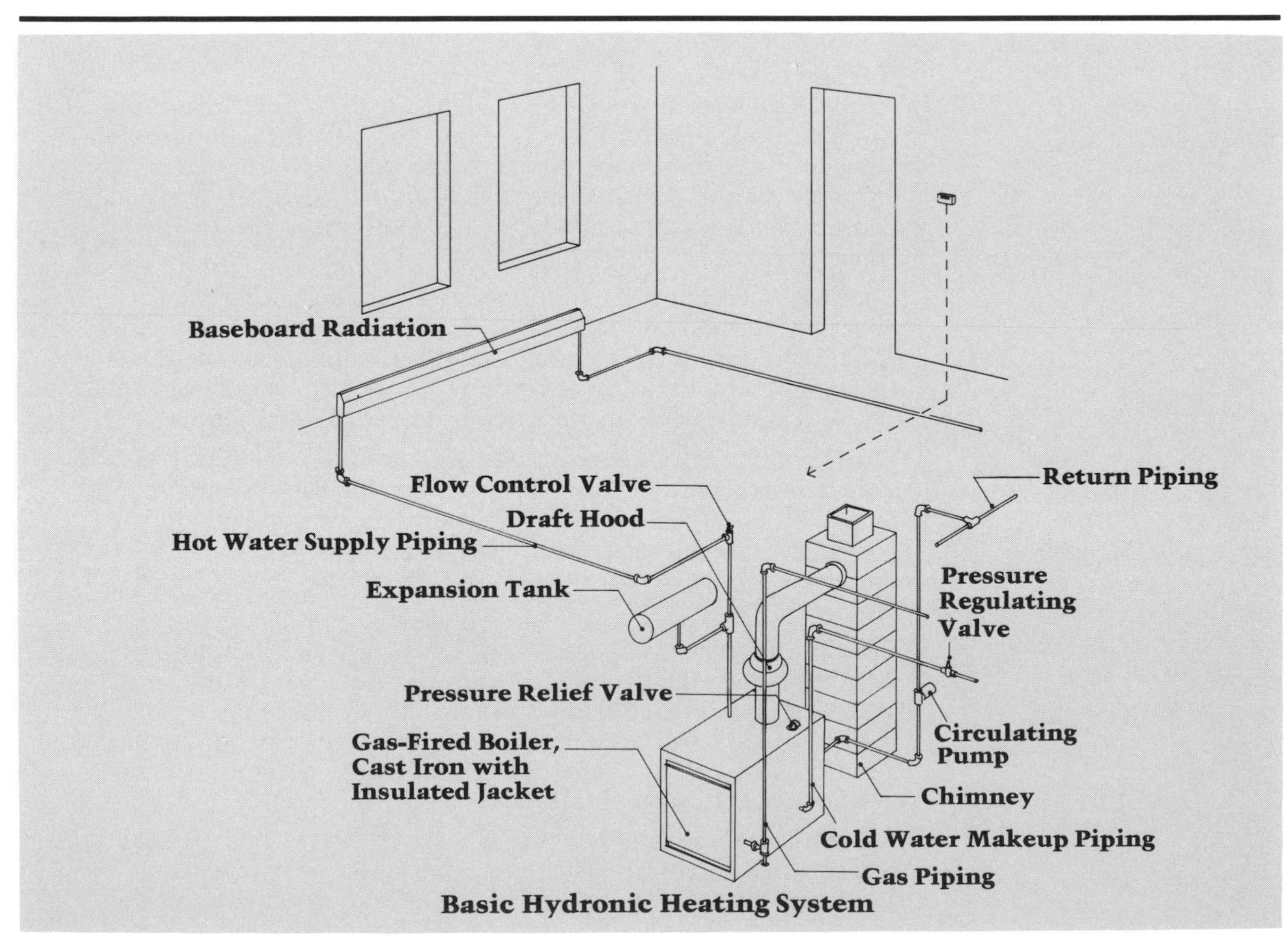

Figure 1.1

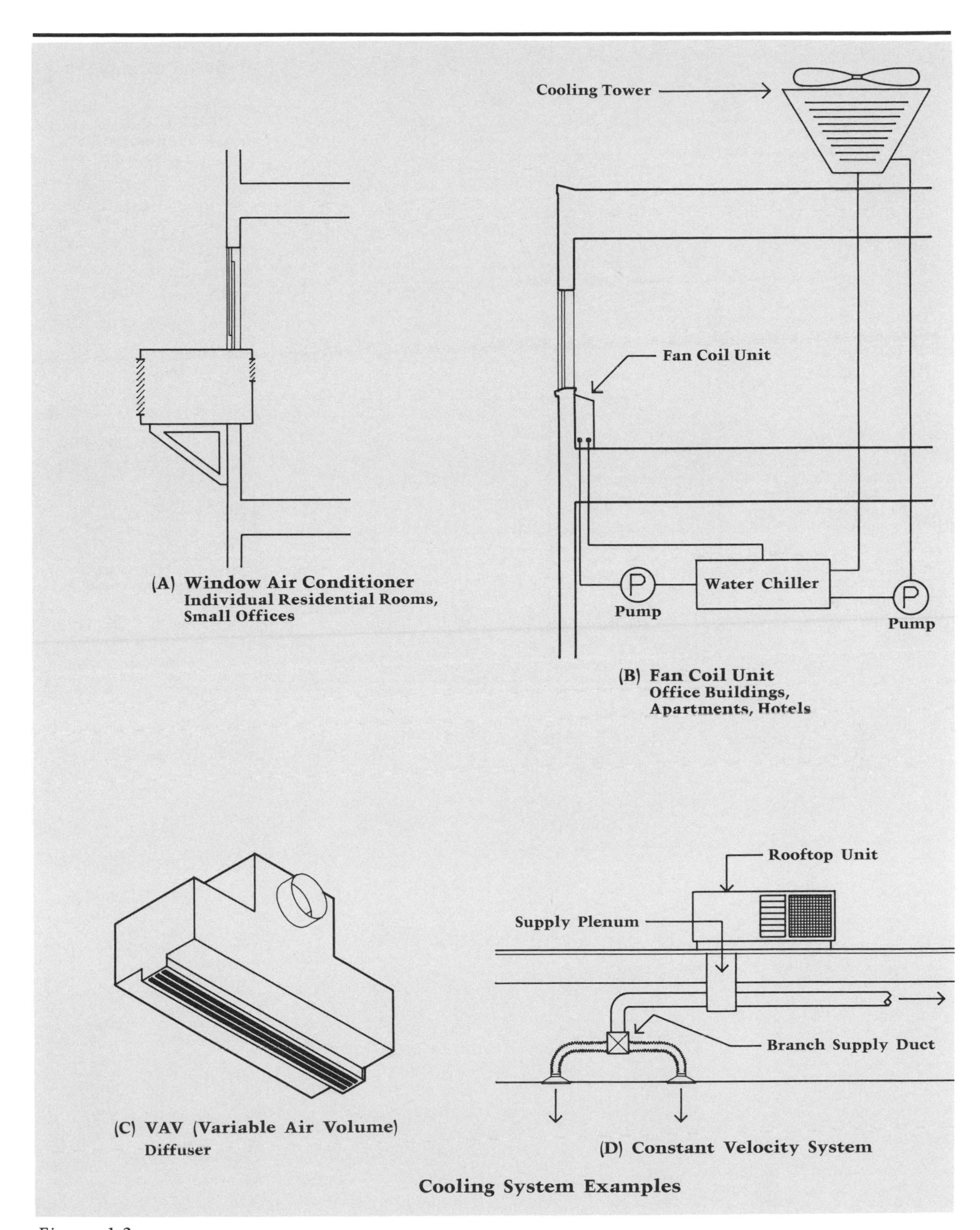
Cooling Tower
Fan Coil Unit
Water Chiller
Pump
Pump
(A) Window Air Conditioner
Individual Residential Rooms,
Small Offices
(B) Fan Coil Unit
Office Buildings,
Apartments, Hotels
Rooftop Unit
Supply Plenum
Branch Supply Duct
(C) VAV (Variable Air Volume)
Diffuser
(D) Constant Velocity System
Cooling System Examples

Figure 1.2

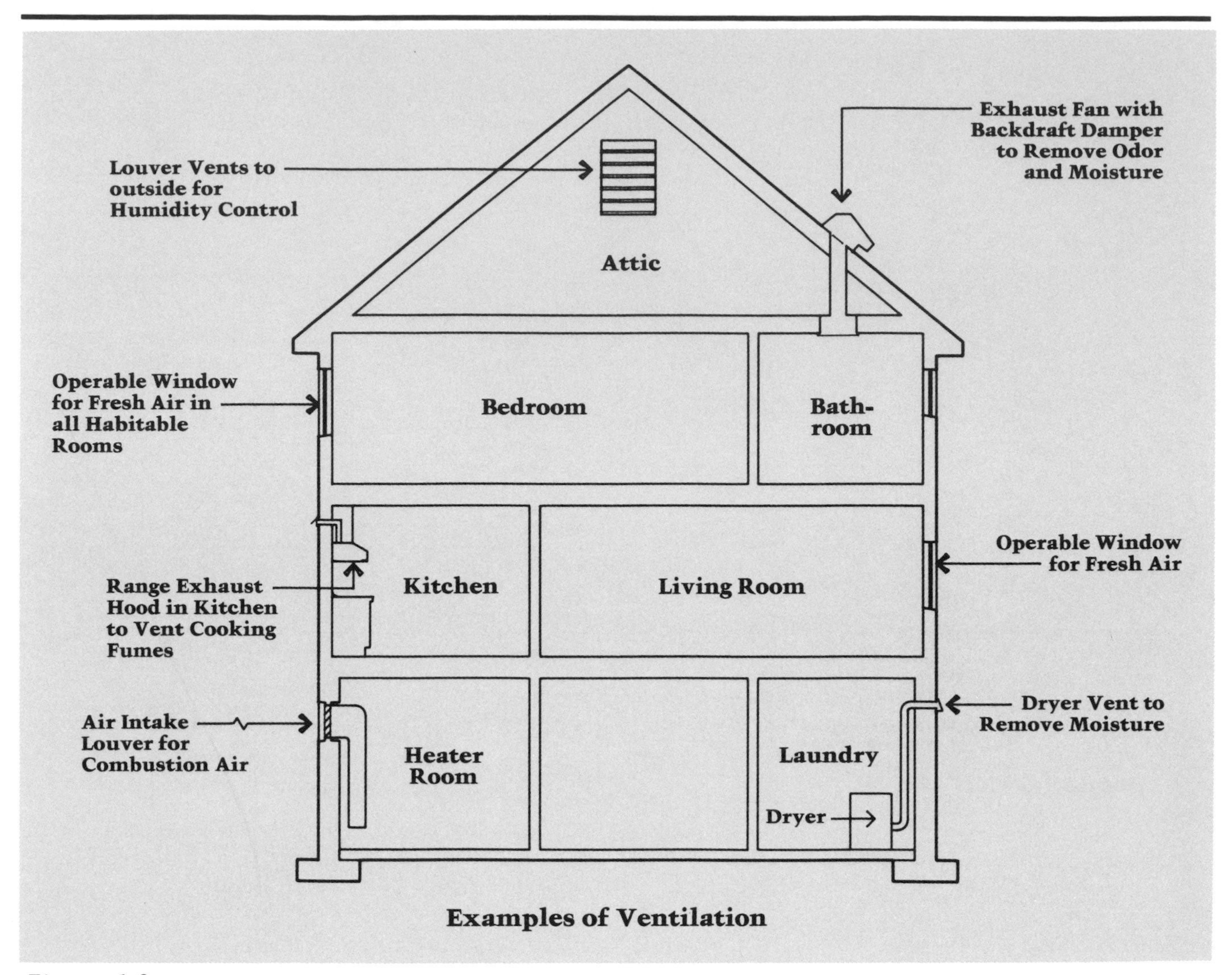
Louver Vents to outside for Humidity Control
Exhaust Fan with Backdraft Damper to Remove Odor and Moisture
Attic
Operable Window for Fresh Air in all Habitable Rooms
Bedroom
Bath-room
Operable Window for Fresh Air
Range Exhaust Hood in Kitchen to Vent Cooking Fumes
Kitchen
Living Room
Air Intake Louver for Combustion Air
Heater Room
Laundry
Dryer Vent to Remove Moisture
Dryer
Examples of Ventilation

Figure 1.3

addition, a distribution system may have valves, dampers, and fittings. (Distribution is further explained in Chapter 9.)

The **terminal units**, located in the conditioned spaces, include convectors (radiators), air diffusers, and fan coil units. These units receive the air or water from the system and utilize it to warm or cool the air in the space. (Terminal units are discussed in detail in Chapter 10.)

Heating

The most common heating systems warm a space by obtaining heat from a source and moving it through the building until it eventually becomes warm air in the space. Purchased steam from a utility company and oil or gas boilers located in a building's boiler room are typical sources of heat. The flow of heat from source to termination is shown in Figure 1.5.

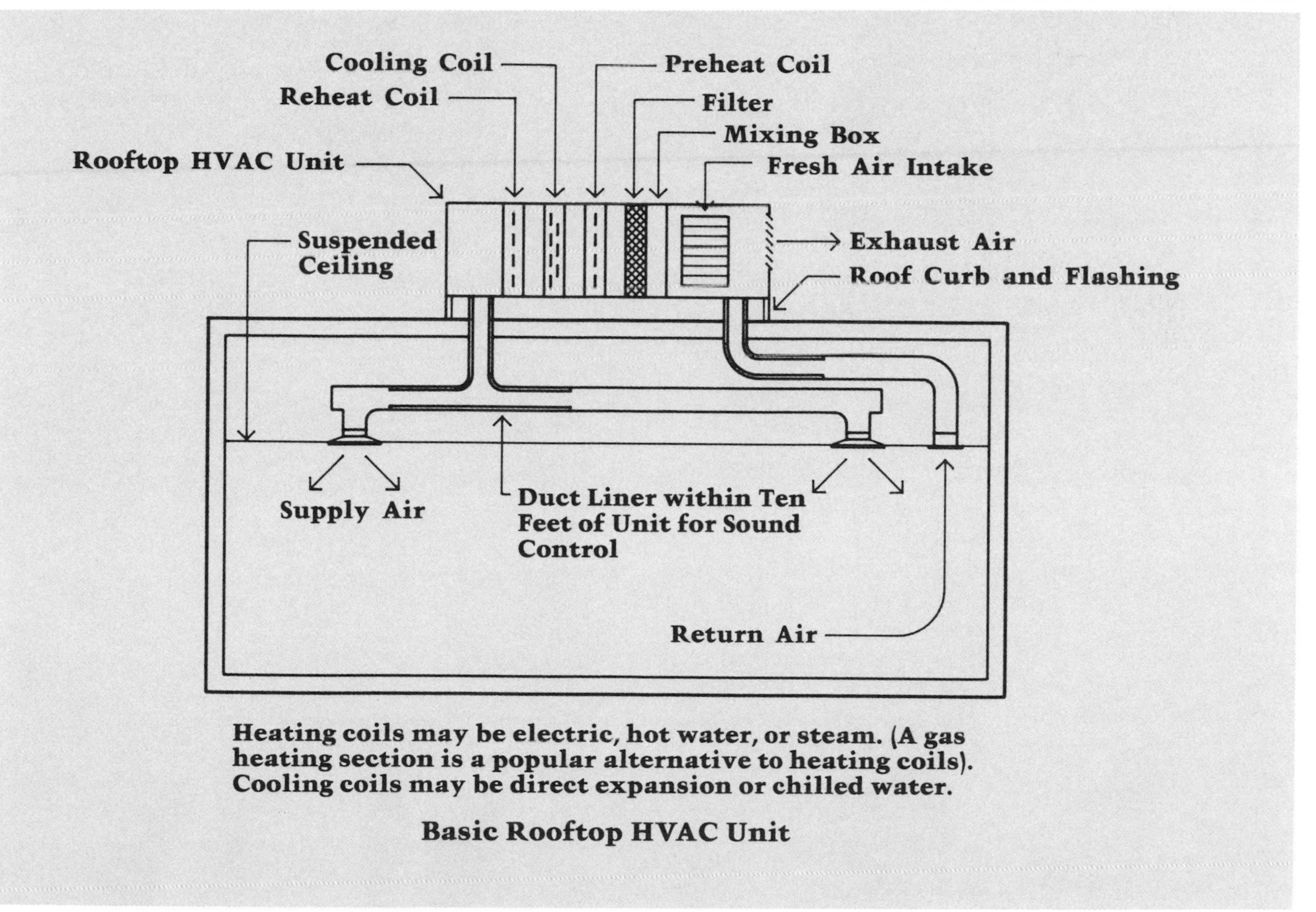

Figure 1.4

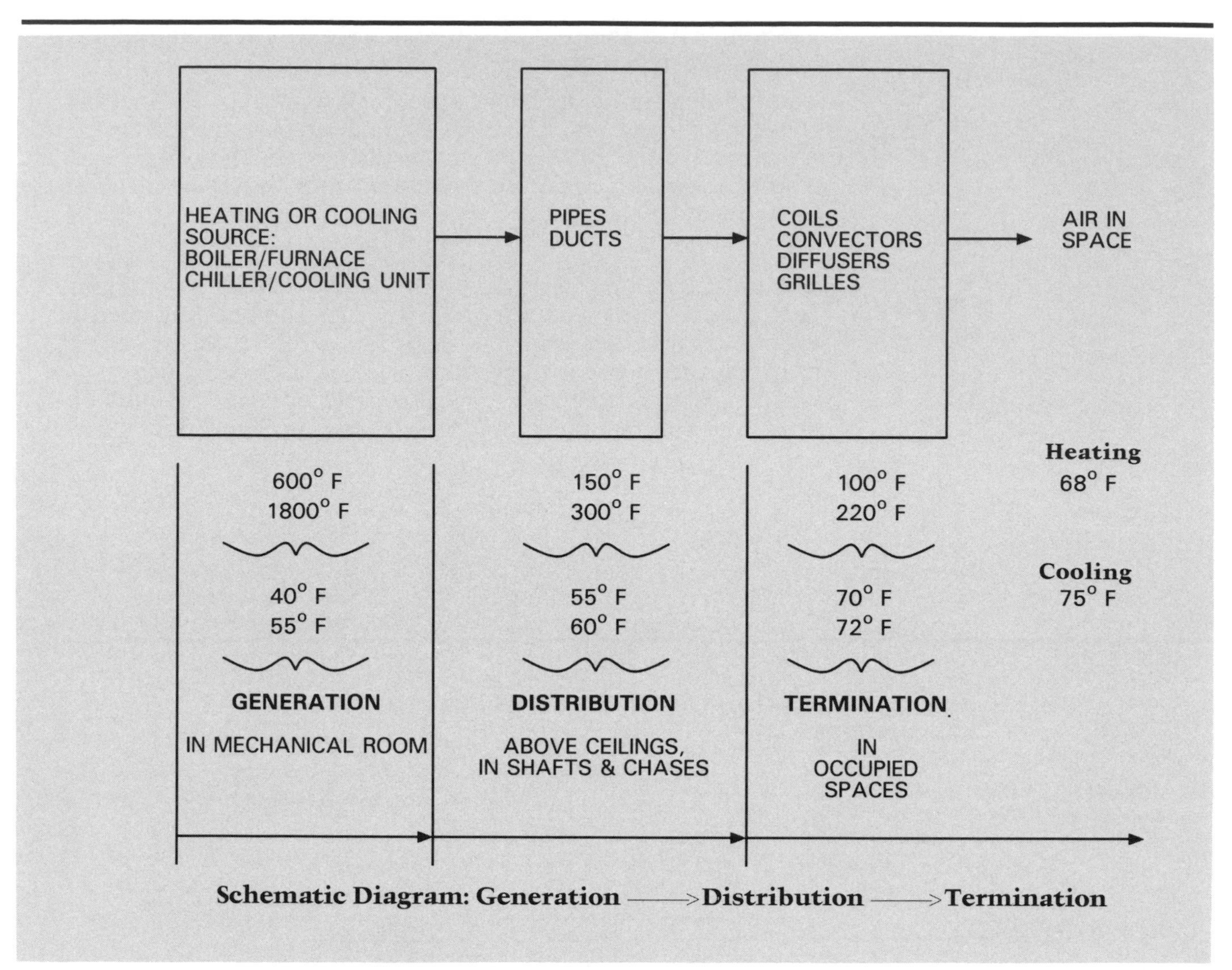
HEATING OR COOLING SOURCE: BOILER/FURNACE CHILLER/COOLING UNIT
PIPES DUCTS
COILS CONVECTORS DIFFUSERS GRILLES
AIR IN SPACE
600° F
1800° F
40° F
55° F
150° F
300° F
55° F
60° F
100° F
220° F
70° F
72° F
Heating
68° F
Cooling
75° F
GENERATION
IN MECHANICAL ROOM
DISTRIBUTION
ABOVE CEILINGS, IN SHAFTS & CHASES
TERMINATION
IN OCCUPIED SPACES
Schematic Diagram: Generation ——> Distribution ——> Termination

Figure 1.5

The unit of measure for heat is the British thermal unit, or Btu. One Btu is the amount of heat necessary to heat one pound of water 1°F. For simplicity, one thousand Btu's per hour is written as 1 MBH. A typical heat loss for a single-family house is often an average of 60,000 Btu's each hour, or 60 MBH.

All systems that move heat downgrade the temperature of the heat as it moves through the system. In the process of moving from a generation source to a terminal unit (in a room), the temperature of heat is reduced from high energy levels (high temperatures) to lower energy levels (lower temperatures). For example, the 1,800°F temperature in the combustion chamber of a boiler becomes 400°F. 150 psi high-pressure steam loses temperature and pressure along the route from source to termination. The heat reaches the terminal unit at about 130°F and finally downgrades to the room air design temperature of 68°F.

The heating ladder shown in Figure 1.6 illustrates the common steps in moving heat from generation to termination. A heating system may use the same medium throughout, such as hot water, or may "step down" from hot water to warm air. On the heating ladder, lateral movement is represented by a straight line, such as from a hot water boiler to a hot water pipe. The line continues to a hot water convector (radiator), which produces warm air for the room. This hot water system does not involve any intermediate heat exchangers (see Figure 1.7).

On the heating ladder it is only possible to move either across or across and down. A system that moves across and down the heating ladder is shown in Figure 1.8. A steam system utilizes steam from a utility company and converts it to hot water in the mechanical room. The hot water is distributed to heating coils that transfer the heat from hot water to warm air, and the warm air is then blown directly into the room. Each time the medium changes, such as from steam to hot water and then from hot water to warm air, a **heat exchanger** and a driving device (pump or fan) are required. Heat exchangers are illustrated in Figure 1.9. (See Chapter 8 for more information on heat exchangers.)

Steam Systems

Steam distribution carries the greatest amount of heat per unit of volume for conventional systems, because each pound of steam delivers approximately 1,000 Btu (1 MBH) as it condenses from steam to water (the latent heat of vaporization). Steam systems do not need pumps or fans to drive them. Instead, the condensing steam creates a vacuum, which draws new steam into the system, and then gravity pushes the condensed steam (hot condensate in water form) back to the boiler or to waste. Condensate pumps and/or vacuum pumps can be used to increase the flow around a system and to improve overall performance. The American Society of Mechanical Engineers (ASME) *Code for Low-Pressure Heating Boilers* limits the working pressure of steam heating boilers to 15 psi and 250°F. Figure 1.10 compares some of the basic characteristics of steam, water, and air systems.

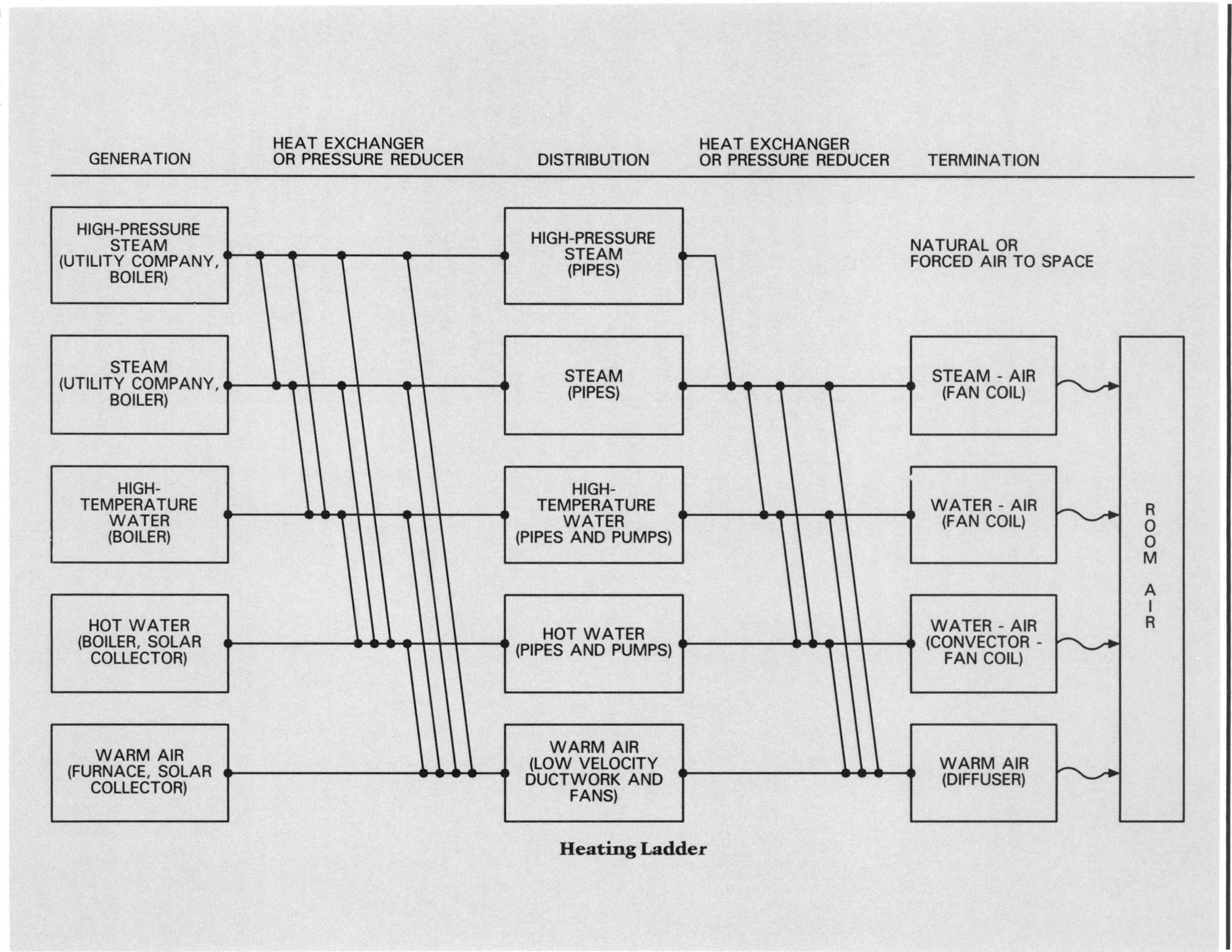

GENERATION
HEAT EXCHANGER OR PRESSURE REDUCER
DISTRIBUTION
HEAT EXCHANGER OR PRESSURE REDUCER
TERMINATION
HIGH-PRESSURE STEAM (UTILITY COMPANY, BOILER)
HIGH-PRESSURE STEAM (PIPES)
NATURAL OR FORCED AIR TO SPACE
STEAM (UTILITY COMPANY, BOILER)
STEAM (PIPES)
STEAM - AIR (FAN COIL)
HIGH-TEMPERATURE WATER (BOILER)
HIGH-TEMPERATURE WATER (PIPES AND PUMPS)
WATER - AIR (FAN COIL)
ROOM AIR
HOT WATER (BOILER, SOLAR COLLECTOR)
HOT WATER (PIPES AND PUMPS)
WATER - AIR (CONVECTOR - FAN COIL)
WARM AIR (FURNACE, SOLAR COLLECTOR)
WARM AIR (LOW VELOCITY DUCTWORK AND FANS)
WARM AIR (DIFFUSER)
Heating Ladder

Figure 1.6

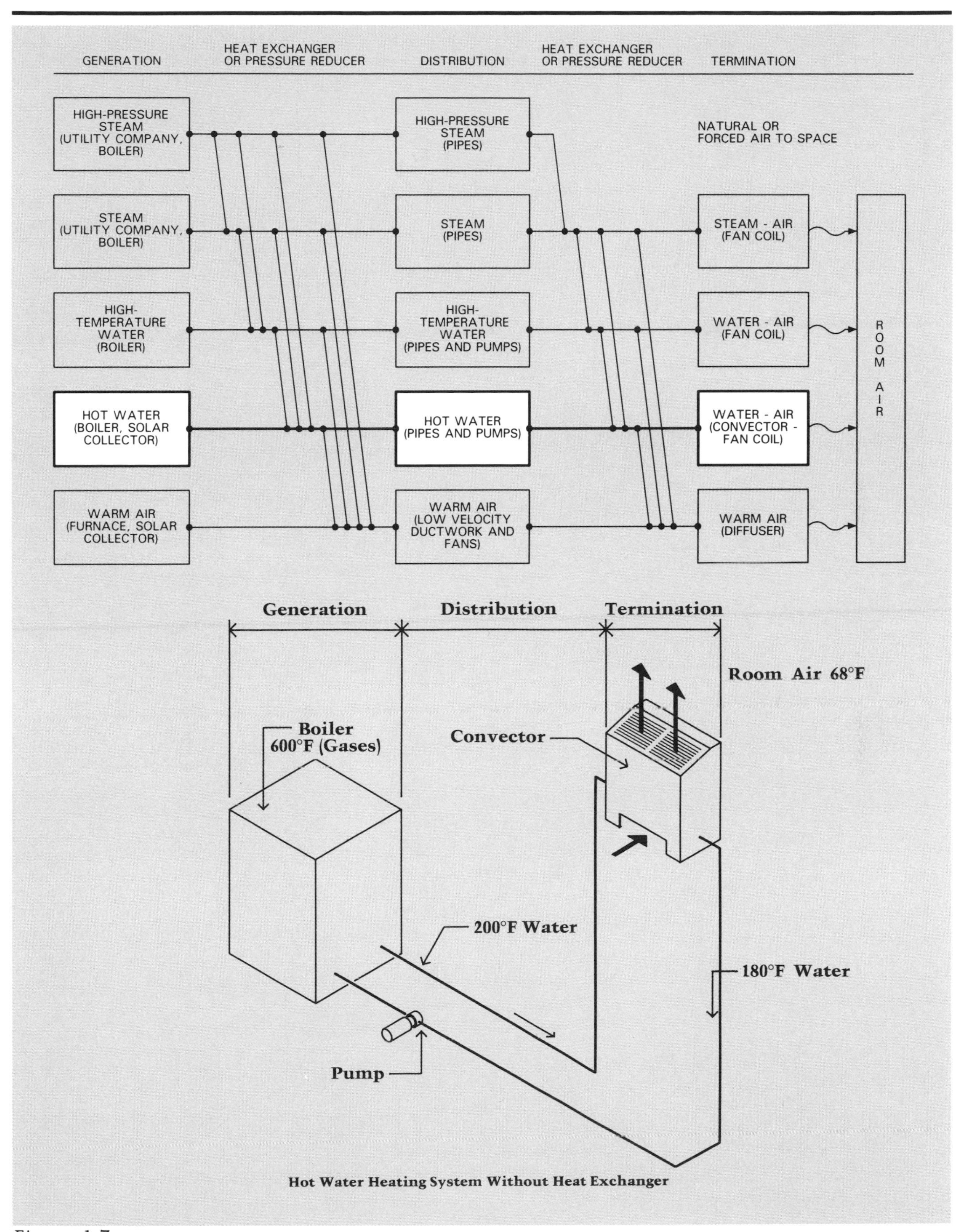

Hot Water Heating System Without Heat Exchanger

Figure 1.7

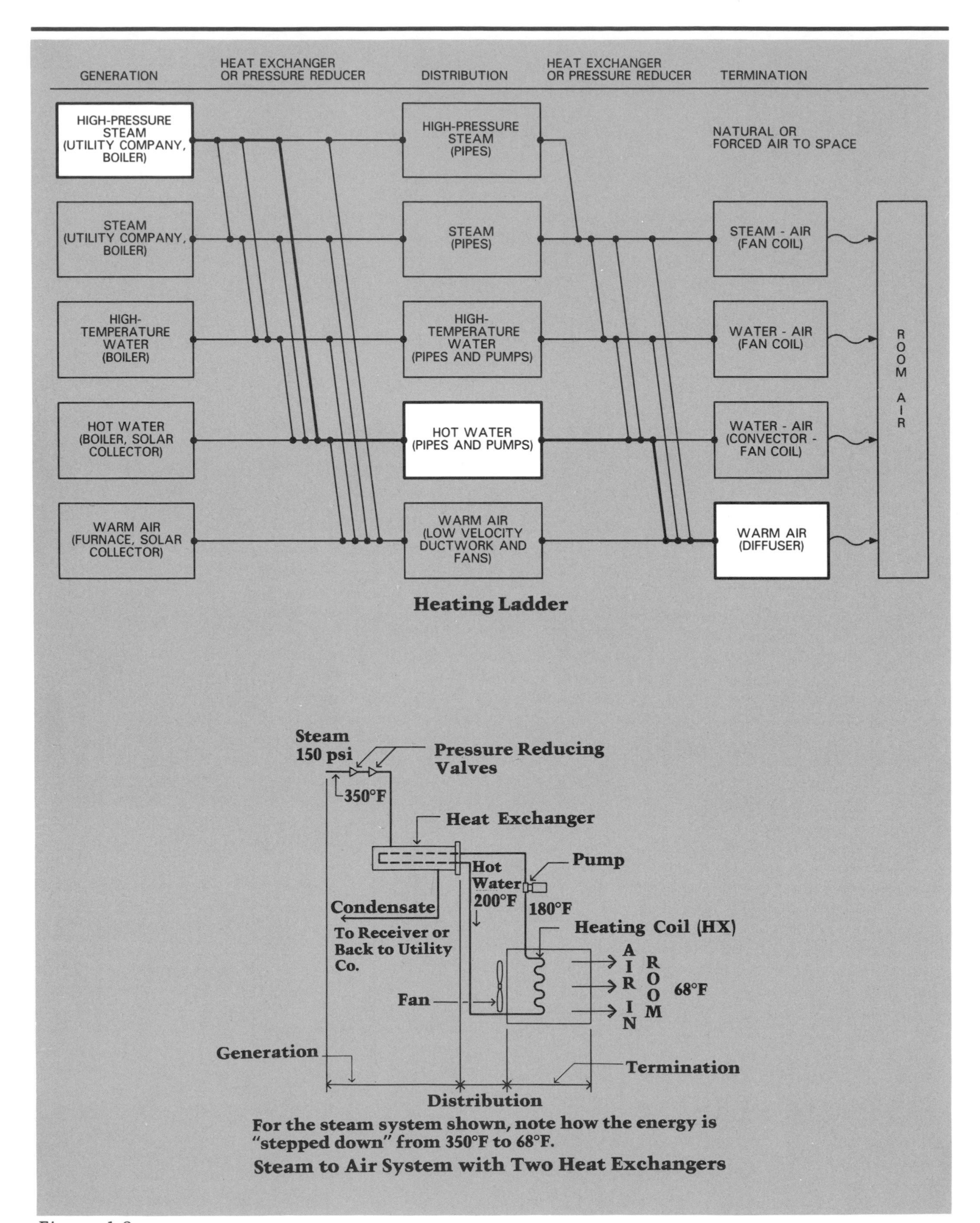

Heating Ladder

For the steam system shown, note how the energy is "stepped down" from 350°F to 68°F.

Steam to Air System with Two Heat Exchangers

Figure 1.8

Steam systems have the advantage of having few mechanical parts. They are well suited to large buildings because of overall competitive installation costs and an ability to supply large quantities of heat.

Hot Water Systems

Hot water systems typically supply water that is between 180°F and 220°F at pressures of 15 to 30 pounds per square inch gauge pressure, or psig. Atmospheric pressure, which is not usually read by the gauge, is 14.7 psi at sea level. When the gauge pressure is combined with the atmospheric pressure, it equals 29.7 psia. This is the *absolute pressure* of the 15 psig reading. Water above the normal boiling temperature of 212°F can be supplied because the boiling point rises with more pressure (water at 15 psig boils at about 250°F). A supply temperature of 200°F is common.

Hot water systems are normally designed for a 20°F temperature drop between the supply water temperature and the return water temperature. The quantities of water required are considerable when compared to steam. One pound of water delivers 20 Btu's, or 1/50 of the heat transmitted in a pound of steam (1/13 of the volume). For this reason, pumps are usually required in hot water systems, which are commonly known as *forced hot water*

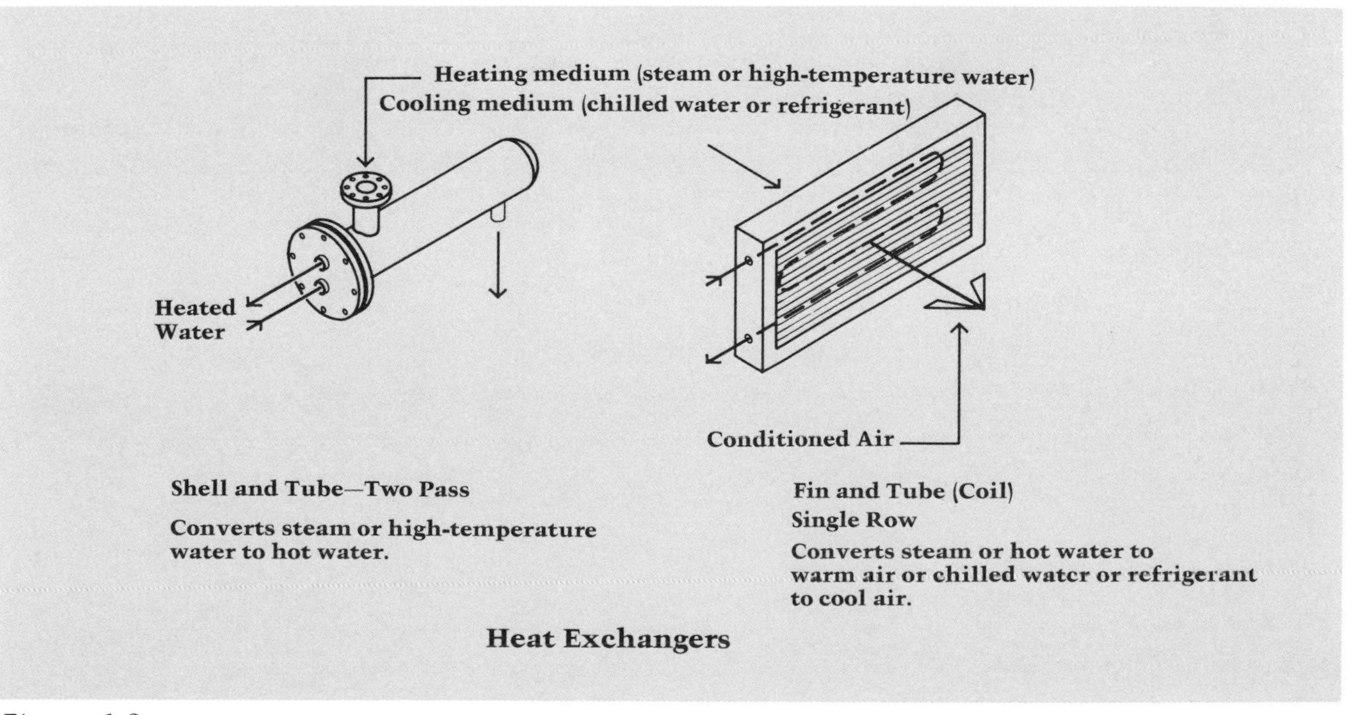

Figure 1.9

systems. The ASME *Code for Low-Pressure Heating Boilers* limits the working pressure of hot-water heating boilers to 30 psi and 250°F.

High-pressure boilers are used for a combination of heating and process loads and are usually designed for 150 psig, found in systems with temperatures in the range of 350°F. The ASME *Code for Power Boilers* is used for boilers operating over 15 psig.

Warm Air Systems

Warm air systems do not have the freeze-up problems that often occur with piping systems. Warm air systems typically supply air from a furnace to the terminal units at 130°F. Because air has the lowest capacity to hold heat, it requires the greatest volume of distribution ductwork. Heating coils in the ductwork continue to warm the air as it passes through the system. The somewhat unwieldy size of the ductwork (when compared to pipe) is often compensated for by the benefits of allowing humidification or dehumidification and the ability to provide fresh outdoor air for ventilation.

Cooling

Cooling is almost universally accomplished by blowing cool air into a room or building. For example, by supplying the proper quantity of air at 60°F to the room temperature of a space, the air will mix and the temperature will be maintained at the design temperature (plus or minus 78°F). The cool air absorbs

Comparison of Steam, Water, and Air Systems

Medium	Heat Delivered Per Pound	Volume Per Pound (Ft.³/lb.)	Volume For 10,000 BTU	Advantages	Disadvantages	Common Uses
Steam 230°F Steam to Condensate	1,000 BTU/lb.	19.0 ft.³/lb. Steam	$\frac{10{,}000}{1{,}000} = 10$ lbs. $10 \text{ lbs.} \times \frac{19\text{ft.}^3}{\text{lbs.}} = 190\text{ft.}^3$	No Pumps	Scalding Risk	Institutional and Commercial
Water 20° ΔT 200° Boiler 180° Return	20 BTU/lb.	0.0166 ft.³/lb. Water	$\frac{10{,}000}{20} = 500$ lbs. 500 lbs. × 0.0166 ft.³/lb. = 83ft.³	Small Pipes	Pump Cost	Residential
Air 60° ΔT 130° Supply 70° Return	15 BTU/lb.	14.5 ft.³/lb. Air	$\frac{10{,}000}{15} = 667$ lbs. 667 lbs. × 145 = 9670ft.³	No Freezing Humidification Filtration and Ventilation Possible	Large Duct Space Required	Combined with Cooling Temporary Heat

Note that water pipes have the smallest, and air ducts have the largest, capacity for transporting heat.

Figure 1.10

the heat gained from transmission through walls, windows, and doors; infiltration of warm air from around doors and windows; solar radiant energy; and internal heat gains from lights, people, and equipment.

The common unit of measure for cooling is the *ton*, which is derived from an earlier period when ice was used for cooling. One ton of cooling is equal to 12,000 Btu's per hour. It takes 144 Btu's to melt one pound of ice at 32°F; when one ton of ice melts in one 24-hour period, it has absorbed 12,000 Btu's per hour.

Just as boilers and furnaces produce heating, a refrigeration system produces cooling. Heating equipment adds heat to a building; cooling equipment, by contrast, subtracts heat, or more simply, pushes heat away from the building. Basic cooling systems are shown on the cooling ladder in Figure 1.11.

As cool air is introduced into a space, it neutralizes the heat gain in order to maintain the desired room temperature. When the 60°F supply air temperature is added to a space, the heat removed from the supply air is rejected to the outside. Some of the return air mixture (78°F) is expelled and some is cooled, mixed with the outside air (88°F, for example), and then redistributed into the system. Figure 1.12 illustrates a basic cooling layout.

There is a significant difference between *making* heat (heating) and *transferring* heat (cooling). The principle of heating is easy to understand. It is just like boiling a cup of water. In heating a building, fuel is burned in the combustion chamber of a boiler or furnace, and the heat energy from the combustion warms the water, steam, or air for the system. Each unit of energy from combustion is added to the system, starting at a high temperature and stepping down along the distribution path.

Cooling is more complicated. It is obviously not a problem to cool a space in the winter — simply opening the windows will cool off a space, as warm air flows "downhill" (outside to colder temperatures) to be replaced by the lower temperature outside air. Opening windows in the summer, however, will obviously not cool a space. The heat inside must be pushed "uphill" to the outdoors. In the summer, the supply air temperature of 60°F is made by taking the heat from the return air mixture and rejecting it to the warmer outdoor air.

This is very similar to what happens with a household refrigerator. The inside of a refrigerator is kept cooler than the surrounding kitchen. To maintain cooling, the refrigeration cycle takes the heat to be removed from the inside of the refrigerator and pumps it outside. The process of taking the heat from the inside of the space and putting it into the refrigerant warms the refrigerant gas. The added energy of the compressor raises the temperature of the refrigerant gas further, to about 110°F. The exposed condenser coils on the back of the refrigerator are cooled by the room air temperature, which turns the refrigerant gas back into a liquid because of the heat reduction. This process keeps the inside of the refrigerator at about 40°F. (The compressor is the only major device that

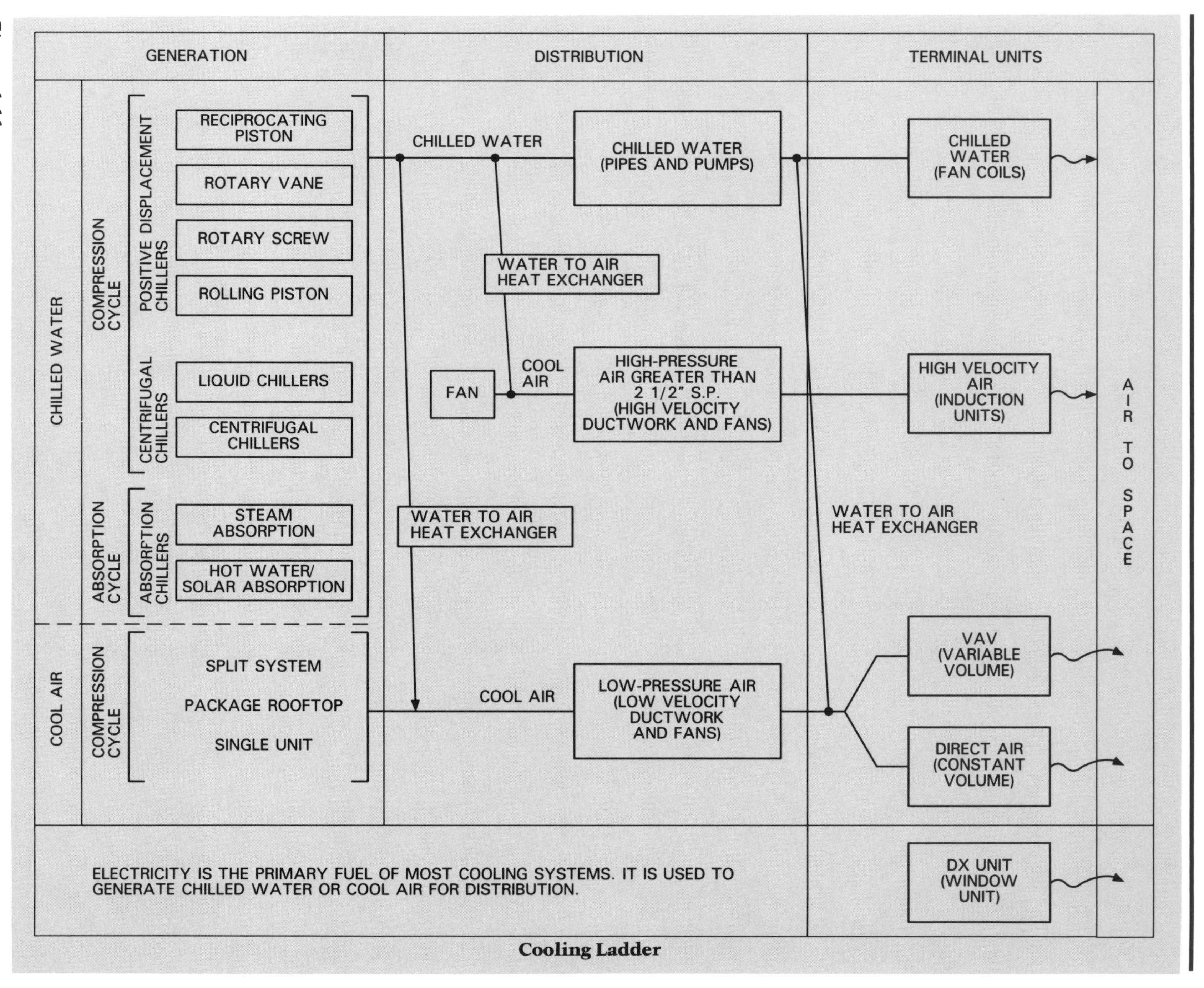

GENERATION
DISTRIBUTION
TERMINAL UNITS
CHILLED WATER
COOL AIR
COMPRESSION CYCLE
ABSORPTION CYCLE
COMPRESSION CYCLE
POSITIVE DISPLACEMENT CHILLERS
CENTRIFUGAL CHILLERS
ABSORPTION CHILLERS
RECIPROCATING PISTON
ROTARY VANE
ROTARY SCREW
ROLLING PISTON
LIQUID CHILLERS
CENTRIFUGAL CHILLERS
STEAM ABSORPTION
HOT WATER/ SOLAR ABSORPTION
SPLIT SYSTEM
PACKAGE ROOFTOP
SINGLE UNIT
CHILLED WATER
CHILLED WATER (PIPES AND PUMPS)
WATER TO AIR HEAT EXCHANGER
FAN
COOL AIR
HIGH-PRESSURE AIR GREATER THAN 2 1/2" S.P. (HIGH VELOCITY DUCTWORK AND FANS)
WATER TO AIR HEAT EXCHANGER
COOL AIR
LOW-PRESSURE AIR (LOW VELOCITY DUCTWORK AND FANS)
WATER TO AIR HEAT EXCHANGER
CHILLED WATER (FAN COILS)
HIGH VELOCITY AIR (INDUCTION UNITS)
VAV (VARIABLE VOLUME)
DIRECT AIR (CONSTANT VOLUME)
DX UNIT (WINDOW UNIT)
AIR TO SPACE
ELECTRICITY IS THE PRIMARY FUEL OF MOST COOLING SYSTEMS. IT IS USED TO GENERATE CHILLED WATER OR COOL AIR FOR DISTRIBUTION.
Cooling Ladder

Figure 1.11

consumes energy in the refrigerator.) The temperatures and layouts for basic heating and cooling systems are shown in Figure 1.13.

In heating, each Btu of heat burned produces nearly one Btu of building heat. In cooling, each Btu to be rejected only requires about one-third to one-fifth of a Btu compressor energy. The effectiveness of cooling is measured by the coefficient of performance. The coefficient of performance equals the amount of cooling energy produced, or heat gain removed (in Btu/hour), divided by the energy consumed to produce the work of the compressor cooling (in Btu/hour).

$$\text{C.O.P.} = \frac{\text{amount of cooling energy produced}}{\text{energy consumed to produce the cooling}}$$

The coefficient of performance generally ranges from 3.5 to 5 for most applications.

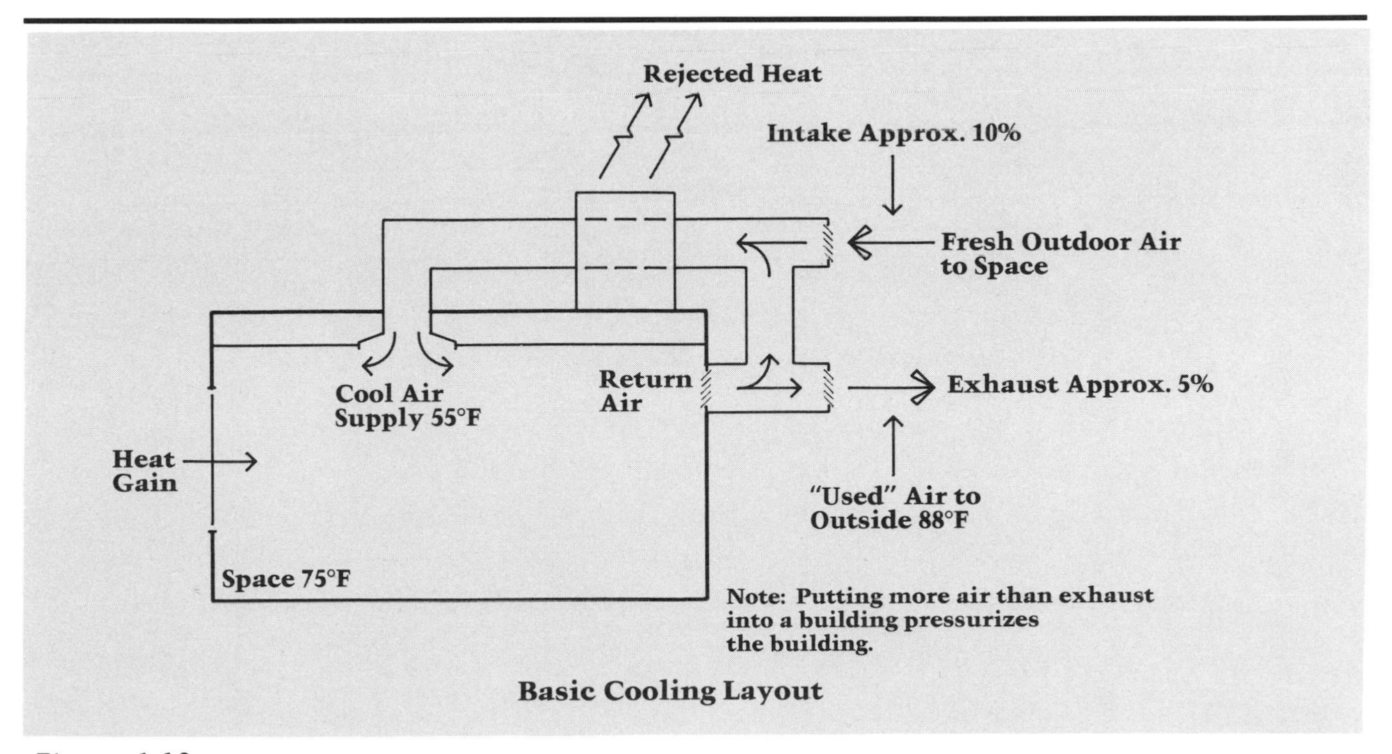

Figure 1.12

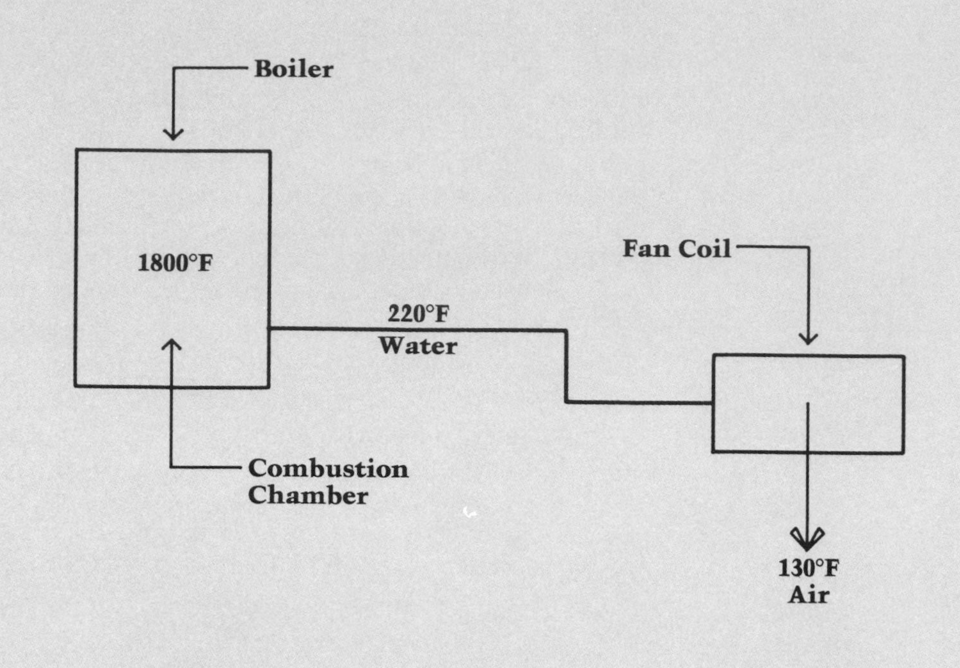

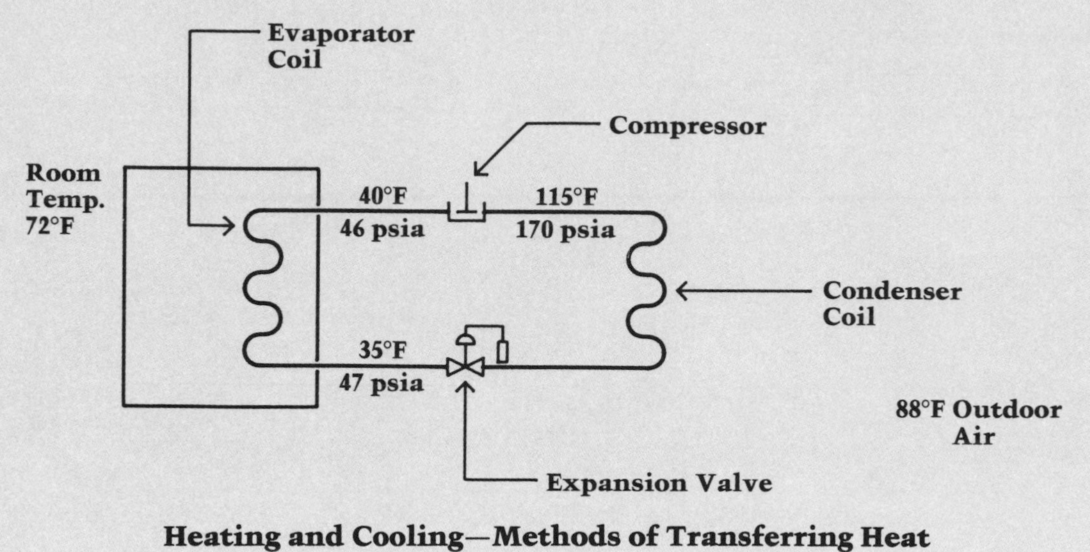

Heating and Cooling—Methods of Transferring Heat

Figure 1.13

Refrigeration Cycles

All cooling systems use refrigeration equipment to produce cooling. There are six refrigeration systems which are listed below.

- Compression cycle
- Absorption cycle
- Steam jet cycle
- Air cycle
- Thermoelectric cycle
- Solar cycle

Since compression and absorption refrigeration systems account for over 98 percent of all installations, this discussion is limited to these two areas. The steam jet, air cycle, thermoelectric cycle, and the solar cycle are rarely installed.

The type of refrigeration cycle selected depends on cost and capacity considerations. Figures 1.14 and 1.15 are selection guides for refrigeration equipment and all types of cooling systems.

Compression Cycle: The most common refrigeration system is the **compression cycle**, which is used in most installations. The compression cycle is the most effective means of heat removal per pound of refrigerant. Figure 1.16 illustrates the compression refrigeration cycle with its four basic pieces of equipment: evaporator, compressor, condenser, and expansion valve. A refrigerant flows through the four pieces of equipment, forming a loop in which the refrigerant changes from a liquid to a gas, and then back to a liquid. This refrigerant has a low boiling point. Nearly all refrigerants boil below 0°F. Beyond the expansion valve is where the boiling (evaporation) takes place. The liquid refrigerant absorbs the heat to be removed from the space surrounding the evaporator. This changes the refrigerant from a liquid to a gas as it absorbs heat and "boils" at room temperature. This process is the reverse of steam heating; the medium in cooling changes from a liquid to a gas, whereas in heating, the medium changes from a gas to a liquid as heat is transferred.

The gas temperature must then be raised significantly above the outdoor temperature by the compressor to allow the outdoor temperature to cool the refrigerant, condense it back to liquid form, and return it for recycling around the loop. The refrigerant has absorbed heat twice: once in the evaporator, where the building heat gain is removed; and again from the compressor, which has added energy to raise the gas temperature above the outdoor air temperature. The total heat absorbed by the refrigerant, from the heat of evaporation and the heat of compression, is finally rejected to the outside by the condenser, and the refrigerant repeats the loop.

The building heat gain is absorbed by the evaporator at no operating cost other than that used by the fan evaporator which, while it may be considerable, is still less than that used by the compressor. It is the compressor that uses the most significant amount of energy and costs the most money to run. As previously explained, the effectiveness of a cooling system is measured by the coefficient of performance. Another term used

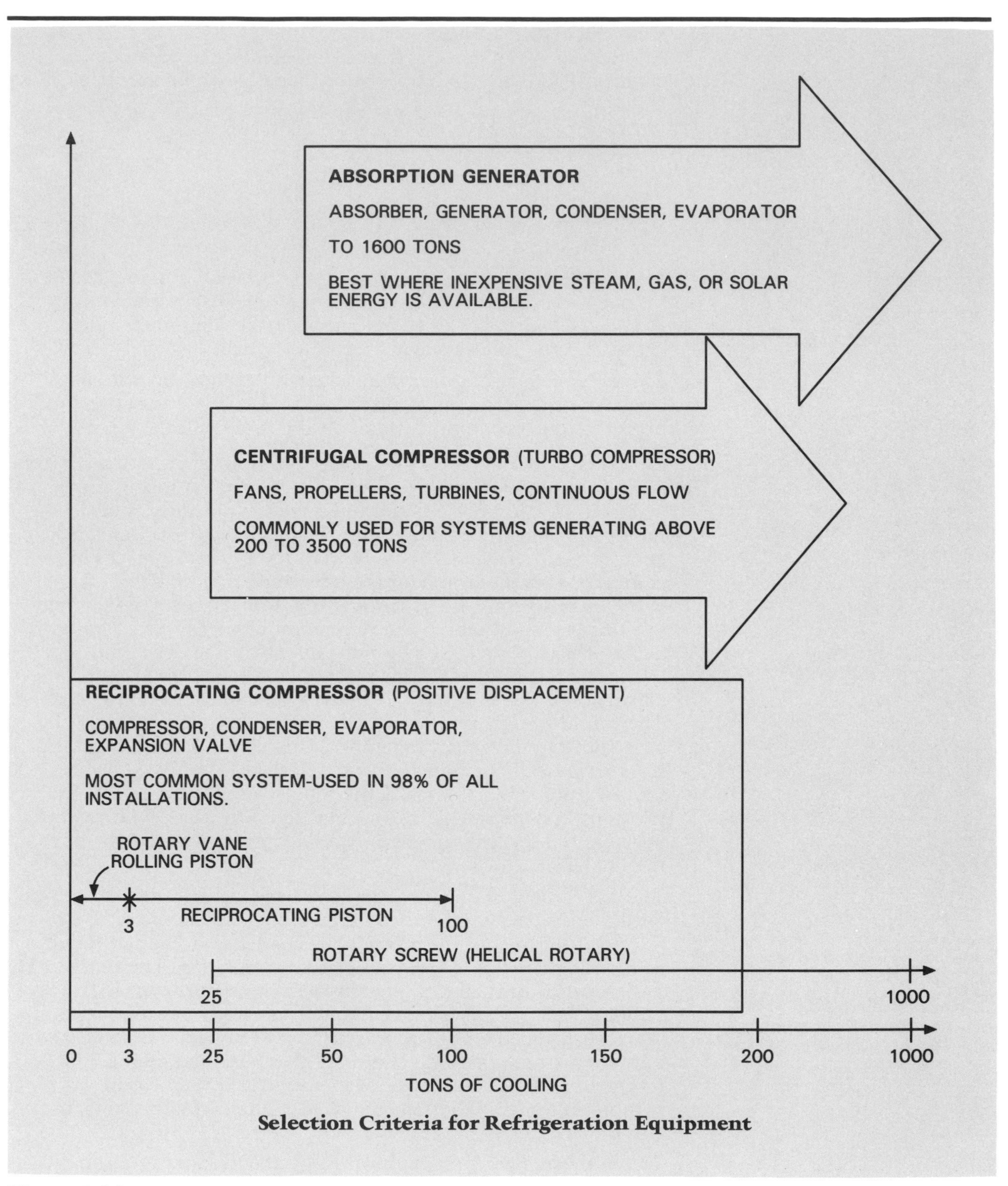

Selection Criteria for Refrigeration Equipment

Figure 1.14

in cooling is the energy efficiency ratio. The energy efficiency ratio equals the heat gain in Btu/hour removed by the equipment, divided by the watts of energy consumed to cool the space.

$$\text{E.E.R.} = \frac{\text{heat gain removed by the equipment (Btu/hr.)}}{\text{watts of energy consumed to cool the space}}$$

Guidelines for Cooling System Selection

System		Subsystem	Range of Tonnage	Common Applications	Remarks
1. Compression Cycle	Positive Displacement	Reciprocating piston	3-100	All buildings	Noise isolation recommended.
	Positive Displacement	Rotary vane	0-3	Refrigerators Window AC units	
	Positive Displacement	Rotary screw	25-1,000	Commercial Institutional	Variable speed possible.
	Positive Displacement	Rolling piston	0-3	Refrigerators Window AC units	
		Centrifugal compressors (turbocompressors)	50-1,000	All buildings	Produce chilled water or brine for circulation to coils and terminal units.
2. Absorption Cycle			3 to 3,500	All buildings	Inexpensive steam, gas, solar most common requirement.
3. Steam Jet Cycle			10-1,000	Foods and chemicals Freeze drying	Special applications.
4. Air Cycle			10-150	Aircraft	Special applications.
5. Thermoelectric Cycle			5-50	Low-rise buildings	Thermoelectric solar panels generate electricity. Good in remote sunny locations.
6. Solar Cycle			50-100	Low-rise buildings	Hot water solar panels supply absorption units.

Figure 1.15

COMPRESSOR
With each stroke of the piston, the pressure of the refrigerant gas is increased as it passes to the condenser. This causes the temperature of the refrigerant to be simultaneously raised and set significantly above the outdoor temperature where it can reject heat.

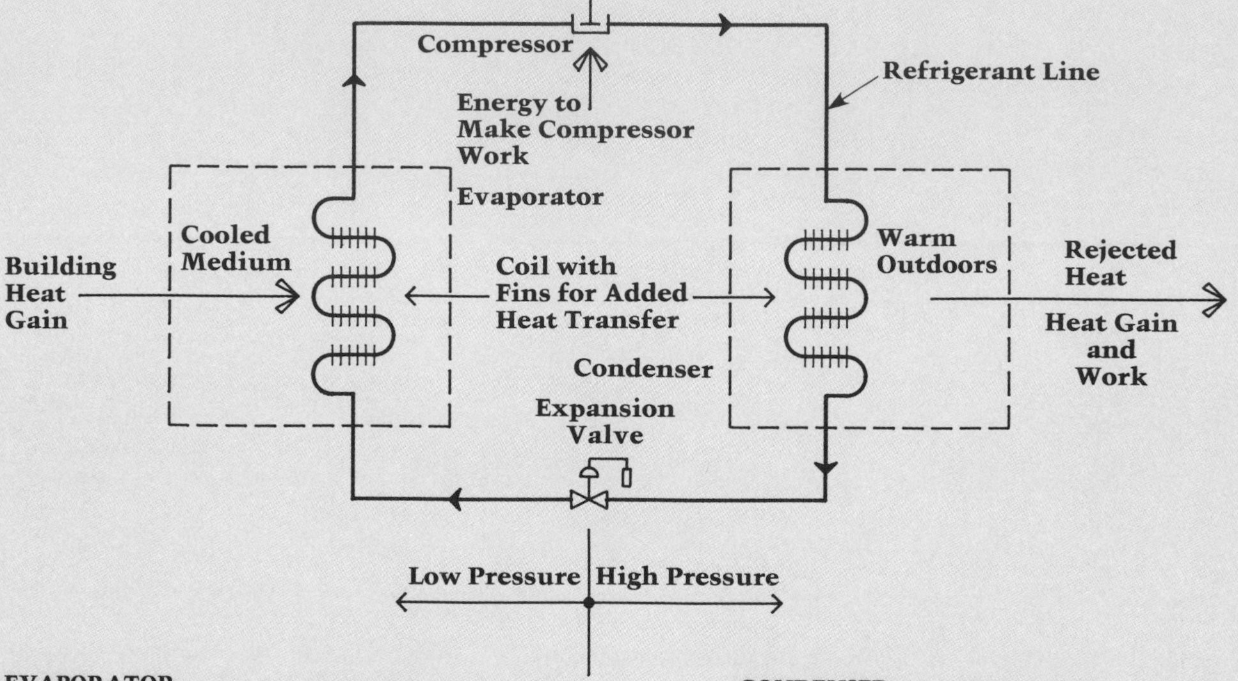

EVAPORATOR
In the evaporator the refrigerant, which has a low boiling point, changes from a liquid to a gas (evaporates, boils) because it absorbs heat from the medium to be cooled. The gas then carries the absorbed heat away from the medium and to the compressor.

CONDENSER
The condenser receives the hot gases from the compressor and cools them with the cooler outdoor air. This causes the gases to liquify or condense and some heat absorbed from the medium and the compressor to be rejected to the outdoors.

EXPANSION VALVE
The hot liquid refrigerant from the condenser, which is at a high pressure level, flows through the orifice of the expansion valve. The orifice causes some of the hot liquid refrigerant to turn to gas at the lower pressure of the evaporator side. This "misting", or boiling, of some of the refrigerant lowers the temperature and pressure of the remaining liquid.

The Compression Refrigeration Cycle

Figure 1.16

Since 1 watt equals 3.41 Btu/hour, the energy efficient ratio equals 3.41 times the value of the coefficient of performance.

1 watt = 3.41 Btu/hr., therefore,

E.E.R. = 3.41 × C.O.P.

Energy efficient ratio values of 9 to 12 are typical, and a value of over 10.5 is considered good.

The type of compression system selected depends on the overall cooling load required, the type of facility, the cost of the system, and the way in which the requirements of each system can be accommodated. Compression cycle systems are either reciprocating compressors (positive displacement) or centrifugal compressors (turbo).

Because there are many criteria that establish system selection, it may happen that more than one type of compressor is adequate for a given application. When this occurs, the final selection is based on either the initial or the life cycle cost of the system.

Reciprocating compressors, or positive displacement compressors, are the most common type of cooling equipment, and are used almost exclusively for systems with up to 50 tons capacity. Positive displacement compressors "squeeze" the refrigerant by using a piston, vane, or screw to compress the gas.

The four different types of reciprocating compressors are listed below. Figure 1.17 illustrates examples of these different types.

- Reciprocating piston
- Rotary vane
- Rotary screw
- Rolling piston

Because they are small, reciprocating compressor units are usually single packaged units, which are air cooled and located on the roof or through a wall.

Some reciprocating compressors are used in split systems, which means that they are divided into two parts. An evaporator, expansion valve, and supply air fan make up one half of the unit, which is located in or near the space to be cooled. The compressor and air cooled condenser (condensing unit) make up the other half of the unit, which is usually located outdoors away from the space, typically on the roof. Reciprocating compressors are best suited to low rise office and residential buildings with many tenants, since each tenant can be held responsible for the individual split system serving the space. The practical distance between the evaporator/valve/fan and the compressor/condenser unit is approximately 50 feet, due to the need to limit pressure and temperature changes in the refrigerant lines connecting the two halves of the system.

Centrifugal compressors are used in larger installations, typically above 200 tons, although some manufacturers have models for loads as low as 50 tons. The centrifugal compressors continually "squeeze" the refrigerant gases by forcing a continual stream of the gas through a fan-like device. This forces the gas into a high-pressure line, which simultaneously raises its temperature. Figure 1.18 illustrates a typical centrifugal chiller.

Absorption Cycle: The absorption cycle is usually used for buildings with cooling loads over 200 tons where the cost of steam or other heat source in summer is inexpensive. The absorption cycle is the next most popular cooling method after the compression cycle. Absorption cycle refrigeration equipment involves the principle that a salt solution spray has an affinity for water. Thus, as a highly concentrated salt solution (anhydrous lithium bromide) is sprayed over a pool of water, some of the water vaporizes and mixes with the salt spray. This happens because chemically, the salt attracts large quantities of water, and also because the concentrated salt spray displaces some of the normally occurring water vapor, causing a partial vacuum of water vapor above the pool of water. These factors combine to draw the water from a liquid to a vapor, which mixes with the salt spray. Whenever water changes from a liquid to a gas, approximately 1000 Btu's (1 MBH) of heat are required to vaporize each pound of water. As the water

Types of Positive Displacement Refrigeration Compressors

Type	Description	Comments
Reciprocating Piston Compressor	A cylinder, piston-type compressor with crankshaft and inlet and outlet valves utilizing the two-cycle method of intake and compression.	• May be electric drive or powered by an internal combustion engine or a steam turbine. • This is the most common compressor used in the 3 to 50 ton range.
Rotating Vane Rotary Compressor	An off-center roller with two oscillating vanes attached rotates within the housing, compressing the gas and discharging through exhaust vane type valves.	• High volumetric efficiency. • Electric drive. • Most common use is 3 tons and smaller.
Rotary Screw (Helical Rotary) Compressor	The single main helical rotor acts against and in conjunction with the two-gate rotors to compress the gas and discharges through vane-type valves.	• Wide load range possible. • Common in 25 to 1,000 ton capacities. • May be electric drive or powered by an internal combustion engine.
Rolling Piston-Type Rotary Compressor	The eccentric shaft and the rotating roller combine with a single oscillating vane (attached to the housing) in compressing the gas and discharging through vane-type valves.	• High volumetric efficiency. • Most common usage is 3 tons and smaller. • Electric drive.

Compressors may be open-type or hermetic. In an open-type, the shaft connects the motor and compressor. The integrity is maintained through a shaft seal. In hermetic design, the motor and compressor are self-contained and are sealed within an air-tight, gas-cooled housing.

Rotary compressors maintain a continuous suction pressure, negating the need for inlet or suction valves.

Halocarbons and ammonia are the two common types of refrigerant used in these compressors.

Figure 1.17

vaporizes, it takes the 1000 Btu/pound from itself and energizes the water vapor. The heat leaves the water pool, chilling it considerably, to below 50°F.

Absorption systems produce chilled water by a different method, but in the same overall tonnage range as centrifugal chillers. Depending on relative installation costs, efficiencies, and operating costs, compression or absorption systems can be interchanged, while leaving the rest of the piping and other HVAC systems intact. The absorption cycle is shown in Figure 1.19.

Absorption equipment is large, heavy, and requires a heat source, which is the energy source for the system. It is most advantageous where the summer cost of the heat source is low. Absorption systems have a natural potential for use with solar energy, because the heat from the sun simultaneously increases the need for cooling and the availability of heat to drive the absorption system.

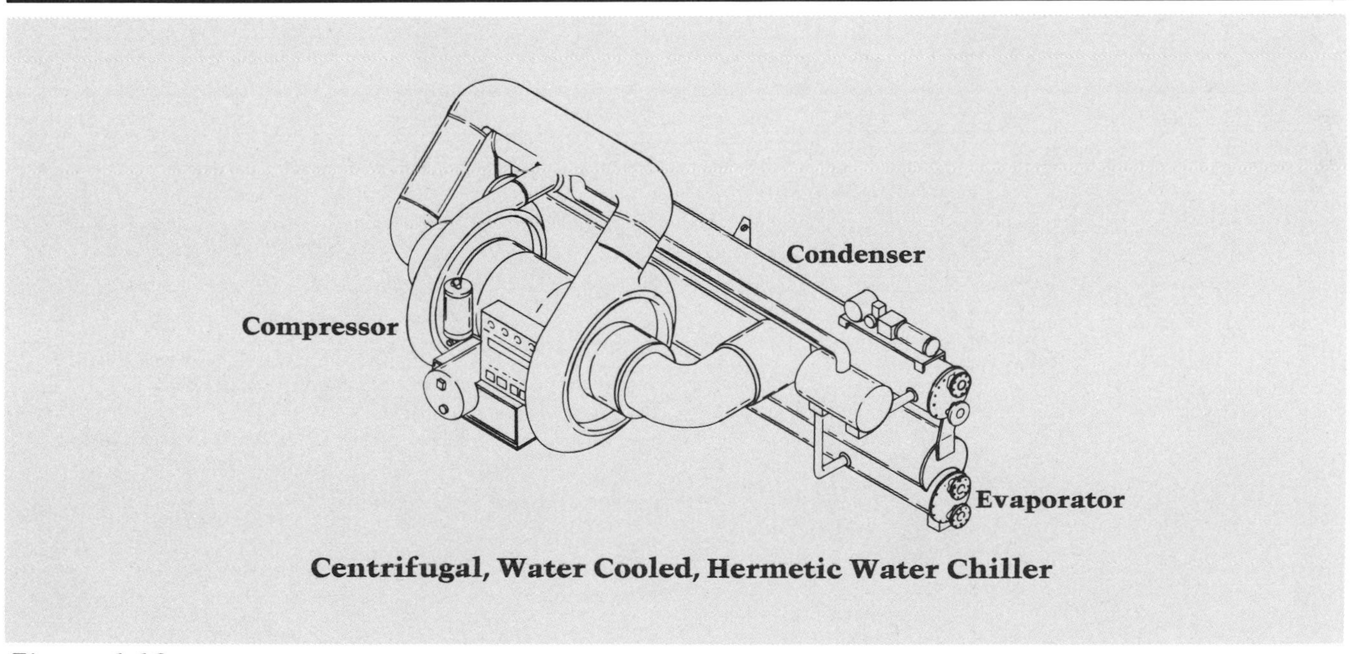

Centrifugal, Water Cooled, Hermetic Water Chiller

Figure 1.18

The selection of the refrigeration cycle usually establishes the principle component parts. The parts are normally furnished by the equipment manufacturer, either as a package unit or in separate, but compatible, component parts.

Condensers

After the type of refrigeration cycle and compressor (where applicable) are chosen, the type of condenser must be selected. There are three principle types of condensers: air cooled condensers, water cooled condensers, and evaporative condensers (see Figure 1.20).

Air cooled condensers are typically used for systems where refrigerant is the medium in the equipment. The refrigerant coils are placed directly in the outdoor air. Mounted outdoors on a roof or through a window, air cooled condensers are suitable

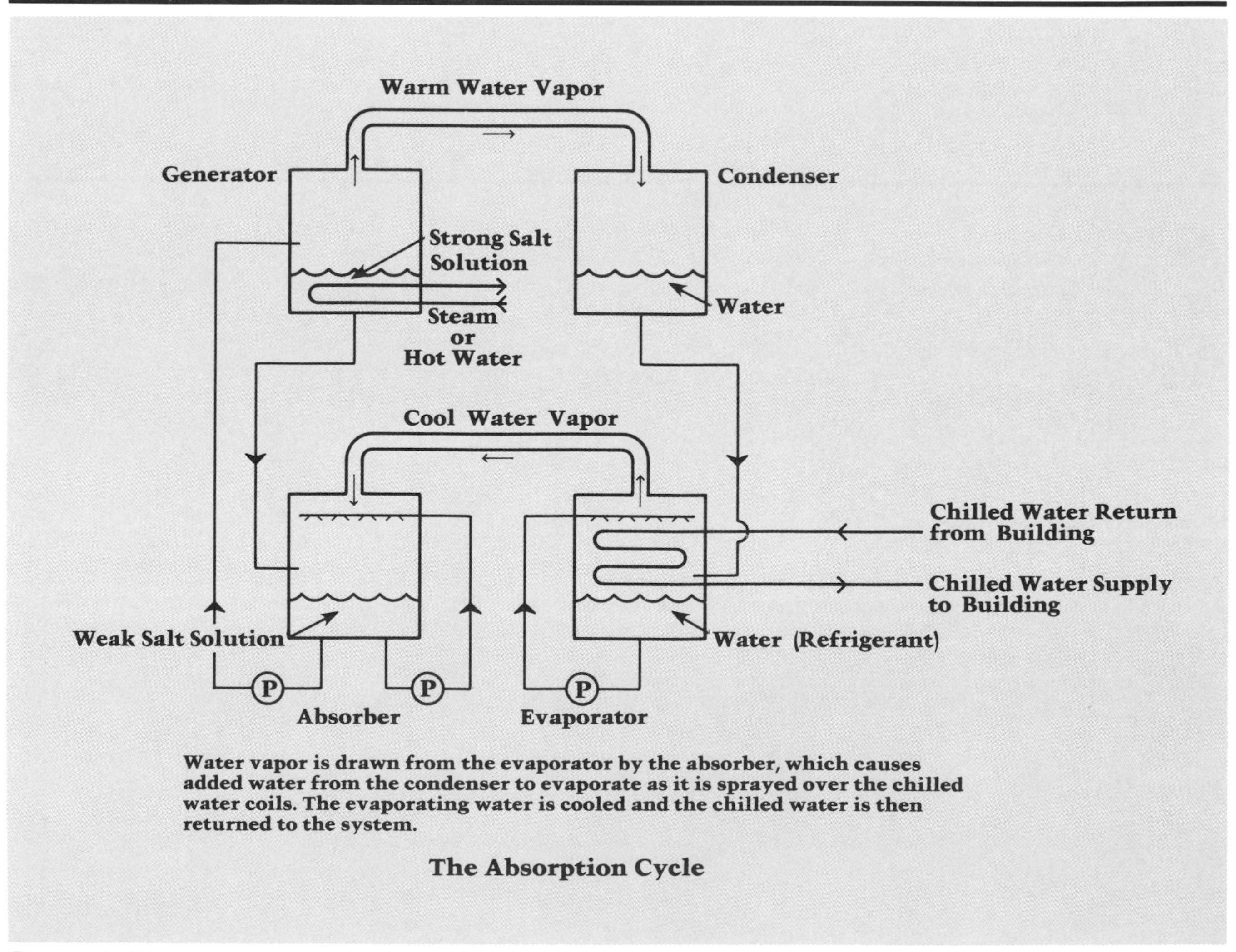

Water vapor is drawn from the evaporator by the absorber, which causes added water from the condenser to evaporate as it is sprayed over the chilled water coils. The evaporating water is cooled and the chilled water is then returned to the system.

The Absorption Cycle

Figure 1.19

Selection Criteria for Condensers				
Type	**Tonnage Range**	**Common Uses**	**Location**	**Special Requirements**
Air Cooled Condenser 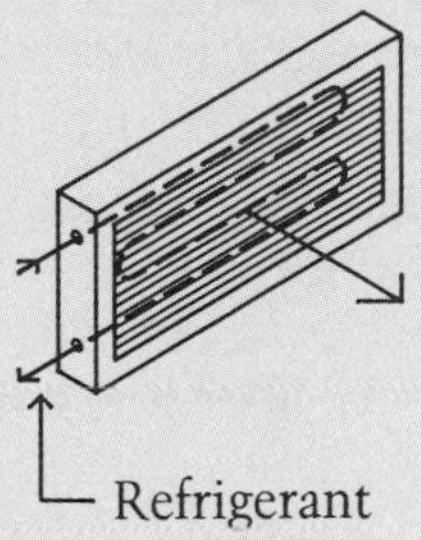 Refrigerant	0–50	Package units using refrigerant only Split system	Rooftop Through wall	Limit run and pipe for split systems.
Water Cooled Condenser 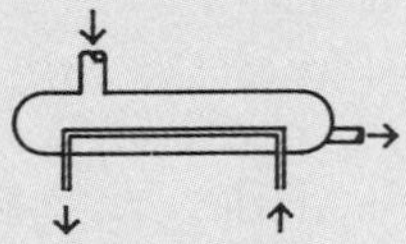	3–150	Small water cooled units using domestic water (where permitted) or water from ponds, lakes, rivers, or groundwater	Rooftop Ground mount	Cooling tower or other water source is required in urban setting.
Cooling Tower 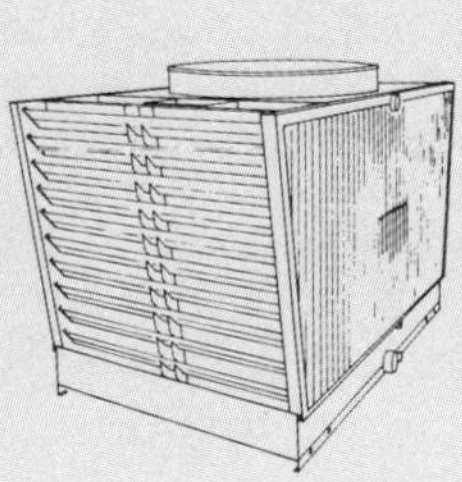	50–1,000	High-rise buildings	Rooftop Ground mount	

Figure 1.20

for loads up to 50 tons. Air around the coils is required for circulation to reject the heat. Fans are usually provided to increase performance. Household refrigerators, common window air conditioners, simple package rooftop units, and split systems all use air cooled condensers.

Closed circuit condenser water coolers are similar in appearance and operation to air cooled condensers. These units cool condenser water or a water glycol solution rather than a refrigerant gas. This type of water cooler is used for winter operation in climates subject to below freezing temperatures.

Water cooled condensers use refrigerant coils that are cooled by water, usually by using a shell and tube heat exchanger. In some cases, river, lake, or groundwater is used. For urban settings, cooling towers which typically receive water at 95°F and return it at 85°F are common. Spray ponds are also sometimes used.

Similar to air cooled condensers, **evaporative condensers** also spray cool water across the outdoor refrigerant coils.

Ductwork

The ideal size and location of the ductwork for air supply systems is different for heating and cooling. In temperate climates, the air ducts for heating and cooling are often the same size, because the heating and cooling loads are roughly equal. For example, a house with a 60,000 Btu/hour heat loss in the winter might have a 50,000 Btu/hour heat gain in the summer. In warmer climates, cooling ducts are usually larger and heating ducts are often not used.

Because warm air rises, heating supply ducts should be placed low and under windows. The rising heated air will stop downdrafts and cold air spill from the windows, will help to prevent condensation on the glass, and will inject the heating supply at the source of greatest heat loss while maintaining overall even room temperature.

By contrast, cooling supply ducts should be located at the ceiling and slightly away from the perimeter of the room in order for the cool air to mix with the warm air. The cool air will drift downward over windows and over the lights and occupants, which are other sources of heat gain.

The ideal sizes and locations of air ducts are illustrated in Figure 1.21.

Using a system that combines heating and cooling in the same air supply ducts involves a compromise due to the inherent conflict between heating and cooling duct sizes and locations. This can be accomplished by larger duct sizes and increased fan capacity. (For more information on ductwork, see Chapter 9.)

Ventilation

Building spaces must be vented for a variety of reasons. The main purpose of ventilation is to provide fresh outdoor air for the occupants. Fresh outdoor air replenishes indoor air for breathing and most noticeably reduces odors, smoke, and fumes caused by people, cooking, and manufacturing processes. All occupied rooms (apartments, offices, stores, schools, hospitals) must be properly ventilated. There are several acceptable methods of providing ventilation: operable windows, fresh outdoor air supply, exhaust air, supply air with exhaust air, and purging. These methods are illustrated in Figure 1.22.

The effect of ventilation on the heating and cooling system is to increase the loads that the heating and cooling system must carry, since the outside air must be treated before it is introduced into the space. The computation of these loads is described in Chapter 2.

Air Conditioning

When a building combines heating and cooling into one system, a rudimentary air conditioning system is established. In addition to heating and cooling, air conditioning also includes humidification and dehumidification, cleaning of air, and providing fresh outdoor air for ventilation. When all of these functions are included in one system, the building can be considered to have a full heating, ventilating, and air conditioning system. An air conditioning ladder is shown in Figure 1.23. This ladder incorporates portions of the heating and cooling ladders previously noted.

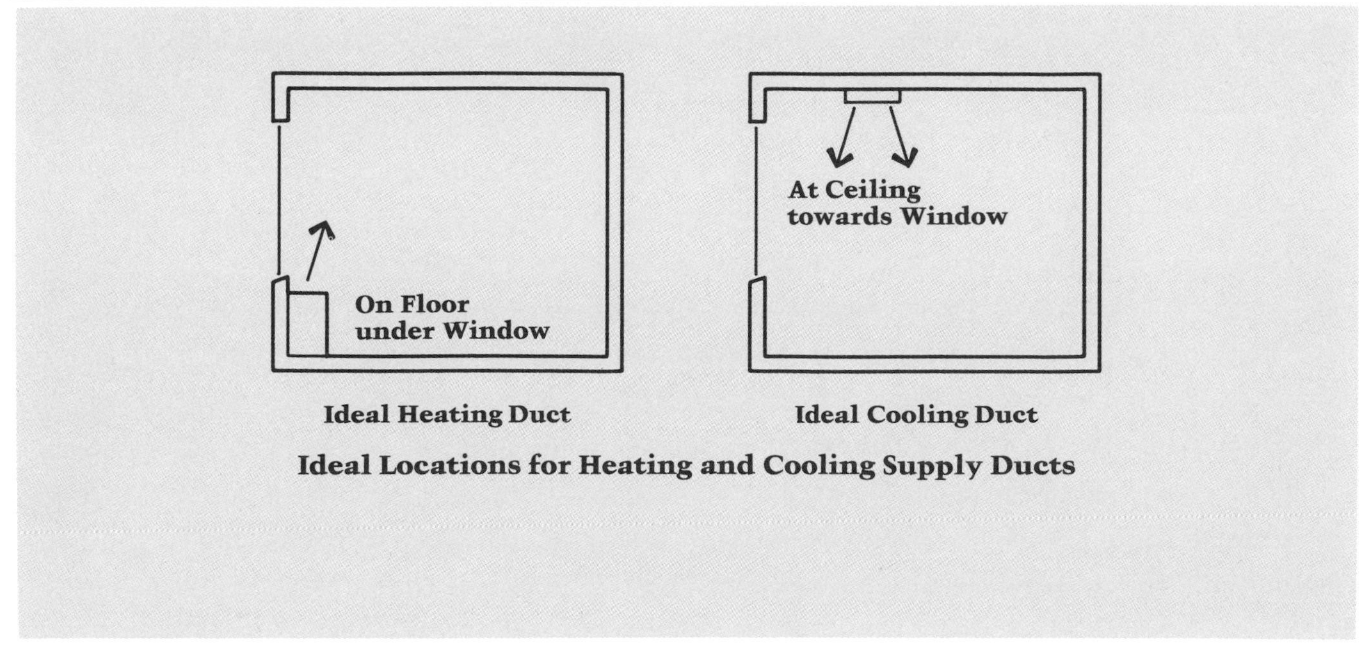

Figure 1.21

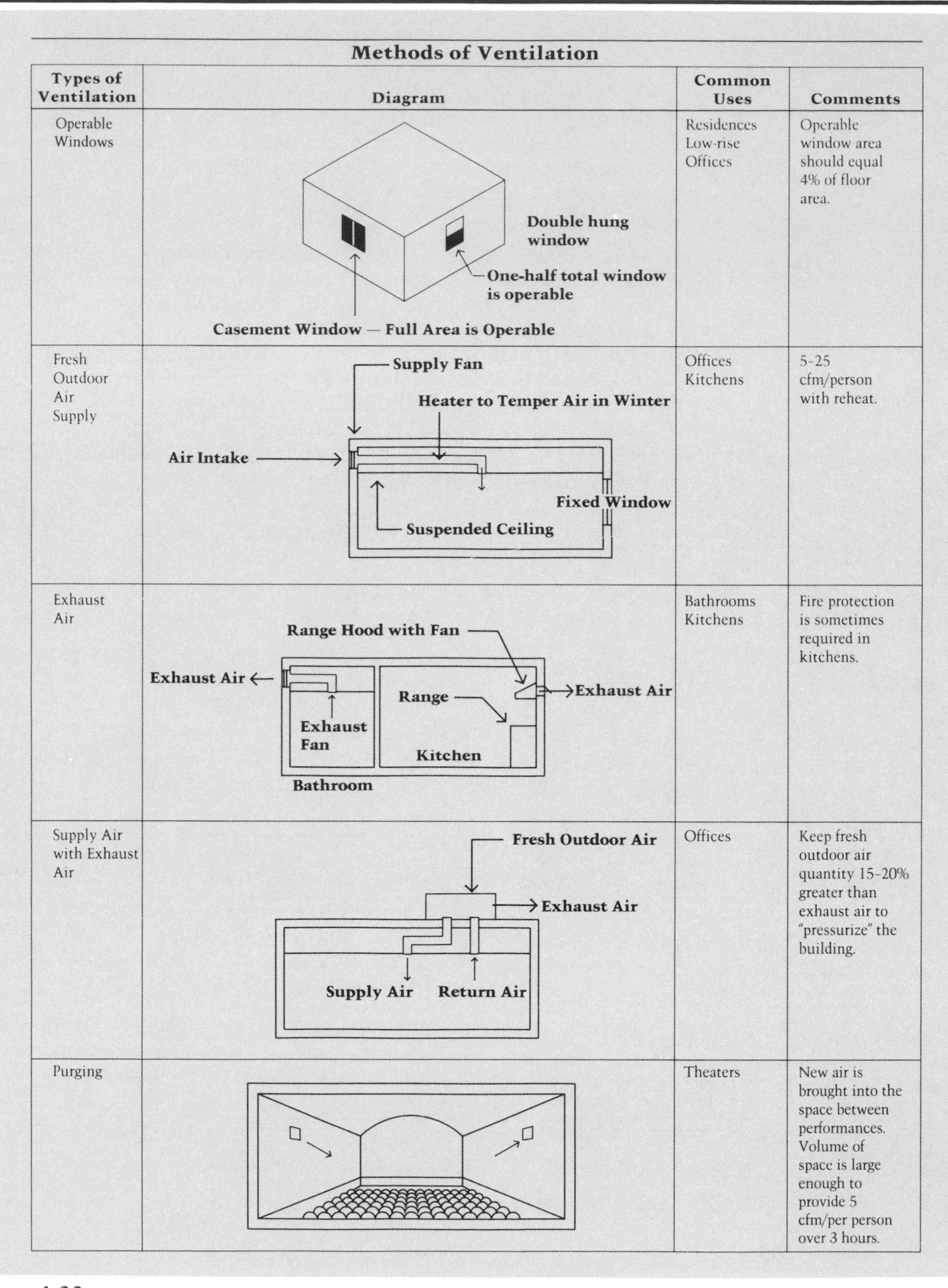

Methods of Ventilation

Types of Ventilation	Diagram	Common Uses	Comments
Operable Windows	Double hung window One-half total window is operable Casement Window — Full Area is Operable	Residences Low-rise Offices	Operable window area should equal 4% of floor area.
Fresh Outdoor Air Supply	Supply Fan Heater to Temper Air in Winter Air Intake Fixed Window Suspended Ceiling	Offices Kitchens	5-25 cfm/person with reheat.
Exhaust Air	Range Hood with Fan Exhaust Air Range Exhaust Air Exhaust Fan Kitchen Bathroom	Bathrooms Kitchens	Fire protection is sometimes required in kitchens.
Supply Air with Exhaust Air	Fresh Outdoor Air Exhaust Air Supply Air Return Air	Offices	Keep fresh outdoor air quantity 15-20% greater than exhaust air to "pressurize" the building.
Purging		Theaters	New air is brought into the space between performances. Volume of space is large enough to provide 5 cfm/per person over 3 hours.

Figure 1.22

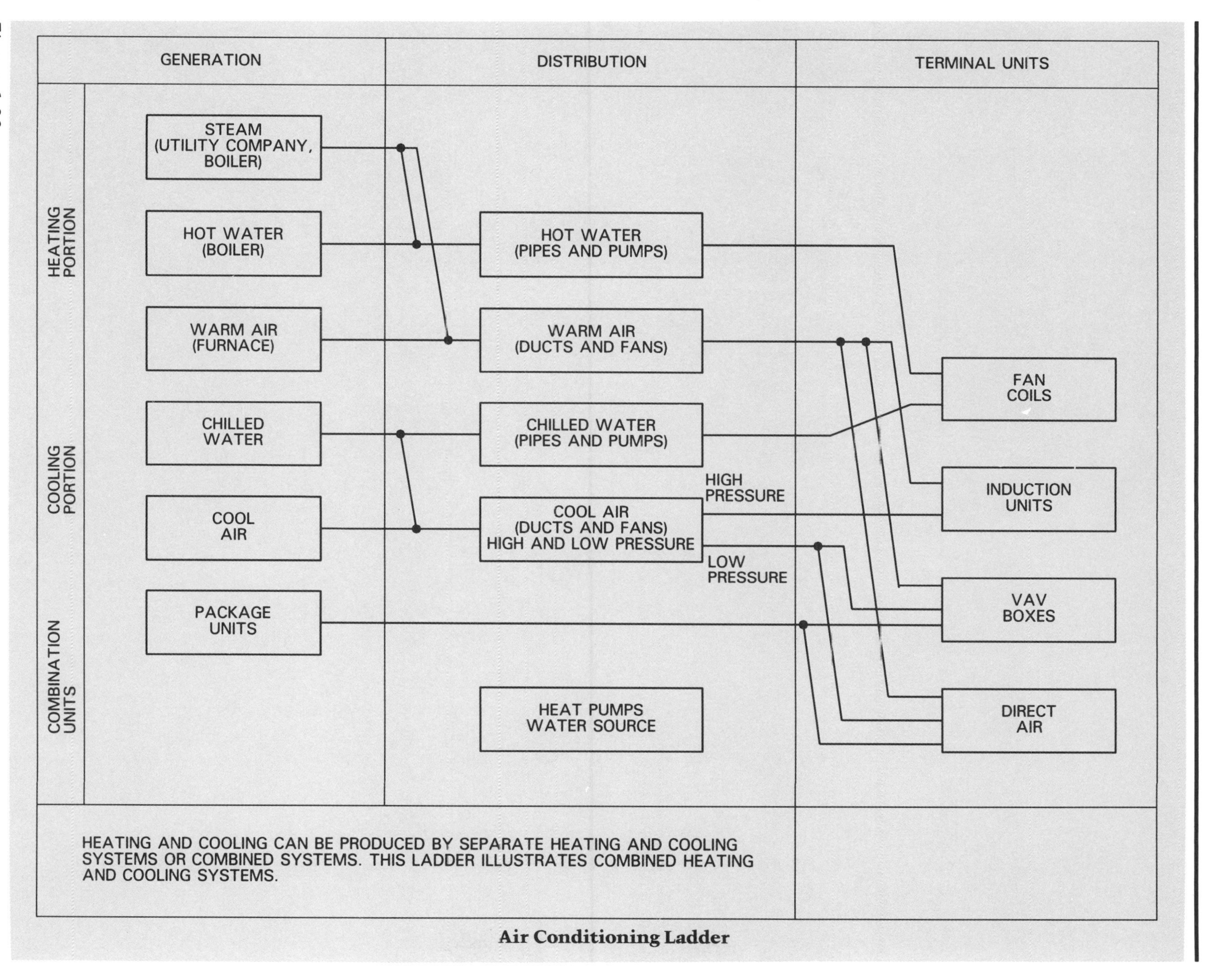

Air Conditioning Ladder

Figure 1.23

CHAPTER TWO

HEATING AND COOLING LOADS

The size of heating and cooling equipment is determined by the "load" it must carry. For a boiler, the load is the amount of heat that must be pushed into the building per hour to keep it warm. For cooling, the load is the amount of heat per hour that must be taken away from the building to keep it cool. Methods for determining both heating and cooling loads are discussed in this chapter.

Heating Loads

Heat is measured in British thermal units, or Btu's. One Btu is equal to the amount of energy necessary to heat one pound of water 1°F, as illustrated in Figure 2.1. Since buildings require large quantities of heat, it is easier to work with smaller units. Therefore, 1,000 Btu's are often written as 1 MBH.

Water

The properties of water are basically constant in the 32–212°F range. Water is easy to work with, because it is a liquid with nearly constant physical characteristics over the range of temperatures normally used in heating, and is readily available and chemically stable. For these reasons, water is the most commonly used heating fluid.

Heating and Sensible Heat

Heating systems are often designed without humidification; all of the heat energy is used to warm the air in a room. When heating systems are designed without humidification, all of the added heat is termed *sensible heat*, because every Btu added to the air is devoted only to raising the air temperature. This heat can be felt or "sensed." When humidification is incorporated into heating systems, the heat load (number of Btu's necessary) increases by approximately 30 percent. (Humidification and latent heat are discussed in more detail in the cooling section of this chapter.)

Heat Loss

The heating load for a building is determined by calculating the amount of heat lost from the building to the outside during the winter design condition. (The common winter design condition is the cold winter night in a particular climate that is surpassed 97½% of the time.)

There are three ways in which heat is lost from a building: *conduction*, *convection*, and *radiation*. The total MBH of these three types of heat loss equals the heat load, or the amount of heat that must be put into the building by the heating system to offset these losses and maintain the proposed comfort zone.

Conduction transfers heat in a chain-like manner. As one molecule heats up, it transfers heat to the molecule next to it. With this type of heat transference, the molecules are stationary and the heat moves, or is conducted, through them. For example, if you hold one end of a steel poker and use the other end to stir a fire, your hand gets warmer as the heat from the fire moves, or is conducted, along the shaft to your hand. Similarly, a building's heat is conducted to the outdoors through the solid surfaces of walls, roofs, floor slabs, glass doors, and windows.

Convection transfers heat as warm molecules actually move from one place to another. If a current of air was passed over the heated poker noted above, the heat of the rod would be transferred to the passing air, or convected. In buildings, heat is convected from the interior to the outdoors by air that leaks, or infiltrates, through cracks around windows and doors, and by the exhaust and ventilation air that moves between the interior and the exterior of the building. Most HVAC distribution systems work by convection. The hot water that moves heat from the boiler through the pipes to the fan coil units and the warm air furnace that distributes heat to the room via ductwork

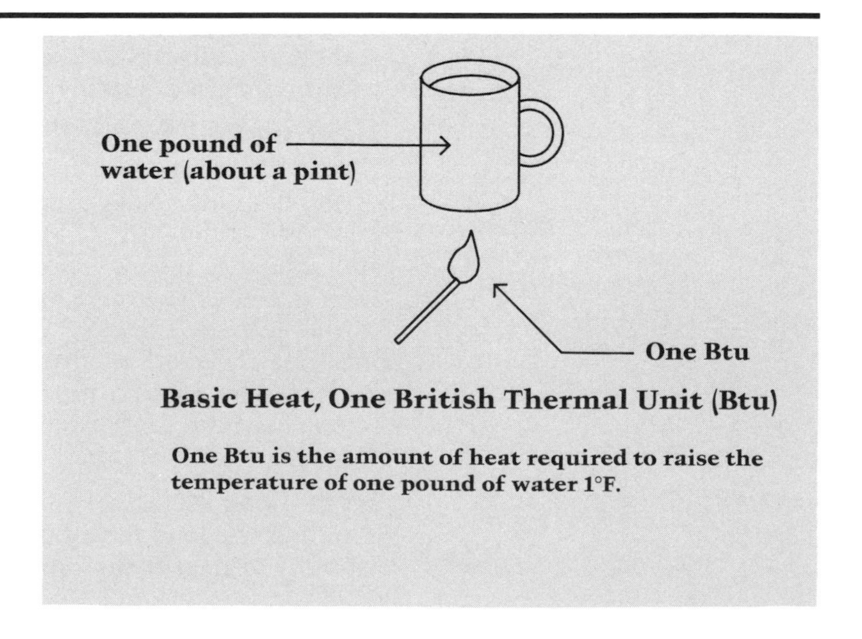

Figure 2.1

are moving heat around the building by convection, either by force or by gravity.

Radiation transfers heat by electromagnetic waves. Sunlight is the most common, and most spectacular, example. Heat from the sun reaches the earth by radiating through 93 million miles of space. In the steel poker example, heat is radiated when it gets "red hot," at which time the heat can be felt several feet away. Heat is radiated onto buildings by the sun during the day. Solar radiation is primarily considered as a factor in calculating the cooling load, where it must be overcome. It is not a significant factor in calculating the heating load.

Figure 2.2 illustrates the basic processes of conduction, convection, and radiation. The heat loads that commonly occur in buildings are illustrated in Figure 2.3. Each type of heat loss is described in the following sections.

Conduction Heat Losses

Conduction heat losses (symbolized by "H_c" in Figure 2.3) are calculated by adding together the conduction heat losses from each separate building material. The formula for computing conduction losses for each material is shown below. Each element is then discussed individually.

$H_c = UA(\Delta T)$

H_c = Conduction heat losses from building materials (Btu/hr.)

U = Overall heat transfer coefficient (Btu/hr. × s.f. × °F) (see Figures 2.5–2.7)

A = Surface area (s.f.)

ΔT = Temperature difference (°F)

Overall Heat Transfer Coefficient: The value of the overall heat transfer coefficient ("U" in the formula for computing conduction heat loss shown above) is based on the type of materials comprising the building enclosure. For example, an uninsulated wood-frame wall has a U value of 0.23 Btu/s.f./hr./°F. This means that 0.23 Btu is conducted through each square foot of wall surface each hour for each 1°F temperature difference between the inside and outside wall surface temperatures. If four inches of insulation are added to this wall, the U value becomes 0.07 Btu/s.f./hr./°F. The value is lowered because less heat is conducted through a better insulated wall. Insulation, which is relatively inexpensive, makes the wall surface more than three times more effective. This is illustrated in Figure 2.4.

The U value is equal to the reciprocal of the sum of the resistances of the components of the building envelope. Computation of the U value is illustrated in Figure 2.5. This formula is shown below.

$$U = \frac{1}{\text{Total resistance}} = \frac{1}{R_T}$$

The **total resistance**, or R_T, equals the sum of the resistances of each individual material making up the building envelope (i.e., $R_T = R_1 + R_2 + R_3 + \ldots$). The resistance of a single material normally varies with its thickness; therefore, in certain cases, the resistance is computed from the material conductivity, or k.

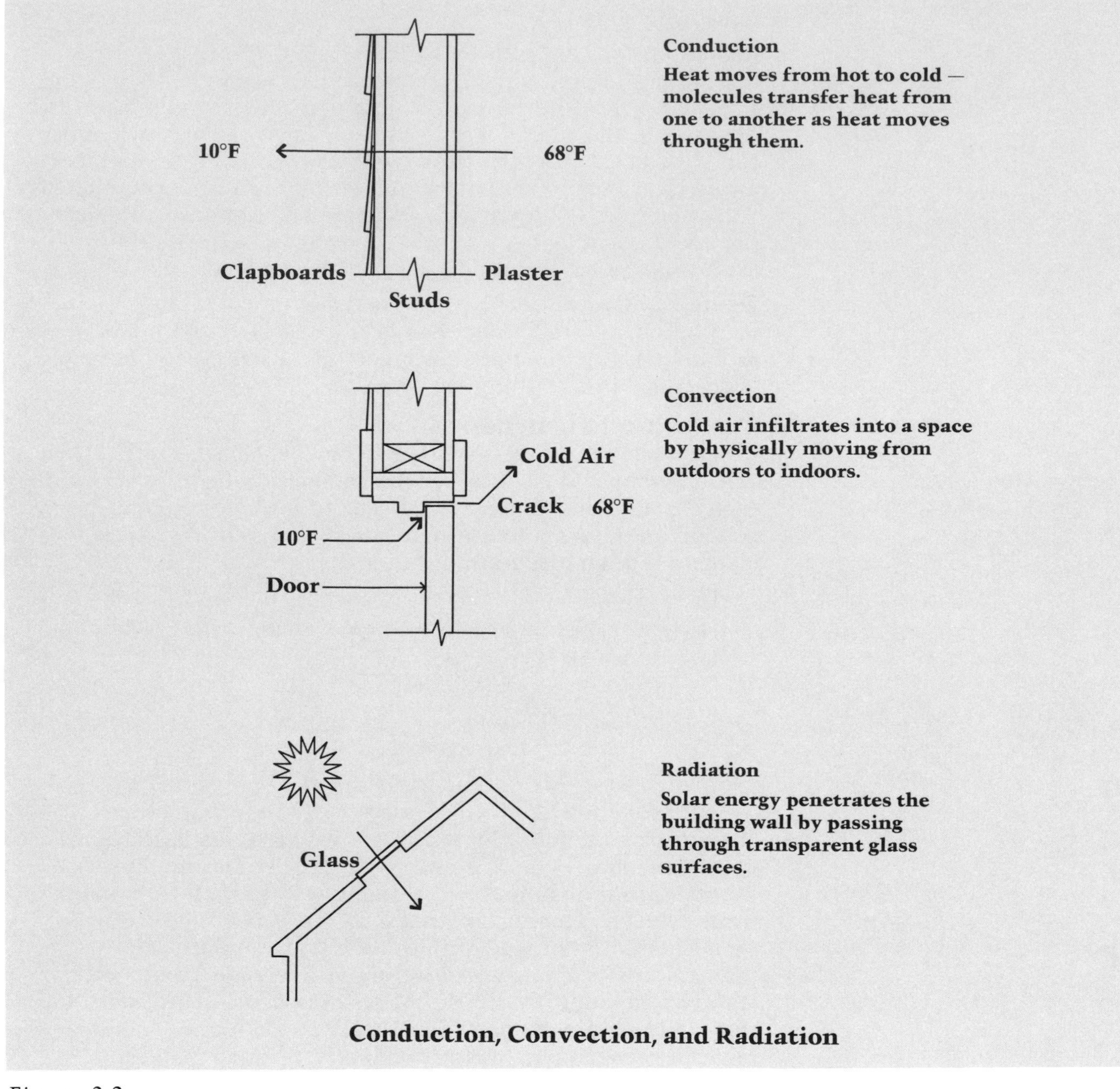
10°F
68°F
Conduction
Heat moves from hot to cold — molecules transfer heat from one to another as heat moves through them.
Clapboards
Plaster
Studs
Cold Air
Convection
Cold air infiltrates into a space by physically moving from outdoors to indoors.
Crack
68°F
10°F
Door
Radiation
Solar energy penetrates the building wall by passing through transparent glass surfaces.
Glass
Conduction, Convection, and Radiation

Figure 2.2

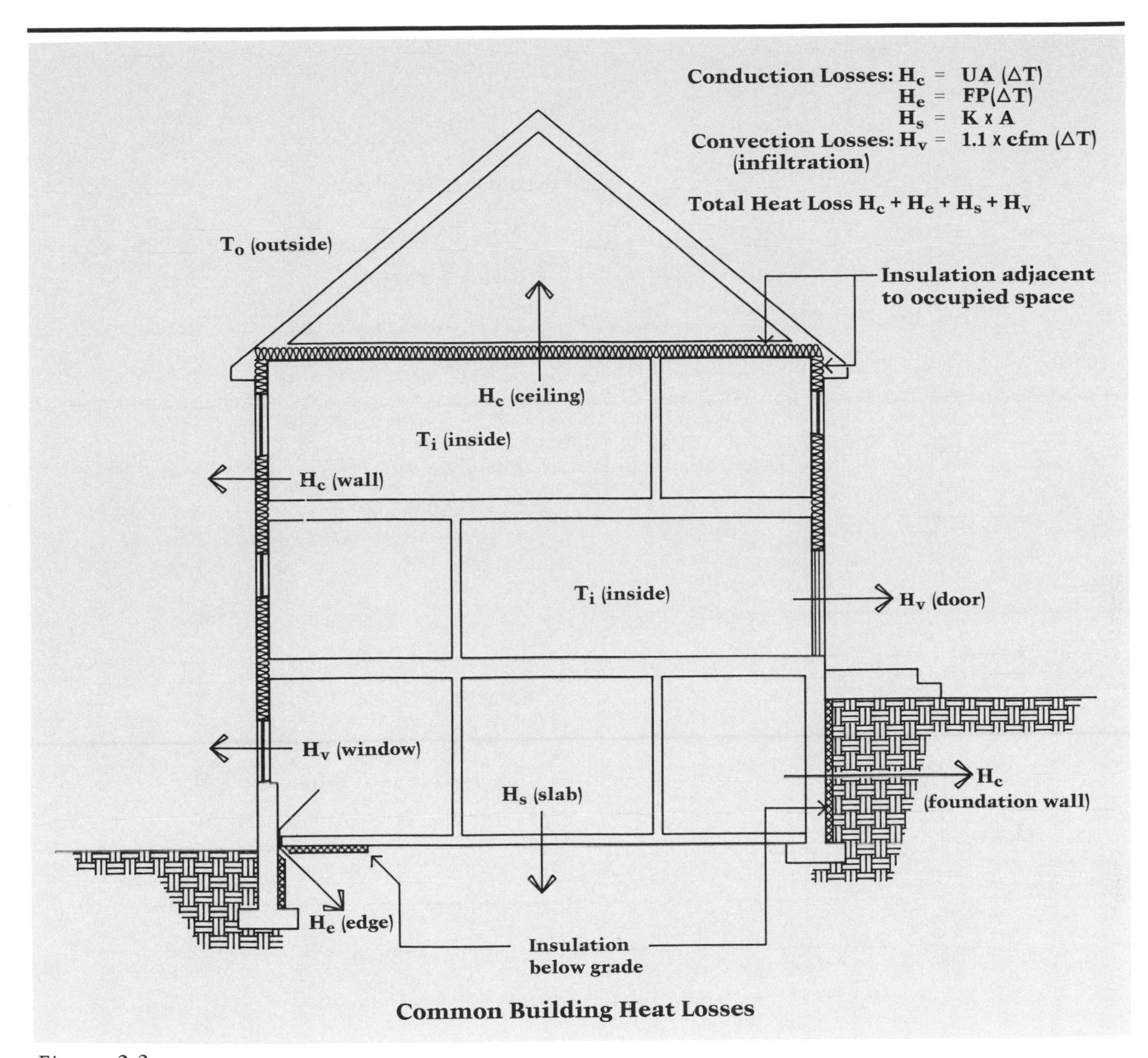

Common Building Heat Losses

Figure 2.3

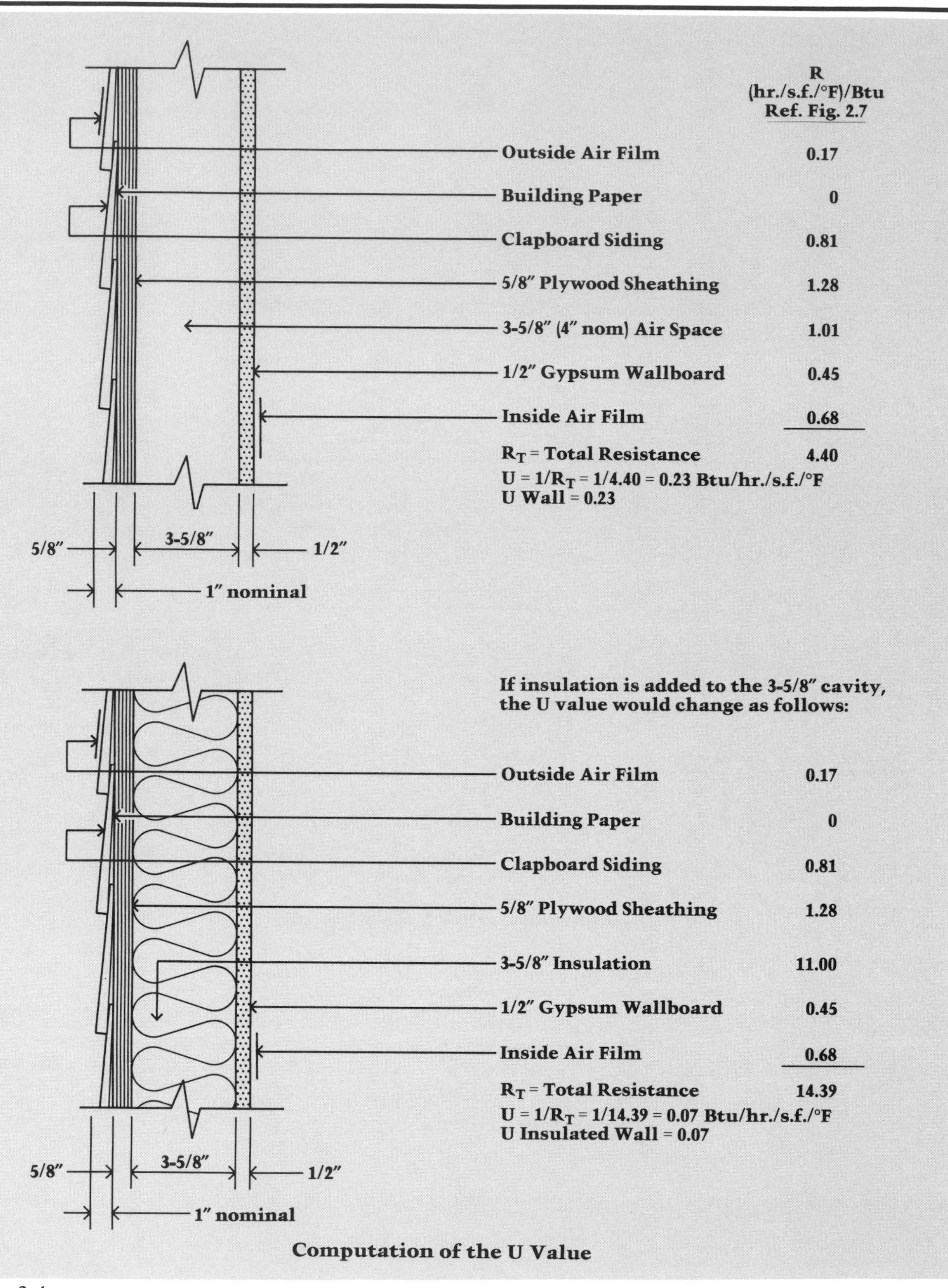

Computation of the U Value

Figure 2.4

Conductivity, Conductance, and the U Value

k - Conductivity Btu/hr./s.f./°F/inch

t = 1″ thick
k = Btu per square foot per hour for each 1°F temperature difference for a 1″ thick material

k is listed for most materials (see Figure 2.7)

T

1′

1′

t

T + 1°F

Heat flow

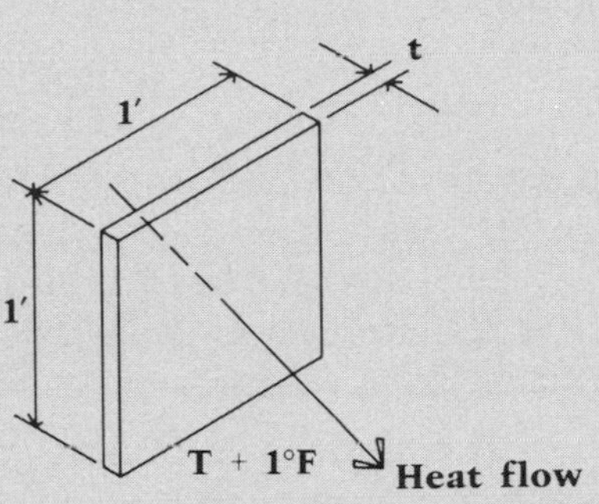

C = Conductance Btu/hr./s.f./°F

C = Btu per square foot per hour for each 1°F temperature difference for the material thickness indicated.

C = k/t

C is listed for common materials (see Figure 2.7)

t = material thickness

T

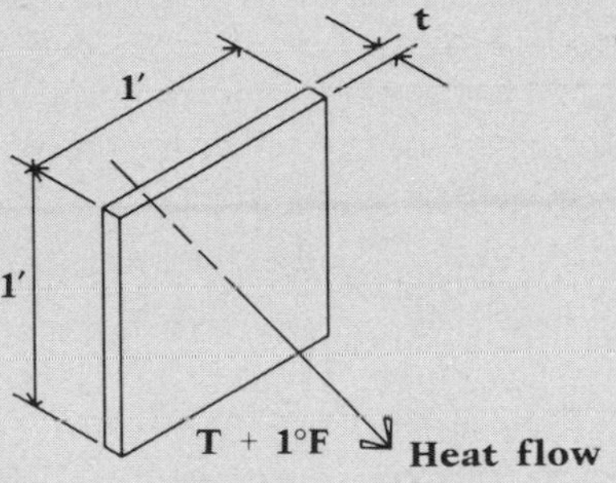

1′

1′

t

T + 1°F

Heat flow

R = Resistance Btu/hr./s.f./°F

R = 1/C or t/k for each material

Air resistances Btu/hr./s.f./°F	Winter	Summer
f_i = Inside air film resistance =	0.68	0.68
f_o = Outside air film resistance =	0.17	0.25
a = Air space (1″ to 4″ nominal) resistance =	0.97	0.86

R_T = Total Resistance

$$R_T = f_i + f_o + R_1 + R_2 + R_3 + \ldots + a \text{ (if applicable)}$$

U = Overall heat transfer coefficient Btu/hr./s.f./°F = $1/R_T$

Figure 2.5

This equals the number of Btu's conducted per square foot per hour for each one degree Fahrenheit temperature difference for a one-inch-thick material. For a material for which the thickness and conductivity is known, the conductance equals the conductivity divided by the thickness, and the resistance equals the reciprocal of the conductance. These formulas are shown below.

$C = k/t$

$R = 1/C$

Thus, for each portion of the building envelope, the U value must be determined. Figure 2.6 lists U values for common building systems and Figure 2.7 lists conductances, conductivities, and resistances (C, k, and R) for common building materials.

Once the overall heat transfer coefficient, U, has been determined for each building envelope system, the total conduction loss is determined by computing the surface area of each material and the change in temperature across each material. Each of these processes is described below.

Surface Area: The surface area, A, for each building material is measured from within the occupied space. Wall areas are measured from the inside finish floor to the underside of the finish ceiling, excluding window and door areas which are computed separately. Slanted ceilings, such as cathedral ceilings, are measured along the slant. Window frames are generally considered as window or glass area.

Temperature Difference (ΔT): The temperature difference, ΔT, is the difference in temperature between the inside and the outside of the wall surface. Figure 2.8 lists indoor design temperatures (T_i) for some typical indoor conditions and Figure 2.9 lists outdoor design temperatures (T_o) for several locations. In winter, this is the difference between the indoor design temperature, 68–72°, and the design outdoor temperature, which varies with location. For example, 9°F is the outdoor design temperature for Boston; thus, the temperature difference is calculated as shown below.

$\Delta T = T_i - T_o$ — in winter; therefore,

$\Delta T = 68 - 9$, or 59°F

In the summer,

$\Delta T = T_o - T_i$

TABLE 23—TRANSMISSION COEFFICIENT U—LIGHT CONSTRUCTION, INDUSTRIAL WALLS*†

FOR SUMMER AND WINTER

Btu/(hr) (sq ft) (deg F temp diff)

All numbers in parentheses indicate weight per sq ft. Total weight per sq ft is sum of wall and finishes.

STUDDING — 3/4" SHEATHING — STUDDING — SHEET IRON

EXTERIOR FINISH	SHEATHING	WEIGHT (lb per sq ft)	INTERIOR FINISH: None	Flat Iron (1)	Insulating Board ½" (2)	Insulating Board 25/32" (3)	Wood ¾" (2)
⅜" Corrugated Transite	None	(1)	1.16	.55	.32	.26	.36
	½" Ins. Board	(2)	.34	.26	.19	.17	.21
	25/32" Ins. Board	(2)	.27	.21	.17	.15	.18
24 Gauge Corrugated Iron	None	(1)	1.40	.60	.33	.27	.38
	½" Ins. Board	(2)	.36	.27	.20	.17	.21
	25/32" Ins. Board	(2)	.28	.22	.17	.15	.18
	¾" Wood	(3)	.46	.33	.22	.19	.24
¾" Wood Siding	None	(2)	.58	.37	.25	.21	.27

1958 ASHAE Guide

Equations: Heat Gain, Btu/hr = (Area, sq ft) × (U value) × (equivalent temp diff, *Table 19*).
Heat Loss, Btu/hr = (Area, sq ft) × (U value) × (outdoor temp − inside temp).

*For addition of air spaces and insulation to walls, refer to *Table 31, page 75*.

†Values apply when sealed with calking compound between sheets, and at ground and roof lines. When sheets are not sealed, increase U factors by 10%. These values may be used for roofs, heat flow up-winter; for heat flow down-summer, multiply above factors by 0.8.

TABLE 24—TRANSMISSION COEFFICIENT U—LIGHTWEIGHT, PREFABRICATED CURTAIN TYPE WALLS*

FOR SUMMER AND WINTER

Btu/(hr) (sq ft) (deg F temp diff)

All numbers in parentheses indicate weight per sq ft. Total weight per sq ft is sum of wall and finishes.

FACING — CORE MATERIAL

INSULATING CORE MATERIAL	DENSITY† (lb/cu ft)	METAL FACING (3) Core Thickness (in.) 1	2	3	4	METAL FACING WITH ¼" AIR SPACE (3) Core Thickness (in.) 1	2	3	4
Glass, Wood, Cotton Fibers	3	.21	.12	.08	.06	.19	.11	.08	.06
Paper Honeycomb	5	.39	.23	.17	.13	.32	.20	.15	.12
Paper Honeycomb with Perlite Fill, Foamglas	9	.29	.17	.12	.09	.25	.15	.11	.09
Fiberboard	15	.36	.21	.15	.12	.29	.19	.14	.11
Wood Shredded (Cemented in Preformed Slabs)	22	.31	.18	.13	.10	.25	.16	.12	.09
Expanded Vermiculite	7	.34	.20	.14	.11	.28	.18	.13	.10
Vermiculite or Perlite Concrete	20	.44	.27	.19	.15	.35	.23	.18	.14
	30	.51	.32	.24	.19	.39	.27	.21	.17
	40	.58	.38	.29	.23	.43	.31	.25	.20
	60	.69	.49	.38	.31	.49	.38	.31	.26

Equations: Heat Gain, Btu/hr = (Area, sq ft) × (U value) × (equivalent temp diff, *Table 19*).
Heat Loss, Btu/hr = (Area, sq ft) × (U value) × (outdoor temp − inside temp).

*For addition of insulation and air spaces to walls, refer to *Table 31, page 75*.

†Total weight per sq ft $= \frac{\text{core density} \times \text{core thickness}}{12} + 3$ lb/sq ft

(Courtesy Carrier Corporation, McGraw-Hill Book Company)

Figure 2.6

TABLE 25—TRANSMISSION COEFFICIENT U—FRAME WALLS AND PARTITIONS*

FOR SUMMER AND WINTER

Btu/(hr) (sq ft) (deg F temp diff)

All numbers in parentheses indicate weight per sq ft. Total weight per sq ft is sum of component materials.

STUDS, EXTERIOR FINISH, PLASTER, PLASTER BASE, SHEATHING

EXTERIOR FINISH	SHEATHING	INTERIOR FINISH								
		None	3/4" Wood Panel (2)	3/8" Gypsum Board (Plaster Board) (2)	Metal Lath Plastered		3/8" Gypsum or Wood Lath Plastered		Insulating Board Plain or Plastered	
					3/4" Sand Plaster(7)	3/4" Lt Wt Plaster(3)	1/2" Sand Plaster(7)	1/2" Lt Wt Plaster(2)	1/2" Board (2)	1" Board (4)
1" Stucco (10)	None, Building Paper	.91	.33	.42	.45	.39	.40	.37	.29	.20
OR Asbestos	5/16" Plywood (1) or 1/2" Gyp (2)	.68	.30	.37	.40	.35	.36	.33	.26	.19
Cement Siding (1)	25/32" Wood & Bldg Paper (2)	.48	.25	.30	.31	.28	.29	.27	.22	.17
OR Asphalt	1/2" Insulating Board (2)	.42	.23	.27	.29	.26	.27	.25	.21	.16
Roll Siding (2)	25/32" Insulating Board (3)	.32	.20	.23	.24	.22	.22	.21	.18	.14
4" Face Brick	None, Building Paper	.73	.30	.37	.40	.35	.36	.33	.26	.19
Veneer (43) OR	5/16" Plywood (1) or 1/2" Gyp (2)	.57	.28	.33	.36	.32	.32	.30	.24	.18
3/8" Plywood (1)	25/32" Wood & Bldg Paper (2)	.42	.23	.27	.29	.26	.27	.25	.21	.16
OR Asphalt	1/2" Insulating Board (2)	.38	.22	.25	.27	.25	.25	.24	.20	.15
Siding (2)	25/32" Insulating Board (3)	.30	.19	.21	.22	.21	.21	.20	.17	.14
Wood Siding (3)	None, Building Paper	.57	.27	.33	.35	.31	.32	.30	.24	.18
OR	5/16" Plywood (1) or 1/2" Gyp (2)	.48	.25	.30	.31	.28	.29	.27	.22	.17
Wood Shingles (2)	25/32" Wood & Bldg Paper	.36	.22	.25	.26	.24	.24	.23	.19	.15
OR 3/4" Wood	1/2" Insulating Board (2)	.33	.20	.23	.24	.22	.23	.22	.18	.14
Panels (3)	25/32" Insulating Board (3)	.27	.18	.20	.21	.19	.19	.19	.16	.13
Wood Shingles	None, Building Paper	.43	.24	.28	.29	.27	.27	.25	.21	.16
Over 5/16" Insul	5/16" Plywood (1) or 1/2" Gyp (2)	.38	.22	.25	.27	.24	.25	.23	.19	.15
Backer Board (3)	25/32" Wood & Bldg Paper	.30	.19	.22	.23	.21	.21	.20	.17	.14
OR Asphalt	1/2" Insulating Board (2)	.28	.18	.20	.21	.20	.20	.19	.16	.13
Insulated Siding (4)	25/32" Insulating Board (3)	.23	.16	.18	.18	.17	.18	.17	.15	.12
Single Partition (Finish on one side only)			.43	.60	.67	.55	.57	.50	.36	.23
Double Partition (Finish on both sides)			.24	.34	.39	.31	.32	.28	.19	.12

1958 ASHAE Guide

Equations: Walls—Heat Gain, Btu/hr = (Area, sq ft) × (U value) × (equivalent temp diff, *Table 19*).
—Heat Loss, Btu/hr = (Area, sq ft) × (U value) × (outdoor temp—inside temp).
Partitions, unconditioned space adjacent—Heat Gain or Loss, Btu/hr = (Area sq ft) × (U value) × (outdoor temp—inside temp—5 F).
Partitions, kitchen or boiler room adjacent—Heat Gain, Btu/hr = (Area sq ft) × (U value)
× (actual temp diff or outdoor temp—inside temp + 15 F to 25 F).

*For addition of insulation and air spaces to partitions, refer to *Table 31, page 75.*

(Courtesy Carrier Corporation, McGraw-Hill Book Company)

Figure 2.6 (cont.)

TABLE 27—TRANSMISSION COEFFICIENT U—FLAT ROOFS COVERED WITH BUILT-UP ROOFING*

FOR HEAT FLOW DOWN—SUMMER. FOR HEAT FLOW UP—WINTER (See Equation at Bottom of Page).

Btu/(hr) (sq ft) (deg F temp diff)

All numbers in parentheses indicate weight per sq ft. Total weight per sq ft is sum of roof, finish and insulation.

TYPE OF DECK	THICKNESS OF DECK (inches) and WEIGHT (lb per sq ft)	CEILING†	INSULATION ON TOP OF DECK, INCHES						
			No Insulation	½ (1)	1 (1)	1½ (2)	2 (3)	2½ (3)	3 (4)
Flat Metal	1 (5)	None or Plaster (6)	.67	.35	.23	.18	.15	.12	.10
		Suspended Plaster (5)	.32	.22	.17	.14	.12	.10	.09
		Suspended Acou Tile (2)	.23	.18	.14	.12	.11	.09	.08
Preformed Slabs—Wood Fiber and Cement Binder	2 (4)	None or Plaster (6)	.20	.16	.13	.11	.10	.09	.08
		Suspended Plaster (5)	.15	.12	.11	.09	.08	.08	.07
		Suspended Acou Tile (2)	.13	.10	.09	.08	.08	.07	.06
	3 (7)	None or Plaster (6)	.14	.11	.10	.09	.08	.08	.07
		Suspended Plaster (5)	.12	.10	.09	.07	.07	.06	.05
		Suspended Acou Tile (2)	.10	.09	.08	.07	.07	.06	.05
Concrete (Sand & Gravel Agg)	4, 6, 8 (47),(70),(93)	None or Plaster (6)	.51	.30	.21	.16	.14	.12	.10
		Suspended Plaster (5)	.28	.20	.16	.13	.12	.10	.09
		Suspended Acou Tile(2)	.21	.16	.13	.11	.10	.09	.08
(Lt Wt Agg on Gypsum Board)	2 (9)	None or Plaster (6)	.27	.20	.15	.13	.11	.10	.08
		Suspended Plaster (5)	.18	.14	.12	.10	.09	.09	.08
		Suspended Acou Tile (2)	.15	.12	.11	.09	.08	.08	.07
	3 (13)	None or Plaster (6)	.21	.16	.13	.11	.10	.09	.08
		Suspended Plaster (5)	.15	.12	.11	.09	.08	.08	.07
		Suspended Acou Tile (2)	.13	.11	.10	.08	.08	.07	.06
	4 (16)	None or Plaster (6)	.17	.14	.11	.10	.09	.08	.07
		Suspended Plaster (5)	.13	.11	.10	.08	.08	.07	.06
		Suspended Acou Tile(2)	.12	.10	.09	.07	.07	.06	.05
Gypsum Slab on ½" Gypsum Board	2 (11)	None or Plaster (6)	.32	.22	.17	.14	.12	.10	.09
		Suspended Plaster (5)	.21	.17	.13	.11	.10	.09	.08
		Suspended Acou Tile (2)	.17	.13	.12	.10	.09	.08	.07
	3 (15)	None or Plaster (6)	.27	.19	.15	.13	.11	.10	.08
		Suspended Plaster (5)	.19	.15	.13	.11	.10	.09	.08
		Suspended Acou Tile (2)	.15	.12	.11	.09	.08	.08	.07
	4 (19)	None or Plaster (6)	.23	.17	.14	.12	.10	.09	.08
		Suspended Plaster (5)	.17	.13	.12	.10	.09	.08	.07
		Suspended Acou Tile (2)	.14	.12	.11	.09	.08	.08	.07
Wood	1 (3)	None or Plaster (6)	.40	.26	.19	.15	.13	.11	.09
		Suspended Plaster (5)	.24	.18	.14	.12	.11	.09	.08
		Suspended Acou Tile (2)	.19	.15	.13	.11	.10	.08	.07
	2 (5)	None or Plaster (6)	.28	.20	.16	.13	.11	.10	.08
		Suspended Plaster (5)	.19	.15	.13	.11	.10	.09	.07
		Suspended Acou Tile (2)	.16	.13	.11	.10	.09	.08	.07
	3 (8)	None or Plaster (6)	.21	.16	.13	.11	.10	.09	.08
		Suspended Plaster (5)	.16	.13	.11	.09	.09	.08	.07
		Suspended Acou Tile (2)	.13	.11	.10	.09	.08	.07	.06

1958 ASHAE Guide

Equations: Summer—(Heat Flow Down) Heat Gain, Btu/hr = (Area, sq ft) × (U value) × (equivalent temp diff, *Table 20*).

Winter—(Heat Flow Up) Heat Loss, Btu/hr = (Area, sq ft) × (U value × 1.1) × (outdoor temp—inside temp).

*For addition of air spaces or insulation to roofs, refer to *Table 31, page 75.*

†For suspended ½" insulation board, plain (.6) or with ½" sand aggregate plaster (5), use values of suspended acou tile.

(Courtesy Carrier Corporation, McGraw-Hill Book Company)

Figure 2.6 (cont.)

TABLE 31—TRANSMISSION COEFFICIENT U—WITH INSULATION & AIR SPACES

SUMMER AND WINTER

Btu/(hr) (sq ft) (deg F temp diff)

U Value Before Adding Insul. Wall, Ceiling, Roof Floor	Addition of Fibrous Insulation — Thickness (Inches)			Add'n of Air Space ¾″ or more *	Addition of Reflective Sheets to Air Space (Aluminum Foil Average Emissivity = .05) — Direction of Heat Flow								
					Winter and Summer Horizontal			Summer Down			Winter Up		
	1	2	3		Added to one or both sides	One sheet in air space	Two sheets in air space	Added to one or both sides	One sheet in air space	Two sheets in air space	Added to one or both sides	One sheet in air space	Two sheets in air space
.60	.19	.11	.08	.38	.34	.18	.11	.12	.06	.05	.36	.20	.14
.58	.19	.11	.08	.37	.33	.18	.11	.12	.06	.05	.36	.20	.14
.56	.18	.11	.08	.36	.32	.18	.11	.11	.06	.05	.35	.20	.14
.54	.18	.11	.08	.36	.31	.17	.11	.11	.06	.05	.34	.19	.14
.52	.18	.11	.08	.35	.30	.17	.10	.11	.06	.05	.33	.19	.14
.50	.18	.11	.08	.34	.29	.17	.10	.11	.06	.05	.32	.19	.13
.48	.17	.11	.08	.33	.28	.16	.10	.11	.06	.04	.31	.18	.13
.46	.17	.10	.08	.32	.28	.16	.10	.11	.06	.04	.30	.18	.13
.44	.17	.10	.07	.31	.27	.16	.10	.11	.06	.04	.29	.18	.13
.42	.16	.10	.07	.30	.26	.15	.10	.11	.06	.04	.28	.17	.13
.40	.16	.10	.07	.29	.26	.15	.10	.10	.06	.04	.27	.17	.12
.38	.16	.10	.07	.28	.25	.15	.09	.10	.06	.04	.26	.17	.12
.36	.15	.10	.07	.27	.24	.14	.09	.10	.06	.04	.25	.16	.12
.34	.15	.10	.07	.26	.23	.14	.09	.10	.06	.04	.24	.16	.12
.32	.15	.10	.07	.25	.22	.13	.09	.10	.05	.04	.23	.15	.11
.30	.14	.09	.07	.23	.21	.13	.09	.10	.05	.04	.22	.15	.11
.28	.14	.09	.07	.22	.20	.13	.08	.09	.05	.04	.20	.14	.10
.26	.13	.09	.07	.21	.19	.12	.08	.09	.05	.04	.19	.13	.10
.24	.13	.09	.07	.20	.17	.12	.08	.09	.05	.04	.18	.13	.10
.22	.12	.08	.06	.18	.16	.11	.08	.08	.05	.04	.16	.12	.09
.20	.12	.08	.06	.17	.15	.10	.07	.08	.05	.04	.15	.11	.09
.18	.11	.08	.06	.15	.14	.10	.07	.08	.05	.04	.14	.11	.08
.16	.10	.07	.06	.14	.12	.09	.07	.07	.05	.04	.13	.10	.08
.14	.09	.07	.05	.12	.11	.08	.06	.07	.04	.04	.12	.09	.07
.12	.08	.06	.05	.11	.10	.08	.06	.06	.04	.03	.10	.08	.07
.10	.07	.06	.05	.09	.08	.07	.05	.06	.04	.03	.09	.07	.06

1958 ASHAE Guide

Insulation Added	Air Space Added	Reflective Sheets Added to One or Both Sides	Reflective Sheet in Air Space	Reflective Sheets in Air Space

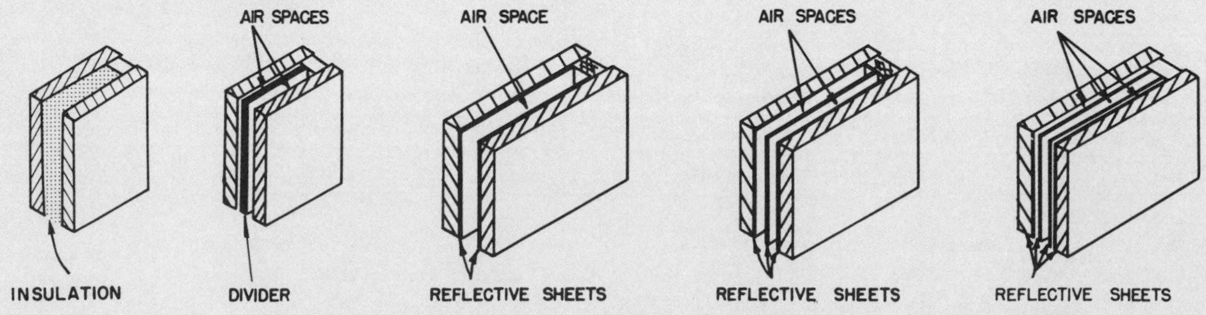

INSULATION
AIR SPACES
DIVIDER
AIR SPACE
REFLECTIVE SHEETS
AIR SPACES
REFLECTIVE SHEETS
AIR SPACES
REFLECTIVE SHEETS

*Checked for summer conditions for up, down and horizontal heat flow. Error from above values is less than 1%.

(Courtesy Carrier Corporation, McGraw–Hill Book Company)

Figure 2.6 (cont.)

TABLE 32—TRANSMISSION COEFFICIENT U—FLAT ROOFS WITH ROOF-DECK INSULATION

SUMMER AND WINTER

Btu/(hr) (sq ft) (deg F temp diff)

U VALUE OF ROOF BEFORE ADDING ROOF DECK INSULATION	Addition of Roof-Deck Insulation Thickness (in.)					
	½	1	1½	2	2½	3
.60	.33	.22	.17	.14	.12	.10
.50	.29	.21	.16	.14	.12	.10
.40	.26	.19	.15	.13	.11	.09
.35	.24	.18	.14	.12	.10	.09
.30	.21	.16	.13	.12	.10	.09
.25	.19	.15	.12	.11	.09	.08
.20	.16	.13	.11	.10	.09	.08
.15	.12	.11	.09	.08	.08	.07
.10	.09	.08	.07	.07	.06	.05

TABLE 33—TRANSMISSION COEFFICIENT U—WINDOWS, SKYLIGHTS, DOORS & GLASS BLOCK WALLS

Btu/(hr) (sq ft) (deg F temp diff)

GLASS

	Vertical Glass							Horizontal Glass			
	Single	Double			Triple			Single		Double (¼″)	
Air Space Thickness (in.)		¼	½	¾-4	¼	½	¾-4	Summer	Winter	Summer	Winter
Without Storm Windows	1.13	0.61	0.55	0.53	0.41	0.36	0.34	0.86	1.40	0.50	0.70
With Storm Windows	0.54							0.43	0.64		

DOORS

Nominal Thickness of Wood (inches)	U Exposed Door	U With Storm Door
1	0.69	0.35
1¼	0.59	0.32
1½	0.52	0.30
1¾	0.51	0.30
2	0.46	0.28
2½	0.38	0.25
3	0.33	0.23
Glass (¾″ Herculite)	1.05	0.43

HOLLOW GLASS BLOCK WALLS

Description*	U
5¾x5¾x3⅞″ Thick—Nominal Size 6x6x4 (*14*)	0.60
7¾x7¾x3⅞″ Thick—Nominal Size 8x8x4 (*14*)	0.56
11¾x11¾x3⅞″ Thick—Nominal Size 12x12x4 (*16*)	0.52
7¾x7¾x3⅞″ Thick with glass fiber screen dividing the cavity (*14*)	0.48
11¾x11¾x3⅞″ Thick with glass fiber screen dividing the cavity (*16*)	0.44

1958 ASHAE Guide

Equation: Heat Gain or Loss, Btu/hr = (Area, sq ft) × (U value) × (outdoor temp — inside temp)

*Italicized numbers in parentheses indicate weight in lb per sq ft.

(Courtesy Carrier Corporation, McGraw-Hill Book Company)

Figure 2.6 (cont.)

Table 3A Thermal Properties of Typical Building and Insulating Materials—Design Values[a]

Description	Density lb/ft³	Conductivity[b] λ Btu•in./h•ft²•F	Conductance (C) Btu/h•ft²•F	Resistance[c] (R) Per inch thickness (1/λ) h•ft²•F/Btu	Resistance[c] (R) For thickness listed (1/C) h•ft²•F/Btu	Specific Heat Btu/lb • deg F
BUILDING BOARD						
Boards, Panels, Subflooring, Sheathing						
Woodboard Panel Products						
Asbestos-cement board	120	4.0	—	*0.25*	—	0.24
Asbestos-cement board 0.125 in.	120	—	33.00	—	*0.03*	
Asbestos-cement board 0.25 in.	120	—	16.50	—	*0.06*	
Gypsum or plaster board 0.375 in.	50	—	3.10	—	*0.32*	0.26
Gypsum or plaster board 0.5 in.	50	—	2.22	—	*0.45*	
Gypsum or plaster board 0.625 in.	50	—	1.78	—	*0.56*	
Plywood (Douglas Fir)[o]	34	0.80	—	*1.25*	—	0.29
Plywood (Douglas Fir) 0.25 in.	34	—	3.20	—	*0.31*	
Plywood (Douglas Fir) 0.375 in.	34	—	2.13	—	*0.47*	
Plywood (Douglas Fir) 0.5 in.	34	—	1.60	—	*0.62*	
Plywood (Douglas Fir) 0.625 in.	34	—	1.29	—	*0.77*	
Plywood or wood panels 0.75 in.	34	—	1.07	—	*0.93*	0.29
Vegetable Fiber Board						
Sheathing, regular density 0.5 in.	18	—	0.76	—	*1.32*	0.31
0.78125 in.	18	—	0.49	—	*2.06*	
Sheathing intermediate density 0.5 in.	22	—	0.82	—	*1.22*	0.31
Nail-base sheathing 0.5 in.	25	—	0.88	—	*1.14*	0.31
Shingle backer 0.375 in.	18	—	1.06	—	*0.94*	0.31
Shingle backer 0.3125 in.	18	—	1.28	—	*0.78*	
Sound deadening board 0.5 in.	15	—	0.74	—	*1.35*	0.30
Tile and lay-in panels, plain or acoustic	18	0.40	—	*2.50*	—	0.14
0.5 in.	18	—	0.80	—	*1.25*	
0.75 in.	18	—	0.53	—	*1.89*	
Laminated paperboard	30	0.50	—	*2.00*	—	0.33
Homogeneous board from repulped paper	30	0.50	—	*2.00*	—	0.28
Hardboard						
Medium density	50	0.73	—	*1.37*	—	0.31
High density, service temp. service underlay	55	0.82	—	*1.22*	—	0.32
High density, std. tempered	63	1.00	—	*1.00*	—	0.32
Particleboard						
Low density	37	0.54	—	*1.85*	—	0.31
Medium density	50	0.94	—	*1.06*	—	0.31
High density	62.5	1.18	—	*0.85*	—	0.31
Underlayment 0.625 in.	40	—	1.22	—	*0.82*	0.29
Wood subfloor 0.75 in.		—	1.06	—	*0.94*	0.33
BUILDING MEMBRANE						
Vapor—permeable felt	—	—	16.70	—	*0.06*	
Vapor—seal, 2 layers of mopped 15-lb felt	—	—	8.35	—	*0.12*	
Vapor—seal, plastic film	—	—	—	—	Negl.	
FINISH FLOORING MATERIALS						
Carpet and fibrous pad	—	—	0.48	—	*2.08*	0.34
Carpet and rubber pad	—	—	0.81	—	*1.23*	0.33
Cork tile 0.125 in.	—	—	3.60	—	*0.28*	0.48
Terrazzo 1 in.	—	—	12.50	—	*0.08*	0.19
Tile—asphalt, linoleum, vinyl, rubber	—	—	20.00	—	*0.05*	0.30
vinyl asbestos						0.24
ceramic						0.19
Wood, hardwood finish 0.75 in.	—	—	1.47	—	*0.68*	
INSULATING MATERIALS						
Blanket and Batt[d]						
Mineral Fiber, fibrous form processed from rock, slag, or glass						
approx.[e] 3–4 in.	0.3-2.0	—	0.091	—	*11*[d]	
approx.[e] 3.5 in.	0.3-2.0	—	0.077	—	13[d]	
approx.[e] 5.5–6.5 in.	0.3-2.0	—	0.053	—	*19*[d]	
approx.[e] 6–7.5 in.	0.3-2.0	—	0.045	—	*22*[d]	
approx.[e] 9–10 in.	0.3-2.0	—	0.033	—	*30*[d]	
approx.[e] 12–13 in.	0.3-2.0	—	0.026	—	*38*[d]	

(Courtesy ASHRAE, 1985 Fundamentals)

Figure 2.7

Table 3A Thermal Properties of Typical Building and Insulating Materials—Design Values[a]

Description	Density lb/ft³	Conductivity[b] λ Btu•in./h•ft²•F	Conductance (C) Btu/h•ft²•F	Resistance[c] (R) Per inch thickness (1/λ) h•ft²•F/Btu	Resistance[c] (R) For thickness listed (1/C) h•ft²•F/Btu	Specific Heat, Btu/lb • deg F
Board and Slabs						
Cellular glass	8.5	0.35	—	*2.86*	—	0.18
Glass fiber, organic bonded	4-9	0.25	—	*4.00*	—	0.23
Expanded perlite, organic bonded	1.0	0.36	—	*2.78*	—	0.30
Expanded rubber (rigid)	4.5	0.22	—	*4.55*	—	0.40
Expanded polystyrene extruded						
Cut cell surface	1.8	0.25	—	*4.00*	—	0.29
Smooth skin surface	1.8-3.5	0.20	—	*5.00*	—	0.29
Expanded polystyrene, molded beads	1.0	0.26	—	*3.85*	—	—
	1.25	0.25	—	*4.00*	—	—
	1.5	0.24	—	*4.17*	—	—
	1.75	0.24	—	*4.17*	—	—
	2.0	0.23	—	*4.35*	—	—
Cellular polyurethane[f] (R-11 exp.)(unfaced)	1.5	0.16	—	*6.25*	—	0.38
Cellular polyisocyanurate[n] (R-11 exp.) (foil faced, glass fiber-reinforced core)	2.0	0.14	—	*7.20*	—	0.22
Nominal 0.5 in.		—	0.278	—	*3.6*	
Nominal 1.0 in.		—	0.139	—	*7.2*	
Nominal 2.0 in.		—	0.069	—	*14.4*	
Mineral fiber with resin binder	15.0	0.29	—	*3.45*	—	0.17
Mineral fiberboard, wet felted						
Core or roof insulation	16-17	0.34	—	*2.94*	—	
Acoustical tile	18.0	0.35	—	*2.86*	—	0.19
Acoustical tile	21.0	0.37	—	*2.70*	—	
Mineral fiberboard, wet molded						
Acoustical tile[h]	23.0	0.42	—	*2.38*	—	0.14
Wood or cane fiberboard						
Acoustical tile[g] 0.5 in.	—	—	0.80	—	*1.25*	0.31
Acoustical tile[g] 0.75 in.	—	—	0.53	—	*1.89*	
Interior finish (plank, tile)	15.0	0.35	—	*2.86*	—	0.32
Cement fiber slabs (shredded wood with Portland cement binder	25-27.0	0.50-0.53	—	*2.0-1.89*	—	—
Cement fiber slabs (shredded wood with magnesia oxysulfide binder)	22.0	0.57	—	*1.75*	—	0.31
LOOSE FILL						
Cellulosic insulation (milled paper or wood pulp)	2.3-3.2	0.27-0.32	—	*3.70-3.13*	—	0.33
Sawdust or shavings	8.0-15.0	0.45	—	*2.22*	—	0.33
Wood fiber, softwoods	2.0-3.5	0.30	—	*3.33*	—	0.33
Perlite, expanded	2.0-4.1	0.27-0.31	—	*3.7-3.3*	—	0.26
	4.1-7.4	0.31-0.36	—	*3.3-2.8*	—	
	7.4-11.0	0.36-0.42	—	*2.8-2.4*	—	
Mineral fiber (rock, slag or glass)						
approx.[e] 3.75-5 in.	0.6-2.0	—	—		11.0	0.17
approx.[e] 6.5-8.75 in.	0.6-2.0	—	—		19.0	
approx.[e] 7.5-10 in.	0.6-2.0	—	—		22.0	
approx.[e] 10.25-13.75 in.	0.6-2.0	—	—		30.0	
Mineral fiber (rock, slag or glass)						
approx.[e] 3.5 in. (closed sidewall application)	2.0-3.5	—	—	—	12.0-14.0	
Vermiculite, exfoliated	7.0-8.2	0.47	—	*2.13*	—	0.32
	4.0-6.0	0.44	—	*2.27*	—	
FIELD APPLIED[q]						
Polyurethane foam	1.5-2.5	0.16-0.18	—	6.25-5.26	—	
Ureaformaldehyde foam	0.7-1.6	0.22-028	—	3.57-4.55	—	
Spray cellulosic fiber base	2.0-6.0	0.24-0.30	—	3.33-4.17	—	
PLASTERING MATERIALS						
Cement plaster, sand aggregate	116	5.0	—	*0.20*	—	0.20
Sand aggregate 0.375 in.	—	—	13.3	—	*0.08*	0.20
Sand aggregate 0.75 in.	—	—	6.66	—	*0.15*	0.20
Gypsum plaster:						
Lightweight aggregate 0.5 in.	45	—	3.12	—	*0.32*	
Lightweight aggregate 0.625 in.	45	—	2.67	—	*0.39*	
Lightweight agg. on metal lath 0.75 in.	—	—	2.13	—	*0.47*	
Perlite aggregate	45	1.5	—	*0.67*	—	0.32

(Courtesy ASHRAE, 1985 Fundamentals)

Figure 2.7 (cont.)

Table 3A Thermal Properties of Typical Building and Insulating Materials—Design Values[a]

Description	Density lb/ft³	Conductivity[b] λ Btu • in./h • ft² • F	Conductance (C) Btu/h • ft² • F	Resistance[c] (R) Per inch thickness (1/λ) h • ft² • F/Btu	Resistance[c] (R) For thickness listed (1/C) h • ft² • F/Btu	Specific Heat, Btu/lb • deg F
PLASTERING MATERIALS						
Sand aggregate	105	5.6	—	*0.18*	—	0.20
Sand aggregate . . . 0.5 in.	105	—	11.10	—	*0.09*	
Sand aggregate . . . 0.625 in.	105	—	9.10	—	*0.11*	
Sand aggregate on metal lath . . . 0.75 in.	—	—	7.70	—	*0.13*	
Vermiculite aggregate	45	1.7	—	*0.59*	—	
MASONRY MATERIALS						
Concretes						
Cement mortar	116	5.0	—	*0.20*	—	
Gypsum-fiber concrete 87.5% gypsum, 12.5% wood chips	51	1.66	—	*0.60*	—	0.21
Lightweight aggregates including expanded shale, clay or slate; expanded slags; cinders; pumice; vermiculite; also cellular concretes	120	5.2	—	*0.19*	—	
	100	3.6	—	*0.28*	—	
	80	2.5	—	*0.40*	—	
	60	1.7	—	*0.59*	—	
	40	1.15	—	*0.86*	—	
	30	0.90	—	*1.11*	—	
	20	0.70	—	*1.43*	—	
Perlite, expanded	40	0.93	—	*1.08*	—	
	30	0.71	—	*1.41*	—	
	20	0.50	—	*2.00*	—	0.32
Sand and gravel or stone aggregate (oven dried)	140	9.0	—	*0.11*	—	0.22
Sand and gravel or stone aggregate (not dried)	140	12.0	—	*0.08*	—	
Stucco	116	5.0	—	*0.20*	—	
MASONRY UNITS						
Brick, common[i]	120	5.0	—	*0.20*	—	0.19
Brick, face[i]	130	9.0	—	*0.11*	—	
Clay tile, hollow:						
1 cell deep . . . 3 in.	—	—	1.25	—	*0.80*	0.21
1 cell deep . . . 4 in.	—	—	0.90	—	*1.11*	
2 cells deep . . . 6 in.	—	—	0.66	—	*1.52*	
2 cells deep . . . 8 in.	—	—	0.54	—	*1.85*	
2 cells deep . . . 10 in.	—	—	0.45	—	*2.22*	
3 cells deep . . . 12 in.	—	—	0.40	—	*2.50*	
Concrete blocks, three oval core:						
Sand and gravel aggregate . . . 4 in.	—	—	1.40	—	*0.71*	0.22
. . . 8 in.	—	—	0.90	—	*1.11*	
. . . 12 in.	—	—	0.78	—	*1.28*	
Cinder aggregate . . . 3 in.	—	—	1.16	—	*0.86*	0.21
. . . 4 in.	—	—	0.90	—	*1.11*	
. . . 8 in.	—	—	0.58	—	*1.72*	
. . . 12 in.	—	—	0.53	—	*1.89*	
Lightweight aggregate . . . 3 in.	—	—	0.79	—	*1.27*	0.21
(expanded shale, clay, slate . . . 4 in.	—	—	0.67	—	*1.50*	
or slag; pumice): . . . 8 in.	—	—	0.50	—	*2.00*	
. . . 12 in.	—	—	0.44	—	*2.27*	
Concrete blocks, rectangular core.[j,k]						
Sand and gravel aggregate						
2 core, 8 in. 36 lb.	—	—	0.96	—	*1.04*	0.22
Same with filled cores[l]	—	—	0.52	—	*1.93*	0.22
Lightweight aggregate (expanded shale, clay, slate or slag, pumice):						
3 core, 6 in. 19 lb.	—	—	0.61	—	*1.65*	0.21
Same with filled cores[l]	—	—	0.33	—	*2.99*	
2 core, 8 in. 24 lb.	—	—	0.46	—	*2.18*	
Same with filled cores[l]	—	—	0.20	—	*5.03*	
3 core, 12 in. 38 lb.	—	—	0.40	—	*2.48*	
Same with filled cores[l]	—	—	0.17	—	*5.82*	
Stone, lime or sand	—	12.50	—	*0.08*	—	0.19
Gypsum partition tile:						
3 • 12 • 30 in. solid	—	—	0.79	—	*1.26*	0.19
3 • 12 • 30 in. 4-cell	—	—	0.74	—	*1.35*	
4 • 12 • 30 in. 3-cell	—	—	0.60	—	*1.67*	

METALS
(See Chapter 39, Table 3)

(Courtesy ASHRAE, 1985 Fundamentals)

Figure 2.7 (cont.)

Table 3A Thermal Properties of Typical Building and Insulating Materials—Design Values[a]

Description	Density lb/ft^3	Conductivity[b] λ Btu•in./$h \cdot ft^2 \cdot F$	Conductance (C) Btu/h•$ft^2 \cdot F$	Resistance[c] (R) Per inch thickness ($1/\lambda$) $h \cdot ft^2 \cdot F$/Btu	Resistance[c] (R) For thickness listed ($1/C$) $h \cdot ft^2 \cdot F$/Btu	Specific Heat, Btu/lb• deg F
ROOFING[h]						
Asbestos-cement shingles	120	—	4.76	—	*0.21*	0.24
Asphalt roll roofing	70	—	6.50	—	*0.15*	0.36
Asphalt shingles	70	—	2.27	—	*0.44*	0.30
Built-up roofing .... 0.375 in.	70	—	3.00	—	*0.33*	0.35
Slate .... 0.5 in.	—	—	20.00	—	*0.05*	0.30
Wood shingles, plain and plastic film faced	—	—	1.06	—	*0.94*	0.31
SIDING MATERIALS (on flat surface)						
Shingles						
Asbestos cement	120	—	4.75	—	*0.21*	
Wood, 16 in., 7.5 exposure	—	—	1.15	—	*0.87*	0.31
Wood, double, 16-in., 12-in. exposure	—	—	0.84	—	*1.19*	0.28
Wood, plus insul. backer board, 0.3125 in.	—	—	0.71	—	*1.40*	0.31
Siding						
Asbestos-cement, 0.25 in., lapped	—	—	4.76	—	*0.21*	0.24
Asphalt roll siding	—	—	6.50	—	*0.15*	0.35
Asphalt insulating siding (0.5 in. bed.)	—	—	0.69	—	*1.46*	0.35
Hardboard siding, 0.4375 in.	40	1.49	—	*0.67*		0.28
Wood, drop, 1 • 8 in.	—	—	1.27	—	*0.79*	0.28
Wood, bevel, 0.5 • 8 in., lapped	—	—	1.23	—	*0.81*	0.28
Wood, bevel, 0.75 • 10 in., lapped	—	—	0.95	—	*1.05*	0.28
Wood, plywood, 0.375 in., lapped	—	—	1.59	—	*0.59*	0.29
Aluminum or Steel[m], over sheathing						
Hollow-backed	—	—	1.61	—	*0.61*	0.29
Insulating-board backed nominal 0.375 in.	—	—	0.55	—	*1.82*	0.32
Insulating-board backed nominal 0.375 in., foil backed			0.34		*2.96*	
Architectural glass	—	—	10.00	—	*0.10*	0.20
WOODS (12% Moisture Content)[o,p]						
Hardwoods						0.39
Oak	41.2 46.8	1.12-1.25	—	*0.89-0.80*	—	
Birch	42.6-45.4	1.16-1.22	—	*0.87-0.82*	—	
Maple	39.8-44.0	1.09-1.19		*0.91-0.88*		
Ash	38.4-41.9	1.06-1.14	—	*0.94-0.88*	—	
Softwoods						0.39
Southern Pine	35.6-41.2	1.00-1.12	—	*1.00-0.89*	—	
Douglas Fir-Larch	33.5 36.3	0.95-1.01	—	*1.06-0.99*	—	
Southern Cypress	31.4-32.1	0.90-0.92	—	*1.11-1.09*	—	
Hem-Fir, Spruce-Pine-Fir	24.5-31.4	0.74-0.90	—	*1.35-1.11*	—	
West Coast Woods, Cedars	21.7-31.4	0.68-0.90	—	*1.48-1.11*	—	
California Redwood	24.5-28.0	0.74-0.82	—	*1.35-1.22*	—	

Notes for Table 3A

[a]Except where otherwise noted, all values are for a mean temperature of 75 F. Representative values for dry materials, selected by ASHRAE TC 4.4, are intended as design (not specification) values for materials in normal use. Insulation materials in actual service may have thermal values that vary from design values depending on their in-situ properties (e.g., density and moisture content). For properties of a particular product, use the value supplied by the manufacturer or by unbiased tests.

[b]To obtain thermal conductivities in But/$h \cdot ft^2 \cdot F$, divide the λ value by 12 in./ft.

[c]Resistance values are the reciprocals of C before rounding off C to two decimal places.

[d]Does not include paper backing and facing, if any. Where insulation forms a boundary (reflective or otherwise) of an air space, see Tables 2A and 2B for the insulating value of an air space with the appropriate effective emittance and temperature conditions of the space.

[e]Conductivity varies with fiber diameter. (See Chapter 20, Thermal Conductivity section.) Insulation is produced in different densities, therefore, there is a wide variation in thickness for the same R-value among manufacturers. No effort should be made to relate any specific R-value to any specific density or thickness.

[f]Values are for aged, unfaced, board stock. For change in conductivity with age of expanded urethane, see Chapter 20, Factors Affecting Thermal Conductivity.

[g]Insulating values of acoustical tile vary, depending on density of the board and on type, size and depth of perforations.

[h]ASTM C 855-77 recognizes the specification of roof insulation on the basis of the C-values shown. Roof insulation is made in thickness to meet these values.

[i]Face brick and common brick do not always have these specific densities. When density differs from that shown, there will be a change in thermal conductivity.

[j]At 45 F mean temperature. Data on rectangular core concrete blocks differ from the above data on oval core blocks, due to core configuration, different mean temperatures, and possibly differences in unit weights. Weight data on the oval core blocks tested are not available.

[k]Weights of units approximately 7.625 in. high and 15.75 in. long. These weights are given as a means of describing the blocks tested, but conductance values are all for 1 ft^2 of area.

[l]Vermiculite, perlite, or mineral wool insulation. Where insulation is used, vapor barriers or other precautions must be considered to keep insulation dry.

[m]Values for metal siding applied over flat surfaces vary widely, depending on amount of ventilation of air space beneath the siding; whether air space is reflective or nonreflective; and on thickness, type, and application of insulating backing-board used. Values given are averages for use as design guides, and were obtained from several guarded hotbox tests (ASTM C236) or calibrated hotbox (ASTM C 976) on hollow-backed types and types made using backing-boards of wood fiber, foamed plastic, and glass fiber. Departures of $\pm$50% or more from the values given may occur.

[n]Time-aged values for board stock with gas-barrier quality (0.001 in. thickness or greater) aluminum foil facers on tow major surfaces.

[o]See Ref. 5.

[p]See Ref. 6, 7, 8 and 9. The conductivity values listed are for heat transfer across the grain. The thermal conductivity of wood varies linearly with the density and the density ranges listed are those normally found for the wood species given. If the density of the wood species is not known, use the mean conductivity value.

(Courtesy ASHRAE, 1985 Fundamentals)

Figure 2.7 (cont.)

TABLE 34—THERMAL RESISTANCES R—BUILDING AND INSULATING MATERIALS

(deg F per Btu) / (hr) (sq ft)

MATERIAL	DESCRIPTION	THICKNESS (in.)	DENSITY (lb per cu ft)	WEIGHT (lb per sq ft)	RESISTANCE R Per Inch Thickness 1/k	RESISTANCE R For Listed Thickness 1/c
BUILDING MATERIALS						
BUILDING BOARD Boards, Panels, Sheathing, etc	Asbestos-Cement Board		120	—	0.25	—
	Asbestos-Cement Board	⅛	120	1.25	—	0.03
	Gypsum or Plaster Board	⅜	50	1.58	—	0.32
	Gypsum or Plaster Board	½	50	2.08	—	0.45
	Plywood		34	—	1.25	—
	Plywood	¼	34	0.71	—	0.31
	Plywood	⅜	34	1.06	—	0.47
	Plywood	½	34	1.42	—	0.63
	Plywood or Wood Panels	¾	34	2.13	—	0.94
	Wood Fiber Board, Laminated or Homogeneous		26	—	2.38	—
			31	—	2.00	—
	Wood Fiber, Hardboard Type		65	—	0.72	—
	Wood Fiber, Hardboard Type	¼	65	1.35	—	0.18
	Wood, Fir or Pine Sheathing	25/32	32	2.08	—	0.98
	Wood, Fir or Pine	1⅝	32	4.34	—	2.03
BUILDING PAPER	Vapor Permeable Felt		—	—	—	0.06
	Vapor Seal, 2 Layers of Mopped 15 lb felt		—	—	—	0.12
	Vapor Seal, Plastic Film		—	—	—	Negl
WOODS	Maple, Oak, and Similar Hardwoods		45	—	0.91	—
	Fir, Pine, and Similar Softwoods		32	—	1.25	—
MASONRY UNITS	Brick, Common	4	120	40	—	.80
	Brick, Face	4	130	43	—	.44
	Clay Tile, Hollow:					
	1 Cell Deep	3	60	15	—	0.80
	1 Cell Deep	4	48	16	—	1.11
	2 Cells Deep	6	50	25	—	1.52
	2 Cells Deep	8	45	30	—	1.85
	2 Cells Deep	10	42	35	—	2.22
	3 Cells Deep	12	40	40	—	2.50
	Concrete Blocks, Three Oval Core Sand & Gravel Aggregate	3	76	19	—	0.40
		4	69	23	—	0.71
		6	64	32	—	0.91
		8	64	43	—	1.11
		12	63	63	—	1.28
	Cinder Aggregate	3	68	17	—	0.86
		4	60	20	—	1.11
		6	54	27	—	1.50
		8	56	37	—	1.72
		12	53	53	—	1.89
	Lightweight Aggregate (Expanded Shale, Clay, Slate or Slag; Pumice)	3	60	15	—	1.27
		4	52	17	—	1.50
		8	48	32	—	2.00
		12	43	43	—	2.27
	Gypsum Partition Tile:					
	3″x12″x30″ solid	3	45	11	—	1.26
	3″x12″x30″ 4-cell	3	35	9	—	1.35
	4″x12″x30″ 3-cell	4	38	13	—	1.67
	Stone, Lime or Sand		150	—	0.08	—

(Courtesy Carrier Corporation, McGraw–Hill Book Company)

Figure 2.7 (cont.)

TABLE 34—THERMAL RESISTANCES R—BUILDING AND INSULATING MATERIALS (Contd)

(deg F per Btu) / (hr) (sq ft)

MATERIAL	DESCRIPTION	THICKNESS (in.)	DENSITY (lb per cu ft)	WEIGHT (lb per sq ft)	RESISTANCE R: Per Inch Thickness 1/k	RESISTANCE R: For Listed Thickness 1/c
	BUILDING MATERIALS, (CONT.)					
MASONRY MATERIALS Concretes	Cement Mortar		116	—	0.20	—
	Gypsum-Fiber Concrete 87½% gypsum, 12½% wood chips		51	—	0.60	—
	Lightweight Aggregates		120	—	0.19	—
	Including Expanded		100	—	0.28	—
	Shale, Clay or Slate		80	—	0.40	—
	Expanded Slag; Cinders		60	—	0.59	—
	Pumice; Perlite; Vermiculite		40	—	0.86	—
	Also, Cellular Concretes		30	—	1.11	—
			20	—	1.43	—
	Sand & Gravel or Stone Aggregate (Oven Dried)		140	—	0.11	—
	Sand & Gravel or Stone Aggregate (Not Dried)		140	—	0.08	—
	Stucco		116	—	0.20	—
PLASTERING MATERIALS	Cement Plaster, Sand Aggregate		116	—	0.20	—
	Sand Aggregate	1/2	116	4.8	—	0.10
	Sand Aggregate	3/4	116	7.2	—	0.15
	Gypsum Plaster:					
	Lightweight Aggregate	1/2	45	1.88	—	0.32
	Lightweight Aggregate	5/8	45	2.34	—	0.39
	Lightweight Aggregate on Metal Lath	3/4	45	2.80	—	0.47
	Perlite Aggregate		45	—	0.67	—
	Sand Aggregate		105	—	0.18	—
	Sand Aggregate	1/2	105	4.4	—	0.09
	Sand Aggregate	5/8	105	5.5	—	0.11
	Sand Aggregate on Metal Lath	3/4	105	6.6	—	0.13
	Sand Aggregate on Wood Lath		105	—	—	0.40
	Vermiculite Aggregate		45	—	0.59	—
ROOFING	Asbestos-Cement Shingles		120	—	—	0.21
	Asphalt Roll Roofing		70	—	—	0.15
	Asphalt Shingles		70	—	—	0.44
	Built-up Roofing	3/8	70	2.2	—	0.33
	Slate	1/2	201	8.4	—	0.05
	Sheet Metal		—	—	Negl	—
	Wood Shingles		40	—	—	0.94
SIDING MATERIALS (On Flat Surface)	Shingles					
	Wood, 16", 7½" exposure		—	—	—	0.87
	Wood, Double, 16", 12" exposure		—	—	—	1.19
	Wood, Plus Insul Backer Board, 5/16"		—	—	—	1.40
	Siding					
	Asbestos-Cement, 1/4" lapped		—	—	—	0.21
	Asphalt Roll Siding		—	—	—	0.15
	Asphalt Insul Siding, 1/2" Board		—	—	—	1.45
	Wood, Drop, 1"x8"		—	—	—	0.79
	Wood, Bevel, 1/2"x8", lapped		—	—	—	0.81
	Wood, Bevel, 3/4"x10", lapped		—	—	—	1.05
	Wood, Plywood, 3/8", lapped		—	—	—	0.59
	Structural Glass		—	—	—	0.10
FLOORING MATERIALS	Asphalt Tile	1/8	120	1.25	—	0.04
	Carpet and Fibrous Pad		—	—	—	2.08
	Carpet and Rubber Pad		—	—	—	1.23
	Ceramic Tile	1	—	—	—	0.08
	Cork Tile		25	—	2.22	—
	Cork Tile	1/8	25	0.26	—	0.28
	Felt, Flooring		—	—	—	0.06
	Floor Tile	1/8	—	—	—	0.05
	Linoleum	1/8	80	0.83	—	0.08
	Plywood Subfloor	5/8	34	1.77	—	0.78
	Rubber or Plastic Tile	1/8	110	1.15	—	0.02
	Terrazzo	1	140	11.7	—	0.08
	Wood Subfloor	25/32	32	2.08	—	0.98
	Wood, Hardwood Finish	3/4	45	2.81	—	0.68

(Courtesy Carrier Corporation, McGraw-Hill Book Company)

Figure 2.7 (cont.)

TABLE 34—THERMAL RESISTANCES R—BUILDING AND INSULATING MATERIALS (Contd)

(deg F per Btu) / (hr) (sq ft)

MATERIAL	DESCRIPTION		THICKNESS (in.)	DENSITY (lb per cu ft)	WEIGHT (lb per sq ft)	RESISTANCE R Per Inch Thickness $\frac{1}{k}$	RESISTANCE R For Listed Thickness $\frac{1}{c}$
INSULATING MATERIALS							
BLANKET AND BATT*	Cotton Fiber			0.8 - 2.0	—	3.85	—
	Mineral Wool, Fibrous Form Processed From Rock, Slag, or Glass			1.5 - 4.0	—	3.70	—
	Wood Fiber			3.2 - 3.6	—	4.00	—
	Wood Fiber, Multi-layer Stitched Expanded			1.5 - 2.0	—	3.70	—
BOARD AND SLABS	Glass Fiber			9.5	—	4.00	—
	Wood or Cane Fiber						
	Acoustical Tile		½	22.4	.93	—	1.19
	Acoustical Tile		¾	22.4	1.4	—	1.78
	Interior Finish (Tile, Lath, Plank)			15.0	—	2.86	—
	Interior Finish (Tile, Lath, Plank)		½	15.0	0.62	—	1.43
	Roof Deck Slab						
	Sheathing (Impreg or Coated)			20.0	—	2.63	—
	Sheathing (Impreg or Coated)		½	20.0	0.83	—	1.32
	Sheathing (Impreg or Coated)		25/32	20.0	1.31	—	2.06
	Cellular Glass			9.0	—	2.50	—
	Cork Board (Without Added Binder)			6.5 - 8.0	—	3.70	—
	Hog Hair (With Asphalt Binder)			8.5	—	3.00	—
	Plastic (Foamed)			1.62	—	3.45	—
	Wood Shredded (Cemented in Preformed Slabs)			22.0	—	1.82	—
LOOSE FILL	Macerated Paper or Pulp Products			2.5 - 3.5	—	3.57	—
	Wood Fiber: Redwood, Hemlock, or Fir			2.0 - 3.5	—	3.33	—
	Mineral Wool (Glass, Slag, or Rock)			2.0 - 5.0	—	3.33	—
	Sawdust or Shavings			8.0 - 15.0	—	2.22	—
	Vermiculite (Expanded)			7.0	—	2.08	—
ROOF INSULATION	All Types						
	Preformed, for use above deck						
	Approximately		½	15.6	.7	—	1.39
	Approximately		1	15.6	1.3	—	2.78
	Approximately		1½	15.6	1.9	—	4.17
	Approximately		2	15.6	2.6	—	5.26
	Approximately		2½	15.6	3.2	—	6.67
	Approximatley		3	15.6	3.9	—	8.33
AIR							
AIR SPACES	POSITION	HEAT FLOW					
	Horizontal	Up (Winter)	¾ - 4	—	—	—	0.85
	Horizontal	Up (Summer)	¾ - 4	—	—	—	0.78
	Horizontal	Down (Winter)	¾	—	—	—	1.02
	Horizontal	Down (Winter)	1½	—	—	—	1.15
	Horizontal	Down (Winter)	4	—	—	—	1.23
	Horizontal	Down (Winter)	8	—	—	—	1.25
	Horizontal	Down (Summer)	¾	—	—	—	0.85
	Horizontal	Down (Summer)	1½	—	—	—	0.93
	Horizontal	Down (Summer)	4	—	—	—	0.99
	Sloping 45°	Up (Winter)	¾ - 4	—	—	—	0.90
	Sloping 45°	Down (Summer)	¾ - 4	—	—	—	0.89
	Vertical	Horiz. (Winter)	¾ - 4	—	—	—	0.97
	Vertical	Horiz. (Summer)	¾ - 4	—	—	—	0.86
AIR FILM	POSITION	HEAT FLOW					
Still Air	Horizontal	Up		—	—	—	0.61
	Sloping 45°	Up		—	—	—	0.62
	Vertical	Horizontal		—	—	—	0.68
	Sloping 45°	Down		—	—	—	0.76
	Horizontal	Down		—	—	—	0.92
15 Mph Wind	Any Position (For Winter)	Any Direction		—	—	—	0.17
7½ Mph Wind	Any Position (For Summer)	Any Direction		—	—	—	0.25

*Includes paper backing and facing if any. In cases where the insulation forms a boundary (highly reflective) of an air space, refer to *Table 31, page 75*

(Courtesy Carrier Corporation, McGraw-Hill Book Company)

Figure 2.7 (cont.)

In some cases, ΔT is measured from another enclosed or unheated space which has a temperature that differs from the outdoor design temperature (T_o). Examples include party walls, attics, basements, attached garages, slabs and walls below grade, or rooms deliberately kept at a special temperature, such as gymnasiums, operating rooms, freezers, or computer rooms. The temperature of the attached space is used instead of the T_o, and in some cases must be estimated. Where the enclosing wall is not an exterior wall, ΔT is equal to the indoor temperature minus the temperature of the adjoining surfaces, or

$$T_i - T_{adj}$$

Figure 2.10 is a guide to establishing ΔT for adjoining surfaces.

Perimeter Losses: When an exterior wall meets the grade, there are additional conductive losses at the edge of the slab.

Table of Indoor Design Temperatures

Space	Design Indoor Temperature °F		Design Relative Humidity % Summer
	Winter	Summer	
Assembly			
Museums	68-72	68-72	40-55
Restaurants	70-74	74-78	50-60
Gymnasiums	55-65	55-65	40-50
Auditoriums	72-76	76-80	50-60
Business	70-74	74-78	40-50
Commercial			
Retail	70 74	72-80*	30-55
Factories and Computer Rooms	68-74	77-85	45-55
High Hazard	special conditions		
Institutional (special applications — hospitals)	74-77	74-79	45-50
Mercantile	70-74	74-78	40-50
Residential			
Apt., House, Hotel	74-77	74-79	45-50
Storage (actual conditions depend upon materials stored)	above 32°	below 95°	N/A

Note: Typical energy code — Winter temp. = 68°F, Summer temp. = 78°F
*Short-term occupancy

Figure 2.8

Weather Data and Design Conditions (winter design @ 97.5% - summer design @ 2.5%)

City	Latitude (1)		Winter Temperatures (1)			Winter Degree Days (2)	Summer (Design Dry Bulb) Temperatures and Relative Humidity		
	0	1'	Med. of Annual Extremes	99%	97½%		1%	2½%	5%
UNITED STATES									
Albuquerque, NM	35	0	6	12	16	4,400	96/61	94/61	92/61
Atlanta, GA	33	4	14	17	22	3,000	95/74	92/74	90/73
Baltimore, MD	39	2	12	14	17	4,600	94/75	92/75	89/74
Birmingham, AL	33	3	17	17	21	2,600	97/74	94/75	93/74
Bismarck, ND	46	5	-31	-23	-19	8,800	95/68	91/68	88/67
Boise, ID	43	3	0	3	10	5,800	96/65	93/64	91/64
Boston, MA	42	2	-1	6	9	5,600	91/73	88/71	85/70
Burlington, VT	44	3	-18	-12	-7	8,200	88/72	85/70	83/69
Charleston, WV	38	2	1	7	11	4,400	92/74	90/73	88/72
Charlotte, NC	35	1	13	18	22	3,200	96/74	94/74	92/74
Casper, WY	42	5	-20	-11	-5	7,400	92/58	90/57	87/57
Chicago, IL	41	5	-5	-3	2	6,600	94/75	91/74	88/73
Cincinnati, OH	39	1	2	1	6	4,400	94/73	92/72	90/72
Cleveland, OH	41	2	-2	1	5	6,400	91/73	89/72	86/71
Columbia, SC	34	0	16	20	24	2,400	98/76	96/75	94/75
Dallas, TX	32	5	14	18	22	2,400	101/75	99/75	97/75
Denver, CO	39	5	-9	-5	1	6,200	92/59	90/59	89/59
Des Moines, IA	41	3	-13	-10	-5	6,600	95/75	92/74	89/73
Detroit, MI	42	2	0	3	6	6,200	92/73	88/72	85/71
Great Falls, MT	47	3	-29	-21	-15	7,800	91/60	88/60	85/59
Hartford, CT	41	5	-4	3	7	6,200	90/74	88/73	85/72
Houston, TX	29	5	24	28	33	1,400	96/77	94/77	92/77
Indianapolis, IN	39	4	-2	-2	2	5,600	93/74	91/74	88/73
Jackson, MS	32	2	17	21	25	2,200	98/76	96/76	94/76
Kansas City, MO	39	1	-2	2	6	4,800	100/75	97/74	94/74
Las Vegas, NV	36	1	18	25	28	2,800	108/66	106/65	104/65
Lexington, KY	38	0	0	3	8	4,600	94/73	92/72	90/72
Little Rock, AR	34	4	13	15	20	3,200	99/76	96/77	94/77
Los Angeles, CA	34	0'	38	41	43	2,000	94/70	90/70	87/69
Memphis, TN	35	0	11	13	18	3,200	98/77	96/76	94/76
Miami, FL	25	5	39	44	47	200	92/77	90/77	89/77
Milwaukee, WI	43	0	-11	-8	-4	7,600	90/74	87/73	84/71
Minneapolis, MN	44	5	-19	-16	-12	8,400	92/75	89/73	86/71
New Orleans, LA	30	0	29	29	33	1,400	93/78	91/78	90/77
New York, NY	40	5	6	11	15	5,000	94/74	91/73	88/72
Norfolk, VA	36	5	18	20	22	3,400	94/77	91/76	89/76
Oklahoma City, OK	35	2	4	9	13	3,200	100/74	97/74	95/73
Omaha, NE	41	2	-12	-8	-3	6,600	97/76	94/75	91/74
Philadelphia, PA	39	5	7	10	14	4,400	93/75	90/74	87/72
Phoenix, AZ	33	3	25	31	34	1,800	108/71	106/71	104/71
Pittsburgh, PA	40	3	1	3	7	6,000	90/72	88/71	85/70
Portland, ME	43	4	-14	-6	-1	7,600	88/72	85/71	81/69
Portland, OR	45	4	17	17	23	4,600	89/68	85/67	81/65
Portsmouth, NH	43	1	-8	-2	2	7,200	88/73	86/71	83/70
Providence, RI	41	4	0	5	9	6,000	89/73	86/72	83/70
Rochester, NY	43	1	-5	1	5	6,800	91/73	88/71	85/70
Salt Lake City, UT	40	5	-2	3	8	6,000	97/62	94/62	92/61
San Francisco, CA	37	5	38	38	40	3,000	80/63	77/62	83/61
Seattle, WA	47	4	22	22	27	5,200	81/68	79/66	76/65
Sioux Falls, SD	43	4	-21	-15	-11	7,800	95/73	92/72	89/71
St. Louis, MO	38	4	1	3	8	5,000	96/75	94/75	92/74
Tampa, FL	28	0	32	36	40	680	92/77	91/77	90/76
Trenton, NJ	40	1	7	11	14	5,000	92/75	90/74	87/73
Washington, DC	38	5	12	14	17	4,200	94/75	92/74	90/74
Wichita, KS	37	4	-1	3	7	4,600	102/72	99/73	96/73
Wilmington, DE	39	4	6	10	14	5,000	93/74	93/74	20/73
ALASKA									
Anchorage	61	1	-29	-23	-18	10,800	73/59	70/58	67/56
Fairbanks	64	5	-59	-51	-47	14,280	82/62	78/60	75/59
CANADA									
Edmonton, Alta.	53	3	-30	-29	-25	11,000	86/66	83/65	80/63
Halifax, N.S.	44	4	-4	1	5	8,000	83/66	80/65	77/64
Montreal, Que.	45	3	-20	-16	-10	9,000	88/73	86/72	84/71
Saskatoon, Sask.	52	1	-35	-35	-31	11,000	90/68	86/66	83/65
St. Johns, Nwf.	47	4	1	3	7	8,600	79/66	77/65	75/64
Saint John, N.B.	45	2	-15	-12	-8	8,200	81/67	79/65	77/64
Toronto, Ont.	43	4	-10	-5	-1	7,000	90/73	87/72	85/71
Vancouver, B.C.	49	1	13	15	19	6,000	80/67	78/66	76/65
Winnipeg, Man.	49	5	-31	-30	-27	10,800	90/73	87/71	84/70

(1) Handbook of Fundamentals, ASHRAE, Inc., NY 1972/1985
(2) Local Climatological Annual Survey, USDC Env. Science Services Administration, Ashville, NC

Table of Outdoor Design Temperatures

Figure 2.9

(These are noted as H_e in Figure 2.3.) Figure 2.11 illustrates perimeter edge slab conditions. The formula for calculating perimeter edge losses is shown below.

H_e = FP(ΔT)
H_e = Heat loss through edge (Btu/hr.)
F = Edge factor (Btu/hr./ft./°F)
P = Perimeter of slab (ft.)
ΔT = Indoor temperature minus outdoor temperature

The edge factor, F, values are shown in Figure 2.11 and vary from 0.47 for edge slabs with perimeter insulation up to 2.74 for uninsulated edge slabs.

Slab Losses: Slabs on grade lose heat to the ground below. The loss is calculated as follows:

H_s = K × A
H_s = Heat loss through slab
K = Constant (Btu/hr./s.f.)
2 for floors and basement walls 4 feet below grade *or*
4 for floors and walls from 0 to 4 feet below grade
A = Area (s.f.)

Convection Losses

Convection losses occur when outside air passes into the building. The two primary sources of convection losses are *infiltration* and *ventilation*. Infiltration results from the leakage

ΔT for Adjoining Unheated Spaces

	Winter	Summer
Adjoining Space	T_i-T_{adj}	T_{adj}-T_i
Party Wall	T_i-T_{adj}	T_{adj}-T_i
Attic (no insulation)*	$(T_i$-$T_o)/2$	95-T_i
Basement (no insulation)*	$(T_i$-$T_o)/2$	0
Attached Garage	T_i-T_o	T_o-T_i
Sun Porch	T_i-T_o	—
Slab on Grade and Walls 0-4′ below Grade	4 Btu/hr./s.f.	0
Slab and Wall below Grade more than 4′	2 Btu/hr./s.f.	0
Crawl Spaces	T_i-T_o	0

T_i = Indoor temperature (°F)
T_o = Outdoor temperature (°F)
T_{adj} = Temperature of adjoining space
*With insulation, the attic and basement temperature will be closer to outdoor temperature.

Figure 2.10

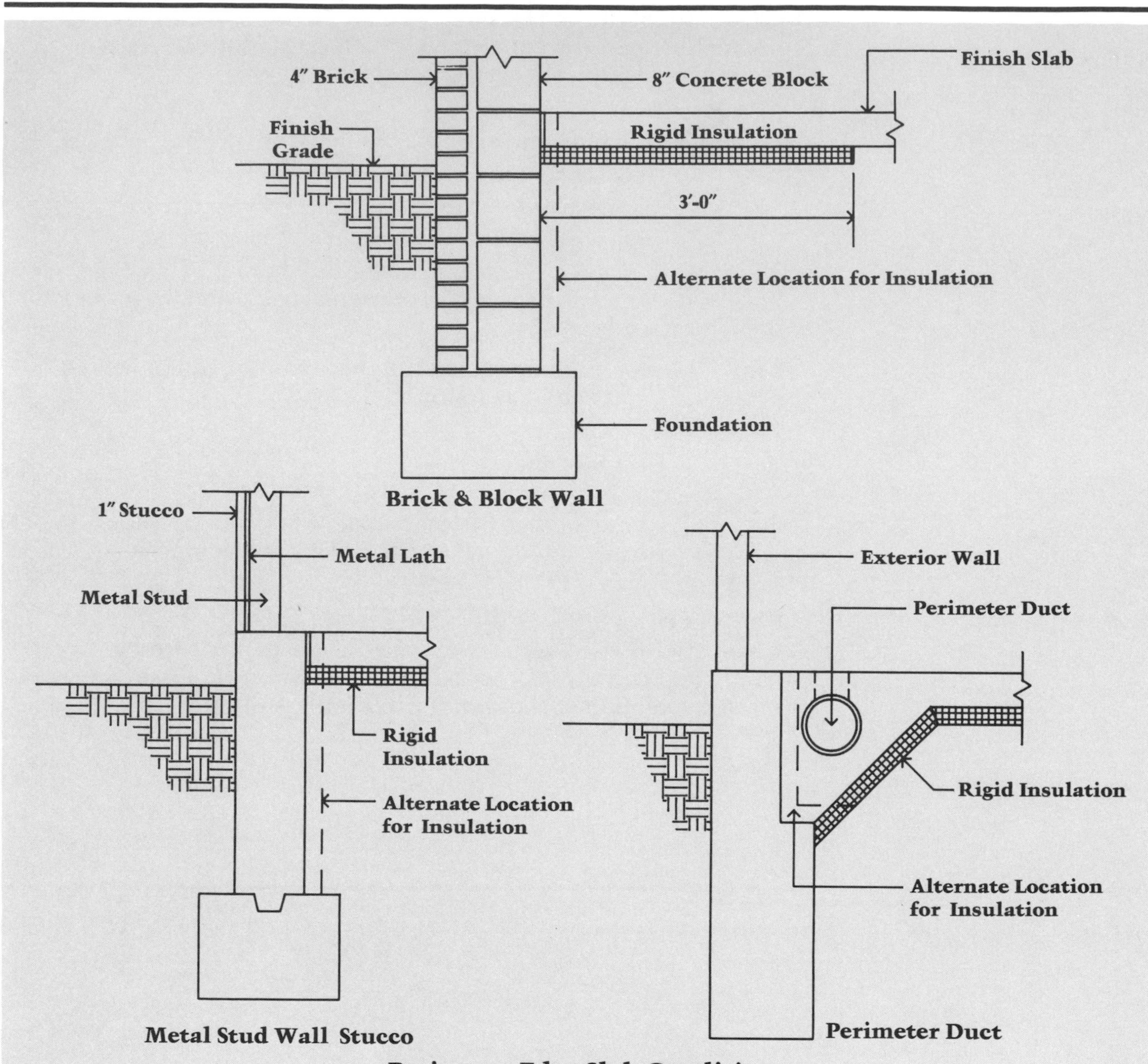

Perimeter Edge Slab Conditions

Table 5 Heat Loss Coefficient of Slab Floor Construction, F_2 (Btu/h·F per ft of perimeter)

Construction	Insulated	Degree Days (65 F Base) 2950	5350	7433
8-in. block wall, brick facing	Uninsulated	0.62	0.68	0.72
	R = 5.4 from edge to footer	0.48	0.50	0.56
4-in block wall, brick facing	Uninsulated	0.80	0.84	0.93
	R = 5.4 from edge to footer	0.47	0.49	0.54
Metal stud wall, stucco	Uninsulated	1.15	1.20	1.34
	R = 5.4 from edge to footer	0.51	0.53	0.58
Poured concrete wall, with duct near perimeter[a]	Uninsulated	1.84	2.12	2.73
	R = 5.4 from edge to footer, 3 ft under floor	0.64	0.72	0.90

[a] Weighted average temperature of the heating duct was assumed at 110 F during the heating season (outdoor air temperature less than 65 F).

(Courtesy ASHRAE, 1985 Fundamentals)

Figure 2.11

of air into the building from cracks around windows, doors, and openings. Ventilation occurs when fresh outdoor air is brought into the building to meet fresh air requirements. In both cases, the cold outdoor air must be heated to bring it to room temperature. The amount of heat necessary to heat the convected air is shown below.

H_v = 1.1 cfm (ΔT)
H_v = Convection heat loss (Btu/hr.)
1.1 = Constant
cfm = Air entering building (cubic feet per minute) (see Figure 2.12)
ΔT = Temperature difference (°F)

The factor 1.1 adjusts for the units as well as for the air density in the winter.

The cubic feet per minute (cfm) of convected air into a building is calculated separately for infiltration and ventilation. For infiltration in the winter, a fifteen mile per hour wind is assumed to be directed at the building. For each crack, a crack coefficient is used to determine how many cfm will infiltrate (per foot of crack) the building.

Figure 2.12 lists crack coefficients for windows and doors. For each opening the air infiltrates, use the equation shown below.

cfm = crack coefficient × length of crack

Infiltration: The wind blows from only one direction at a time. The cold air that infiltrates one side of a building usually forces warm air to exfiltrate the other side. Although it is impossible for air to infiltrate a building from all four sides at once, infiltration calculations are based on the cracks on all four sides of the building. Most designers correct for this inconsistency by sizing the distribution and terminal units for the full load and reducing the load from infiltration on the generating system by one-third to one-half. In this way, the boiler heats only the actual imposed load, while the piping and radiators are properly sized for their maximum possible load. The infiltration diversity is shown in Figure 2.13.

Ventilation: The other source of convection heat losses in buildings is ventilation air. For buildings with heating only, operable windows often satisfy the ventilation requirements. Ventilation exhaust fans used for kitchens and bathrooms operate intermittently and do not exceed the infiltrated air convection losses. Ventilated air that does exceed infiltrated air should be added to the convected cfm quantity to be heated. In large buildings, the amount of ventilated air typically exceeds the infiltrated air. In small buildings the reverse is common.

Radiation Losses

Most heat losses are calculated based on the coldest winter night for a given climate. In the daytime during the winter, there is a solar heat gain, which some designers account for by partially offsetting the heat loss. Factors such as electric lights and body heat are also incorporated by some designers in comprehensive energy designs.

pressure, tends to flow down thru the building and out the doors on the street level, thereby offsetting some of the infiltration thru them.

In low buildings, air infiltrates thru open doors on the windward side unless sufficient outdoor air is introduced thru the air conditioning equipment to offset it; refer to *"Offsetting Infiltration with Outdoor Air."*

With *doors on opposite walls,* the infiltration can be considerable if the two are open at the same time.

Basis of Table 41
— Infiltration thru Windows and Doors, Summer

The data in *Tables 41a, b and c* is based on a wind velocity of 7.5 mph blowing directly at the window or door, and on observed crack widths around typical windows and doors. This data is derived from *Table 44* which lists infiltration thru cracks around windows and doors as established by ASHAE tests.

Table 41d shows values to be used for doors on opposite walls for various percentages of time that each door is open.

The data in *Table 41e* is based on actual tests of typical applications.

Use of Table 41
— Infiltration thru Windows and Doors, Summer

The data in *Table 41* is used to determine the infiltration thru windows and doors on the windward side with the wind blowing directly at them. When the wind direction is oblique to the windows or doors, multiply the values in *Tables 41a, b, c, d,* by 0.60 and apply to total areas. For specific locations, adjust the values in *Table 41* to the design wind velocity; refer to *Table 1, page 10.*

During the summer, infiltration is calculated for the windward side(s) only, because stack effect is small and, therefore, causes the infiltration air to flow in a downward direction in tall buildings (over 100 ft). Some of the air infiltrating thru the windows will exfiltrate thru the windows on the leeward side(s), while the remaining infiltration air flows out the doors, thus offsetting some of the infiltration thru the doors. To determine the net infiltration thru the doors, determine the infiltration thru the windows on the windward side, multiply this by .80, and subtract from the door infiltration. For low buildings the door infiltration on the windward side should be included in the estimate.

TABLE 41—INFILTRATION THRU WINDOWS AND DOORS—SUMMER*

7.5 mph Wind Velocity†

TABLE 41a—DOUBLE HUNG WINDOWS‡

DESCRIPTION	CFM PER SQ FT SASH AREA					
	Small—30" x 72"			Large—54" x 96"		
	No W-Strip	W-Strip	Storm Sash	No W-Strip	W-Strip	Storm Sash
Average Wood Sash	.43	.26	.22	.27	.17	.14
Poorly Fitted Wood Sash	1.20	.37	.60	.76	.24	.38
Metal Sash	.80	.35	.40	.51	.22	.25

TABLE 41b—CASEMENT TYPE WINDOWS‡

DESCRIPTION	CFM PER SQ FT SASH AREA									
	Percent Openable Area									
	0%	25%	33%	40%	45%	50%	60%	66%	75%	100%
Rolled Section—Steel Sash										
Industrial Pivoted	.33	.72	—	.99	—	—	—	1.45	—	2.6
Architectural Projected	—	.39	—	—	—	.55	.74	—	—	—
Residential	—	—	.28	—	—	.49	—	—	—	.63
Heavy Projected	—	—	—	—	.23	—	—	.32	.39	—
Hollow Metal—Vertically Pivoted	.27	.58	—	.82	—	—	—	1.2	—	2.2

REPRESENTATIVE TYPES OF WINDOWS
(VIEWED FROM OUTSIDE)

(Courtesy Carrier Corporation, McGraw-Hill Book Company)

Figure 2.12

TABLE 41—INFILTRATION THRU WINDOWS AND DOORS—SUMMER* (Contd)

7.5 mph Wind Velocity†

TABLE 41c—DOORS ON ONE OR ADJACENT WALLS, FOR CORNER ENTRANCES

DESCRIPTION	CFM PER SQ FT AREA**		CFM	
	No Use	Average Use	Standing Open	
			No Vestibule	Vestibule
Revolving Doors—Normal Operation	.8	5.2	—	—
Panels Open	—	—	1,200	900
Glass Door—3/16″ Crack	4.5	10.0	700	500
Wood Door (3′ x 7′)	1.0	6.5	700	500
Small Factory Door	.75	6.5	—	—
Garage & Shipping Room Door	2.0	4.5	—	—
Ramp Garage Door	2.0	6.75	—	—

TABLE 41d—SWINGING DOORS ON OPPOSITE WALLS

% Time 2nd Door is Open	CFM PER PAIR OF DOORS				
	% Time 1st Door is Open				
	10	25	50	75	100
10	100	250	500	750	1,000
25	250	625	1250	1875	2,500
50	500	1250	2500	3750	5,000
75	750	1875	3750	5625	7,500
100	1000	2500	5000	7500	10,000

TABLE 41e—DOORS

APPLICATION	CFM PER PERSON IN ROOM PER DOOR		
	72″ Revolving Door	36″ Swinging Door	
		No Vestibule	Vestibule
Bank	6.5	8.0	6.0
Barber Shop	4.0	5.0	3.8
Candy and Soda	5.5	7.0	5.3
Cigar Store	20.0	30.0	22.5
Department Store (Small)	6.5	8.0	6.0
Dress Shop	2.0	2.5	1.9
Drug Store	5.5	7.0	5.3
Hospital Room	—	3.5	2.6
Lunch Room	4.0	5.0	3.8
Men's Shop	2.7	3.7	2.8
Restaurant	2.0	2.5	1.9
Shoe Store	2.7	3.5	2.6

*All values in *Table 41* are based on the wind blowing directly at the window or door. When the wind direction is oblique to the window or door, multiply the above values by 0.60 and use the total window and door area on the windward side(s).

†Based on a wind velocity of 7.5 mph. For design wind velocities different from the base, multiply the above values by the ratio of velocities.

‡Includes frame leakage where applicable.

**Vestibules may decrease the infiltration as much as 30% when the door usage is light. When door usage is heavy, the vestibule is of little value for reducing infiltration.

Example 1 — Infiltration in Tall Buildings, Summer

Given:

A 20-story building in New York City oriented true north. Building is 100 ft long and 100 ft wide with a floor-to-floor height of 12 ft. Wall area is 50% residential casement windows having 50% fixed sash. There are ten 7 ft x 3 ft swinging glass doors on the street level facing south.

Find:

Infiltration into the building thru doors and windows, disregarding outside air thru the equipment and the exhaust air quantity.

Solution:

The prevailing wind in New York City during the summer is south, 13 mph *(Table 1, page 10).*

(Courtesy Carrier Corporation, McGraw-Hill Book Company)

Figure 2.12 (cont.)

TABLE 43—INFILTRATION THRU WINDOWS AND DOORS—WINTER*

15 mph Wind Velocity†

TABLE 43a—DOUBLE HUNG WINDOWS ON WINDWARD SIDE‡

DESCRIPTION	CFM PER SQ FT AREA					
	Small—30″ x 72″			Large—54″ x 96″		
	No W-Strip	W-Strip	Storm Sash	No W-Strip	W-Strip	Storm Sash
Average Wood Sash	.85	.52	.42	.53	.33	.26
Poorly Fitted Wood Sash	2.4	.74	1.2	1.52	.47	.74
Metal Sash	1.60	.69	.80	1.01	.44	.50

NOTE: W-Strip denotes weatherstrip.

TABLE 43b—CASEMENT TYPE WINDOWS ON WINDWARD SIDE‡

DESCRIPTION	CFM PER SQ FT AREA									
	Percent Ventilated Area									
	0%	25%	33%	40%	45%	50%	60%	66%	75%	100%
Rolled Section—Steel Sash										
Industrial Pivoted	.65	1.44	—	1.98	—	—	—	2.9	—	5.2
Architectural Projected	—	.78	—	—	—	1.1	1.48	—	—	—
Residential	—	—	.56	—	—	.98	—	—	—	1.26
Heavy Projected	—	—	—	—	.45	—	—	.63	.78	—
Hollow Metal—Vertically Pivoted	.54	1.19	—	1.64	—	—	—	2.4	—	4.3

TABLE 43c—DOORS ON ONE OR ADJACENT WINDWARD SIDES‡

DESCRIPTION	CFM PER SQ FT AREA**				
	Infrequent Use	Average Use			
		1 & 2 Story Bldg.	Tall Building (ft)		
			50	100	200
Revolving Door	1.6	10.5	12.6	14.2	17.3
Glass Door—(3/16″ Crack)	9.0	30.0	36.0	40.5	49.5
Wood Door 3′ x 7′	2.0	13.0	15.5	17.5	21.5
Small Factory Door	1.5	13.0			
Garage & Shipping Room Door	4.0	9.0			
Ramp Garage Door	4.0	13.5			

*All values in *Table 43* are based on the wind blowing directly at the window or door. When the prevailing wind direction is oblique to the window or doors, multiply the above values by 0.60 and use the total window and door area on the windward side(s).

†Based on a wind velocity of 15 mph. For design wind velocities different from the base, multiply the table values by the ratio of velocities.

‡Stack effect in tall buildings may also cause infiltration on the leeward side. To evaluate this, determine the equivalent velocity (V_e) and subtract the design velocity (V). The equivalent velocity is:

$V_e = \sqrt{V^2 - 1.75a}$ (upper section)

$V_e = \sqrt{V^2 + 1.75b}$ (lower section)

Where a and b are the distances above and below the mid-height of the building, respectively, in ft.

Multiply the table values by the ratio $(V_e - V)/15$ for doors and one half of the windows on the leeward side of the building. (Use values under "1 and 2 Story Bldgs" for doors on leeward side of tall buildings.)

**Doors on opposite sides increase the above values 25%. Vestibules may decrease the infiltration as much as 30% when door usage is light. If door usage is heavy, the vestibule is of little value in reducing infiltration. Heat added to the vestibule will help maintain room temperature near the door.

(Courtesy Carrier Corporation, McGraw-Hill Book Company)

Figure 2.12 (cont.)

INFILTRATION — CRACK METHOD (Summer or Winter)

The crack method of evaluating infiltration is more accurate than the area methods. It is difficult to establish the exact crack dimensions but, in certain close tolerance applications, it may be necessary to evaluate the load accurately. The crack method is applicable both summer and winter.

Basis of Table 44
— Infiltration thru Windows and Doors, Crack Method

The data on windows in *Table 44* are based on ASHAE tests. These test results have been reduced 20% because, as infiltration occurs on one side, a certain amount of pressure builds up in the building, thereby reducing the infiltration. The data on glass and factory doors has been calculated from observed typical crack widths.

Use of Table 44
— Infiltration thru Windows and Doors, Crack Method

Table 44 is used to determine the infiltration thru the doors and windows listed. This table does not take into account winter stack effect which must be evaluated separately, using the equivalent wind velocity formulas previously presented.

Example 5 — Infiltration thru Windows, Crack Method

Given:

A 4 ft x 7 ft residential casement window facing south.

Find:

The infiltration thru this window.

Solution:

Assume the crack widths are measured as follows:
Window frame — none, well sealed
Window openable area — 1/32 in. crack; length, 20 ft

Assume the wind velocity is 30 mph due south.
Infiltration thru window = 20 × 2.1 = 42 cfm *(Table 44)*

TABLE 44—INFILTRATION THRU WINDOWS AND DOORS—CRACK METHOD—SUMMER—WINTER*

TABLE 44a—DOUBLE HUNG WINDOWS—UNLOCKED ON WINDWARD SIDE

TYPE OF DOUBLE HUNG WINDOW	CFM PER LINEAR FOOT OF CRACK — Wind Velocity—Mph											
	5		10		15		20		25		30	
	No W-Strip	W-Strip	No W-Strip	W-Strip	No W-Strip	W-Strip	No W-Strip	W-Strip	No W-Strip	W-Strip	No W-Strip	W-Strip
Wood Sash												
Average Window	.12	.07	.35	.22	.65	.40	.98	.60	1.33	.82	1.73	1.05
Poorly Fitted Window	.45	.10	1.15	.32	1.85	.57	2.60	.85	3.30	1.18	4.20	1.53
Poorly Fitted—with Storm Sash	.23	.05	.57	.16	.93	.29	1.30	.43	1.60	.59	2.10	.76
Metal Sash	.33	.10	.78	.32	1.23	.53	1.73	.77	2.3	1.00	2.8	1.27

TABLE 44b—CASEMENT TYPE WINDOWS ON WINDWARD SIDE

TYPE OF CASEMENT WINDOW AND TYPICAL CRACK SIZE		CFM PER LINEAR FOOT OF CRACK — Wind Velocity—Mph					
		5	10	15	20	25	30
Rolled Section—Steel Sash							
Industrial Pivoted	1/16" crack	.87	1.80	2.9	4.1	5.1	6.2
Architectural Projected	1/32" crack	.25	.60	1.03	1.43	1.86	2.3
Architectural Projected	3/64" crack	.33	.87	1.47	1.93	2.5	3.0
Residential Casement	1/64" crack	.10	.30	.55	.78	1.00	1.23
Residential Casement	1/32" crack	.23	.53	.87	1.27	1.67	2.10
Heavy Casement Section Projected	1/64" crack	.05	.17	.30	.43	.58	.80
Heavy Casement Section Projected	1/32" crack	.13	.40	.63	.90	1.20	1.53
Hollow Metal—Vertically Pivoted		.50	1.46	2.40	3.10	3.70	4.00

*Infiltration caused by stack effect must be calculated separately during the winter.
†No allowance has been made for usage. See *Table 43* for infiltration due to usage.

(Courtesy Carrier Corporation, McGraw-Hill Book Company)

Figure 2.12 (cont.)

**TABLE 44—INFILTRATION THRU WINDOWS AND DOORS—CRACK METHOD—SUMMER—WINTER*
(Contd)**

TABLE 44c—DOORS† ON WINDWARD SIDE

TYPE OF DOOR	CFM PER LINEAR FOOT OF CRACK					
	Wind Velocity— mph					
	5	10	15	20	25	30
Glass Door—Herculite						
Good Installation ⅛″ crack	3.2	6.4	9.6	13.0	16.0	19.0
Average Installation ³⁄₁₆″ crack	4.8	10.0	14.0	20.0	24.0	29.0
Poor Installation ¼″ crack	6.4	13.0	19.0	26.0	26.0	38.0
Ordinary Wood or Metal						
Well Fitted—W-Strip	.45	.60	.90	1.3	1.7	2.1
Well Fitted—No W-Strip	.90	1.2	1.8	2.6	3.3	4.2
Poorly Fitted—No W-Strip	.90	2.3	3.7	5.2	6.6	8.4
Factory Door ⅛″ crack	3.2	6.4	9.6	13.0	16.0	19.0

VENTILATION

VENTILATION STANDARDS

The introduction of outdoor air for ventilation of conditioned spaces is necessary to dilute the odors given off by people, smoking and other internal air contaminants.

The amount of ventilation required varies primarily with the total number of people, the ceiling height and the number of people smoking. People give off body odors which require a minimum of 5 cfm per person for satisfactory dilution. Seven and one half cfm per person is recommended. This is based on a population density of 50 to 75 sq ft per person and a typical ceiling height of 8 ft. With greater population densities, the ventilation quantity should be increased. When people smoke, the additional odors given off by cigarettes or cigars require a minimum of 15 to 25 cfm per person. In special gathering rooms with heavy smoking, 30 to 50 cfm per person is recommended.

SCHEDULED VENTILATION

In comfort applications, where local codes permit, it is possible to reduce the capacity requirements of the installed equipment by reducing the ventilation air quantity at the time of peak load. This quantity can be reduced at times of peak to, in effect, minimize the outdoor air load. At times other than peak load, the calculated outdoor air quantity is used. Scheduled ventilation is recommended *only for installations operating more than 12 hours or 3 hours longer than occupancy,* to allow some time for flushing out the building when no odors are being generated. It has been found, by tests, that few complaints of stuffiness are encountered when the outdoor air quantity is reduced for short periods of time, provided the flushing period is available. It is recommended that the outdoor air quantity be reduced to no less than 40% of the recommended quantity.

(Courtesy Carrier Corporation, McGraw-Hill Book Company)

Figure 2.12 (cont.)

Heating Load Summary

The maximum design heating load is usually based on the demand of a cold winter night in a given climate. This is true in a residence, a hospital, or an auditorium where the facility is typically in use at night. For other building types, such as office buildings, two different design conditions are used. One condition is daytime-occupied use and the other is unoccupied (setback) night period.

The outdoor design temperature depends on the geographic location (see Figure 2.9). The inside design temperature is usually 68°F. The heating load consists of conduction and convection heat losses (see Figures 2.3 and 2.14).

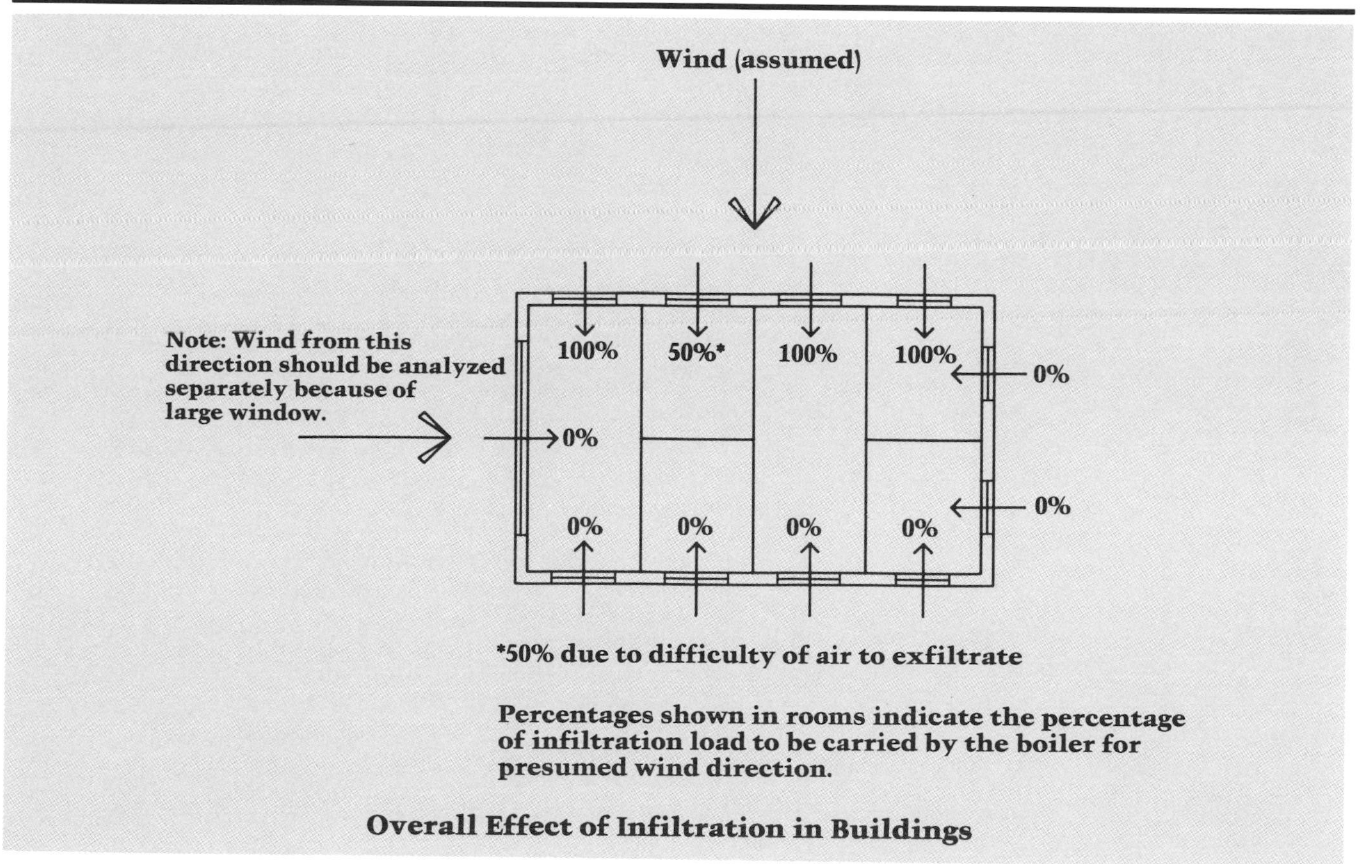

Figure 2.13

Cooling Loads

The size of a cooling system is based on the load it must carry. In cooling, the load is expressed in Btu/hour or tons. One ton of cooling equals 12,000 Btu/hour, or 12 MBH. The cooling load represents the amount of heat that must be *removed* from a building, usually referred to as heat gain. In addition to the heat gain from conduction and convection (computed similarly to heat losses), cooling loads must take into account radiation and internal heat gain from lights, appliances, power, and people. All of these factors represent additional loads to a cooling system.

Heat Loss Formulas

Conduction Heat Losses

$H_c = UA\ (\Delta T)$

H_c = Conduction loss (Btu/hr.)
U = Overall heat transfer coefficient (Btu/hr. × ft.^{2} × °F) (see Figures 2.5–2.7)
A = Surface area (s.f.)
ΔT = Temperature difference (°F)

$H_e = FP\ (\Delta T)$

H_e = Heat loss through edge (Btu/hr.)
F = Edge factor (Btu/hr./ft./°F)
P = Perimeter of slab (ft.)
ΔT = Indoor temperature minus outdoor temperature

Convection Heat Losses

$H_s = K \times A$ *or*
= 2 × Area for floors and basement walls 4 feet below grade *or*
= 4 × Area for floors and walls from 0 to 4 feet below grade

H_s = Heat loss through slab
K = Constant (Btu /hr./s.f.)
A = Area (s.f.)

$H_v = 1.1\ \text{cfm}\ (\Delta T)$

H_v = Convection loss (Btu/hr.)
1.1 = Constant
cfm = Air entering building (cubic feet per minute) (see Figure 2.12)
ΔT = Temperature difference (°F)

Total Heat Loss (Btu/hr.)

$H_t = H_c + H_e + H_v + H_s$

H_c = Conduction losses (walls, windows, floor, roof, attics, crawl spaces, garages, basements)
H_e = Edge losses, when applicable
H_s = Slab losses (grades, foundations, walls)
H_v = Infiltration and ventilation

Figure 2.14

Cooling loads are based on the statistics for a hot summer's day that is only surpassed 2-½% of the time in a particular climate. The critical time and date (which varies with each building) is determined by designers, because cooling load calculations are more complicated than heating load calculations. The primary reason for this complexity is the solar energy heat gain factor, computations for which are elaborate. The solar orientation of the building and the amount of glass receiving radiant energy varies not only from building to building, but also day by day, and even hour by hour. Furthermore, the heat "stored" in a building depends on the overall mass of the building, the hours of operation, and the length of time that lights are on in the building each day. The color of the building and type of sunscreen are also important factors.

The actual cooling load calculation can involve many iterations or tests during the cooling season. The resulting design conditions may include many variations. For example, one room facing east may have its heaviest heat gain at 10:00 a.m. on June 19, while another room facing west may have its design load (heaviest heat gain) at 5:00 p.m. on August 21; the maximum overall load for the building as a whole may occur at 3:00 p.m. on July 20. The interior temperature may swing with the varied loads as the cooling equipment responds.

In addition to sensible heat (a change in the air temperature that is felt or "sensed"), cooling systems must also account for *latent heat gains*. Latent heat is the energy that is required to change a solid to a liquid or a liquid to a gas; no temperature change occurs during this process. In cooling, latent heat is the energy resulting from condensation of moisture in the air. Condensing water vapor is the reverse of boiling. It takes approximately 1,000 Btu's, or 1 MBH, to boil or condense a pound of water. Boiling normally occurs at 212°F. However, it is possible to vaporize 50°F water or to pass humid air at 80°F over a cooling coil that produces 50°F air, which condenses some of the moisture in the air. The portion of the cooling energy that changes the 50°F condensate is the latent heat. Latent heat for cooling occurs when outdoor air (which contains moisture) is introduced into a building. People and cooking equipment also add moisture (latent heat) to a space. Sensible and latent heat gains are illustrated in Figure 2.15. Figure 2.16 illustrates the five categories of common building heat gains, which are listed below.

- Conduction heat gain
- Convection heat gain
- Solar heat gain
- Internal heat gain (lights, motors, cooking, etc.)
- People

Conduction Heat Gain

Conduction heat gain ("H_c" in Figure 2.16) is calculated in a manner similar to heat loss. The conductive heat gain through each envelope system (roof, walls, windows, and doors) is calculated separately, as shown in the following formula.

H_c = UA (CLTD)

H_c = Conduction heat gain (Btu/hr.)

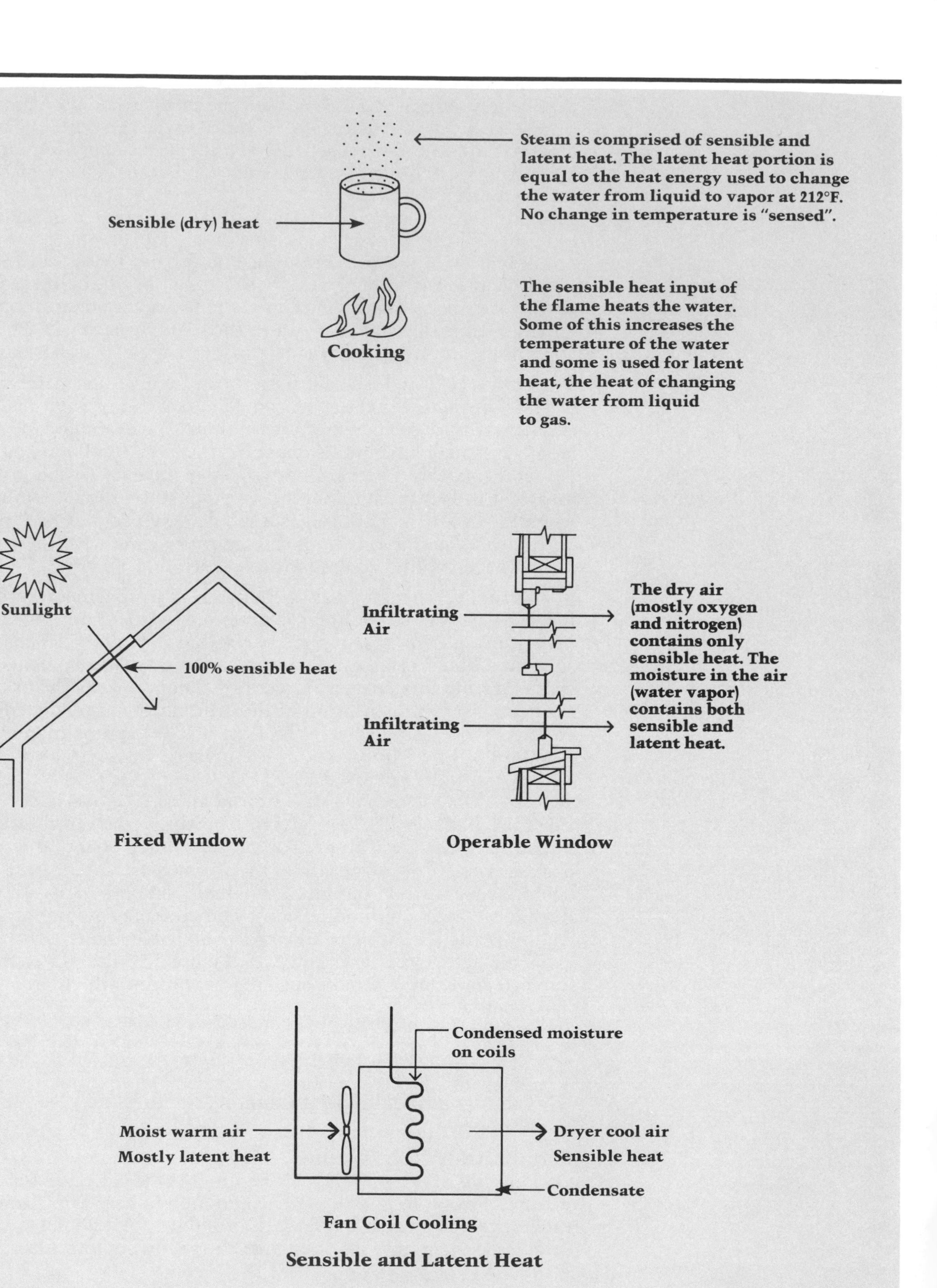

Sensible and Latent Heat

Figure 2.15

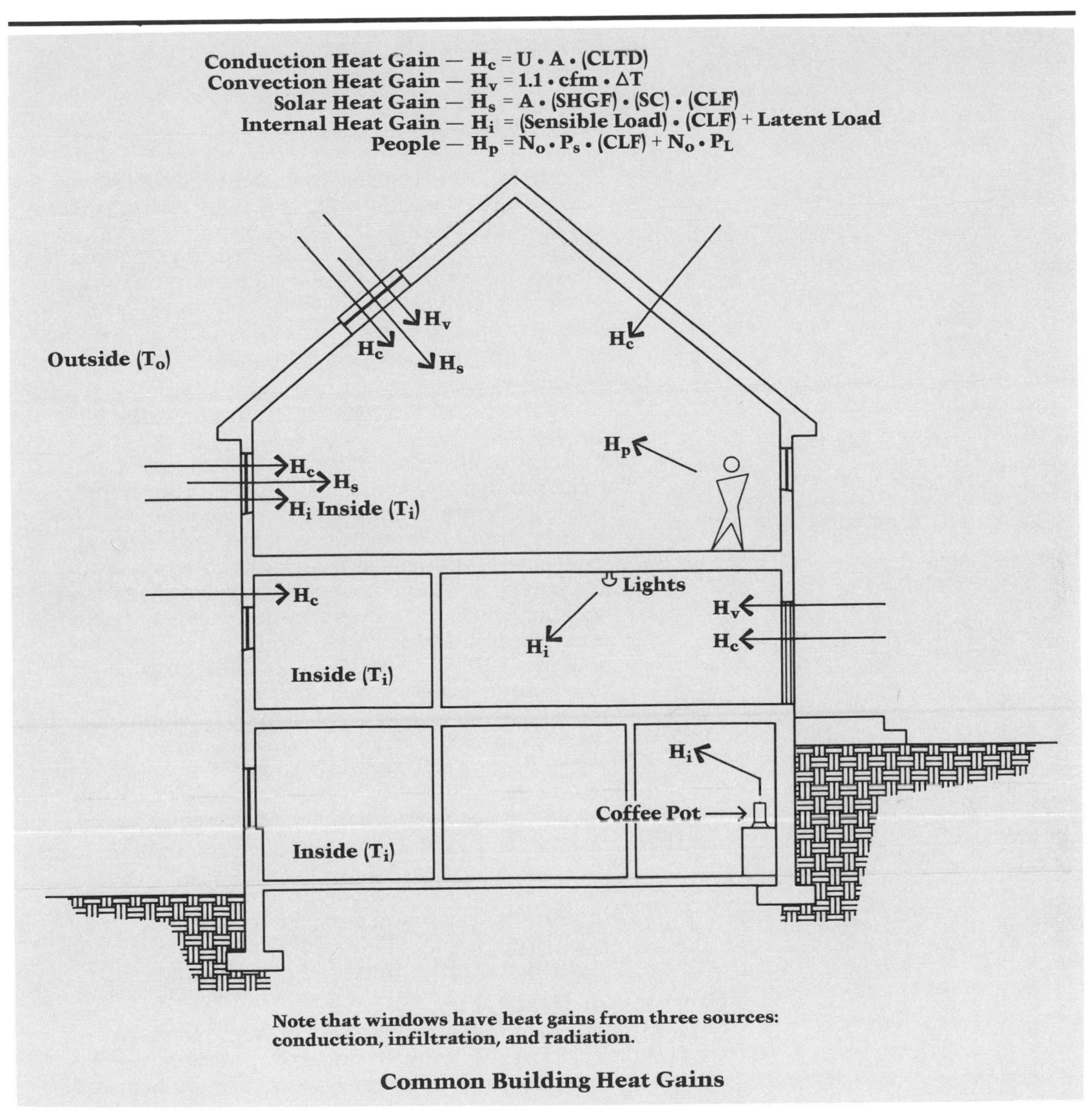

Common Building Heat Gains

Figure 2.16

U	=	Overall heat transfer coefficient (Btu/s.f./°F) (see Figures 2.5–2.7)
A	=	Surface area (s.f.)
CLTD	=	Cooling load temperature difference (°F) (see Figure 2.17)

U The overall heat transfer coefficient for cooling is computed in the same way as it is for heating. The U values are used for both heating and cooling. Figures 2.5, 2.6, and 2.7 can be used to compute the U value for cooling. Because the wind velocity and air density are both lower in the summer, some designers choose to recompute the cooling U values by adjusting for the changes in f_i and f_o, shown in Figure 2.5.

A The surface area is computed in the same manner as for heating.

CLTD Calculating the temperature difference (between indoor design and actual outdoor conditions) for cooling is more complicated than the procedure for heating. Interior design temperatures can be read from the recommended indoor design temperatures in Figure 2.8. The temperature on the outside of the building envelope, however, is not necessarily the outdoor temperature. For example, a tar and gravel roof may reach a temperature of 250°F on a summer day, substantially warmer than the ambient 88°F outside design temperature. Similarly, wall color, the daily temperature range, the inside design temperature, the weight of construction, the latitude, orientation, time of day, and month all affect the actual outside surface temperature. (See Figures 2.8 and 2.9 for T_i and T_o.) For glass, CLTD equals $T_o - T_i$. The American Society of Heating, Refrigeration, and Air Conditioning Engineers (ASHRAE) has a more complete method for calculating the CLTD of glass, but many designers prefer the slightly conservative value stated here.

Convection Heat Gain

Convection heat gain is calculated in a manner similar to convection heat loss. Both infiltration and ventilation contribute to heat gain. The formula for convection heat gain is shown below.

$$H_v = \underbrace{1.1 \text{ cfm } (\Delta T)}_{\text{Sensible heat gain}} + \underbrace{4840 \text{ cfm } (\Delta W)}_{\text{Latent heat gain}}$$

The cfm of air entering the building from infiltration is computed in the same way as infiltration affecting heating, except that for cooling, a 7.5 mph wind velocity is used (see Figure 2.12). When an air handling system is used to pressurize the building, the infiltration will equal zero, since no air can infiltrate to a higher pressure. In this situation, outside air is used both to ventilate and pressurize the building; the amount of air required for ventilation is used for the cfm value.

26.8 **CHAPTER 26** **1985 Fundamentals Handbook**

Table 5 Cooling Load Temperature Differences for Calculating Cooling Load from Flat Roofs

Roof No	Description of Construction	Weight lb/ft²	U-value Btu/(h·ft²·F)	1	2	3	4	5	6	7	8	9	10	11	12	13	14	15	16	17	18	19	20	21	22	23	24	Hour of Maximum CLTD	Minimum CLTD	Maximum CLTD	Difference CLTD
				Solar Time, h																											
	Without Suspended Ceiling																														
1	Steel sheet with 1-in. (or 2-in.) insulation	7 (8)	0.213 (0.124)	1	−2	−3	−3	−5	−3	6	19	34	49	61	71	78	79	77	70	59	45	30	18	12	8	5	3	14	−5	79	84
2	1-in. wood with 1-in. insulation	8	0.170	6	3	0	−1	−3	−3	−2	4	14	27	39	52	62	70	74	74	70	62	51	38	28	20	14	9	16	−3	74	77
3	4-in. l.w concrete	18	0.213	9	5	2	0	−2	−3	−3	1	9	20	32	44	55	64	70	73	71	66	57	45	34	25	18	13	16	−3	73	76
4	2-in. h.w. concrete with 1-in. (or 2-in.) insulation	29	0.206 (0.122)	12	8	5	3	0	−1	−1	3	11	20	30	41	51	59	65	66	66	62	54	45	36	29	22	17	16	−1	67	68
5	1-in. wood with 2-in. insulation	9	0.109	3	0	−3	−4	−5	−7	−6	−3	5	16	27	39	49	57	63	64	62	57	48	37	26	18	11	7	16	−7	64	71
6	6-in. l.w. concrete	24	0.158	22	17	13	9	6	3	1	1	3	7	15	23	33	43	51	58	62	64	62	57	50	42	35	28	18	1	64	63
7	2.5-in. wood with 1-in. insulation	13	0.130	29	24	20	16	13	10	7	6	6	9	13	20	27	34	42	48	53	55	56	54	49	44	39	34	19	6	56	50
8	8-in. l.w. concrete	31	0.126	35	30	26	22	18	14	11	9	7	7	9	13	19	25	33	39	46	50	53	54	53	49	45	40	20	7	54	47
9	4-in. h.w. concrete with 1-in. (or 2-in.) insulation	52 (52)	0.200 (0.120)	25	22	18	15	12	9	8	8	10	14	20	26	33	40	46	50	53	53	52	48	43	38	34	30	18	8	53	45
10	2.5-in. wood with 2-in. insulation	13	0.093	30	26	23	19	16	13	10	9	8	9	13	17	23	29	36	41	46	49	51	50	47	43	39	35	19	8	51	43
11	Roof terrace system	75	0.106	34	31	28	25	22	19	16	14	13	13	15	18	22	26	31	36	40	44	45	46	45	43	40	37	20	13	46	33
12	6-in. h.w. concrete with 1-in. (or 2-in.) insulation	(75) 75	0.192 (0.117)	31	28	25	22	20	17	15	14	14	16	18	22	26	31	36	40	43	45	45	44	42	40	37	34	19	14	45	31
13	4-in. wood with 1-in. (or 2-in) insulation	17 (18)	0.106 (0.078)	38	36	33	30	28	25	22	20	18	17	16	17	18	21	24	28	32	36	39	41	43	43	42	40	22	16	43	27
	With Suspended Ceiling																														
1	Steel Sheet with 1-in. (or 2-in.) insulation	9 (10)	0.134 (0.092)	2	0	−2	−3	−4	−4	−1	9	23	37	50	62	71	77	78	74	67	56	42	28	18	12	8	5	15	−4	78	82
2	1-in. wood with 1-in. insulation	10	0.115	20	15	11	8	5	3	2	3	7	13	21	30	40	48	55	60	62	61	58	51	44	37	30	25	17	2	62	60
3	4-in. l.w. concrete	20	0.134	19	14	10	7	4	2	0	0	4	10	19	29	39	48	56	62	65	64	61	54	46	38	30	24	17	0	65	65
4	2-in. h.w. concrete with 1-in. insulation	30	0.131	28	25	23	20	17	15	13	13	14	16	20	25	30	35	39	43	46	47	46	44	41	38	35	32	18	13	47	34
5	1-in. wood with 2-in. insulation	10	0.083	25	20	16	13	10	7	5	5	7	12	18	25	33	41	48	53	57	57	56	52	46	40	34	29	18	5	57	52
6	6-in. l.w. concrete	26	0.109	32	28	23	19	16	13	10	8	7	8	11	16	22	29	36	42	48	52	54	54	51	47	42	37	20	7	54	47
7	2.5-in. wood with 1-in. insulation	15	0.096	34	31	29	26	23	21	18	16	15	15	16	18	21	25	30	34	38	41	43	44	44	42	40	37	21	15	44	29
8	8-in. l.w. concrete	33	0.093	39	36	33	29	26	23	20	18	15	14	14	15	17	20	25	29	34	38	42	45	46	45	44	42	21	14	46	32
9	4-in. h.w. concrete with 1-in. (or 2-in.) insulation	53 (54)	0.128 (0.090)	30	29	27	26	24	22	21	20	20	21	22	24	27	29	32	34	36	38	38	38	37	36	34	33	19	20	38	18
10	2.5-in. wood with 2-in.insulation	15	0.072	35	33	30	28	26	24	22	20	18	18	18	20	22	25	28	32	35	38	40	41	41	40	39	37	21	18	41	23
11	Roof terrace system	77	0.082	30	29	28	27	26	25	24	23	22	22	22	23	23	25	26	28	29	31	32	33	33	33	33	32	22	22	33	11
12	6-in. h.w. concrete with 1-in. (or 2-in) insulation	77 (77)	0.125 (0.088)	29	28	27	26	25	24	23	22	21	21	22	23	25	26	28	30	32	33	34	34	34	33	32	31	20	21	34	13
13	4-in. wood with 1-in (or 2-in.) insulation	19 (20)	0.082 (0.064)	35	34	33	32	31	29	27	26	24	23	22	21	22	22	24	25	27	30	32	34	35	36	37	36	23	21	37	16

(1) *Direct Application of Table 5 Without Adjustments:*

Values in Table 5 were calculated using the following conditions:

- Dark flat surface roof ("dark" for solar radiation absorption)
- Indoor temperature of 78 F
- Outdoor maximum temperature of 95 F with outdoor mean temperature of 85 F and an outdoor daily range of 21 deg F
- Solar radiation typical of 40 deg North latitude on July 21
- Outside surface resistance, R_o = 0.333 ft² • F • h/Btu
- Without and with suspended ceiling, but no attic fans or return air ducts in suspended ceiling space
- Inside surface resistance, R_i = 0.685 ft² • F • h/Btu

(2) *Adjustments to Table 5 Values:*

The following equation makes adjustments for deviations of design and solar conditions from those listed in (1) above.

$$CLTD_{corr} = [(CLTD + LM) \cdot K + (78 - T_R) + (T_o - 85)] \cdot f$$

where CLTD is from this table

(a) LM is latitude-month correction from Table 9 for a horizontal surface,

(b) K is a color adjustment factor applied after first making month-latitude adjustments. Credit should not be taken for a light-colored roof except where permanence of light color is established by experience, as in rural areas or where there is little smoke

K = 1.0 if dark colored or light in an industrial area

K = 0.5 if permanently light-colored (rural area)

(c) $(78 - T_R)$ is indoor design temperature correction

(d) $(T_o - 85)$ is outdoor design temperature correction, where T_o is the average outside temperature on design day

(e) f is a factor for attic fan and or ducts above ceiling applied after all other adjustments have been made

f = 1.0 no attic or ducts

f = 0.75 positive ventilation

Values in Table 5 were calculated without and with a suspended ceiling, but made no allowances for positive ventilation or return ducts thru the space. If ceiling is insulated and a fan is used between ceiling and roof, CLTD may be reduced by 25% (f = 0.75). Use of the suspended ceiling space for a return air plenum or with return air ducts should be analyzed separately.

(3) *Roof Constructions Not Listed in Table:*

The U-Values listed are to be used only as guides. The actual value of U as obtained from tables such as Tables 3 and 4, Chapter 23 or as calculated for the actual roof construction should be used.

(Courtesy ASHRAE, 1985 Fundamentals)

Figure 2.17

An actual roof construction not in this table would be thermally similar to a roof in the table, if it has similar mass, lb_m/ft^2, and similar heat capacity $Btu/ft^2 \cdot F$. In this case, use the CLTD from this table as corrected by Note (2) above.

Example: A flat roof without a suspended ceiling has mass properties = 18.0 lb/ft^2,
U = 0.20 $Btu/h \cdot ft^2 \cdot F$, and heat capacity = 9.5 $Btu/ft^2 \cdot F$.
Use $CLTD_{uncorr}$ from Roof No. 13, to obtain $CLTD_{corr}$ and use the actual U value to calculate $q/A = U\ (CLTD_{corr}) = 0.20\ (CLTD_{corr})$.

(4) *Additional Insulation*

For each R-7 increase in R-value from insulation added to the roof structure, use a CLTD for a roof whose weight and heat capacity are approximately the same, but whose CLTD has a maximum value 2 h later. If this is not possible, because a roof with longest time lag has already been selected, use an effective CLTD in cooling load calculation equal to 29 deg F.

Table 6 Wall Construction Group Description

Group No.	Description of Construction	Weight (lb/ft^2)	U-Value ($Btu/h \cdot ft^2 \cdot F$)	Code Numbers of Layers (see Table 8)
4-in. Face Brick + (*Brick*)				
C	Air Space + 4-in. Face Brick	83	0.358	A0, A2, B1, A2, E0
D	4-in. Common Brick	90	0.415	A0, A2, C4, E1, E0
C	1-in. Insulation or Air Space + 4-in. Common Brick	90	0.174-0.301	A0, A2, C4, B1/B2, E1, E0
B	2-in. Insulation + 4-in. Common Brick	88	0.111	A0, A2, B3, C4, E1, E0
B	8-in. Common Brick	130	0.302	A0, A2, C9, E1, E0
A	Insulation or Air Space + 8-in. Common brick	130	0.154-0.243	A0, A2, C9, B1/B2, E1, E0
4-in. Face Brick + (*H.W. Concrete*)				
C	Air Space + 2-in. Concrete	94	0.350	A0, A2, B1, C5, E1, E0
B	2-in. Insulation + 4-in. Concrete	97	0.116	A0, A2, B3, C5, E1, E0
A	Air Space or Insulation + 8-in. or more Concrete	143-190	0.110-0.112	A0, A2, B1, C10/11, E1, E0
4-in. Face Brick + (*L.W. or H.W. Concrete Block*)				
E	4-in. Block	62	0.319	A0, A2, C2, E1, E0
D	Air Space or Insulation + 4-in. Block	62	0.153-0.246	A0, A2, C2, B1/B2, E1, E0
D	8-in. Block	70	0.274	A0, A2, C7, A6, E0
C	Air Space or 1-in. Insulation + 6-in. or 8-in. Block	73-89	0.221-0.275	A0, A2, B1, C7/C8, E1, E0
B	2-in. Insulation + 8-in. Block	89	0.096-0.107	A0, A2, B3, C7/C8, E1, E0
4-in. Face Brick + (*Clay Tile*)				
D	4-in. Tile	71	0.381	A0, A2, C1, E1, E0
D	Air Space + 4-in. Tile	71	0.281	A0, A2, C1, B1, E1, E0
C	Insulation + 4-in. Tile	71	0.169	A0, A2, C1, B2, E1, E0
C	8-in. Tile	96	0.275	A0, A2, C6, E1, E0
B	Air Space or 1-in. Insulation + 8-in. Tile	96	0.142-0.221	A0, A2, C6, B1/B2, E1, E0
A	2-in. Insulation + 8-in. Tile	97	0.097	A0, A2, B3, C6, E1, E0
H.W. Concrete Wall + (*Finish*)				
E	4-in. Concrete	63	0.585	A0, A1, C5, E1, E0
D	4-in. Concrete + 1-in. or 2-in. Insulation	63	0.119-0.200	A0, A1, C5, B2/B3, E1, E0
C	2-in. Insulation + 4-in. Concrete	63	0.119	A0, A1, B6, C5, E1, E0
C	8-in. Concrete	109	0.490	A0, A1, C10, E1, E0
B	8-in. Concrete + 1-in. or 2-in. Insulation	110	0.115-0.187	A0, A1, C10, B5/B6, E1, E0
A	2-in. Insulation + 8-in. Concrete	110	0.115	A0, A1, B3, C10, E1, E0
B	12-in. Concrete	156	0.421	A0, A1, C11, E1, E0
A	12-in. Concrete + Insulation	156	0.113	A0, C11, B6, A6, E0
L. W. and H.W. Concrete Block + (*Finish*)				
F	4-in. Block + Air Space/Insulation	29	0.161-0.263	A0, A1, C2, B1/B2, E1, E0
E	2-in. Insulation + 4-in. Block	29-37	0.105-0.114	A0, A1, B3, C2/C3, E1, E0
E	8-in. Block	47-51	0.294-0.402	A0, A1, C7/C8, E1, E0
D	8-in. Block + Air Space/Insulation	41-57	0.149-0.173	A0, A1, C7/C8, B1/B2, E1, E0
Clay Tile + (*Finish*)				
F	4-in. Tile	39	0.419	A0, A1, C1, E1, E0
F	4-in. Tile + Air Space	39	0.303	A0, A1, C1, B1, E1, E0
E	4-in. Tile + 1-in. Insulation	39	0.175	A0, A1, C1, B2, E1, E0
D	2-in. Insulation + 4-in. Tile	40	0.110	A0, A1, B3, C1, E1, E0
D	8-in. Tile	63	0.296	A0, A1, C6, B1/B2, E1, E0
C	8-in. Tile + Air Space/1-in. Insulation	63	0.151-0.231	A0, A1, C6, B1/B2, E1, E0
B	2-in. Insulation + 8-in. Tile	63	0.099	A0, A1, B3, C6, E1, E0
Metal Curtain Wall				
G	With/without air Space + 1- in./ 2-in. 3-in. Insulation	5-6	0.091-0.2 30	A0, A3, B5/B6/B12, A3, E0
Frame Wall				
G	1-in. to 3-in. Insulation	16	0.081-0.1 78	A0, A1, B1, B2/B3/B4, E1, E0

(Courtesy ASHRAE, 1985 Fundamentals)

Figure 2.17 (cont.)

Table 7 Cooling Load Temperature Differences for Calculating Cooling Load from Sunlit Walls

North Latitude Wall Facing	Solar Time, h 0100	0200	0300	0400	0500	0600	0700	0800	0900	1000	1100	1200	1300	1400	1500	1600	1700	1800	1900	2000	2100	2200	2300	2400	Hr of Maximum CLTD	Minimum CLTD	Maximum CLTD	Difference CLTD
Group A Walls																												
N	14	14	14	13	13	13	12	12	11	11	10	10	10	10	10	10	11	11	12	12	13	13	14	14	2	10	14	4
NE	19	19	19	18	17	17	16	15	15	15	15	15	16	16	17	18	18	18	19	19	20	20	20	20	22	15	20	5
E	24	24	23	23	22	21	20	19	19	18	19	19	20	21	22	23	24	24	25	25	25	25	25	25	22	18	25	7
SE	24	23	23	22	21	20	20	19	18	18	18	18	18	19	20	21	22	23	23	24	24	24	24	24	22	18	24	6
S	20	20	19	19	18	18	17	16	16	15	14	14	14	14	14	15	16	17	18	19	19	20	20	20	23	14	20	6
SW	25	25	25	24	24	23	22	21	20	19	19	18	17	17	17	17	18	19	20	22	23	24	25	25	24	17	25	8
W	27	27	26	26	25	24	24	23	22	21	20	19	19	18	18	18	18	19	20	22	23	25	26	26	1	18	27	9
NW	21	21	21	20	20	19	19	18	17	16	16	15	15	14	14	14	15	15	16	17	18	19	20	21	1	14	21	7
Group B Walls																												
N	15	14	14	13	12	11	11	10	9	9	9	8	9	9	9	10	11	12	13	14	14	15	15	15	24	8	15	7
NE	19	18	17	16	15	14	13	12	12	13	14	15	16	17	18	19	19	20	20	21	21	21	20	20	21	12	21	9
E	23	22	21	20	18	17	16	15	15	15	17	19	21	22	24	25	26	26	27	27	26	26	25	24	20	15	27	12
SE	23	22	21	20	18	17	16	15	14	14	15	16	18	20	21	23	24	25	26	26	26	26	25	24	21	14	26	12
S	21	20	19	18	17	15	14	13	12	11	11	11	11	12	14	15	17	19	20	21	22	22	22	21	23	11	22	11
SW	27	26	25	24	22	21	19	18	16	15	14	14	13	13	14	15	17	20	22	25	27	28	28	28	24	13	28	15
W	29	28	27	26	24	23	21	19	18	17	16	15	14	14	14	15	17	19	22	25	27	29	29	30	24	14	30	16
NW	23	22	21	20	19	18	17	15	14	13	12	12	12	11	12	12	13	15	17	19	21	22	23	23	24	11	23	9
Group C Walls																												
N	15	14	13	12	11	10	9	8	8	7	7	8	8	9	10	12	13	14	15	16	17	17	17	16	22	7	17	10
NE	19	17	16	14	13	11	10	10	11	13	15	17	19	20	21	22	22	23	23	23	23	22	21	20	20	10	23	13
E	22	21	19	17	15	14	12	12	14	16	19	22	25	27	29	29	30	30	30	29	28	27	26	24	18	12	30	18
SE	22	21	19	17	15	14	12	12	12	13	16	19	22	24	26	28	29	29	29	29	28	27	26	24	19	12	29	17
S	21	19	18	16	15	13	12	10	9	9	9	10	11	14	17	20	22	24	25	26	25	25	24	22	20	9	26	17
SW	29	27	25	22	20	18	16	15	13	12	11	11	11	13	15	18	22	26	29	32	33	33	32	31	22	11	33	22
W	31	29	27	25	22	20	18	16	14	13	12	12	12	13	14	16	20	24	29	32	35	35	35	33	22	12	35	23
NW	25	23	21	20	18	16	14	13	11	10	10	10	10	11	12	13	15	18	22	25	27	27	27	26	22	10	27	17
Group D Walls																												
N	15	13	12	10	9	7	6	6	6	6	6	7	8	10	12	13	15	17	18	19	19	19	18	16	21	6	19	13
NE	17	15	13	11	10	8	7	8	10	14	17	20	22	23	23	24	24	25	25	24	23	22	20	18	19	7	25	18
E	19	17	15	13	11	9	8	9	12	17	22	27	30	32	33	33	32	32	31	30	28	26	24	22	16	8	33	25
SE	20	17	15	13	11	10	8	8	10	13	17	22	26	29	31	32	32	32	31	30	28	26	24	22	17	8	32	24
S	19	17	15	13	11	9	8	7	6	6	7	9	12	16	20	24	27	29	29	29	27	26	24	22	19	6	29	23
SW	28	25	22	19	16	14	12	10	9	8	8	8	10	12	16	21	27	32	36	38	38	37	34	31	21	8	38	30
W	31	27	24	21	18	15	13	11	10	9	9	9	10	11	14	18	24	30	36	40	41	40	38	34	21	9	41	32
NW	25	22	19	17	14	12	10	9	8	7	7	8	9	10	12	14	18	22	27	31	32	32	30	27	22	7	32	25
Group E Walls																												
N	12	10	8	7	5	4	3	4	5	6	7	9	11	13	15	17	19	20	21	23	20	18	16	14	20	3	22	19
NE	13	11	9	7	6	4	5	9	15	20	24	25	25	26	26	26	26	26	25	24	22	19	17	15	16	4	26	22
E	14	12	10	8	6	5	6	11	18	26	33	36	38	37	36	34	33	32	30	28	25	22	20	17	13	5	38	33
SE	15	12	10	8	7	5	5	8	12	19	25	31	35	37	37	36	34	33	31	28	26	23	20	17	15	5	37	32
S	15	12	10	8	7	5	4	3	4	5	9	13	19	24	29	32	34	33	31	29	26	23	20	17	17	3	34	31
SW	22	18	15	12	10	8	6	5	5	6	7	9	12	18	24	32	38	43	45	44	40	35	30	26	19	5	45	40
W	25	21	17	14	11	9	7	6	6	6	7	9	11	14	20	27	36	43	49	49	45	40	34	29	20	6	49	43
NW	20	17	14	11	9	7	6	5	5	5	6	8	10	13	16	20	26	32	37	38	36	32	28	24	20	5	38	33
Group F Walls																												
N	8	6	5	3	2	1	2	4	6	7	9	11	14	17	19	21	22	23	24	23	20	16	13	11	19	1	23	23
NE	9	7	5	3	2	1	5	14	23	28	30	29	28	27	27	27	27	26	24	22	19	16	13	11	11	1	30	29
E	10	7	6	4	3	2	6	17	28	38	44	45	43	39	36	34	32	30	27	24	21	17	15	12	12	2	45	43
SE	10	7	6	4	3	2	4	10	19	28	36	41	43	42	39	36	34	31	28	25	21	18	15	12	13	2	43	41
S	10	8	6	4	3	2	1	1	3	7	13	20	27	34	38	39	38	35	31	26	22	18	15	12	16	1	39	38
SW	15	11	9	6	5	3	2	2	4	5	8	11	17	26	35	44	50	53	52	45	37	28	23	18	18	2	53	48
W	17	13	10	7	5	4	3	3	4	6	8	11	14	20	28	39	49	57	60	54	43	34	27	21	19	3	60	57
NW	14	10	8	6	4	3	2	2	3	5	8	10	13	15	21	27	35	42	46	43	35	28	22	18	19	2	46	44
Group G Walls																												
N	3	2	1	0	−1	2	7	8	9	12	15	18	21	23	24	24	25	26	22	15	11	9	7	5	18	−1	26	27
NE	3	2	1	0	−1	9	27	36	39	35	30	26	26	27	27	26	25	22	18	14	11	9	7	5	9	−1	39	40
E	4	2	1	0	−1	11	31	47	54	55	50	40	33	31	30	29	27	24	19	15	12	10	8	6	10	−1	55	56
SE	4	2	1	0	−1	5	18	32	42	49	51	48	42	36	32	30	27	24	19	15	12	10	8	6	11	−1	51	52
S	4	2	1	0	−1	0	1	5	12	22	31	39	45	46	43	37	31	25	20	15	12	10	8	5	14	−1	46	47
SW	5	4	3	1	0	0	2	5	8	12	16	26	38	50	59	63	61	52	37	24	17	13	10	8	16	0	63	63
W	6	5	3	2	1	1	2	5	8	11	15	19	27	41	56	67	72	67	48	29	20	15	11	8	17	1	72	71
NW	5	3	2	1	0	0	2	5	8	11	15	18	21	27	37	47	55	55	41	25	17	13	10	7	18	0	55	55

(1) *Direct Application of the Table Without Adjustments:*

Values in the Table were calculated using the same conditions for walls as outlined for the roof CLTD table, Table 5. These values may be used for all normal air-conditioning estimates usually without correction (except as noted below) when the load is calculated for the hottest weather.

For totally shaded walls use the North orientation values.

(2) *Adjustments to Table Values:*

The following equation makes adjustments for conditions other than those listed in Note (1).

$$CLTD_{corr} = (CI\ TD + LM) \times K + (78 - T_R) + (T_o - 85)$$

where CLTD is from Table 7 at the wall orientation.

(a) LM is the latitude-month correction from Table 9.

(b) K is a color adjustment factor applied after first making month-latitude adjustment

K = 1.0 if dark colored or light in an industrial area

K = 0.83 if permanently medium-colored (rural area)

K = 0.65 if permanently light-colored (rural area)

Credit should not be taken for wall color other than dark except where permanence of color is established by experience, as in rural areas or where there is little smoke.

(Courtesy ASHRAE, 1985 Fundamentals)

Figure 2.17 (cont.)

Colors: Light — Cream

Medium — Medium blue, medium green, bright red, light brown, unpainted wood and natural color concrete

Dark — Dark blue, red, brown and green

(c) $(78 - T_R)$ is indoor design temperature correction

(d) $(T_o - 85)$ is outdoor design temperature correction, where T_o is the average outside temperature on design day.

(3) *Wall Construction Not Listed:*

The U-Values listed are to be used only as guides. The actual value of U as obtained from tables such as Tables 3 and 4, Chapter 23 or as calculated for the actual wall structure should be used.

An actual wall construction not listed in this table (or Table 6) would be thermally similar to a wall in the table, if it has similar mass, lb/ft^2, and similar heat capacity Btu/ft^2 • F. In that case, use the CLTD from this table as corrected by Note (2) above.

(4) *Additional Insulation:*

For each 7 increase in R-value from insulation added to the wall structures in Table 6, use the CLTD for the wall group with the next higher letter in the alphabet (e.g., C →A). For example, move to a group B wall when the initial wall group is C. When the insulation is added to the exterior of the construction rather than the interior, use the CLTD for the wall group two letters higher. If this is not possible, because a wall on Group A has already been selected, use an effective CLTD in the load calculation as given in the following table.

CLTD, Uncorrected, When Vertical Wall Structure is "Thermally" Heavier than Group A because of Added Insulation

N	NE	E	SE	S	SW	W	NW
11	17	22	21	17	21	22	17

Table 8 Thermal Properties and Code Numbers of Layers Used in Calculation of Coefficients for Roof and Wall

Description	Code Number	Thickness and Thermal Properties[a]					
		L	*k*	*D*	*SH*	*R*	Mass
Outside surface resistance	A0					0.333	
1-in. Stucco (asbestos cement or wood siding plaster, etc)	A1	0.0833	0.4	116	0.20	0.208	9.66
4-in. facebrick (dense concrete)	A2	0.3333	0.75	130	0.22	0.444	43.3
Steel siding (aluminum or other lightweight cladding)	A3	0.0050	26.0	480	0.10	0.000	2.40
Outside surface resistance						0.333	
0.5-in. slag, membrane	A4	0.0417	0.83	55	0.40		
0.375-in. felt		0.0313	0.11	70	0.40		
Finish	A6	0.0417	0.24	78	0.26	0.174	3.25
4-in. facebrick	A7	0.3333	0.77	125	0.22	0.433	41.6
Air Space Resistance	B1					0.91	
1-in. insulation	B2	0.0833	0.025	2.0	0.2	332	0.17
2-in. insulation	B3	0.1667	0.025	2.0	0.2	6.68	0.33
3-in. insulation	B4	0.2500	0.025	2.0	0.2	10.03	0.50
1-in. insulation	B5	0.0833	0.025	5.7	0.2	3.33	0.47
2-in. insulation	B6	0.1667	0.025	5.7	0.2	6.68	0.95
1-in. wood	B7	0.0833	0.067	37.0	0.6	1.19	3.08
2.5-in. wood	B8	0.2083	0.067	37.0	0.6	2.98	7.71
4-in. wood	B9	0.3333	0.067	37.0	0.6	4.76	2.3
2-in. wood	B10	0.1667	0.067	37.0	0.6	2.39	6.18
3-in. wood	B11	0.2500	0.067	37.0	0.6	3.58	9.25
3-in. insulation	B12	0.2500	0.025	5.7	0.2	10.00	1.42
4-in. insulation	B13	0.3333	0.025	5.7	0.2	13.33	1.90
5-in. insulation	B14	0.4167	0.025	5.7	0.2	16.67	2.38
6-in. insulation	B15	0.5000	0.025	5.7	0.2	20.00	2.85
4-in. clay tile	C1	0.3333	0.33	70.0	0.2	1.01	23.3
4-in. l.w. concrete block	C2	0.3333	0.22	38.0	0.2	1.51	12.7
4-in. l.w. concrete block	C3	0.3333	0.47	61.0	0.2	0.71	20.3
4-in. common brick	C4	0.3333	0.42	120.0	0.2	0.79	40.0
4-in. l.w. concrete	C5	0.3333	1.00	140.0	0.2	0.333	46.6
8-in. clay tile	C6	0.6667	0.33	70.0	0.2	2.02	46.7
8-in. l.w. concrete block	C7	0.6667	0.33	38.0	0.2	2.02	25.4
8-in. l.w. concrete block	C8	0.6667	0.6	61.0	0.2	1.11	40.7
8-in. common brick	C9	0.6667	0.42	120.0	0.2	1.59	80.0
8-in. l.w. concrete	C10	0.6667	1.00	140.0	0.2	0.667	93.4
12-in. l.w. concrete	C11	1.0000	1.00	140.0	0.2	1.08	140.0
2-in. l.w. concrete	C12	1.6667	1.00	140.0	0.2	1.167	23.4
6-in. l.w. concrete	C13	0.5000	1.00	140.0	0.500	0.2	70.0
4-in. l.w. concrete	C14	0.3333	0.10	40.0	0.2	3.333	13.3
6-in. l.w. concrete	C15	0.5000	0.10	40.0	0.2	5.000	20.0
8-in. l.w. concrete	C16	0.6667	0.10	40.0	0.2	6.667	26.7
8-in. l.w. concrete block (filled insulation)	C17	0.6667	0.08	18.0	0.2	9.00	12.0
8-in. l.w. concrete block (filled insulation)	C18	0.6667	0.34	53.0	0.2	1.98	35.4
12-in. l. w. concrete block (filled insulation)	C19	1.0000	0.08	19.0	0.2	13.5	19.0
12-in. l.w. concrete block (filled insulation)	C20	1.0000	0.39	56.0	0.2	2.59	56.0
Inside surface resistance	E0					.685	
0.75-in. plaster; 0.75-in. gypsum or other similar finishing layer	E1	0.0625	0.42	100	0.2	0.149	6.25
0.5-in. slag or stone	E2	0.0417	0.83	55	0.40	0.050	2.29
0.375-in. felt & membrane	E3	0.0313	0.11	70	0.40	0.285	2.19
Ceiling air space	E4					1.0	
Acoustic tile	E5	0.0625	0.035	30	0.20	1.786	1.88

[a]Units: L = ft; k = Btu/h•ft•F; D = lb/ft^3; SH = Btu/lb•F; R = ft^2•F•h/Btu; WT = lb/ft^2.

(Courtesy ASHRAE, 1985 Fundamentals)

Figure 2.17 (cont.)

26.12 CHAPTER 26 1985 Fundamentals Handbook

Table 9 CLTD Correction For Latitude and Month Applied to Walls and Roofs, North Latitudes

Lat.	Month	N	NNE NNW	NE NW	ENE WNW	E W	ESE WSW	SE SW	SSE SSW	S	HOR
0	Dec	−3	−5	−5	−5	−2	0	3	6	9	−1
	Jan/Nov	−3	−5	−4	−4	−1	0	2	4	7	−1
	Feb/Oct	−3	−2	−2	−2	−1	−1	0	−1	0	0
	Mar/Sept	−3	0	1	−1	−1	−3	−3	−5	−8	0
	Apr/Aug	5	4	3	0	−2	−5	−6	−8	−8	−2
	May/Jul	10	7	5	0	−3	−7	−8	−9	−8	−4
	Jun	12	9	5	0	−3	−7	−9	−10	−8	−5
8	Dec	−4	−6	−6	−6	−3	0	4	8	12	−5
	Jan/Nov	−3	−5	−6	−5	−2	0	3	6	10	−4
	Feb/Oct	−3	−4	−3	−3	−1	−1	1	2	4	−1
	Mar/Sept	−3	−2	−1	−1	−1	−2	−2	−3	−4	0
	Apr/Aug	2	2	2	0	−1	−4	−5	−7	−7	−1
	May/Jul	7	5	4	0	−2	−5	−7	−9	−7	−2
	Jun	9	6	4	0	−2	−6	−8	−9	−7	−2
16	Dec	−4	−6	−8	−8	−4	−1	4	9	13	−9
	Jan/Nov	−4	−6	−7	−7	−4	−1	4	8	12	−7
	Feb/Oct	−3	−5	−5	−4	−2	0	2	5	7	−4
	Mar/Sept	−3	−3	−2	−2	−1	−1	0	0	0	−1
	Apr/Aug	−1	0	−1	−1	−1	−3	−3	−5	−6	0
	May/Jul	4	3	3	0	−1	−4	−5	−7	−7	0
	Jun	6	4	4	1	−1	−4	−6	−8	−7	0
24	Dec	−5	−7	−9	−10	−7	−3	3	9	13	−13
	Jan/Nov	−4	−6	−8	−9	−6	−3	3	9	13	−11
	Feb/Oct	−4	−5	−6	−6	−3	−1	3	7	10	−7
	Mar/Sept	−3	−4	−3	−3	−1	−1	1	2	4	−3
	Apr/Aug	−2	−1	0	−1	−1	−2	−1	−2	−3	0
	May/Jul	1	2	2	0	0	−3	−3	−5	−6	1
	Jun	3	3	3	1	0	−3	−4	−6	−6	1
32	Dec	−5	−7	−10	−11	−8	−5	2	9	12	−17
	Jan/Nov	−5	−7	−9	−11	−8	−4	2	9	12	−15
	Feb/Oct	−4	−6	−7	−8	−4	−2	4	8	11	−10
	Mar/Sept	−3	−4	−4	−4	−2	−1	3	5	7	−5
	Apr/Aug	−2	−2	−1	−2	0	−1	0	1	1	−1
	May/Jul	1	1	1	0	0	−1	−1	−3	−3	1
	Jun	1	2	2	1	0	−2	2	−4	−4	2
40	Dec	−6	−8	−10	−13	−10	−7	0	7	10	−21
	Jan/Nov	−5	−7	−10	−12	−9	−6	1	8	11	−19
	Feb/Oct	−5	−7	−8	−9	−6	−3	3	8	12	−14
	Mar/Sept	−4	−5	−5	−6	−3	−1	4	7	10	−8
	Apr/Aug	−2	−3	−2	−2	0	0	2	3	4	−3
	May/Jul	0	0	0	0	0	0	0	0	1	1
	Jun	1	1	1	0	1	0	0	−1	−1	2
48	Dec	−6	−8	−11	−14	−13	−10	−3	2	6	−25
	Jan/Nov	−6	−8	−11	−13	−11	−8	−1	5	8	−24
	Feb/Oct	−5	−7	−10	−11	−8	−5	1	8	11	−18
	Mar/Sept	−4	−6	−6	−7	−4	−1	4	8	11	−11
	Apr/Aug	−3	−3	−3	−3	−1	0	4	6	7	−5
	May/Jul	0	−1	0	0	1	1	3	3	4	0
	Jun	1	1	2	1	2	1	2	2	3	2
56	Dec	−7	−9	−12	−16	−16	−14	−9	−5	−3	−28
	Jan/Nov	−6	−8	−11	−15	−14	−12	−6	−1	2	−27
	Feb/Oct	−6	−8	−10	−12	−10	−7	0	6	9	−22
	Mar/Sept	−5	−6	−7	−8	−5	−2	4	8	12	−15
	Apr/Aug	−3	−4	−4	−4	−1	1	5	7	9	−8
	May/Jul	0	0	0	0	2	2	5	6	7	−2
	Jun	2	1	2	1	3	3	4	5	6	1
64	Dec	−7	−9	−12	−16	−17	−18	−16	−14	−12	−30
	Jan/Nov	−7	−9	−12	−16	−16	−16	−13	−10	−8	−29
	Feb/Oct	−6	−8	−11	−14	−13	10	−4	1	4	−26
	Mar/Sept	−5	−7	−9	−10	−7	−4	2	7	11	−20
	Apr/Aug	−3	−4	−4	−4	−1	1	5	9	11	−11
	May/Jul	1	0	1	0	3	4	6	8	10	−3
	Jun	2	2	2	2	4	4	6	7	9	0

(1) Corrections in this table are in degrees F. The correction is applied directly to the CLTD for a wall or roof as given in Tables 5A and 7A.
(2) The CLTD correction given in this table is *not* applicable to Table 10, Cooling Load Temperature Differences for Conduction through Glass.
(3) For South latitudes, replace Jan. through Dec. by July through June.

(Courtesy ASHRAE, 1985 Fundamentals)

Figure 2.17 (cont.)

Latent Heat Loads (ΔW)

The same cfm that causes sensible heat gain also carries moisture that adds to the latent heat load. Latent heat, or ΔW, is the change in moisture measured in pounds of moisture per pound of dry air (see Figure 2.18). For example, the amount of latent heat for outdoor air at 88°F/50 percent relative humidity that enters an indoor condition of 78°F/90 percent relative humidity has a ΔW of:

$$0.015 - 0.008 = 0.007 \text{ lbs. moisture/lb. dry air}$$

Radiation Heat Gain

Solar heat gain that enters transparent surfaces varies with location, time of day, day of the year, shading, orientation, building color, mass, hours of operation, and daily temperature range. The basic formula for computing radiant energy heat gain is shown below.

H_s = A · SHGF · SC · CLF
H_s = Radiant heat gain (Btu/hr.)
A = Glass area (s.f.)
SGHF = Solar heat gain factor (Btu/hr./s.f.)
SC = Shading coefficient (see Figure 2.20)
CLF = Cooling load factor (see Figure 2.21)

A Area — This is the glass area for each window, door, skylight, or other opening. The area is measured perpendicular to the ground for windows and doors, and parallel to the ground for skylights and roof openings.

SHGF Solar Heat Gain Factor — This figure represents the amount of heat gain (measured in Btu/hour) which enters the building through each square foot of glass due to solar radiation through an unshaded single pane of flat glass. The value of the SHGF depends on the orientation, latitude, month, and time of day. Values for SHGF are listed in Figure 2.19.

SC Shading Coefficient — For unshaded single-pane flat glass, the shading coefficient, or SC, equals 1.0. As the type of glass varies — reflective glass or double-pane, for example — less energy can enter into the building and the overall heat gain due to solar radiation is reduced. Similarly, exterior awnings and interior shades or drapes reduce the SC. Figure 2.20 is a table of SC values to be used in the formula for radiant heat gain.

CLF Cooling Load Factor — Figure 2.21 lists cooling load factors to be used. For each pane of glass, the value is read based on the orientation and time of day.

AIR CONDITIONING | C8.4-000 | General

Table 8:4-003 Psychrometric Table

Dewpoint or Saturation Temperature (F)

Relative humidity (%)	32	35	40	45	50	55	60	65	70	75	80	85	90	95	100
100	32	35	40	45	50	55	60	65	70	75	80	85	90	95	100
90	30	33	37	42	47	52	57	62	67	72	77	82	87	92	97
80	27	30	34	39	44	49	54	58	64	68	73	78	83	88	93
70	24	27	31	36	40	45	50	55	60	64	69	74	79	84	88
60	20	24	28	32	36	41	46	51	55	60	65	69	74	79	83
50	16	20	24	28	33	36	41	46	50	55	60	64	69	73	78
40	12	15	18	23	27	31	35	40	45	49	53	58	62	67	71
30	8	10	14	18	21	25	29	33	37	42	46	50	54	59	62
20	6	7	8	9	13	16	20	24	28	31	35	40	43	48	52
10	4	4	5	5	6	8	9	10	13	17	20	24	27	30	34

Dry bulb temperature (F) (column headings)

This table shows the relationship between RELATIVE HUMIDITY, DRY BULB TEMPERATURE AND DEWPOINT. As an example, assume that the thermometer in a room reads 75°F, and we know that the relative humidity is 50%. The chart shows the dewpoint temperature to be 55°. That is, any surface colder than 55°F will "sweat" or collect condensing moisture. This surface could be the outside of an uninsulated chilled water pipe in the summertime, or the inside surface of a wall or deck in the wintertime. After determining the extreme ambient parameters. The table at the left is useful in determining which surfaces need insulation or vapor barrier protection.

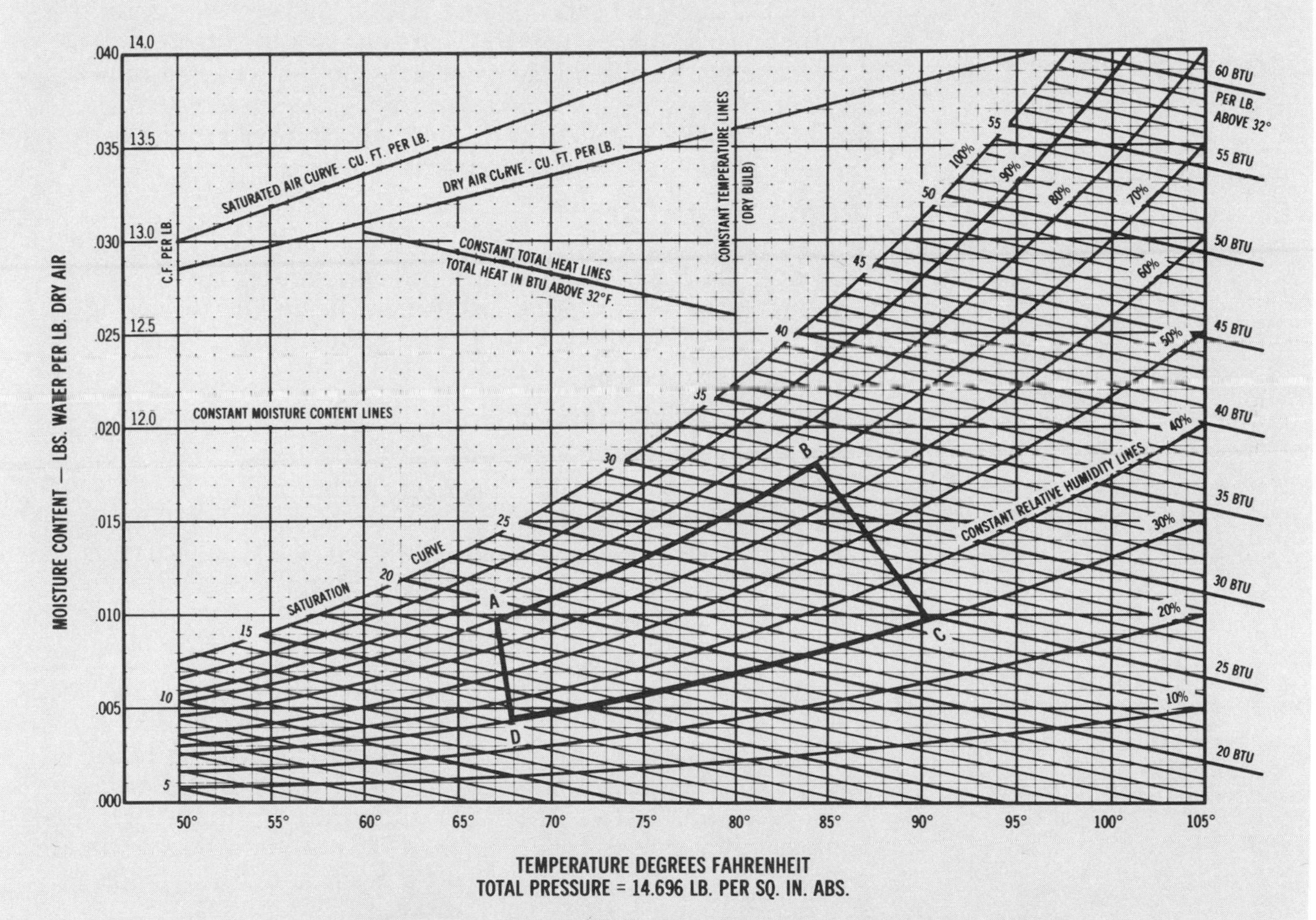

Psychrometric chart showing different variables based on one pound of dry air. Space marked A B C D is temperature-humidity range which is most comfortable for majority of people.

Figure 2.18

Internal Heat Gain

All energy consumed in a building eventually results in a form of heat gain. Lights, office equipment, appliances, and computers all consume power that generates heat in the building and adds to the heat gain. Some of this internal heat gain is "dry," which means that it contains no moisture and only produces sensible heat gains. Examples of heat gains that are sensible only are lights, computers, copy machines, and televisions. Other pieces of equipment are "wet," meaning they produce some moisture, which adds approximately 1,000 Btu's of latent heat per pound of moisture. Examples of equipment that produces both sensible (dry) and latent (wet) loads include cooking equipment, hair dryers, instrument sterilizers, coffee makers, and indoor fountains. The following formula is used for computing internal heat gain:

H_i = (Sensible load) CLF + (Latent load)

H_i = Internal heat gain (Btu/hr.)

CLF = Cooling load factor

Figure 2.22 lists the sensible and latent loads for common internal heat gains. Figure 2.21 lists cooling load factors. To determine the sensible load of electric appliances and lights not listed, multiply the rated wattage by 3.4 to convert watts to Btu/hour. Add 25 percent if fluorescent fixtures are used to account for energy consumed by the ballasts.

People Heat Gain

The building's occupants represent a special type of internal heat gain. Each person in a building contributes a certain amount to the sensible and latent heat load. The following formula is used for computing heat gained from people:

H_p = N_o P_s (CLF) + N_o P_L

H_p = People heat gain (Btu/hr.)

N_o = Number of occupants

P_s = Sensible heat gain per person (Btu/hr.) (see Figure 2.23)

CLF = Cooling load factor, usually 1.0 (see Figure 2.21)

P_L = Latent heat gain per person (Btu/hr.) (see Figure 2.23)

The number of occupants, N_o, is determined by the use of each space. For auditoriums, restaurants, theaters, and similar spaces, the occupancy equals the number of seats. Other building uses assign population on a square foot basis. The cooling load factor usually equals 1.0 except in low density applications. Refer to Figure 2.21 or ASHRAE guidelines for a more detailed analysis. Figure 2.23 lists the sensible and latent heat gains for people. The heat gain depends on the amount and type of activity taking place.

Cooling Load Summary

The maximum cooling load is based on the greatest anticipated heat gain during the cooling season. Generally, conditions for 3:00 p.m. on a typical August day are used as one of several conditions tested. Because so many variable conditions are involved, many designers employ sophisticated computer

Table 11 Maximum Solar Heat Gain Factor, Btu/h·ft² for Sunlit Glass, North Latitudes

0° N Lat

	N	NNE/ NNW	NE/ NW	ENE/ WNW	E/ W	ESE/ WSW	SE/ SW	SSE/ SSW	S	HOR
Jan.	34	34	88	177	234	254	235	182	118	296
Feb.	36	39	132	205	245	247	210	141	67	306
Mar.	38	87	170	223	242	223	170	87	38	303
Apr.	71	134	193	224	221	184	118	38	37	284
May	113	164	203	218	201	154	80	37	37	265
June	129	173	206	212	191	140	66	37	37	255
July	115	164	201	213	195	149	77	38	38	260
Aug.	75	134	187	216	212	175	112	39	38	276
Sept.	40	84	163	213	231	213	163	84	40	293
Oct.	37	40	129	199	236	238	202	135	66	299
Nov.	35	35	88	175	230	250	230	179	117	293
Dec.	34	34	71	164	226	253	240	196	138	288

4° N Lat

	N	NNE/ NNW	NE/ NW	ENE/ WNW	E/ W	ESE/ WSW	SE/ SW	SSE/ SSW	S	HOR
Jan.	33	33	79	170	229	252	237	193	141	286
Feb.	35	35	123	199	242	248	215	152	88	301
Mar.	38	77	163	219	242	227	177	96	43	302
Apr.	55	125	189	223	223	190	126	43	38	287
May	93	154	200	220	206	161	89	38	38	272
June	110	164	202	215	196	147	73	38	38	263
July	96	154	197	215	200	156	85	39	38	267
Aug.	59	124	184	215	214	181	120	42	40	279
Sep.	39	75	156	209	231	216	170	93	44	293
Oct.	36	36	120	193	234	239	207	148	86	294
Nov.	34	34	79	168	226	248	232	190	139	284
Dec.	33	33	62	157	221	250	242	206	160	277

8° N Lat

	N	NNE/ NNW	NE/ NW	ENE/ WNW	E/ W	ESE/ WSW	SE/ SW	SSE/ SSW	S	HOR
Jan.	32	32	71	163	224	250	242	203	162	275
Feb.	34	34	114	193	239	248	219	165	110	294
Mar.	37	67	156	215	241	230	184	110	55	300
Apr.	44	117	184	221	225	195	134	53	39	289
May	74	146	198	220	209	167	97	39	38	277
June	90	155	200	217	200	141	82	39	39	269
July	77	145	195	215	204	162	93	40	39	272
Aug.	47	117	179	214	216	186	128	51	41	282
Sep.	38	66	149	205	230	219	176	107	56	290
Oct.	35	35	112	187	231	239	211	160	108	288
Nov.	33	33	71	161	220	245	233	200	160	273
Dec.	31	31	55	149	215	246	247	215	179	265

12° N Lat

	N	NNE/ NNW	NE/ NW	ENE/ WNW	E/ W	ESE/ WSW	SE/ SW	SSE/ SSW	S	HOR
Jan.	31	31	63	155	217	246	247	212	182	262
Feb.	34	34	105	186	235	248	226	177	133	286
Mar.	36	58	148	210	240	233	190	124	73	297
Apr.	40	108	178	219	227	200	142	64	40	290
May	60	139	194	220	212	173	106	40	40	280
June	75	149	198	217	204	161	90	40	40	274
July	63	139	191	215	207	168	102	41	41	275
Aug.	42	109	174	212	218	191	135	62	142	282
Sep.	37	57	142	201	229	222	182	121	73	287
Oct.	34	34	103	180	227	238	219	172	130	280
Nov.	32	32	63	153	214	241	243	209	179	260
Dec.	30	30	47	141	207	242	251	223	197	250

16° N Lat

	N	NNE NNW	NE/ NW	ENE WNW	E/ W	ESE/ WSW	SE/ SW	SSE/ SSW	S	HOR
Jan.	30	30	55	147	210	244	251	223	199	248
Feb.	33	33	96	180	231	247	233	188	154	275
Mar.	35	53	140	205	239	235	197	138	93	291
Apr.	39	99	172	215	227	204	150	77	45	289
May	52	132	189	218	215	179	115	45	41	282
June	66	142	194	217	207	167	99	41	41	277
July	55	132	187	214	210	174	111	44	42	277
Aug.	41	100	168	209	219	196	143	74	46	282
Sep.	36	50	134	196	227	224	191	134	93	282
Oct.	33	33	95	174	223	237	225	183	150	270
Nov.	30	30	55	145	206	241	247	220	196	246
Dec.	29	29	41	132	198	241	254	233	212	234

20° N Lat

	N	NNE/ NNW	NE/ NW	ENE/ WNW	E/ W	ESE/ WSW	SE/ SW	SSE/ SSW	S	HOR
Jan.	29	29	48	138	201	243	253	233	214	232
Feb.	31	31	88	173	226	244	238	201	174	263
Mar.	34	49	132	200	237	236	206	152	115	284
Apr.	38	92	166	213	228	208	158	91	58	287
May	47	123	184	217	217	184	124	54	42	283
June	59	135	189	216	210	173	108	45	42	279
July	48	124	182	213	212	179	119	53	43	278
Aug.	40	91	162	206	220	200	152	88	57	280
Sep.	36	46	127	191	225	225	199	148	114	275
Oct.	32	32	87	167	217	236	231	196	170	258
Nov.	29	29	48	136	197	239	249	229	211	230
Dec.	27	27	35	122	187	238	254	241	226	217

24° N. Lat

	N	NNE/ NNW	NE/ NW	ENE/ WNW	E/ W	ESE/ WSW	SE/ SW	SSE/ SSW	S	HOR
Jan.	27	27	41	128	190	240	253	241	227	214
Feb.	30	30	80	165	220	244	243	213	192	249
Mar.	34	45	124	195	234	237	214	168	137	275
Apr.	37	88	159	209	228	212	169	107	75	283
May	43	117	178	214	218	190	132	67	46	282
June	55	127	184	214	212	179	117	55	43	279
July	45	116	176	210	213	185	129	65	46	278
Aug.	38	87	156	203	220	204	162	103	72	277
Sep.	35	42	119	185	222	225	206	163	134	266
Oct.	31	31	79	159	211	237	235	207	187	244
Nov.	27	27	42	126	187	236	249	237	224	213
Dec.	26	26	29	112	180	234	247	247	237	199

28° N. Lat

	N (Shade)	NNE/ NNW	NE/ NW	ENE/ WNW	E/ W	ESE/ WSW	SE/ SW	SSE/ SSW	S	HOR
Jan.	25	25	35	117	183	235	251	247	238	196
Feb.	29	29	72	157	213	244	246	224	207	234
Mar.	33	41	116	189	231	237	221	182	157	265
Apr.	36	84	151	205	228	216	178	124	94	278
May	40	115	172	211	219	195	144	83	58	280
June	51	125	178	211	213	184	128	68	49	278
July	41	114	170	208	215	190	140	80	57	276
Aug.	38	83	149	199	220	207	172	120	91	272
Sep.	34	38	111	179	219	226	213	177	154	256
Oct.	30	30	71	151	204	236	238	217	202	229
Nov.	26	26	35	115	181	232	247	243	235	195
Dec.	24	24	24	99	172	227	248	251	246	179

32° N. Lat

	N (Shade)	NNE/ NW	NE/ NW	ENE/ WNW	E/ W	ESE/ WSW	SE/ SW	SSE/ SSW	S	HOR
Jan.	24	24	29	105	175	229	249	250	246	176
Feb.	27	27	65	149	205	242	248	232	221	217
Mar.	32	37	107	183	227	237	227	195	176	252
Apr.	36	80	146	200	227	219	187	141	115	271
May	38	111	170	208	220	199	155	99	74	277
June	44	122	176	208	214	189	139	83	60	276
July	40	111	167	204	215	194	150	96	72	273
Aug.	37	79	141	195	219	210	181	136	111	265
Sept.	33	35	103	173	215	227	218	189	171	244
Oct.	28	28	63	143	195	234	239	225	215	213
Nov.	24	24	29	103	173	225	245	246	243	175
Dec.	22	22	22	84	162	218	246	252	252	158

36° N. Lat

	N (Shade)	NNE/ NNW	NE/ NW	ENE/ WNW	E/ W	ESE/ WSW	SE/ SW	SSE/ SSW	S	HOR
Jan.	22	22	24	90	166	219	247	252	252	155
Feb.	26	26	57	139	195	239	248	239	232	199
Mar.	30	33	99	176	223	238	232	206	192	238
Apr.	35	76	144	196	225	221	196	156	135	262
May	38	107	168	204	220	204	165	116	93	272
June	47	118	175	205	215	194	150	99	77	273
July	39	107	165	201	216	199	161	113	90	268
Aug.	36	75	138	190	218	212	189	151	131	257
Sep.	31	31	95	167	210	228	223	200	187	230
Oct.	27	27	56	133	187	230	239	231	225	195
Nov.	22	22	24	87	163	215	243	248	248	154
Dec.	20	20	20	69	151	204	241	253	254	136

(Courtesy ASHRAE, 1985 Fundamentals)

Figure 2.19

Table 11 Maximum Solar Heat Gain Factor, Btu/h·ft² for Sunlit Glass, North Latitudes (concluded)

40° N. Lat										
	N (Shade)	NNE/ NNW	NE/ NW	ENE/ WNW	E/ W	ESE/ WSW	SE/ SW	SSE/ SSW	S	HOR
Jan.	20	20	20	74	154	205	241	252	254	133
Feb.	24	24	50	129	186	234	246	244	241	180
Mar.	29	29	93	169	218	238	236	216	206	223
Apr.	34	71	140	190	224	223	203	170	154	252
May	37	102	165	202	220	208	175	133	113	265
June	48	113	172	205	216	199	161	116	95	267
July	38	102	163	198	216	203	170	129	109	262
Aug.	35	71	135	185	216	214	196	165	149	247
Sep.	30	30	87	160	203	227	226	209	200	215
Oct.	25	25	49	123	180	225	238	236	234	177
Nov.	20	20	20	73	151	201	237	248	250	132
Dec.	18	18	18	60	135	188	232	249	253	113

44° N. Lat										
	N (Shade)	NNE/ NNW	NE/ NW	ENE/ WNW	E/ W	ESE/ WSW	SE/ SW	SSE/ SSW	S	HOR
Jan.	17	17	17	64	138	189	232	248	252	109
Feb.	22	22	43	117	178	227	246	248	247	160
Mar.	27	27	87	162	211	236	238	224	218	206
Apr.	33	66	136	185	221	224	210	183	171	240
May	36	96	162	201	219	211	183	148	132	257
June	47	108	169	205	215	203	171	132	115	261
July	37	96	159	198	215	206	179	144	128	254
Aug.	34	66	132	180	214	215	202	177	165	236
Sep.	28	28	80	152	198	226	227	216	211	199
Oct.	23	23	42	111	171	217	237	240	239	157
Nov.	18	18	18	64	135	186	227	244	248	109
Dec.	15	15	15	49	115	175	217	240	246	89

48° N. Lat										
	N (Shade)	NNE/ NNW	NE/ NW	ENE/ WNW	E/ W	ESE/ WSW	SE/ SW	SSE/ SSW	S	HOR
Jan.	15	15	15	53	118	175	216	239	245	85
Feb.	20	20	36	103	168	216	242	249	250	138
Mar.	26	26	80	154	204	234	239	232	228	188
Apr.	31	61	132	180	219	225	215	194	186	226
May	35	97	158	200	218	214	192	163	150	247
June	46	110	165	204	215	206	180	148	134	252
July	37	96	156	196	214	209	187	158	146	244
Aug.	33	61	128	174	211	216	208	188	180	223
Sep.	27	27	72	144	191	223	228	223	220	182
Oct.	21	21	35	96	161	207	233	241	242	136
Nov.	15	15	15	52	115	172	212	234	240	85
Dec.	13	13	13	36	91	156	195	225	233	65

52° N. Lat										
	N (Shade)	NNE/ NNW	NE/ NW	ENE/ WNW	E/ W	ESE/ WSW	SE/ SW	SSE/ SSW	S	HOR
Jan.	13	13	13	39	92	155	193	222	230	62
Feb.	18	18	29	85	156	202	235	247	250	115
Mar.	24	24	73	145	196	230	239	238	236	169
Apr.	30	56	128	177	215	224	220	204	199	211
May	34	98	154	198	217	217	199	175	167	235
June	45	111	161	202	214	210	188	162	152	242
July	36	97	152	194	213	212	195	171	163	233
Aug.	32	56	124	169	208	216	212	197	193	208
Sep.	25	25	65	136	182	218	228	228	227	163
Oct.	19	19	28	80	148	192	225	238	240	114
Nov.	13	13	13	39	90	152	189	217	225	62
Dec.	10	10	10	19	73	127	172	199	209	42

56° N. Lat										
	N (Shade)	NNE/ NNW	NE/ NW	ENE/ WNW	E/ W	ESE/ WSW	SE/ SW	SSE/ SSW	S	HOR
Jan.	10	10	10	21	74	126	169	194	205	40
Feb.	16	16	21	71	139	184	223	239	244	91
Mar.	22	22	65	136	185	224	238	241	241	149
Apr.	28	58	123	173	211	223	223	213	210	195
May	36	99	149	195	215	218	206	187	181	222
June	53	111	160	199	213	213	196	174	168	231
July	37	98	147	192	211	214	201	183	177	221
Aug.	30	56	119	165	203	216	215	206	203	193
Sep.	23	23	58	126	171	211	227	230	231	144
Oct.	16	16	20	68	132	176	213	229	234	91
Nov.	10	10	10	21	72	122	165	190	200	40
Dec.	7	7	7	7	47	92	135	159	171	23

60° N. Lat										
	N (Shade)	NNE/ NNW	NE/ NW	ENE/ WNW	E/ W	ESE/ WSW	SE/ SW	SSE/ SSW	S	HOR
Jan.	7	7	7	7	46	88	130	152	164	21
Feb.	13	13	13	58	118	168	204	225	231	68
Mar.	20	20	56	125	173	215	234	241	242	128
Apr.	27	59	118	168	206	222	225	220	218	178
May	43	98	149	192	212	220	211	198	194	208
June	58	110	162	197	213	215	202	186	181	217
July	44	97	147	189	208	215	206	193	190	207
Aug.	28	57	114	161	199	214	217	213	211	176
Sep.	21	21	50	115	160	202	222	229	231	123
Oct.	14	14	14	56	111	159	193	215	221	67
Nov.	7	7	7	7	45	86	127	148	160	22
Dec.	4	4	4	4	16	51	76	100	107	9

64° N. Lat										
	N (Shade)	NNE/ NNW	NE/ NW	ENE/ WNW	E/ W	ESE/ WSW	SE/ SW	SSE/ SSW	S	HOR
Jan.	3	3	3	3	15	45	67	89	96	8
Feb.	11	11	11	43	89	144	177	202	210	45
Mar.	18	18	47	113	159	203	226	236	239	105
Apr.	25	59	113	163	201	219	225	225	224	160
May	48	97	150	189	211	220	215	207	204	192
June	62	114	162	193	213	216	208	196	193	203
July	49	96	148	186	207	215	211	202	200	192
Aug.	27	58	109	157	193	211	217	217	217	159
Sept.	19	19	43	103	148	189	213	224	227	101
Oct.	11	11	11	40	83	135	167	191	199	46
Nov.	4	4	4	4	15	44	66	87	93	8
Dec.	0	0	0	0	1	5	11	14	15	1

Table 12 Maximum Solar Heat Gain Factor For Externally Shaded Glass, Btu/h·ft² (Based on Ground Reflectance of 0.2)

Use for latitudes 0-24 deg.
For latitudes greater than 24, use north orientation, Table 11A.
For horizontal glass in shade, use the tabulated values for all latitudes

	N	NNE/ NNW	NE/ NW	ENE/ WNW	E/ W	ESE/ WSW	SE/ SW	SSE/ SSW	S	(ALL LAT.) HOR
Jan.	31	31	31	32	34	36	37	37	38	16
Feb.	34	34	34	35	36	37	38	38	39	16
Mar.	6	36	37	38	39	40	40	39	39	19
Apr.	40	40	41	42	42	42	41	40	40	24
May	43	44	45	46	45	43	41	40	40	28
June	45	46	47	47	46	44	41	40	40	31
July	45	45	46	47	47	45	42	41	41	31
Aug.	42	42	43	45	46	45	43	42	42	28
Sept.	37	37	38	40	41	42	42	41	41	23
Oct.	34	34	34	36	38	39	40	40	40	19
Nov.	32	32	32	32	34	36	38	38	39	17
Dec.	30	30	30	31	32	34	36	37	37	15

CALCULATING SPACE COOLING LOAD FROM HEAT GAIN THROUGH INTERIOR PARTITIONS, CEILINGS AND FLOORS

Whenever a conditioned space is adjacent to a space in which a different temperature prevails, transfer of heat through the separating structural section must be considered. The heat transfer rate q in Btu/h, is given by:

$$q = UA\,(t_b - t_i) \tag{13}$$

where

U = coefficient of overall heat transfer between the adjacent and the conditioned space, Btu/h·ft²·F
A = area of separating section concerned, ft²
t_b = average air temperature in adjacent space, F
t_i = air temperature in conditioned space, F

(Courtesy ASHRAE, 1985 Fundamentals)

Figure 2.19 (cont.)

CHAPTER 4. SOLAR HEAT GAIN THRU GLASS 1–57

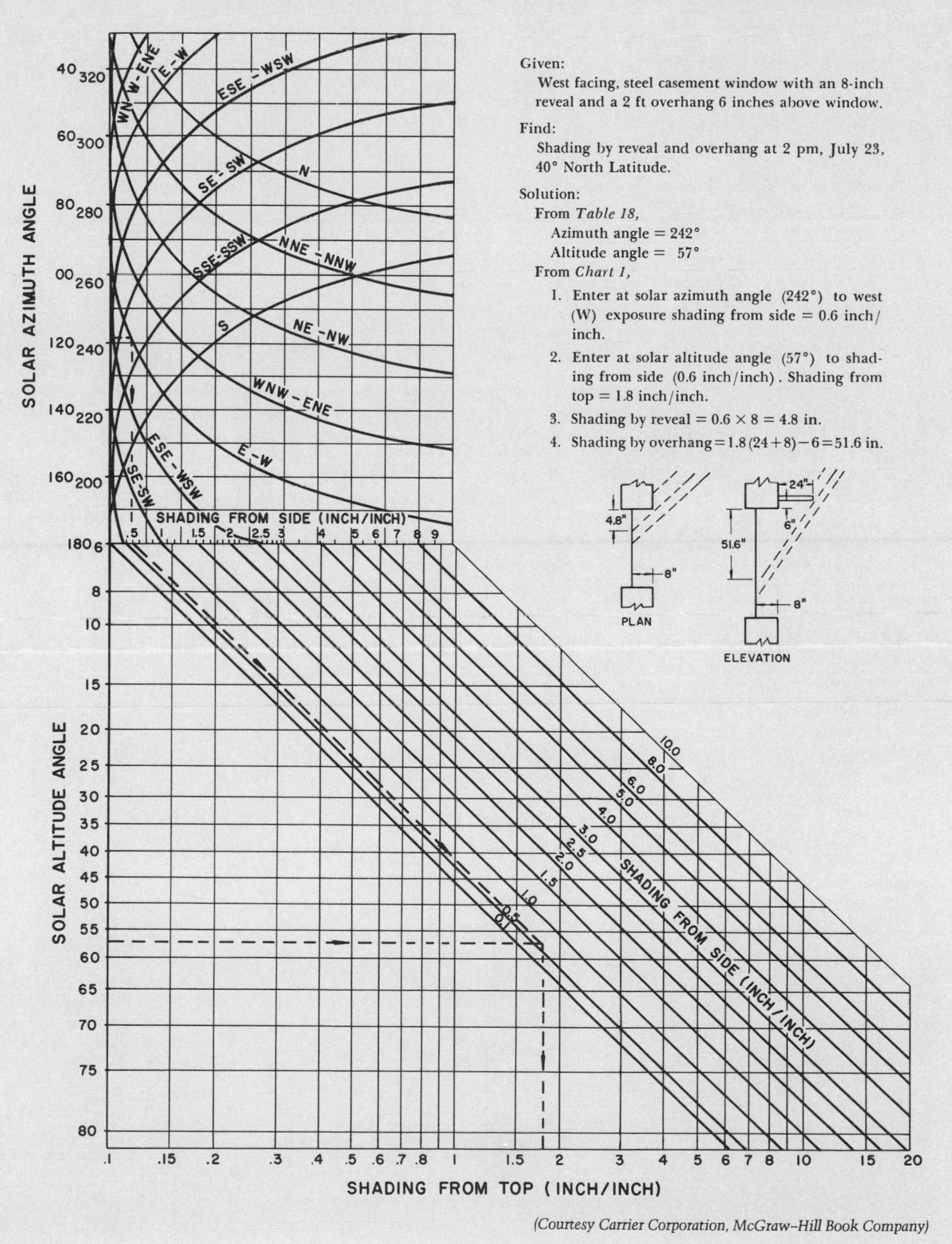
CHART 1 — SHADING FROM REVEALS, OVERHANGS, FINS AND ADJACENT BUILDINGS

Given:
West facing, steel casement window with an 8-inch reveal and a 2 ft overhang 6 inches above window.

Find:
Shading by reveal and overhang at 2 pm, July 23, 40° North Latitude.

Solution:
From *Table 18,*
Azimuth angle = 242°
Altitude angle = 57°
From *Chart 1,*

1. Enter at solar azimuth angle (242°) to west (W) exposure shading from side = 0.6 inch/inch.
2. Enter at solar altitude angle (57°) to shading from side (0.6 inch/inch). Shading from top = 1.8 inch/inch.
3. Shading by reveal = 0.6 × 8 = 4.8 in.
4. Shading by overhang = 1.8 (24 + 8) − 6 = 51.6 in.

(Courtesy Carrier Corporation, McGraw-Hill Book Company)

Figure 2.20

NOTE: Actually the reaction on the solar heat reflected back through the glass from the blind is not always identical to the first pass as assumed in this example. The first pass through the glass filters out most of solar radiation that is to be absorbed in the glass, and the second pass absorbs somewhat less. For simplicity, the reaction is assumed identical, since the quantities are normally small on the second pass.

Basis of Table 16
Over-all Factors for Solar Heat Gain thru Glass, With and Without Shading Devices

The factors in *Table 16* are based on:

1. An outdoor film coefficient of 2.8 Btu/(hr) (sq ft) (deg F) at 5 mph wind velocity.
2. An inside film coefficient of 1.8 Btu/(hr)(sq ft) (deg F), 100-200 fpm. This is not 1.47 as normally used, since the present practice in well designed systems is to sweep the window with a stream of air.
3. A 30° angle of incidence which is the angle at which most exposures peak. The 30° angle of incidence is approximately the balance point on reduction of solar heat coming through the atmosphere and the decreased transmissibility of glass. Above the 30° angle the transmissibility of glass decreases, and below the 30° angle the atmosphere absorbs or reflects more.
4. All shading devices fully drawn, except roller shades. Experience indicates that roller shades are *seldom fully drawn,* so the factors have been slightly increased.
5. Venetian blind slats horizontal at 45° and shading screen slats horizontal at 17°.
6. Outdoor canvas awnings ventilated at sides and top. (See *Table 16* footnote.)
7. Since *Table 15* is based on the net solar heat gain thru ordinary glass, all calculated solar heat factors are divided by .88 *(Fig. 12).*
8. The average absorptivity, reflectivity and transmissability for common glass and shading devices at a 30° angle of incidence along with shading factors appear in the table below.

Use of Table 16
— Over-all Factors for Solar Heat Gain thru Glass, With and Without Shading Devices

The factors in *Table 16* are multiplied by the values in *Table 15* to determine the solar heat gain thru different combinations of glass and shading devices. The correction factors listed under *Table 15* are to be used if applicable. Transmission due to temperature difference between the inside and outdoor air must be added to the solar heat gain to determine total gain thru glass.

Example 3 — Partially Drawn Shades

Occasionally it is necessary to estimate the cooling load in a building where the blinds are not to be fully drawn. The procedure is illustrated in the following example:

Given:
West exposure, 40° North latitude
Thermopane window with white venetian blind on inside, ¾ drawn.

Find:
Peak solar heat gain.

Solution:
By inspection of *Table 15,* the boxed boldface values for peak solar heat gain, occurring at 4:00 p.m. on July 23
= 164 Btu/(hr)(sq ft)

TYPES OF GLASS OR SHADING DEVICES*	Absorptivity (a)	Reflectivity (r)	Transmissibility (t)	Solar Factor†
Ordinary Glass	.06	.08	.86	1.00
Regular Plate, ¼"	.15	.08	.77	.94
Glass, Heat Absorbing	by mfg.	.05	(1 − .05 − a)	—
Venetian Blind, Light Color	.37	.51	.12	.56‡
Medium Color	.58	.39	.03	.65‡
Dark Color	.72	.27	.01	.75‡
Fiberglass Cloth, Off White (5.72 - 61/58)	.05	.60	.35	.48‡
Cotton Cloth, Beige (6.18 - 91/36)	.26	.51	.23	.56‡
Fiberglass Cloth, Light Gray	.30	.47	.23	.59‡
Fiberglass Cloth, Tan (7.55 - 57/29)	.44	.42	.14	.64‡
Glass Cloth, White, Golden Stripes	.05	.41	.54	.65‡
Fiberglass Cloth, Dark Gray	.60	.29	.11	.75‡
Dacron Cloth, White (1.8 - 86/81)	.02	.28	.70	.76‡
Cotton Cloth, Dark Green, Vinyl Coated (similar to roller shade)	.85	.15	.00	.88‡
Cotton Cloth, Dark Green (6.06 - 91/36)	.02	.28	.70	.76‡

*Factors for various draperies are given for guidance only since the actual drapery material may be different in color and texture; figures in parentheses are ounces per sq yd, and yarn count warp/filling. Consult manufacturers for actual values.

†Compared to ordinary glass.

‡For a shading device in combination with ordinary glass.

(Courtesy Carrier Corporation, McGraw–Hill Book Company)

Figure 2.20 (cont.)

Values of U can be obtained from Chapter 23. Temperature t_b may have any value over a considerable range according to conditions in the adjacent space. The temperature in a kitchen or boiler room may be as much as 15 to 50 deg F above the outdoor air temperature. It is recommended that actual temperatures in adjoining spaces be measured wherever practicable; where nothing is known except that the adjacent space is of conventional construction and contains no heat sources, $t_b - t_i$ should be considered the difference between the outdoor air and conditioned-space design dry- bulb temperatures *minus* 5 deg F. In some cases, the air temperature in the adjacent space will always correspond to the outdoor air temperature.

For floors directly in contact with the ground, or over an underground basement that is neither ventilated nor warmed, heat transfer may be neglected for cooling load estimates.

Table 13 Cooling Load Factors for Glass without Interior Shading, North Latitudes

Fenes-tration Facing	Room Con-struction	Solar Time, h																							
		1	2	3	4	5	6	7	8	9	10	11	12	13	14	15	16	17	18	19	20	21	22	23	24
N (Shaded)	L	0.17	0.14	0.11	0.09	0.08	0.33	0.42	0.48	0.56	0.63	0.71	0.76	0.80	0.82	0.82	0.79	0.75	0.84	0.61	0.48	0.38	0.31	0.25	0.20
	M	0.23	0.20	0.18	0.16	0.14	0.34	0.41	0.46	0.53	0.59	0.65	0.70	0.73	0.75	0.76	0.74	0.75	0.79	0.61	0.50	0.42	0.36	0.31	0.27
	H	0.25	0.23	0.21	0.20	0.19	0.38	0.45	0.49	0.55	0.60	0.65	0.69	0.72	0.72	0.72	0.70	0.70	0.75	0.57	0.46	0.39	0.34	0.31	0.28
NNE	L	0.06	0.05	0.04	0.03	0.03	0.26	0.43	0.47	0.44	0.41	0.40	0.39	0.39	0.38	0.36	0.33	0.30	0.26	0.20	0.16	0.13	0.10	0.08	0.07
	M	0.09	0.08	0.07	0.06	0.06	0.24	0.38	0.42	0.39	0.37	0.37	0.36	0.36	0.36	0.34	0.33	0.30	0.27	0.22	0.18	0.16	0.14	0.12	0.10
	H	0.11	0.10	0.09	0.09	0.08	0.26	0.39	0.42	0.39	0.36	0.35	0.34	0.34	0.33	0.32	0.31	0.28	0.25	0.21	0.18	0.16	0.14	0.13	0.12
NE	L	0.04	0.04	0.03	0.02	0.02	0.23	0.41	0.51	0.51	0.45	0.39	0.36	0.33	0.31	0.28	0.26	0.23	0.19	0.15	0.12	0.10	0.08	0.06	0.05
	M	0.07	0.06	0.06	0.05	0.04	0.21	0.36	0.44	0.45	0.40	0.36	0.33	0.31	0.30	0.28	0.26	0.23	0.21	0.17	0.15	0.13	0.11	0.09	0.08
	H	0.09	0.08	0.08	0.07	0.07	0.23	0.37	0.44	0.44	0.39	0.34	0.31	0.29	0.27	0.26	0.24	0.22	0.20	0.17	0.14	0.13	0.12	0.11	0.10
ENE	L	0.04	0.03	0.03	0.02	0.02	0.21	0.40	0.52	0.57	0.53	0.45	0.39	0.34	0.31	0.28	0.25	0.22	0.18	0.14	0.12	0.09	0.08	0.06	0.05
	M	0.07	0.06	0.05	0.05	0.04	0.20	0.35	0.45	0.49	0.47	0.41	0.36	0.33	0.30	0.28	0.26	0.23	0.20	0.17	0.14	0.12	0.11	0.09	0.08
	H	0.09	0.09	0.08	0.07	0.07	0.22	0.36	0.46	0.49	0.45	0.38	0.33	0.30	0.27	0.25	0.23	0.21	0.19	0.16	0.14	0.13	0.12	0.11	0.10
E	L	0.04	0.03	0.03	0.02	0.02	0.19	0.37	0.51	0.57	0.57	0.50	0.42	0.37	0.32	0.29	0.25	0.22	0.19	0.15	0.12	0.10	0.08	0.06	0.05
	M	0.07	0.06	0.06	0.05	0.05	0.18	0.33	0.44	0.50	0.51	0.46	0.39	0.35	0.31	0.29	0.26	0.23	0.21	0.17	0.15	0.13	0.11	0.10	0.08
	H	0.09	0.09	0.08	0.08	0.07	0.20	0.34	0.45	0.49	0.49	0.43	0.36	0.32	0.29	0.26	0.24	0.22	0.19	0.17	0.15	0.13	0.12	0.11	0.10
ESE	L	0.05	0.04	0.03	0.03	0.02	0.17	0.34	0.49	0.58	0.61	0.57	0.48	0.41	0.36	0.32	0.28	0.24	0.20	0.16	0.13	0.10	0.09	0.07	0.06
	M	0.08	0.07	0.06	0.05	0.05	0.16	0.31	0.43	0.51	0.54	0.51	0.44	0.39	0.35	0.32	0.29	0.26	0.22	0.19	0.16	0.14	0.12	0.11	0.09
	H	0.10	0.09	0.09	0.08	0.08	0.19	0.32	0.43	0.50	0.52	0.49	0.41	0.36	0.32	0.29	0.26	0.24	0.21	0.18	0.16	0.14	0.13	0.12	0.11
SE	L	0.05	0.04	0.04	0.03	0.03	0.13	0.28	0.43	0.55	0.62	0.63	0.57	0.48	0.42	0.37	0.33	0.28	0.24	0.19	0.15	0.12	0.10	0.08	0.07
	M	0.09	0.08	0.07	0.06	0.05	0.14	0.26	0.38	0.48	0.54	0.56	0.51	0.45	0.40	0.36	0.33	0.29	0.25	0.21	0.18	0.16	0.14	0.12	0.10
	H	0.11	0.10	0.10	0.09	0.08	0.17	0.28	0.40	0.49	0.53	0.53	0.48	0.41	0.36	0.33	0.30	0.27	0.24	0.20	0.18	0.16	0.14	0.13	0.12
SSE	L	0.07	0.05	0.04	0.04	0.03	0.06	0.15	0.29	0.43	0.55	0.63	0.64	0.60	0.52	0.45	0.40	0.35	0.29	0.23	0.18	0.15	0.12	0.10	0.08
	M	0.11	0.09	0.08	0.07	0.06	0.08	0.16	0.26	0.38	0.48	0.55	0.57	0.54	0.48	0.43	0.39	0.35	0.30	0.25	0.21	0.18	0.16	0.14	0.12
	H	0.12	0.11	0.11	0.10	0.09	0.12	0.19	0.29	0.40	0.49	0.54	0.55	0.51	0.44	0.39	0.35	0.31	0.27	0.23	0.20	0.18	0.16	0.15	0.13
S	L	0.08	0.07	0.05	0.04	0.04	0.06	0.09	0.14	0.22	0.34	0.48	0.59	0.65	0.65	0.59	0.50	0.43	0.36	0.28	0.22	0.18	0.15	0.12	0.10
	M	0.12	0.11	0.09	0.08	0.07	0.08	0.11	0.14	0.21	0.31	0.42	0.52	0.57	0.58	0.53	0.47	0.41	0.36	0.29	0.25	0.21	0.18	0.16	0.14
	H	0.13	0.12	0.12	0.11	0.10	0.11	0.14	0.17	0.24	0.33	0.43	0.51	0.56	0.55	0.50	0.43	0.37	0.32	0.26	0.22	0.20	0.18	0.16	0.15
SSW	L	0.10	0.08	0.07	0.06	0.05	0.06	0.09	0.11	0.15	0.19	0.27	0.39	0.52	0.62	0.67	0.65	0.58	0.46	0.36	0.28	0.23	0.19	0.15	0.12
	M	0.14	0.12	0.11	0.09	0.08	0.09	0.11	0.13	0.15	0.18	0.25	0.35	0.46	0.55	0.59	0.59	0.53	0.44	0.35	0.30	0.25	0.22	0.19	0.16
	H	0.15	0.14	0.13	0.12	0.11	0.12	0.14	0.16	0.18	0.21	0.27	0.37	0.46	0.53	0.57	0.55	0.49	0.40	0.32	0.26	0.23	0.20	0.18	0.16
SW	L	0.12	0.10	0.08	0.06	0.05	0.06	0.08	0.10	0.12	0.14	0.16	0.24	0.36	0.49	0.60	0.66	0.66	0.58	0.43	0.33	0.27	0.22	0.18	0.14
	M	0.15	0.14	0.12	0.10	0.09	0.09	0.10	0.12	0.13	0.15	0.17	0.23	0.33	0.44	0.53	0.58	0.59	0.53	0.41	0.33	0.28	0.24	0.21	0.18
	H	0.15	0.14	0.13	0.12	0.11	0.12	0.13	0.14	0.16	0.17	0.19	0.25	0.34	0.44	0.52	0.56	0.56	0.49	0.37	0.30	0.25	0.21	0.19	0.17
WSW	L	0.12	0.10	0.08	0.07	0.05	0.06	0.07	0.09	0.10	0.12	0.13	0.17	0.26	0.40	0.52	0.62	0.66	0.61	0.44	0.34	0.27	0.22	0.18	0.15
	M	0.15	0.13	0.12	0.10	0.09	0.09	0.10	0.11	0.12	0.13	0.14	0.17	0.24	0.35	0.46	0.54	0.58	0.55	0.42	0.34	0.28	0.24	0.21	0.18
	H	0.15	0.14	0.13	0.12	0.11	0.11	0.12	0.13	0.14	0.15	0.16	0.19	0.26	0.36	0.46	0.53	0.56	0.51	0.38	0.30	0.25	0.21	0.19	0.17
W	L	0.12	0.10	0.08	0.06	0.05	0.06	0.07	0.08	0.10	0.11	0.12	0.14	0.20	0.32	0.45	0.57	0.64	0.61	0.44	0.34	0.27	0.22	0.18	0.14
	M	0.15	0.13	0.11	0.10	0.09	0.09	0.09	0.10	0.11	0.12	0.13	0.14	0.19	0.29	0.40	0.50	0.56	0.55	0.41	0.33	0.27	0.23	0.20	0.17
	H	0.14	0.13	0.12	0.11	0.10	0.11	0.12	0.13	0.14	0.14	0.15	0.16	0.21	0.30	0.40	0.49	0.54	0.52	0.38	0.30	0.24	0.21	0.18	0.16
WNW	L	0.12	0.10	0.08	0.06	0.05	0.06	0.07	0.09	0.10	0.12	0.13	0.15	0.17	0.26	0.40	0.53	0.63	0.62	0.44	0.34	0.27	0.22	0.18	0.14
	M	0.15	0.13	0.11	0.10	0.09	0.09	0.10	0.11	0.12	0.13	0.14	0.15	0.17	0.24	0.35	0.47	0.55	0.55	0.41	0.33	0.27	0.23	0.20	0.17
	H	0.14	0.13	0.12	0.11	0.10	0.11	0.12	0.13	0.14	0.15	0.16	0.17	0.18	0.25	0.36	0.46	0.53	0.52	0.38	0.30	0.24	0.20	0.18	0.16
NW	L	0.11	0.09	0.08	0.06	0.05	0.06	0.08	0.10	0.12	0.14	0.16	0.17	0.19	0.23	0.33	0.47	0.59	0.60	0.42	0.33	0.26	0.21	0.17	0.14
	M	0.14	0.12	0.11	0.09	0.08	0.09	0.10	0.11	0.13	0.14	0.16	0.17	0.18	0.21	0.30	0.42	0.51	0.54	0.39	0.32	0.26	0.22	0.19	0.16
	H	0.14	0.12	0.11	0.10	0.10	0.10	0.12	0.13	0.15	0.16	0.18	0.18	0.19	0.22	0.30	0.41	0.50	0.51	0.36	0.29	0.23	0.20	0.17	0.15
NNW	L	0.12	0.09	0.08	0.06	0.05	0.07	0.11	0.14	0.18	0.22	0.25	0.27	0.29	0.30	0.33	0.44	0.57	0.62	0.44	0.33	0.26	0.21	0.17	0.14
	M	0.15	0.13	0.11	0.10	0.09	0.10	0.12	0.15	0.18	0.21	0.23	0.26	0.27	0.28	0.31	0.39	0.51	0.56	0.41	0.33	0.27	0.23	0.20	0.17
	H	0.14	0.13	0.12	0.11	0.10	0.12	0.15	0.17	0.20	0.23	0.25	0.26	0.28	0.28	0.31	0.38	0.49	0.53	0.38	0.30	0.25	0.21	0.18	0.16
HOR	L	0.11	0.09	0.07	0.06	0.05	0.07	0.14	0.24	0.36	0.48	0.58	0.66	0.72	0.74	0.73	0.67	0.59	0.47	0.37	0.29	0.24	0.19	0.16	0.13
	M	0.16	0.14	0.12	0.11	0.09	0.11	0.16	0.24	0.33	0.43	0.52	0.59	0.64	0.67	0.66	0.62	0.56	0.47	0.38	0.32	0.28	0.24	0.21	0.18
	H	0.17	0.16	0.15	0.14	0.13	0.15	0.20	0.28	0.36	0.45	0.52	0.59	0.62	0.64	0.62	0.58	0.51	0.42	0.35	0.29	0.26	0.23	0.21	0.19

L = Light construction: frame exterior wall, 2-in. concrete floor slab, approximately 30 lb of material/ft^2 of floor area.
M = Medium construction: 4-in. concrete exterior wall, 4-in. concrete floor slab, approximately 70 lb of building material/ft^2 of floor area.
H = Heavy construction: 6-in. concrete exterior wall, 6-in. concrete floor slab, approximately 130 lb of building materials/ft^2 of floor area.

(Courtesy ASHRAE, 1985 Fundamentals)

Figure 2.21

Table 14 Cooling Load Factors for Glass with Interior Shading, North Latitudes (All Room Constructions)

Fenes-tration Facing	Solar Time, h																							
	0100	0200	0300	0400	0500	0600	0700	0800	0900	1000	1100	1200	1300	1400	1500	1600	1700	1800	1900	2000	2100	2200	2300	2400
N	0.08	0.07	0.06	0.06	0.07	0.73	0.66	0.65	0.73	0.80	0.86	0.89	0.89	0.86	0.82	0.75	0.78	0.91	0.24	0.18	0.15	0.13	0.11	0.10
NNE	0.03	0.03	0.02	0.02	0.03	0.64	0.77	0.62	0.42	0.37	0.37	0.37	0.36	0.35	0.32	0.28	0.23	0.17	0.08	0.07	0.06	0.05	0.04	0.04
NE	0.03	0.02	0.02	0.02	0.02	0.56	0.76	0.74	0.58	0.37	0.29	0.27	0.26	0.24	0.22	0.20	0.16	0.12	0.06	0.05	0.04	0.04	0.03	0.03
ENE	0.03	0.02	0.02	0.02	0.02	0.52	0.76	0.80	0.71	0.52	0.31	0.26	0.24	0.22	0.20	0.18	0.15	0.11	0.06	0.05	0.04	0.04	0.03	0.03
E	0.03	0.02	0.02	0.02	0.02	0.47	0.72	0.80	0.76	0.62	0.41	0.27	0.24	0.22	0.20	0.17	0.14	0.11	0.06	0.05	0.05	0.04	0.03	0.03
ESE	0.03	0.03	0.02	0.02	0.02	0.41	0.67	0.79	0.80	0.72	0.54	0.34	0.27	0.24	0.21	0.19	0.15	0.12	0.07	0.06	0.05	0.04	0.04	0.03
SE	0.03	0.03	0.02	0.02	0.02	0.30	0.57	0.74	0.81	0.79	0.68	0.49	0.33	0.28	0.25	0.22	0.18	0.13	0.08	0.07	0.06	0.05	0.04	0.04
SSE	0.04	0.03	0.03	0.03	0.02	0.12	0.31	0.54	0.72	0.81	0.81	0.71	0.54	0.38	0.32	0.27	0.22	0.16	0.09	0.08	0.07	0.06	0.05	0.04
S	0.04	0.04	0.03	0.03	0.03	0.09	0.16	0.23	0.38	0.58	0.75	0.83	0.80	0.68	0.50	0.35	0.27	0.19	0.11	0.09	0.08	0.07	0.06	0.05
SSW	0.05	0.04	0.04	0.03	0.03	0.09	0.14	0.18	0.22	0.27	0.43	0.63	0.78	0.84	0.80	0.66	0.46	0.25	0.13	0.11	0.09	0.08	0.07	0.06
SW	0.05	0.05	0.04	0.04	0.03	0.07	0.11	0.14	0.16	0.19	0.22	0.38	0.59	0.75	0.83	0.81	0.69	0.45	0.16	0.12	0.10	0.09	0.07	0.06
WSW	0.05	0.05	0.04	0.04	0.03	0.07	0.10	0.12	0.14	0.16	0.17	0.23	0.44	0.64	0.78	0.84	0.78	0.55	0.16	0.12	0.10	0.09	0.07	0.06
W	0.05	0.05	0.04	0.04	0.03	0.06	0.09	0.11	0.13	0.15	0.16	0.17	0.31	0.53	0.72	0.82	0.81	0.61	0.16	0.12	0.10	0.08	0.07	0.06
WNW	0.05	0.05	0.04	0.03	0.03	0.07	0.10	0.12	0.14	0.16	0.17	0.18	0.22	0.43	0.65	0.80	0.84	0.66	0.16	0.12	0.10	0.08	0.07	0.06
NW	0.05	0.04	0.04	0.03	0.03	0.07	0.11	0.14	0.17	0.19	0.20	0.21	0.22	0.30	0.52	0.73	0.82	0.69	0.16	0.12	0.10	0.08	0.07	0.06
NNW	0.05	0.05	0.04	0.03	0.03	0.11	0.17	0.22	0.26	0.30	0.32	0.33	0.34	0.34	0.39	0.61	0.82	0.76	0.17	0.12	0.10	0.08	0.07	0.06
HOR.	0.06	0.05	0.04	0.04	0.03	0.12	0.27	0.44	0.59	0.72	0.81	0.85	0.85	0.81	0.71	0.58	0.42	0.25	0.14	0.12	0.10	0.08	0.07	0.06

CALCULATING SPACE COOLING LOAD FROM HEAT SOURCES WITHIN THE CONDITIONED SPACE

Lighting

An accurate estimate of the space cooling load imposed by lighting, often the major space load component, is essential in air-conditioning system design. Calculation of this load component is not straightforward; the rate of heat gain to the air caused by lights can be quite different from the power supplied to the lights.

Some of the energy emanating from lights is in the form of radiation that only affects the air after it has been absorbed by walls, floors and furniture, and has warmed them to a temperature higher than the air temperature. This absorbed energy, stored by the structure, contributes to the space cooling load after a time lag, and is present after the lights are switched off. The time lag effect (Fig. 2) should be taken into account when calculating the cooling load, since the actual load is lower than the instantaneous heat gain and the peak load may be affected significantly.

Generally, the instantaneous rate of heat gain from electric lighting in watts can be calculated from:

$$q = \text{total light wattage} \cdot \text{use factor} \cdot \text{special allowance factor} \qquad (14)$$

The rate of heat gain can be expressed in units of Btu/h by multiplying Eq.(14) by 3.413. The *total light wattage* is obtained from the ratings of all fixtures installed for general illumination or special use. The *use factor* is the ratio of the wattage in use, for the conditions under which the load estimate is being made, to the total installed wattage. For commercial applications such as stores, the use factor is generally unity. The *special allowance factor* is introduced for fluorescent fixtures and fixtures requiring more energy than their rated wattage. For fluorescent fixtures, the special allowance factor accounts for ballast losses, which can be as high as 2.19 for 32-W single lamp high-output fixtures on 277 V. Rapid-start, 40-W lamp fixture allowance factors vary from a low of 1.18 for two lamps on 277 V to a high of 1.30 for one lamp on 118 V. For other industrial fixtures, such as sodium lamp fixtures, special allowance factors may vary from 1.04 to 1.37, depending on the manufacturer.

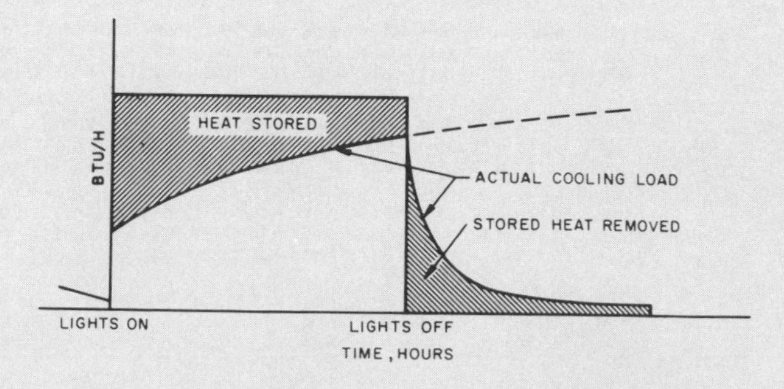

Fig. 2 Thermal Storage Effect in Cooling Load from Lights

(Courtesy ASHRAE, 1985 Fundamentals)

Figure 2.21 (cont.)

Table 20 Recommended Rate of Heat Gain from Commercial Cooking Appliances Located in the Air-Conditioned Area[a]

Appliance	Capacity	Overall Dim., Inches Width • Depth • Height	Miscellaneous Data (Dimensions in Inches)	Manufacturer's Input Rating		Probable Max. Hourly Input Btuh	Recommended Rate of Heat Gain, Btuh			
				Boiler hp or Watts	Btuh		Without Hood			With Hood[b]
							Sensible	Latent	Total	All Sensible
Gas-Burning, Counter Type										
Broiler-griddle		31 • 20 • 18			36,000	18,000	11,700	6,300	18,000	3,600
Coffee brewer per burner			With *warm* position		5,500	2,500	1,750	750	2,500	500
Water heater burner			With storage tank		11,000	5,000	3,850	1,650	5,500	1,100
Coffee urn	3 gal	12-inch dia.			10,000	5,000	3,500	1,500	5,000	1,000
	5 gal.	14-inch dia.			15,000	7,500	5,250	2,250	7,500	1,500
	8 gal. twin	25-inch wide			20,000	10,000	7,000	3,000	10,000	2,000
Deep fat fryer	15 lb fat	14 • 21 • 15			30,000	15,000	7,500	7,500	15,000	3,000
Dry food warmer per sq ft of top					1,400	700	560	140	700	140
Griddle, frying per sq ft of top					15,000	7,500	4,900	2,600	7,500	1,500
Short order stove, per burner			Open grates		10,000	5,000	3,200	1,800	5,000	1,000
Steam table per sq ft of top					2,500	1,250	750	500	1,250	250
Toaster, continuous	360 slices/hr	19 • 16 • 30	2 slices wide		12,000	6,000	3,600	2,400	6,000	1,200
	720 slices/hr	24 • 16 • 30	4 slices wide		20,000	10,000	6,000	4,000	10,000	2,000
Gas-Burning, Floor Mounted Type										
Broiler, unit		24 • 26 grid	Same burner heats oven		70,000	35,000				7,000
Deep fat fryer	32 lb fat		14-in. kettle		65,000	32,500				6,500
	56 lb fat		18-in. kettle		100,000	50,000				10,000
Oven, deck, per sq ft of hearth area			Same for 7and 12 high decks		4,000	2,000				400
Oven, roasting		32 • 32 • 60	Two ovens—24 • 28 • 15		80,000	40,000	Exhaust hood required	Exhaust hood required	Exhaust hood required	8,000
Range, heavy duty		32 • 42 • 33								
Top section			32 wide • 39 deep		64,000	32,000				6,400
Oven			25 • 28 • 15		40,000	20,000				4,000
Range, jr., heavy duty		31 • 35 • 33								
Top section			31 wide • 32 deep		45,000	22,500				4,500
Oven			24 • 28 • 15		35,000	17,500				3,500
Range, restaurant type										
Per 2-burner sect.			12 wide • 28 deep		24,000	12,000				2,400
Per oven			24 • 22 • 14		30,000	15,000				3,000
Per broiler-griddle			24 wide • 26 deep		35,000	17,500				3,500
Electric, Counter Type										
Coffee brewer										
per burner				625	2,130	1,000	770	230	1,000	340
per warmer				160	545	300	230	70	300	90
automatic	240 cups per hr	27 • 21 • 22	4-burner + water htr.	5,000	17,000	8,500	6,500	2,000	8,500	1,700
Coffee urn	3 gal.			2,000	6,800	3,400	2,550	850	3,400	1,000
	5 gal.			3,000	10,200	5,100	3,850	1,250	5,100	1,600
	8 gal. twin			4,000	13,600	6,800	5,200	1,600	6,800	2,100
Deep fat fryer	14 lb fat	13 • 22 • 10		5,500	18,750	9.400	2,800	6,600	9,400	3,000
	21 lb fat	16 • 22 • 10		8,000	27,300	13,700	4,100	9,600	13,700	4,300
Dry food warmer, per sq ft of top				240	820	400	320	80	400	130
Egg boiler	2 cups	10 • 13 • 25		1,100	3,750	1,900	1,140	760	1,900	600
Griddle, frying, per sq ft of top				2,700	9,200	4,600	3,000	1,600	4,600	1,500
Griddle-Grill		18 • 20 • 13	Grid, 200 sq in.	6,000	20,400	10,200	6,600	3,600	10,200	3,200
Hotplate		18 • 20 • 13	2 heating units	5,200	17,700	8,900	5,300	3,600	8,900	2,800
Roaster		18 • 20 • 13		1,650	5,620	2,800	1,700	1,100	2,800	900
Roll warmer		18 • 20 • 13		1,650	5,620	2,800	2,600	200	2,800	900
Toaster, continuous	360 slices/hr	15 • 15 • 28	2 slices wide	2,200	7,500	3,700	1.960	1,740	3,700	1,200
	720 slices/hr	20 • 15 • 28	4 slices wide	3,000	10,200	5,100	2,700	2,400	5,100	1,600
Toaster, pop-up	4 slice	12 • 11 • 9		2,540	8,350	4,200	2,230	1.970	4,200	1,300
Waffle iron		18 • 20 • 13	2 grids	1,650	5,620	2,800	1,680	1,120	2,800	900

(Courtesy ASHRAE, 1985 Fundamentals)

Figure 2.22

Table 20 Recommended Rate of Heat Gain from Commercial Cooking Appliances Located in the Air-Conditioned Area[a] (Continued)

Appliance	Capacity	Overall Dim., Inches Width • Depth • Height	Miscellaneous Data (Dimensions in Inches)	Manufacturer's Input Rating: Boiler hp or Watts	Manufacturer's Input Rating: Btuh	Probable Max. Hourly Input Btuh	Recommended Rate of Heat Gain, Btuh — Without Hood: Sensible	Without Hood: Latent	Without Hood: Total	With Hood[b]: All Sensible
Electric, Floor Mounted Type										
Griddle[c]		36 • 32 • 37	36 • 25 cooking surface	16,800	57,300					2,060
Broiler, no oven			23 wide • 25 deep grid	12,000	40,900	20,500				6,500
with oven			23 • 27 • 12 oven	18,000	61,400	30,700				9,800
Broiler, single deck[c]		36 • 36 • 54		16,000	54,600					10,800
Deep fat fryer	28 lb fat	20 • 38 • 36	14 wide • 15 deep kettle	12,000	40,900	20,500				6,500
	60 lb fat	24 • 36 • 36	20 wide • 20 deep kettle	18,000	61,400	30,700				9,800
Fryer[c]		15 • 32 • 36	13 • 23 cooking surface	22,000	75,000					730
Oven, baking, per sq ft of hearth			Compartment 8-in. high	500	1,700	850	Exhaust hood required	Exhaust hood required	Exhaust hood required	270
Oven, roasting, per sq ft of hearth			Compartment 12-in. high	900	3,070	1,500				490
Range, heavy duty[c]										
Top section		38 • 36 • 37	36 • 24 cooking surface	15,000	51,200					19,100
Oven				6,700	22,900					1,700
Range, medium duty		30 • 32 • 36								
Top section				8,000	27,300	13,600				4,300
Oven				3,600	12,300	6,200				1,900
Range, light duty		30 • 29 • 36								
Top section				6,600	22,500	11,200				3,600
Oven				3,000	10,200	5,100				1,600
Convection Oven[c]		38 • 36 • 55		11,000	37,500					1,540
Charbroiler[c]		36 • 24 • 34	30 • 18 cooking surface	16,500	56,300					4,320
Steam cooker, two sections[c]		36 • 29 • 64		24,000	81,900					3,140
Steam Heated										
Coffee urn	3 gal.			0.2	6,600	3,300	2,180	1,120	3.300	1,000
	5 gal.			0.3	10,000	5,000	3,300	1,700	5,000	1,600
	8 gal. twin			0.4	13,200	6,600	4,350	2,250	6,600	2,100
Steam table per sq ft of top			With insets	0.05	1,650	825	500	325	825	260
Bain marie per sq ft of top			Open Tank	0.10	3,300	1,650	825	825	1,650	520
Oyster steamer				0.5	16,500	8,250	5,000	3,250	8,250	2,600
Steam kettles per gal. capacity			Jacketed type	0.06	2,000	1,000	600	400	1,000	320
Compartment steamer per compartment		24 • 25 • 12 compartment	Floor mounted	1.2	40,000	20,000	12,000	8,000	20,000	6,400
Compartment steamer	3 pans 12 • 20 • 2.5		Single counter unit	0.5	16,500	8,250	5,000	3,250	8,250	2,600
Plate warmer per cu ft				0.05	1,650	825	550	275	825	260

[a]The data in this table (except as noted in c below) was determined by assuming the hourly heat input was 0.50 times the manufacturer's energy input rating. This is conservative on the average but could result in heat gain estimates higher or lower than actual heat gains depending on the appliance. Consult the text for additional discussion.

[b]For poorly designed or undersized exhaust systems the heat gains in this column should be doubled and half of the increase assumed as latent heat.

[c]Based on measured heat gain at typical idle conditions. For open island canopies mutiply values by 1.32. For additional information, see Ref. 41.

Table 21 Rate of Heat Gain from Miscellaneous Appliances

Appliance	Miscellaneous Data	Manufacturer's Rating: Watts	Manufacturer's Rating: Btuh	Recommended Rate of Heat Gain, Btuh: Sensible	Latent	Total
Electrical Appliances						
Hair dryer	Blower type	1580	5400	2300	400	2700
Hair dryer	Helmet type	705	2400	1870	330	2200
Permanent wave machine	60 heaters @25 W, 36 in normal use	1500	5000	850	150	1000
Neon sign, per linear ft of tube	0.5 in., dia			30		30
	0.375 in., dia			60		60
Sterilizer, instrument		1100	3750	650	1200	1850
Gas-Burning Appliances						
Lab burners Bunsen	0.4375-in. barrel		3000	1680	420	2100
Fishtail	1.5-in. wide		5000	2800	700	3500
Meeker	1-in. diameter		6000	3360	840	4200
Gas light, per burner	Mantle type		2000	1800	200	2000
Cigar lighter	Continuous flame		2500	900	100	1000

(Courtesy ASHRAE, 1985 Fundamentals)

Figure 2.22 (cont.)

Table 17E Cooling Load Factors When Lights Are on for 16 Hours

"a" Coefficients	"b" Classification	Number of hours after lights are turned on 0	1	2	3	4	5	6	7	8	9	10	11	12	13	14	15	16	17	18	19	20	21	22	23
	A	0.12	0.54	0.63	0.70	0.76	0.81	0.85	0.88	0.90	0.92	0.94	0.95	0.96	0.97	0.97	0.98	0.98	0.54	0.43	0.35	0.28	0.23	0.18	0.15
	B	0.23	0.66	0.69	0.72	0.75	0.78	0.80	0.82	0.84	0.85	0.87	0.88	0.89	0.90	0.91	0.92	0.93	0.49	0.44	0.39	0.35	0.32	0.29	0.26
0.45	C	0.29	0.72	0.74	0.75	0.77	0.78	0.80	0.81	0.82	0.83	0.84	0.85	0.86	0.87	0.88	0.88	0.89	0.45	0.42	0.39	0.37	0.35	0.33	0.31
	D	0.31	0.75	0.76	0.77	0.77	0.78	0.79	0.79	0.80	0.81	0.81	0.82	0.82	0.83	0.83	0.84	0.84	0.40	0.39	0.37	0.36	0.35	0.34	0.33
	A	0.10	0.63	0.70	0.76	0.81	0.84	0.87	0.90	0.92	0.93	0.95	0.96	0.97	0.97	0.98	0.98	0.99	0.44	0.35	0.28	0.23	0.18	0.15	0.12
	B	0.19	0.72	0.75	0.77	0.80	0.82	0.84	0.85	0.87	0.88	0.89	0.90	0.91	0.92	0.93	0.94	0.94	0.40	0.36	0.32	0.29	0.26	0.24	0.21
0.55	C	0.24	0.77	0.79	0.80	0.81	0.82	0.83	0.84	0.85	0.86	0.87	0.88	0.88	0.89	0.90	0.90	0.91	0.37	0.34	0.32	0.30	0.29	0.27	0.25
	D	0.26	0.80	0.80	0.81	0.82	0.82	0.83	0.83	0.84	0.84	0.85	0.85	0.86	0.86	0.86	0.87	0.87	0.33	0.32	0.31	0.30	0.29	0.28	0.27
	A	0.07	0.71	0.77	0.81	0.85	0.88	0.90	0.92	0.94	0.95	0.96	0.97	0.97	0.98	0.98	0.99	0.99	0.34	0.27	0.22	0.18	0.14	0.12	0.09
	B	0.15	0.78	0.81	0.82	0.84	0.86	0.87	0.88	0.90	0.91	0.92	0.92	0.93	0.94	0.94	0.95	0.96	0.31	0.28	0.25	0.23	0.20	0.18	0.16
0.65	C	0.18	0.82	0.83	0.84	0.85	0.86	0.87	0.88	0.89	0.89	0.90	0.90	0.91	0.92	0.92	0.93	0.93	0.28	0.27	0.25	0.24	0.22	0.21	0.20
	D	0.20	0.84	0.85	0.85	0.86	0.86	0.87	0.87	0.87	0.88	0.88	0.88	0.89	0.89	0.89	0.90	0.90	0.25	0.25	0.24	0.23	0.22	0.22	0.21
	A	0.05	0.79	0.83	0.87	0.89	0.91	0.93	0.94	0.95	0.96	0.97	0.98	0.98	0.98	0.99	0.99	0.99	0.24	0.20	0.16	0.13	0.10	0.08	0.07
	B	0.11	0.85	0.86	0.87	0.89	0.90	0.91	0.92	0.93	0.93	0.94	0.95	0.95	0.96	0.96	0.96	0.97	0.22	0.20	0.18	0.16	0.15	0.13	0.12
0.75	C	0.13	0.87	0.88	0.89	0.89	0.90	0.91	0.91	0.92	0.92	0.93	0.93	0.94	0.94	0.94	0.95	0.95	0.20	0.19	0.18	0.17	0.16	0.15	0.14
	D	0.14	0.89	0.89	0.89	0.90	0.90	0.90	0.91	0.91	0.91	0.91	0.92	0.92	0.92	0.92	0.93	0.93	0.18	0.18	0.17	0.17	0.16	0.16	0.15

Table 18 Rates of Heat Gain from Occupants of Conditioned Spaces[a]

Degree of Activity	Typical Application	Total Heat Adults, Male Btu/h	Total Heat Adjusted[b] Btu/h	Sensible Heat Btu/h	Latent Heat Btu/h
Seated at rest	Theater, movie	400	350	210	140
Seated, very light work writing	Offices, hotels, apts	480	420	230	190
Seated, eating	Restaurant[c]	520	580 [c]	255	325
Seated, light work, typing	Offices, hotels, apts	640	510	255	255
Standing, light work or walking slowly	Retail Store, bank	800	640	315	325
Light bench work	Factory	880	780	345	435
Walking, 3 mph, light machine work	Factory	1040	1040	345	695
Bowling[d]	Bowling alley	1200	960	345	615
Moderate dancing	Dance hall	1360	1280	405	875
Heavy work, heavy machine work, lifting	Factory	1600	1600	565	1035
Heavy work, athletics	Gymnasium	2000	1800	635	1165

[a]Note; Tabulated values are based on 78 F room dry-bulb temperature. For 80 F room dry-bulb, the total heat remains the same, but the sensible heat value should be decreased by approximately 8% and the latent heat values increased accordingly.

[b]Adjusted total heat gain is based on normal percentage of men, women and children for the application listed, with the postulate that the gain from an adult female is 85% of that for an adult male, and that the gain from a child is 75% of that for an adult male.

[c]Adjusted total heat value for eating in a restaurant, includes 60 Btu/h for food per individual (30 Btu/h sensible and 30 Btu/h latent).

[d]For bowling figure one person per alley actually bowling, and all others as sitting 400 Btu/h or standing and walking slowly 790 Btu/h.

Also refer to Tables 4 and 7, Chapter 8.
All values rounded to nearest 10 Btu/h.

Table 19 Sensible Heat Cooling Load Factors for People

Total Hours in Space	Hours after Each Entry Into Space 1	2	3	4	5	6	7	8	9	10	11	12	13	14	15	16	17	18	19	20	21	22	23	24
2	0.49	0.58	0.17	0.13	0.10	0.08	0.07	0.06	0.05	0.04	0.04	0.03	0.03	0.02	0.02	0.02	0.02	0.01	0.01	0.01	0.01	0.01	0.01	0.01
4	0.49	0.59	0.66	0.71	0.27	0.21	0.16	0.14	0.11	0.10	0.08	0.07	0.06	0.06	0.05	0.04	0.04	0.03	0.03	0.03	0.02	0.02	0.02	0.01
6	0.50	0.60	0.67	0.72	0.76	0.79	0.34	0.26	0.21	0.18	0.15	0.13	0.11	0.10	0.08	0.07	0.06	0.06	0.05	0.04	0.04	0.03	0.03	0.03
8	0.51	0.61	0.67	0.72	0.76	0.80	0.82	0.84	0.38	0.30	0.25	0.21	0.18	0.15	0.13	0.12	0.10	0.09	0.08	0.07	0.06	0.05	0.05	0.04
10	0.53	0.62	0.69	0.74	0.77	0.80	0.83	0.85	0.87	0.89	0.42	0.34	0.28	0.23	0.20	0.17	0.15	0.13	0.11	0.10	0.09	0.08	0.07	0.06
12	0.55	0.64	0.70	0.75	0.79	0.81	0.84	0.86	0.88	0.89	0.91	0.92	0.45	0.36	0.30	0.25	0.21	0.19	0.16	0.14	0.12	0.11	0.09	0.08
14	0.58	0.66	0.72	0.77	0.80	0.83	0.85	0.87	0.89	0.90	0.91	0.92	0.93	0.94	0.47	0.38	0.31	0.26	0.23	0.20	0.17	0.15	0.13	0.11
16	0.62	0.70	0.75	0.79	0.82	0.85	0.87	0.88	0.90	0.91	0.92	0.93	0.94	0.95	0.95	0.96	0.49	0.39	0.33	0.28	0.24	0.20	0.18	0.16
18	0.66	0.74	0.79	0.82	0.85	0.87	0.89	0.90	0.92	0.93	0.94	0.94	0.95	0.96	0.96	0.97	0.97	0.97	0.50	0.40	0.33	0.28	0.24	0.21

(Courtesy ASHRAE, 1985 Fundamentals)

Figure 2.23

programs that analyze a building hour-by-hour to produce the design conditions for the entire building. This analysis can also be done on a room-by-room basis. Cooling loads in buildings consist of conduction, convection, radiation, internal, and people heat gains. This information is summarized in Figure 2.24.

Figure 2.25 lists typical building cooling loads from *Means Mechanical Cost Data* as an overall guide to calculations.

Heat Gain Formulas

Conduction Heat Gain

H_c = UA (CLTD)

H_c = Conduction heat gain (Btu/hr.)
U = Overall heat transfer coefficient (Btu/s.f./°F) (see Figures 2.5–2.7)
A = Surface area (s.f.)
CLTD = Cooling load temperature difference (°F) (see Figure 2.8)

Convection Heat Gain

H_v = 1.1 cfm (ΔT) + 4840 cfm (ΔW)

Sensible heat gain: H_v = 1.1 cfm (ΔT)
Latent heat gain: 4840 cfm (ΔW)

Radiation Heat Gain

H_s = A (SHGF) (SC) (CLF)

H_s = Radiant heat gain (Btu/hr.)
A = Glass area (s.f.)
SGHF = Solar heat gain factor (Btu/hr./s.f.)
SC = Shading coefficient (see Figure 2.20)
CLF = Cooling load factor (see Figure 2.21)

Internal Heat Gain

H_i = (sensible load) CLF + (latent load)

H_i = Internal heat gain (Btu/hr.)
CLF = Cooling load factor

People Heat Gain

$H_p = N_oP_sCLF + N_oP_L$

H_p = People heat gain (Btu/hr.)
N_o = Number of occupants
P_s = Sensible heat gain per person (Btu/hr.) (see Figure 2.23)
CLF = Cooling load factor, usually 1.0 (see Figure 2.21)
P_L = Latent heat gain per person (Btu/hr.) (see Figure 2.23)

Total Heat Gain

$H_t = H_c + H_v + H_s + H_i + H_p$

Figure 2.24

AIR CONDITIONING | C8.4-000 | General

Table 8.4-001

General: The purpose of air conditioning is to control the environment of a space so that comfort is provided for the occupants and/or conditions are suitable for the processes or equipment contained therein. The several items which should be evaluated to define system objectives are:

Temperature Control
Humidity Control
Cleanliness
Odor, smoke and fumes
Ventilation

Efforts to control the above parameters must also include consideration of the degree or tolerance of variation, the noise level introduced, the velocity of air motion and the energy requirements to accomplish the desired results.

The variation in **temperature** and **humidity** is a function of the sensor and the controller. The controller reacts to a signal from the sensor and produces the appropriate suitable response in either the terminal unit, the conductor of the transporting medium (air, steam, chilled water, etc.), or the source (boiler, evaporating coils, etc.).

The **noise level** is a by-product of the energy supplied to moving components of the system. Those items which usually contribute the most noise are pumps, blowers, fans, compressors and diffusers. The level of noise can be partially controlled through use of vibration pads, isolators, proper sizing, shields, baffles and sound absorbing liners.

Some **air motion** is necessary to prevent stagnation and stratification. The maximum acceptable velocity varies with the degree of heating or cooling which is taking place. Most people feel air moving past them at velocities in excess of 25 FPM as an annoying draft, however, velocities up to 45 FPM may be acceptable in certain cases. Ventilation, expressed as air changes per hour and percentage of fresh air, is usually an item regulated by local codes.

Selection of the system to be used for a particular application is usually a trade-off. In some cases the building size, style, or room available for mechanical use limits the range of possibilities. Prime factors influencing the decision are first cost and total life (operating, maintenance and replacement costs). The accuracy with which each parameter is determined will be an important measure of the reliability of the decision and subsequent satisfactory operation of the installed system.

Heat delivery may be desired from an air conditioning system. Heating capability usually is added as follows: A gas fired burner or hot water/steam/electric coils may be added to the air handling unit directly and heat all air equally. For limited or localized heat requirements the water/steam/electric coils may be inserted into the duct branch supplying the cold areas. Gas fired duct furnaces are also available. **Note:** when water or steam coils are used the cost of piping and boiler must also be added. For a rough estimate use the cost per square foot of the appropriate sized hydronic system with unit heater. This will give the boiler, piping, and the unit heaters which would approximate the cost of the heating coils.The installed cost of electric and gas heaters, boilers and other heat related items on a unit basis may be located in Section 15.5 of *Means Mechanical Cost Data.*

Table 8.4-002 Air Conditioning Requirements

BTU's Per Hour Per S.F. of Floor Area and S.F. Per Ton of Air Conditioning

Type Building	BTU per S.F.	S.F. per Ton	Type Building	BTU per S.F.	S.F. per Ton	Type Building	BTU per S.F.	S.F. per Ton
Apartments, Individual	26	450	Dormitory, Rooms	40	300	Libraries	50	240
Corridors	22	550	Corridors	30	400	Low Rise Office, Exterior	38	320
Auditoriums & Theaters	40	18*/300	Dress Shops	43	280	Interior	33	360
Banks	50	240	Drug Stores	80	150	Medical Centers	28	425
Barber Shops	48	250	Factories	40	300	Motels	28	425
Bars & Taverns	133	90	High Rise Office-Ext. Rms.	46	263	Office (small suite)	43	280
Beauty Parlors	66	180	Interior Rooms	37	325	Post Office, Individual Office	42	285
Bowling Alleys	68	175	Hospitals, Core	43	280	Central Area	46	260
Churches	36	20*/330	Perimeter	46	260	Residences	20	600
Cocktail Lounges	68	175	Hotel, Guest Rooms	44	275	Restaurants	60	200
Computer Rooms	141	85	Public Spaces	55	220	Schools & Colleges	46	260
Dental Offices	52	230	Corridors	30	400	Shoe Stores	55	220
Dept. Stores, Basement	34	350	Industrial Plants, Offices	38	320	Shop'g. Ctrs., Super Markets	34	350
Main Floor	40	300	General Offices	34	350	Retail Stores	48	250
Upper Floor	30	400	Plant Areas	40	300	Specialty Shops	60	200

*Persons per ton

12,000 BTU = 1 ton of air conditioning

Figure 2.25

CHAPTER THREE

CODES, REGULATIONS, AND STANDARDS

Every building in a city, town, or municipality is governed by a building code. Building codes are minimum construction standards established to protect public health, safety, and welfare. Typically, a local building code establishes minimum design requirements for all building systems, including HVAC systems and components. The owner is legally obligated to establish and install heating, ventilating, and air conditioning systems that conform to the governing code.

The local building codes regulate design, methods of construction, quality of material, and building use for the structures within a political jurisdiction. The local code may be state or city drafted, or it may be adapted from one of several recognized standards, such as the Building Official's and Code Administrator's International (BOCA) Codes. BOCA is an international professional organization that researches and has established a model building code.

Issues such as fire protection, material quality, and structural integrity are addressed in detail in each local building code. The aspects of HVAC design which relate to these issues must be considered in order to ensure the public's safety. For example, since HVAC involves combustion, proper construction of flues and ductwork is essential. Procedures for ongoing inspection and maintenance are also important. The weight of HVAC equipment must be checked against structural limits. Almost every chapter of the building code contains material that has some relevance to HVAC design.

Energy Conservation

Since the oil crisis of the 1970's, most states have incorporated into their codes a chapter on energy conservation. While there are other factors that contribute to energy use, such as lighting and electrical loads, HVAC is clearly a major focus for energy conservation. As a result, there are extensive code requirements for this aspect of HVAC design. The design conditions established by the code are fundamental to its energy conservation requirements. The heating system must be able to achieve certain health and comfort temperatures and specific temperatures prescribed by code; for example, generally about 70°F design temperature for heating and 78°F design temperature for cooling. Factors that influence the temperature include room type (living room or gymnasium), building type (hospital or office building), and building location (New England or the South, for example). Many other aspects of building design contribute to the achievement of the required temperatures. Some of these factors include:

- the insulation of the building envelope which, if adequate, makes the work of heating and cooling the building easier;
- lighting, which may carry its own heat; and
- the use pattern of the building.

Although the code establishes the specific design temperature that the heating system must be able to maintain, heating is a dynamic system and the HVAC design must allow for some flexibility in loads. This flexibility is characteristic of a good design.

Energy conservation codes also cover the energy consumption levels of HVAC equipment. Both HVAC systems and related components are governed by certain established criteria. The coefficients of the performance and energy efficiency ratio are listed in the energy conservation article of the code. Equipment must conform to these criteria. This should be checked with the manufacturer — not only for code conformance, but also to compare cost savings in energy consumption.

Cooling

Cooling is not mandated by code. However, there are certain accepted design standards for cooling systems. For example, a typical summer outdoor maximum design temperature is 88°F. The range of humidity allowed is also established, since humidity levels play a major role in the comfort of people in a given room temperature. Indoor design relative humidity for heating is approximately 40 percent.

Controls

HVAC controls are typically covered by code articles on energy conservation. Complex HVAC systems (as determined by building use) must be capable of meeting the individual requirements for each building area. This diversity in temperature within a single building is achieved by controls that monitor heating and cooling temperatures. Another type of control saves energy by setting back the temperature at the end of the day or by shutting off the equipment when it is not needed. Criteria for these controls are established by the energy conservation articles included in most codes.

Professional Organizations

The background research for energy conservation standards is rarely performed by each individual municipality. Instead, it is done by several organizations, one of which is the American Society of Heating, Refrigeration, and Air Conditioning Engineers (ASHRAE); the information gathered is incorporated into the building code. What the local code does not cover is referenced by the standards of the various professional organizations in the construction field. These references are usually listed in the back of each code and should be consulted for complete information. For HVAC, the standards are based on information supplied by ASHRAE and SMACNA (Sheet Metal and Air Conditioning Contractors National Association), among many others. These organizations form a vast information network that covers accepted engineering and industry practice and goes beyond the local code's mandate of protecting the public's health, safety, and welfare. To this extent, the criteria established are not required by law, but represent practitioners' wisdom.

The material controlled by such standards covers several different categories. It is produced by professional societies (such as ASHRAE), and is written at a certain level with appropriate detail, as represents the organization. ASME, the American Society of Mechanical Engineers, is similar in the way that they present their material. The "Boiler and Pressure Vessel Code," researched and established by ASME, is an important reference tool for HVAC designers.

Contractors also have their organizations. SMACNA is a good example. Contractor's associations have a different emphasis than engineering associations. The key issues tend to be accepted industry practice for equipment and installation, rather than concern with research and engineering formulas and derivations for calculations. Each type of material has its place in the information network and is useful in its own way.

Materials Testing

All construction materials, including those for HVAC (particularly pipe and metal for ductwork), are covered in the publications of several national organizations. These include the American Society of Testing and Materials (ASTM), Underwriters Laboratories (UL), and the American National Standards Institute (ANSI). Each of these organizations also has a unique place in the information network. ASTM represents an engineering society whose main concern is to establish the characteristics of engineering materials. ASTM also conducts the testing required to establish the performance of material under different temperatures and stresses.

Underwriters Laboratories (UL) is a private testing company that tests equipment under different circumstances for various criteria. The UL label has become nationally known as an indication that a piece of equipment or material meets the high standards UL requires. The testing criteria that UL uses varies, but it is usually from accepted industry sources, such as ASHRAE or ASTM. UL actually performs the tests on a given model of the equipment as submitted by the manufacturer. It then produces manuals listing accepted equipment. Each piece of equipment which has been built to the performance criteria accepted by UL will also be allowed to carry a UL label.

The American National Standards Institute (ANSI) is a national organization which brings together the variety of organizations evaluating materials and equipment in the United States. ANSI does not conduct its own tests or establish material quality. Instead, it incorporates those standards produced by other organizations. By coordinating this information, ANSI attempts to establish one system for evaluating materials in equipment.

Fire Safety Standards

An important element in both the governing codes and good engineering practice is the code and research produced by the National Fire Protection Association (NFPA). Since HVAC involves combustion, NFPA requirements are extensive. NFPA research has established standards for flammable liquids and the installation of oil-burning, gas, and air conditioning and ventilation equipment. NFPA has also established the National Electric Code and National Fuel Gas Code.

Typical Requirements

The material presented in this chapter is representative of the standards and information sources that govern and assist the HVAC designer. Together, these resources form a matrix which can be used in developing HVAC systems and components.

The following items contain typical code requirements for HVAC systems. Each of these items should be reviewed and verified for a particular building and locality.

Drawings:

These are typically prepared by a registered architect or licensed professional engineer in all buildings other than one- or two-family residences.

Design Conditions:

	Summer		**Winter**	
Temperature:	*Indoor*	*Outdoor*	*Indoor*	*Outdoor*
	75–78°F	88°F (varies)	68°F	10–40°F
Humidity: (varies)	50%			

U Values:

Building Enclosure "U" Values Minimum (Btu/hr./s.f./°F)

Walls	0.08
Roof	0.06
Windows	0.53 (double insulated)

Lighting Level:

3 watts per square foot maximum

Ventilation:

2 air changes allowed per hour – 25 percent fresh outdoor air or 5 cfm per person. Natural light and air in residential spaces except bathrooms and kitchens.

Equipment Oversizing:

25 percent maximum

Figure 3.1 lists the regulatory agency codes, regulations, and standards referenced in the Building Officials and Code Administrators International (BOCA) *National Mechanical Code 1987.*

Reference Standards

American National Standards Institute, Inc. (ANSI)
1430 Broadway
New York, New York 10018

Title	Standard Reference Number
Ductile-Iron Pipe, Centrifugally Cast, in Metal Molds or Sand-Lined Molds for Gas	A21.52-82
Cast Iron Pipe Flanges and Flanged Fittings, Class 25, 125, 250 and 800	B16.1-75
Pipe Flanges and Flanged Fittings, Steel Nickel Alloy and other Special Alloys	B16.5-81
Factory-Made Wrought Steel Buttwelding Fittings 1981 Supplement B16.9a	B16.9-78
Forged Steel Fittings, Socket-Welding and Threaded	B16.11-80
Cast Copper Alloy Solder Joint Pressure Fittings	B16.18-84
Wrought Copper and Copper Alloy Solder Joint Pressure Fittings	B16.22-80
Cast Copper Alloy Solder Joint Drainage Fittings DWV	B16.23-84
Bronze Pipe Flanges and Flanged Fittings, Class 150 and 300	B16.24-79
Cast Copper Alloy Fittings for Flared Copper Tubes	B16.26-83
Wrought Steel Buttwelding Short Radius Elbows and Returns	B16.28-78
Wrought Copper and Wrought Copper Alloy Solder Joint Drainage Fittings - DWV	B16.29-80

Air Conditioning and Refrigeration Institute (ARI)
1501 Wilson Blvd.
Suite 600
Arlington, Virginia 22209

Title	Standard Reference Number
Unitary Air-Conditioning Equipment	210-81

American Society of Heating, Refrigerating and Air Conditioning Engineers, Inc. (ASHRAE)
1791 Tullie Circle, N.E.
Atlanta, Georgia 30329

Title	Standard Reference Number
Safety Code for Mechanical Refrigeration	15-78
Number Designation of Refrigerants	34-78
Thermal Environmental Conditions for Human Occupancy	55-81
Energy Conservation in New Building Design	90A-80
Handbook, Fundamentals Volume	ASHRAE-85

American Society of Mechanical Engineers (ASME)
United Engineering Center
345 East 47th Street
New York, New York 10017

Title	Standard Reference Number
Boiler and Unfired Pressure Vessel Code, Section VIII, Division 1 & 2 Summer 83, Winter 83, Summer 84, Winter 84, Summer 85, Winter 85 and Summer 86 Addenda	ASME-83
Pipe Threads, General Purpose (inch)	B1.20.1-83
Malleable-Iron Threaded Fittings, Classes 150 and 300	B16.3-85
Cast Bronze Threaded Fittings, Class 125 and 250	B16.15-85

Figure 3.1

Reference Standards (continued)	
American Society for Testing and Materials (ASTM) 1916 Race Street Philadelphia, Pennsylvania 19103	
Title	**Standard Reference Number**
Pipe, Steel, Black and Hot Dipped Zinc Coated, Welded and Seamless	A53-86
Seamless Carbon Steel Pipe for High Temperature Service	A106-86
Pipe, Steel, Black and Hot-Dipped Zinc-Coated (Galvanized) Welded and Seamless, for Ordinary Uses	A120-84
Gray Iron Castings for Valves, Flanges and Pipe Fittings	A126-84
Copper Brazed Steel Tubing	A254-84
Gray Iron and Ductile Iron Pressure Pipe	A377-84
Piping Fittings of Wrought Carbon Steel and Alloy Steel for Low Temperature Service	A420-85A
Steel Sheet, Zinc-Coated (Galvanized) by the Hot-Dip Process — General Requirements	A525-86
Electric-Resistance-Welded Coiled Steel Tubing for Gas and Fuel Oil Lines	A539-85
Solder Metal	B32-83
Standard Sizes of Seamless Copper Pipe	B42-85
Standard Sizes of Seamless Red Brass Pipe	B43-86
Seamless Copper Tube	B75-86
Seamless Copper Water Tube	B88-86
Seamless Brass Tube	B135-86
Aluminum-Alloy Drain Seamless Tubes	B210-82A
Aluminum-Alloy Seamless Pipe and Seamless Extruded Tube	B241-83A
Wrought Seamless Copper and Copper-Alloy Tube	B251-86
Seamless Copper Tube for Air Conditioning and Refrigeration Field Service	B280-83
Threadless Copper Pipe	B302-85
Refractories for Incinerators and Boilers	C64-77
Ground Fire Clay as a Refractory Mortar for Laying Up Fireclay Brick	C105-81
Classification of Insulating Fire Brick	C155-84
Clay Flue Linings	C315-83
Hot Surface Performance of High Temperature Thermal Insulation — Test for	C411-82
Mineral Fiber Block and Board Thermal Insulation	C612-83
Flash Point by Tag Closed Tester — Test for	D56-82
Flash Point by Pensky — Martens Closed Tester — Method of Test	D93-85
Acrylonitrile-Butadiene-Styrene (ABS) Plastic Pipe Schedules 40 and 80	D1527-82
Poly (Vinyl Chloride) (PVC) Plastic Pipe, Schedules 40, 80 and 120	D1785-83
Solvent Cement for Acrylonitrile-Butadiene-Styrene (ABS) Plastic Pipe and Fittings	D2235-81
Poly (Vinyl Chloride) (PVC) Plastic Pipe (SDR-PR)	D2241-86
Acrylonitrile-Butadiene-Styrene (ABS) Plastic Pipe, (SDR-PR)	D2282-82
Poly (Vinyl Chloride) (PVC) Plastic Pipe Fittings, Schedule 40	D2466-78
Socket-Type Poly (Vinyl Chloride) (PVC) Plastic Pipe Fittings, Schedule 80	D2467-76A
Acrylonitrile-Butadiene-Styrene (ABS) Plastic Pipe Fittings, Schedule 40	D2468-80
Socket-Type Acrylonitrile-Butadiene-Styrene (ABS) Plastic Pipe Fittings, Schedule 80	D2469-76
Thermoplastic Gas Pressure Pipe, Tubing and Fittings	D2513-85A
Reinforced Epoxy Resin Gas Pressure Pipe and Fittings	D2517-81
Solvent Cements for Poly (Vinyl Chloride) (PVC) Plastic Pipe and Fittings	D2564-84
Chlorinated Poly (Vinyl Chloride) (CPVC) Plastic Hot- and Cold-Water Distribution Systems	D2846-82

Figure 3.1 (cont.)

Reference Standards (continued)

Title	Standard Reference Number
Polybutylene (PB) Plastic Hot-Water Distribution Systems	D3309-85B
Surface Burning Characteristics of Building Materials	E84-84
Behavior of Materials in a Vertical Tube Furnace @ 750° C. — Standard Test Method	E136-82
Safe Handling of Solvent Cements used for Joining Thermoplastic Pipe and Fittings — Recommended Practice	F402-80
Socket-Type Chlorinated Poly (Vinyl Chloride) (CPVC) Plastic Pipe Fittings, Schedule 40	F438-82
Socket-Type Chlorinated Poly (Vinyl Chloride) (CPVC) Plastic Pipe Fittings, Schedule 80	F439-82
Chlorinated Poly (Vinyl Chloride) (CPVC) Plastic Pipe, Schedules 40 and 80	F441-86
Chlorinated Poly (Vinyl Chloride) (CPVC) Plastic Pipe (SDR-PR)	F442-85
Solvent Cements for Chlorinated Poly (Vinyl Chloride) (CPVC) Plastic Pipe and Fittings	F493-85
Field Measurement of Soil Resistivity Using the Wenner Four-Electrode Method — Field Measurement	G57-84

American Welding Society (AWS)
P.O. Box 351040
Miami, Florida 33135

Title	Standard Reference Number
Brazing Filler Metal	A5.8-81

Building Officials and Code Administrators International (BOCA)
4051 West Flossmoor Road
Country Club Hills, Illinois 60477-5795

Title	Standard Reference Number
National Mechanical Code	NMC-87
National Building Code	NBC-87
National Fire Prevention Code	NFPC-87
National Plumbing Code	NPC-87

United States Department of Energy (DOE)
Washington, D.C. 20545

Title	Standard Reference Number
Furnaces and Boilers — U.S. Department of Energy Test Procedures	Federal Register Volume 49, FR12148, No. 61-84

Federal Specifications General Service Administration (FS)
7th and D Streets
Specifications Section
Room 6039
Washington, D.C. 20407

Title	Standard Reference Number
Pipe, Bends, Traps, Caps and Plugs (for Industrial Pressure and Soil and Waste Applications)	WW-P-325B-76

Figure 3.1 (cont.)

Reference Standards (continued)

International Institute of Ammonia Refrigeration (IIAR)
111 E. Wacker Drive
Chicago, IL 60601

Title	Standard Reference Number
Equipment, Design and Installation of Ammonia Mechanical Refrigeration Systems	2-84

Manufacturers Standardization Society of the Valve and Fittings Industry, Inc. (MSS)
127 Park Street, N.E.
Vienna, Virginia 22180

Title	Standard Reference Number
Pipe Hangers and Supports — Selection and Application	SP-69-83

National Association of Corrosion Engineers (NACE)
P.O. Box 218340
Houston, TX 77218

Title	Standard Reference Number
Control of External Corrosion of Underground or Submerged Metallic Piping Systems	RP-01-69-83

National Fire Protection Association (NFPA)
Batterymarch Park
Quincy, Massachusetts 02269

Title	Standard Reference Number
Flammable and Combustible Liquids Code	30-84
Automotive and Marine Service Station Code	30A-84
Storage and Handling of Cellulose Nitrate Motion Picture Film	40-82
Storage and Handling of Liquefied Petroleum Gases	58-83
Explosion Venting Guide	68-78
National Electrical Code	70-84
Repair Garages — Standard	88B-85

Sheet Metal and Air Conditioning Contractors National Association, Inc. (SMACNA)
8224 Old Courthouse Road
Vienna, Virginia 22180

Title	Standard Reference Number
Fibrous Glass Ducts — Construction Standards	SMACNA-79
Metallic Ducts — HVAC Duct Construction Standards	SMACNA-85

Figure 3.1 (cont.)

Reference Standards (continued)

Underwriters Laboratories, Inc. (UL)
333 Pfingsten Road
Northbrook, Illinois 60062

Title	Standard Reference Number
Flammability of Plastic Materials for Parts in Devices and Appliances	94-85
Chimneys, Factory-Built, Residential Type and Building Heating Appliance	103-86
Factory-Built Fireplaces	127-86
Pressure Regulating Valves for LP Gas	144-85
Factory Made Air Ducts and Connectors	181-81
Gas Vents	441-86
Dampers, Fire and Ceiling Dampers — 1986 Supplement	555-79
Type L Low-Temperature Venting Systems	641-86
Grease Extractors for Exhaust Ducts	710-83
Fireplace Stoves	737-86
Fire and Smoke Characteristics of Electrical and Optical-Fiber Cables Used in Air-Handling Spaces — Test Method	910-85
Medium Heat Appliance, Factory Built Chimneys	959-86
Room Heaters, Solid Fuel Type	1482-86

Office of Technical Information Bureau of Mines (USBM)
Department of the Interior
2401 E. Street, N.W.
Washington, D.C. 20241

Title	Standard Reference Number
Ringelmann Smoke Chart	USBM Circular No. 8333 May, 1967

Adapted from the Building Officials and Code Administrators International (BOCA) *National Mechanical Code 1987.*

Figure 3.1 (cont.)

CHAPTER FOUR

HOW TO SELECT A SYSTEM

The three previous chapters were designed to familiarize the reader with the principles behind heating and cooling, how to calculate loads, and the codes and regulations that should be followed in designing HVAC systems. This chapter is an introduction to system selection and contains guidelines for using the information presented in the remaining chapters of the book.

Selection Criteria

Certain criteria must be determined for every building in order to select the appropriate HVAC system. Many of these criteria are established by the owner. They influence what components the building will have and the range and quality of the selected components. Basic system selection criteria include building life, cost, technical feasibility, overall fit, and whether the building is new or is being remodeled. Each of these criteria is described in the following sections.

Building Life

The anticipated life of a building is an important factor to consider when designing an HVAC system. Commercial office buildings, for example, have a financial life of ten to twenty years. Many developers typically sell a property within seven years of building it. Institutional buildings are often expected to have a life span greater than fifty years. Hospitals have certain components that undergo upgrading every three to ten years. Clearly, the effort and cost for a short-lived project are different than those for a building that should last over fifty years. The useful life of the equipment in the building should, accordingly, be selected to match the anticipated building life span.

Cost

The costs for a project can be evaluated in two ways. First, the cost of *initial construction* must be considered. This is the cost to complete the project, including land, legal, design, marketing, and construction costs. Most projects are approved or not approved on the basis of the anticipated initial cost. This is also the easier cost to establish. The *life cycle cost* of the facility is not so easy to anticipate. This is the total cost of a building over its entire useful life. This is a more accurate, but more complex, financial analysis. Projects are not paid for all at once. In reality, money is borrowed for a project's development costs and paid off over the term of the note. In addition, there are operating and maintenance costs each year that are added to the annual principal and interest charges along with all other expenses, such as taxes, replacement, and broker fees. If the overall life cycle costs are less than the income the building is expected to produce, the project is considered profitable. A careful life cycle analysis will typically recommend more efficient, but more expensive, equipment than an initial cost analysis, because the "bottom line" is that over each year of the building's life, the maintenance and operation of such equipment is less costly than the initially less expensive items.

Technical Feasibility

In selecting a system, it is useful to determine as early as possible any technical limitations. Is gas available? Can the building have a basement? Will the structure hold the added weight of the intended equipment? Do local environmental considerations prohibit certain types of equipment? Is the existing plant large enough for the new load?

Many initial choices are eliminated because they are not possible or permitted. If these limitations can be identified early, there will be more time to design for the appropriate choices.

Overall Fit

An HVAC system must be designed to fit into the building. The actual working out of a system that threads its way through a building takes a considerable amount of time and is often one of the major challenges a designer faces. Some items that should be determined early in a project include:

- Space requirements for mechanical equipment, including the size and availability of a mechanical equipment room.
- Location and sizes for chimney and shafts.
- Routes and space requirements for pipes and ducts.
- Places where mechanical systems cross other systems and may require larger chases or lower ceilings.

Close cooperation between design team members is essential to produce an efficiently designed building.

Renovation Projects

The charts and recommendations in this book can be used as a starting point for new construction. In existing buildings, however, the generation equipment is replaced after fifteen to forty years of use. For existing buildings, the decisions regarding new equipment are guided by the following additional considerations:

- **Energy audit recommendations:** An energy audit is usually performed by a professional engineer and results in recommended changes based on an evaluation of existing systems. The cost of replacement is weighed against the annual savings it will bring. Pay-backs under three years are usually implemented.

- **Condition of existing equipment:** Worn out or inefficient equipment is often replaced as part of a capital improvement program. Savings in repairs and tax benefits are part of the economic benefits listed.
- **Operation and maintenance cost of existing equipment:** Equipment that is costly to run or repair may be completely replaced.
- **Cost and disruption of any replacement:** When replacing equipment, it is important to maintain continuity of service. This may require off-hour premium time, which can increase the replacement cost.
- **Increased income generated by the improvements:** New equipment, however expensive, may open the building to new, higher paying markets. Rent projections may show a positive cash flow for improvements.
- **Feasibility:** The overall ability to implement the work must be verified. Considerations such as the capacity of the roof to support a cooling tower and access to adequate electrical power are crucial to any plan.

Each building type has a typical set of general requirements for common heating and cooling systems. For most building types, there is more than one choice for a suitable HVAC system. Selecting an HVAC generating system follows a logical decision-making process, illustrated in Figure 4.1.

There are four basic selections to make in order to establish the type of HVAC system:

1. Select the type of system (heating and/or cooling).
2. Select the fuel.
3. Select the distribution system and the terminal units.
4. Select the generation equipment and accessories.

Once the above steps are completed, the basic characteristics of the entire system can be established.

The following discussion applies to both new and renovated systems. However, when a worn out or inefficient piece of equipment is to be replaced, the new equipment must be compatible with the systems that remain.

The four basic selections are discussed in the following sections. Each section demonstrates how each selection is made. Following the discussion, three separate examples—one for heating only, one for cooling only, and one combined system—are presented. Once the basic characteristics of a system are selected, the remaining portions of the book should be used to size each separate part of the entire system, including generation, distribution, terminal units, controls, and accessories.

Figure 4.2 lists building types, along with comments on the typical types of heating and cooling systems available. This chart also notes whether the heating and cooling systems are separate or combined.

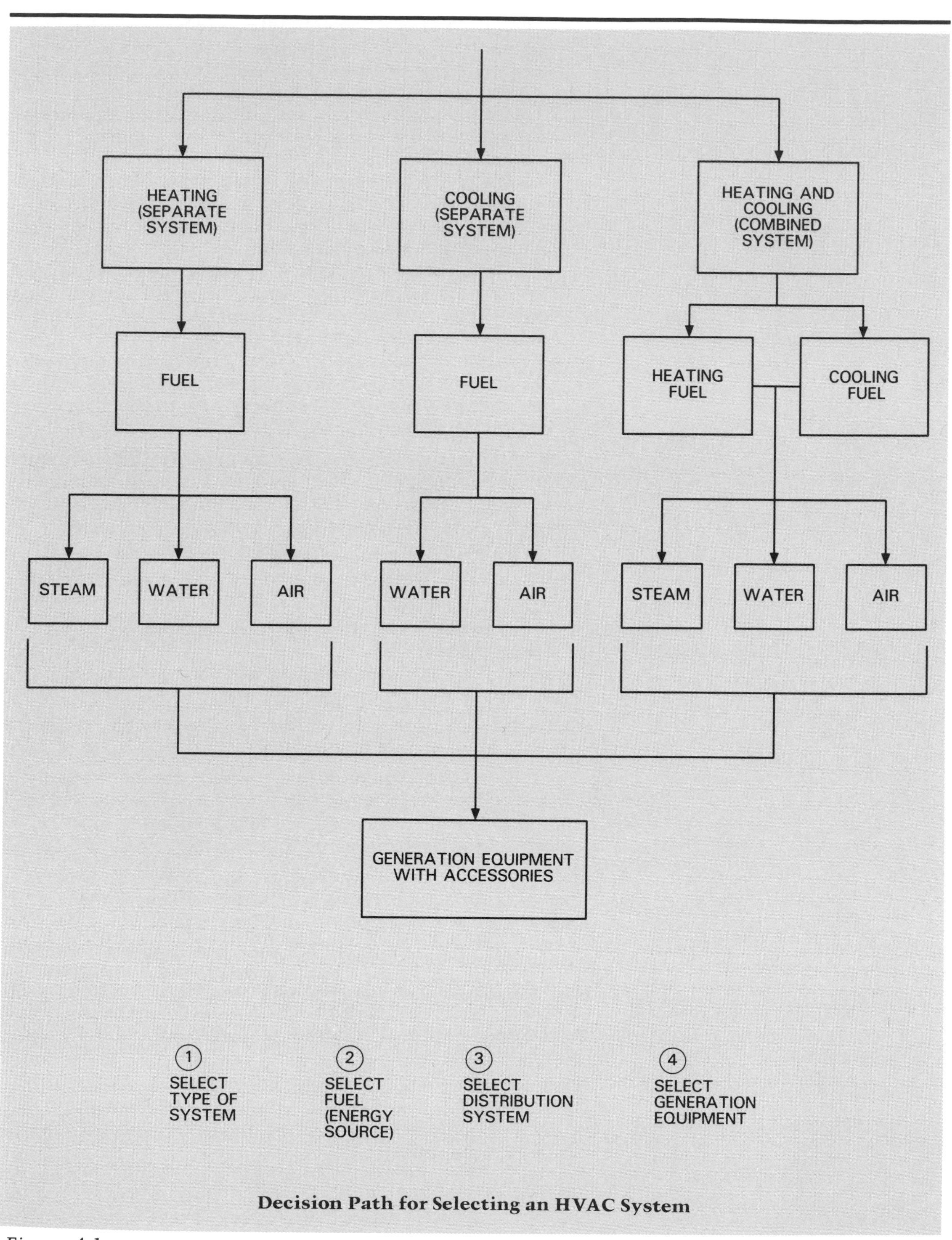

Decision Path for Selecting an HVAC System

Figure 4.1

Types of Heating and Cooling Systems for Buildings

Building Type	Heating — Separate (Fig. 1.6)	Cooling — Separate (Fig. 1.11)	Heating and Cooling Combined (Fig. 1.23)	Remarks
Assembly				
Movie Theaters, Cinemas	—	—	All air central system	Large make-up air consideration. Check local codes.
Nightclubs, Restaurants	—	—	All air central system	Smoke and humidity removal requires large make-up air consideration. Negative pressure in kitchens for odor removal.
Libraries, Museums	—	—	All air central system	Humidity control.
Churches	Perimeter radiation	DX	All air central system	—
Schools	Thru-the-wall fan coils Perimeter radiation	—	—	Local codes regulate outdoor air make-up.
Swimming Pools	—	—	All air central system	Humidity control.
Business				
Offices	Perimeter radiation	Central air	Package-Multizone	—
Bank	Perimeter radiation	—	All air central system	—
Courthouse	Perimeter radiation	Local units	All air central system	—
Multi-story	Perimeter-hot water	—	All air central system	—
Lobby, Hallways	—	—	Separate units	Separate hours.
Typical Office Floor	Reheat coils, electric or hot water	Fan coils, VAV units	Central air with economizer, heat pumps	VAV, induction, dual duct, fan coil.
Computer Rooms	—	Tenant supplied unit	—	Tie to cooling tower.
Lobby	—	—	—	Set 5°-8° buffer.
Commercial/ Retail				
Department Store	Reheat coils as required	—	All air central system	—
One-story Stores	Supplemental radiation at perimeter	—	Rooftop HVAC unit	—
Supermarkets	—	—	Rooftop HVAC unit	Reclaim heat from compressors.
Factory	Electric, gas-fired space heaters or hydronic	Special conditions only	All air central system	—
Hazardous	Hot water	Chilled water	Hydronic	Non-sparking motors and controls.
Institutional				
Hospitals	Steam-combined with hot water	Central	All air-dual duct	Special ventilation standards.
Penal	Hydronic, concealed or remote	—	—	Special security standards.

Figure 4.2

Selecting the Type of System

The decision process shown in Figure 4.1 illustrates the basic choices to be made when selecting an HVAC system. Heating is mandatory in most buildings in temperate climates for reasons of human health and comfort as well as for public safety (for example, to prevent sprinkler systems from freezing). The decision to be made is whether or not the building requires cooling. Cooling is generally optional. However, most new and renovated buildings are expected to provide cooling.

Figure 4.2 lists common HVAC systems for particular building types. For example, for a mid-rise middle income residential building, a combined HVAC rooftop unit is listed as the common system. In addition to the recommendations provided in this chart, current practice in the area for similar buildings should be considered. Another factor to be considered in choosing the HVAC system is the effect that the type of cooling system chosen will have on the overall marketing of the property.

Types of Heating and Cooling Systems for Buildings (continued)

Building Type	Heating — Separate (Fig. 1.6)	Cooling — Separate (Fig. 1.11)	Heating and Cooling Combined (Fig. 1.23)	Remarks
Residential				
1 & 2 Family	Warm air system or hydronic with radiators	Split system or window units	Heat pump (air-to-air) requires booster or with electric reheat	Operable windows and bathroom exhaust fan.
Multi-family Residences and Elderly Housing	Electric or hydronic	Window units	Heat pump (water-to-air) fan coil units	Fresh air to corridors — bathroom and kitchen exhaust.
Hotels	—	—	Hydronic or electric fan coil units	Same as above.
Storage Warehouses	Electric unit, gas or hydronic unit heaters	Walk-in coolers Built-up direct expansion systems	— —	Check requirements for material stored.
Existing Buildings Renovation	Note the type of system presently in the building and compare with the above table. Review the recommendations in the book. See discussion on installing new systems versus renovation. When the renovated use is different from the existing system, check for overall feasibility.			

Figure 4.2 (cont.)

Based on the above factors, select from among the following types of HVAC systems:

- separate heating,
- separate cooling, and
- combined heating and cooling.

The possible systems for each of these choices are illustrated on the heating ladder (Figure 1.6), the cooling ladder (Figure 1.11), and the air conditioning ladder (Figure 1.23), presented in Chapter 1.

Selecting the Type of Fuel

Before selecting any HVAC equipment, the type of fuel must be chosen. The fuel for heating systems is selected independent of cooling systems and is usually chosen based on economy and availability. Figure 4.3 lists the basic properties and costs of common fuels.

The basic fuels are natural gas, propane gas, oil, steam, electricity, solar energy, and coal. The following sections contain comments on the use of each type.

Electricity

Electricity is universal to all buildings because it is necessary for lighting and power. It is the most commonly used fuel for air

Basic Properties and Costs of Fuels

Fuel Type	Unit	Heating Value BTU/Unit	Cost per Unit (cents)*	Overall System Efficiency	Net Cost per MBH* (cents)	Remarks
Electricity	Kilowatt	3,413	12–15	100	3.5–4.4	
Steam	Pounds (at atmospheric pressure)	1,000	0.8–1.3	60–75	1.1–2.1	
Oil #2	Gallon	138,500	70–130	60–88	0.6–1.5	
Oil #4	Gallon	145,000	70–130	60–88	0.7–1.5	
Oil #6	Gallon	152,000	60–120	60–80	—	Preheat
Natural Gas	CCF (100 C.F.)**	103,000	80–100	65–92	.84–1.49	
Propane	Gallon	95,500	80–110	65–90	—	
Coal	Pound	13,000	7–12	45–75	.71–2.05	
Solar	S.F. of collector — varies with location and collector					

*Northeastern U.S., 1987
**Note: 1 therm = 1.013 CCF

Figure 4.3

conditioning since the majority of refrigeration is achieved with compressors, which require electricity to operate. When compared with other fuels, electricity has a high cost and alternatives are generally sought. There are, however, some particular advantages to using electricity that often outweigh its high cost per Btu.

- Electrical systems can be individually metered, making them preferable for multi-tenant buildings, such as residential condominiums.
- Electrical systems are free of freezing problems and do not leak. As long as plumbing lines are heat-traced electrically, the heating system can be completely turned off, a particular advantage for spaces that may be unoccupied for long periods of time. (Air systems have the similar advantage of being freezeproof and not leaking.)
- Electrical systems can be regulated by individual room control thermostats. This means that unused rooms can be turned down or off, thus compensating for higher fuel costs.
- Electrical systems do not need chimneys, air intake louvers, or pumps. They have few (if any) parts that wear out and are generally less expensive to install and maintain.
- The efficiency of electric heat is 100 percent. All the consumed electrical energy is delivered to the space as heat.

Steam

When steam is available from the utility in the street or from a central plant, it has the particular advantage of not requiring a boiler or chimney. Pressure-reducing stations are used to reduce the high-pressure steam, as required, to low or medium pressure. Heat exchangers are utilized in producing hot water from the utility steam.

Oil

Fuel oil is one of the most commonly used heating fuels. Fuel oil is graded from #1, highly refined, to #6, crude. The refined oils are more expensive and have a lower heat content, but are chosen because they flow at room temperatures. #2 oil is commonly used in heating applications with small burners; #4 oil is commonly used for larger burners. #5 and #6 are nearly solid at room temperature and must be preheated in order to get them to flow.

Oil is normally stored in tanks, which may be exposed or buried in the ground. Current laws may require containment and/or leak detection systems. Tanks may be made of fiberglass or steel, single or double wall. Transfer pumps that supply oil to the burner reservoir or day tank are usually required on long runs.

Gas

There are several gases available for heating. They include natural gas, propane (often called bottled gas), butane, methane, hydrogen, and carbon monoxide.

Natural gas and propane are the most common types of fuel used. When available from the street, natural gas has the advantage of not requiring a storage tank. Natural gas is manufactured from coke or oil and obtained directly from the earth. Propane is most often used in rural areas where natural gas lines are not available. Gas can also be used for cooking and is often selected for this reason.

Coal

Coal is a solid fuel and requires special consideration to feed the boiler (stoker) and remove the ash. It is no longer practical to perform these procedures manually; larger coal furnaces and boilers have been developed with automatic stoking and ash removal. Nevertheless, the practical considerations of delivery, storage, and handling of solid fuel, in addition to its sulphur content (which has been associated with environmental problems), have restricted its use to large industrial or campus plants.

Coal is classified as either anthracite or bituminous—hard or soft—in a series of ratings with a relatively even heat value (12,000 – 14,000 Btu/lb.).

Solar

Solar energy can be used to generate domestic hot water, warm air, and hot water for absorption chillers. The costs vary with each application. Among other issues, the roof structure must be checked to ensure that it can support the solar panels for both gravity and uplift forces.

Selection: The selection of fuel should be based not only on the cost per MBH, but also on any special installation costs or fuel conversion charges. Oil and propane, for example, require fuel tanks, while steam, electricity, and natural gas involve utility connection charges.

In urban areas, steam is a preferred heating fuel, because it brings no combustible fuel into the building—a great asset from the viewpoint of the fire department and insurance companies—and it requires no boiler or chimney. Therefore, steam systems occupy less of a building's valuable floor space—an asset to owners and developers. Steam heat is almost always used when it is available in urban high-rise buildings.

With the exception of steam, gas, or hot water-fed absorption chillers, virtually all cooling equipment runs by electricity.

In existing buildings, the practicality of continuing with the fuel currently used should be examined. In many cases, conversion to a different fuel is more economical despite the costs of demolition and installation of a new system.

Selecting the Distribution System

For a complete description of distribution systems, refer to Chapter 9, "Distribution and Driving Systems." The following is a brief discussion on the distribution system selection process for renovation projects.

Renovation Projects

It is common to retain and re-use existing distribution systems when the same type of system (separate, or combined heating and/or cooling) will be used. Other considerations are whether the existing distribution system is in good condition, and if it can serve the same overall layout or use after renovation. When it is practical to retain the existing distribution system, the result is often considerable cost savings, when compared with the cost of demolition and installation of a new system.

When the type of system or use of the renovated building will change, the general guidelines for new construction are followed to select the type of distribution system.

When the condition of the distribution system is poor, it should be replaced with a new system. Repairs to existing systems—particularly older steam or water systems—can be very difficult and often never-ending.

Figure 4.4 illustrates the distribution system selection process. This chart can be used to determine the appropriate type of distribution system for the building.

Selecting the Generation Equipment

Once the choices for the type of overall building system, the type of fuel, and the type of distribution system have been made, the type of generation equipment and its accessories can be determined. Figure 4.5 shows portions of the heating ladder, the cooling ladder, and the air conditioning ladder, which can be used to determine the type of generating equipment required. (See Figures 1.6, 1.11, and 1.23 in Chapter 1 for more detail on the heating ladder, the cooling ladder, and the air conditioning ladder, respectively.)

Preliminary Selection of Distribution Systems

Type of Building System	Type of Distribution System Conditions	Select
Separate Heating	Individual room control and/or tenant billing required	Electric
	Steam heating and other steam uses—industrial	Steam
	Continuous use buildings	Hot water
	Intermittent use with freezing potential	Warm air
Separate Cooling	One- and two-story buildings	Air
Heating and Cooling	High-rise office — fan coil units	Water
	— induction units/central air	Air and reheat
	— Variable air volume	Air and reheat

Figure 4.4

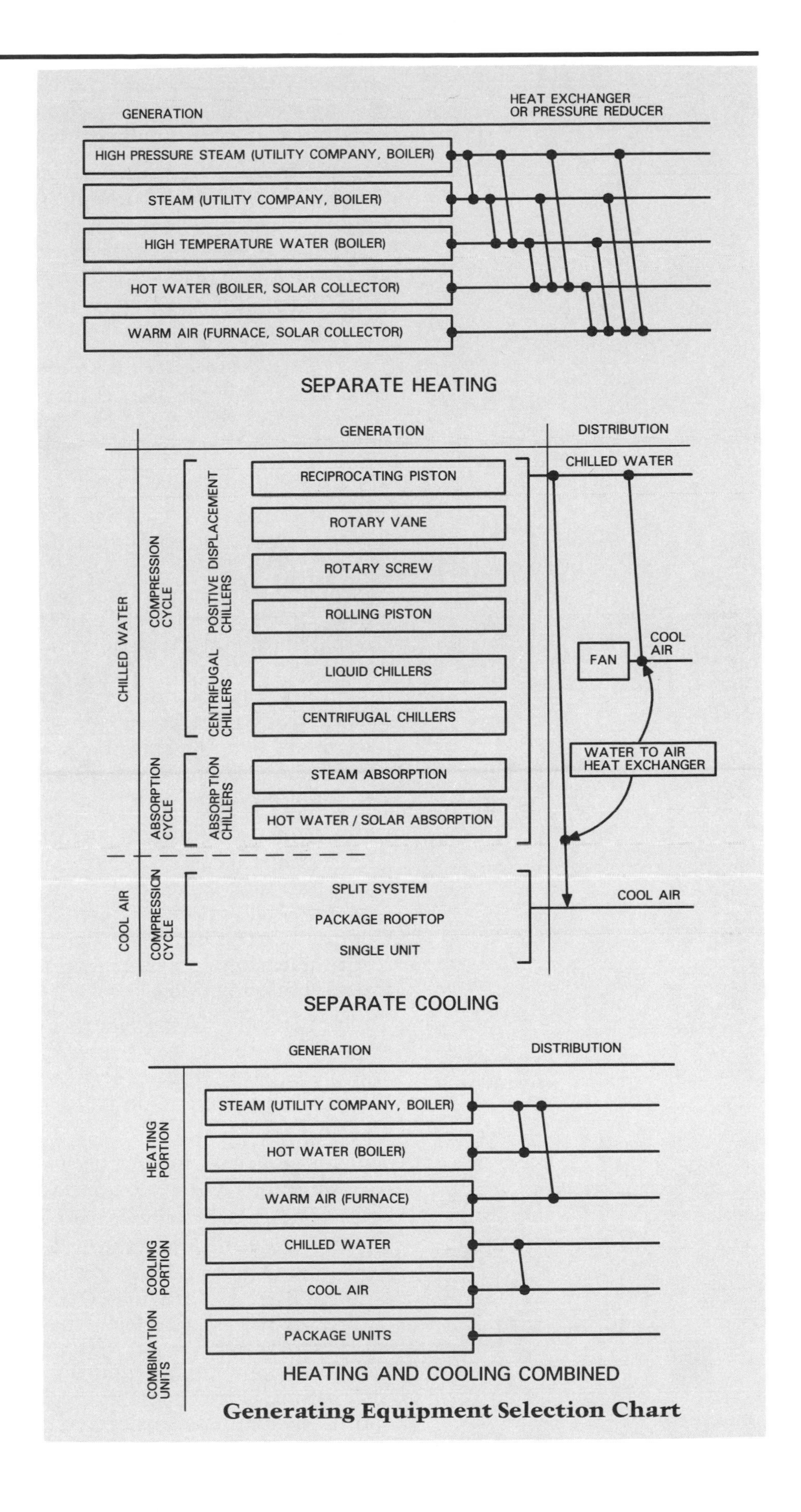

Generating Equipment Selection Chart

Figure 4.5

Equipment Selection Examples

The following examples illustrate the selection process for separate heating, separate cooling, and combined heating and cooling systems. The principles previously explained are illustrated in the following three examples.

Selection of a Separate Heating System

For this example, the building is a mid-rise housing complex in a suburban area. The heat load for the complex is 1500 MBH. According to the selection process shown in Figure 4.1, decisions are made in the following order.

Type of System: From Figure 4.2, the residential building, select a separate heating system.

Type of Fuel: From Figure 4.3, #2 oil is selected. In a suburban area, steam and natural gas may not be available. Electricity, depending on cost, could have been selected, especially if individual heat is to be metered and paid by each tenant.

Type of Distribution System: From Figure 4.4, select hot water.

Type of Generating Equipment (with accessories): The main heating generating equipment is an oil-fired hot water boiler (#2 oil). (The Means line number for this equipment is 155-120. Equipment data sheets for generation equipment appear at the end of Chapter 5.) The path of this example is traced on Figure 4.6, part of the heating ladder. Some designers recommend installing two boilers in each building. Because the maximum design load occurs only a few days each year, sizing one boiler for one-third and the second boiler for two-thirds of the load maximizes overall efficiency, cuts down on wear and tear, and provides a backup for heating when one boiler is not operative. For simplicity, one boiler is used for this example.

Selection of a Separate Cooling System

The building used for this example is an existing four-story office building. It has an existing centralized steam heating system using perimeter convectors/radiators in good condition. Cooling is to be provided. The cooling load is 40 tons. According to the selection process shown in Figure 4.1, decisions are made in the following order.

Type of System: Using Figure 4.2, check the "existing buildings" entry. Because the new use will be the same as the old use and the existing heating system is in good condition, leave the heating alone and design a separate "stand-alone" cooling system.

Type of Fuel: From Figure 1.11, it can be seen that electricity is the common fuel for cooling except where cheap steam, gas, or solar energy is available to run large absorption machines.

Type of Distribution System: Because radiation already lines the perimeter of the building, cooling is best provided from the ceiling. Since existing (older) buildings generally have high ceilings, an air system is the first choice because it will not interfere with the steam system; it will fit into the high ceiling; and it will provide distribution in the ceiling, the ideal location for cooling. Since the heating system is centralized, a centralized cooling system is also selected. (A different choice would be made if each floor or tenant were paying individually for

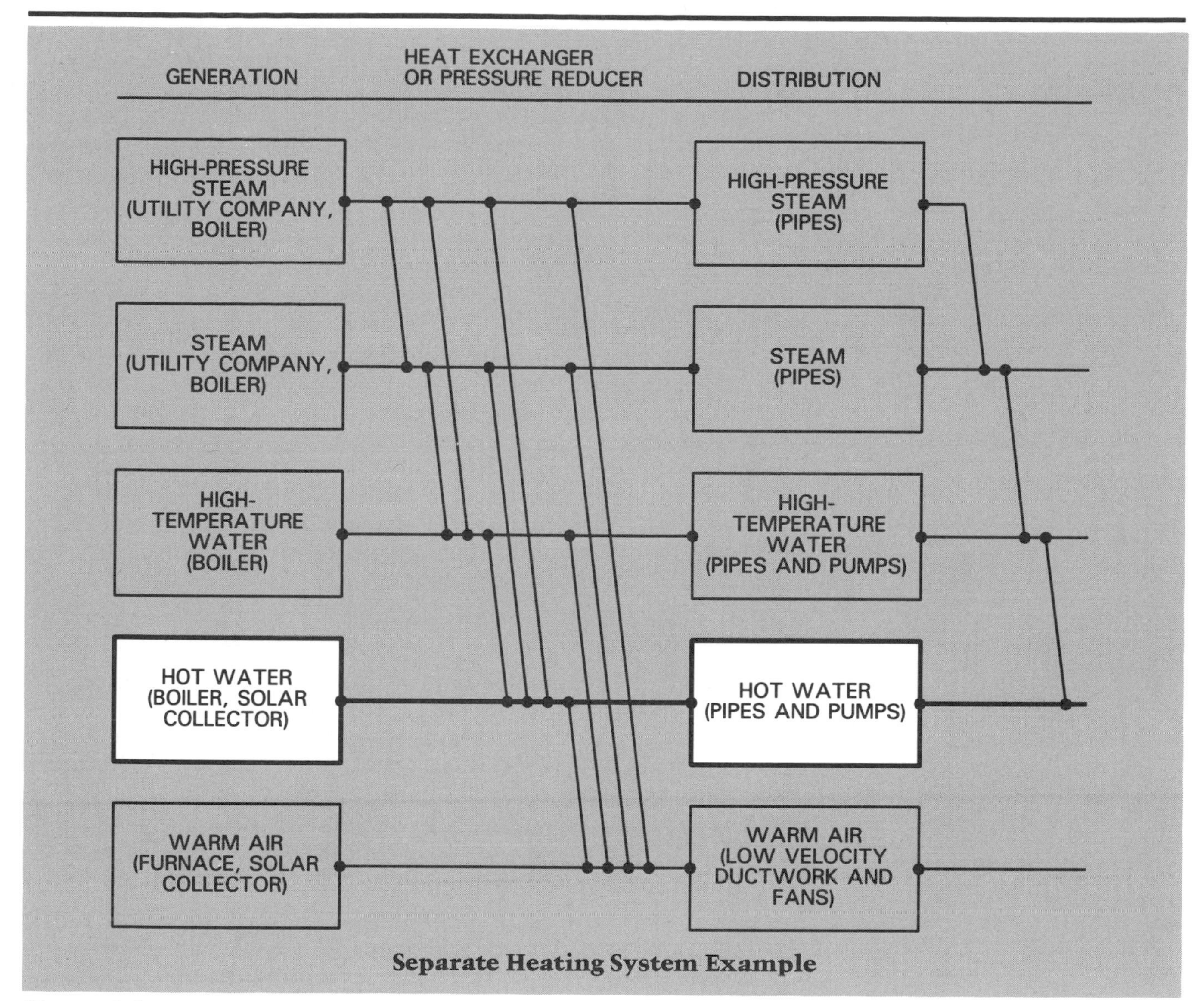

Separate Heating System Example

Figure 4.6

electricity.) A central system requires a piece of equipment to reject the heat from the building to the outside. This device might be an air cooled condenser, a cooling tower, an evaporative condenser, or a rooftop unitary piece of equipment. The choice of this equipment should be reviewed by a structural engineer to make sure that the weight of the equipment can be supported by the particular type of roof framing.

Type of Generating Equipment (with accessories): The main cooling generating equipment is an electric package unit mounted on the roof and zoned to each tenant. Figure 4.7, part of the cooling ladder, traces the selection path for this example.

Selection of a Combined Heating and Cooling System

In this example, the building is a 15-story office building with 8000 square feet per floor. It includes a ground floor retail area and is located in a downtown section of the city. Hook-ups to all major services and utilities are available, including steam. The heating design load is 2,500 MBH and the cooling load is 300 tons. Cooling is required year round for the interior of the building. According to the selection process shown in Figure 4.1, decisions are made in the following order.

Type of System: From Figure 4.2, select a combined heating and cooling system.

Type of Fuel: For combined systems, the heating and cooling fuels are selected independently. Steam is selected as the heat source because it is available; reasonably competitive (see Figure 4.3); generates savings; and because boilers, tanks, and chimneys are not necessary.

Electricity is selected as the energy source for cooling because the building has sufficient electrical capacity available. (An absorption unit is also worthy of consideration in this case.)

Type of Distribution System: The possible mediums for the distribution system are air and water. The basic choices are a centralized plant that services the entire building or a decentralized system of smaller units, down to one or more for each tenant. Usually, individual tenant systems may be separately metered.

A building's HVAC system is selected after many meetings with the owner, developer, architect, and HVAC engineer. For this building, it is assumed that after several trial runs and cost estimates, a centralized system is selected. For other projects in different cities with varying costs, building designs, marketing strategies, and tenant requirements, the choices will be different. To make the best possible choice, it is necessary to combine experience with trial selections of full systems. For each trial selected, compare the advantages and disadvantages, including the cost of the mechanical system. The impact of these choices on the overall building should also be considered.

With the centralized system selected, the possible mediums for heating and cooling can be analyzed. The layout of centralized distribution systems is shown in Figure 4.8.

Cooling the building interior is required year round, because the heat gain from lights, people, and equipment contributes to an

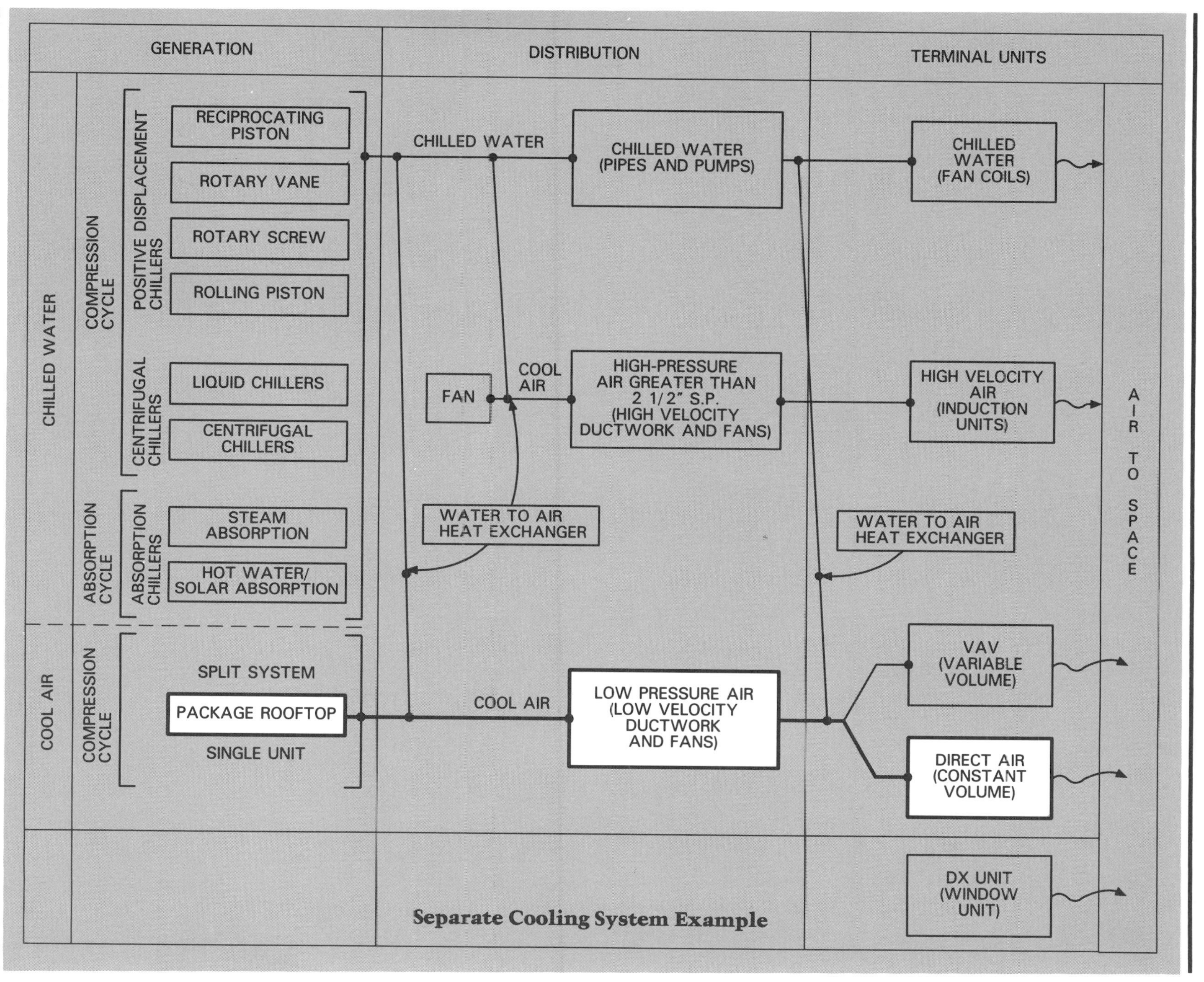

Separate Cooling System Example

Figure 4.7

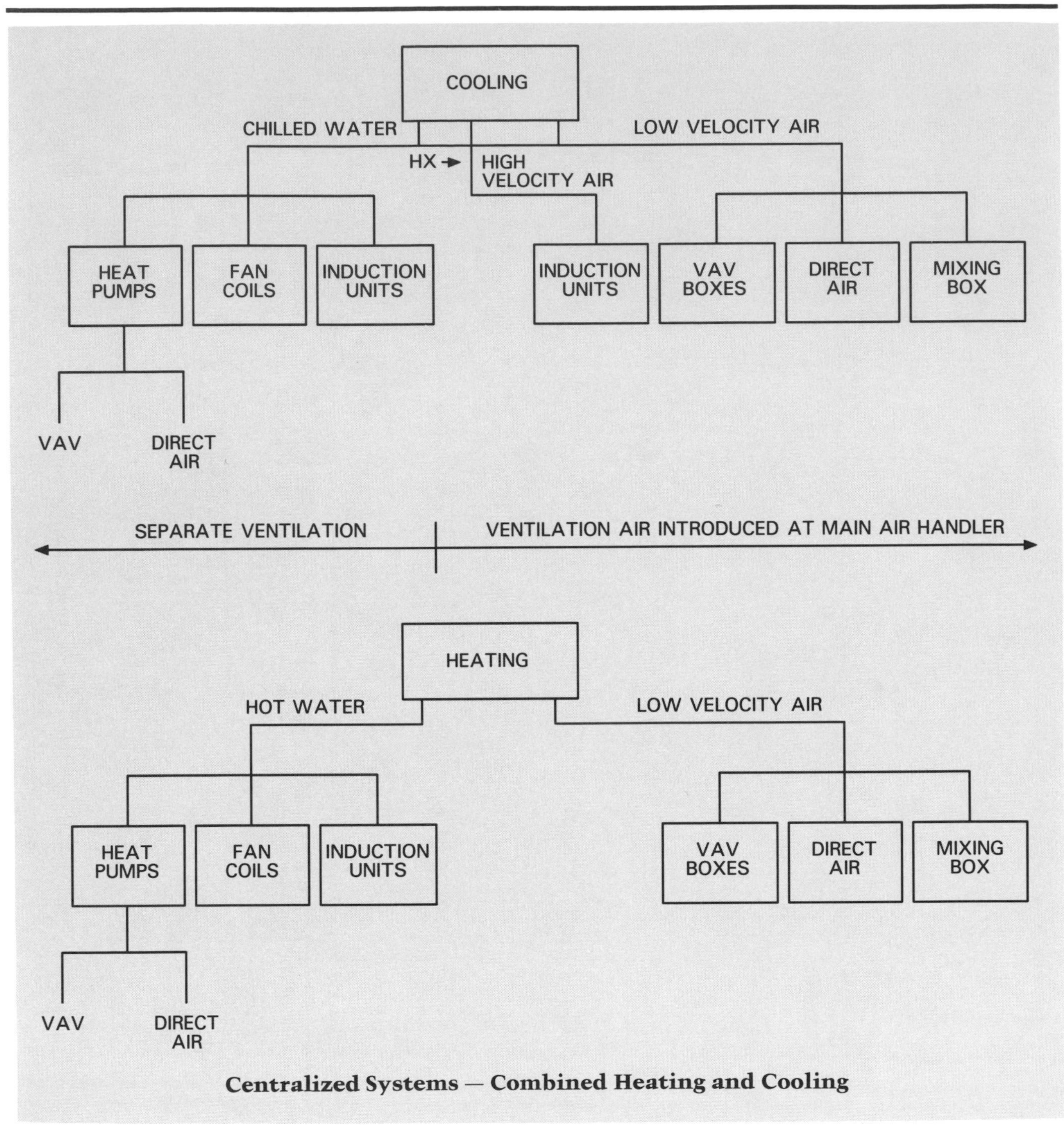

Centralized Systems — Combined Heating and Cooling

Figure 4.8

excess temperature that must be removed from the interior, even in winter. By contrast, because the perimeter of a typical high-rise building loses heat through the window wall, heating of the window wall surfaces is required in winter, even though the interior zone is usually always cooled. The need to simultaneously heat and cool office buildings in winter and cool the interior during the cooler spring and fall months has resulted in the development of two major energy conservation systems: free cooling and heat pumps (see Chapter 5, "Generation Equipment").

Now the distribution system is chosen. It is assumed that there is an available central core in the interior for supply and return air, and that the owner wishes to be responsible for and maintain as few pieces of equipment as possible. Based on the chart shown in Figure 4.8, an economizer cycle, low-velocity air system to variable air volume (VAV) boxes is selected for cooling the interior. This system will provide ventilation air and individual control for interior spaces.

Heating and cooling the perimeter can be achieved in several ways. Fan coils are selected for this example, because they do not occupy any more floor or ceiling space for air ducts; they can be placed under windows, an ideal location for heat and condensation control; they can be individually controlled; and they can accommodate the different volumes of air required for heating and cooling.

Generation Equipment: The primary generating equipment for this combined system is as follows:

- **Cooling:** an electric 300-ton chiller and cooling tower tied to an air handler with economizer and perimeter fan coils.
- **Heating:** utility steam to pressure-reducing station, heat exchangers for hot water distribution to perimeter fan coils, and freeze protection coils for air handler.

Figure 4.9 illustrates the path this system follows on the cooling ladder, and Figure 4.10, the heating ladder, for this combined heating and cooling system.

Summary

By using the basic outline presented, the overall system can be selected. Equipment data sheets are provided at the end of the chapters on generation, distribution, terminal units, controls, and accessory equipment to further specify individual pieces of equipment. Each piece of equipment is identified by a Means line number in order to determine the cost in the current annual edition of *Means Mechanical Cost Data*. Where more than one choice is possible, a separate analysis can be performed for each, and the advantages weighed against the disadvantages.

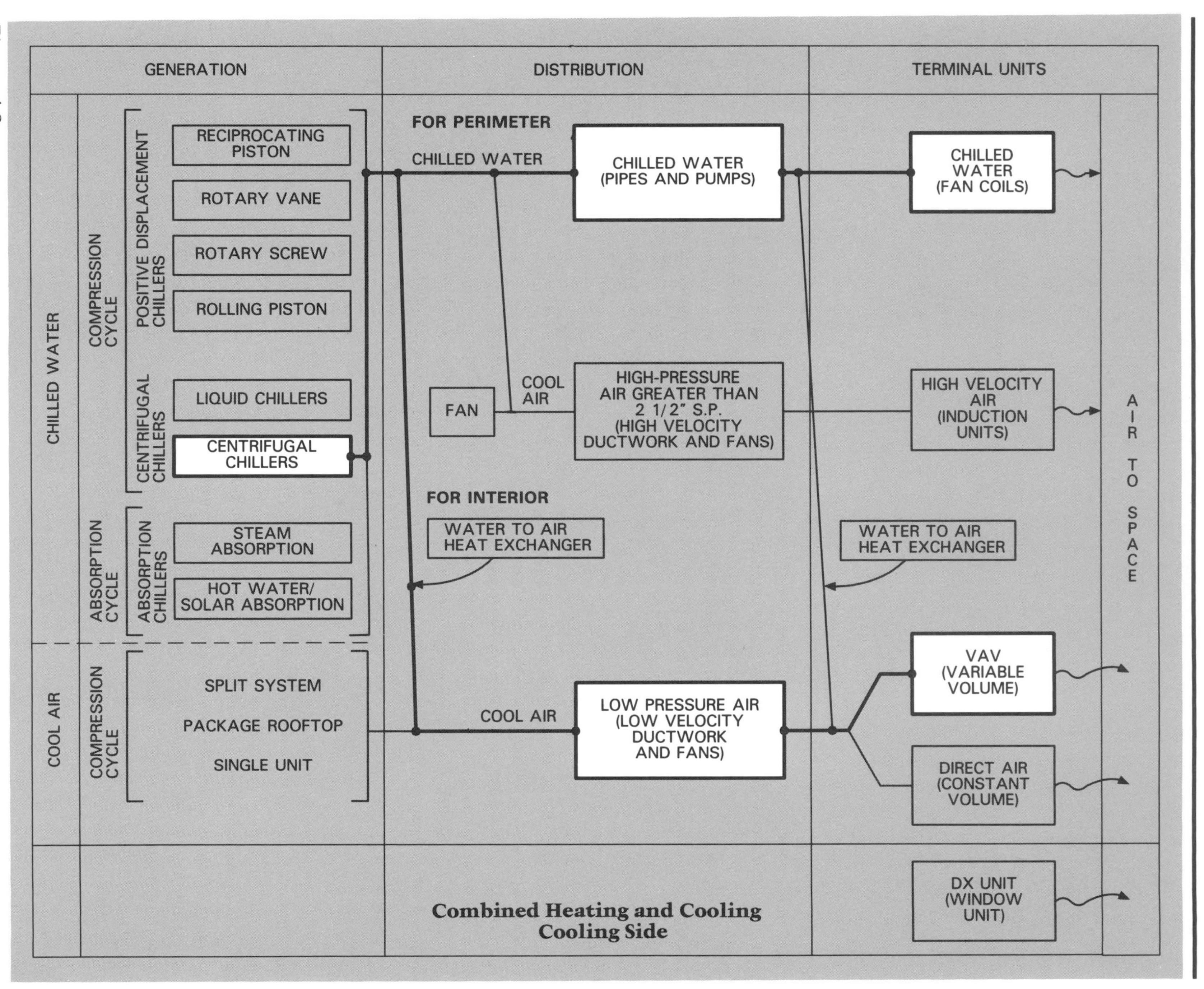

GENERATION
DISTRIBUTION
TERMINAL UNITS
CHILLED WATER
COMPRESSION CYCLE
POSITIVE DISPLACEMENT CHILLERS
RECIPROCATING PISTON
ROTARY VANE
ROTARY SCREW
ROLLING PISTON
CENTRIFUGAL CHILLERS
LIQUID CHILLERS
CENTRIFUGAL CHILLERS
ABSORPTION CYCLE
ABSORPTION CHILLERS
STEAM ABSORPTION
HOT WATER/ SOLAR ABSORPTION
COOL AIR
COMPRESSION CYCLE
SPLIT SYSTEM
PACKAGE ROOFTOP
SINGLE UNIT
FOR PERIMETER
CHILLED WATER
CHILLED WATER (PIPES AND PUMPS)
FAN
COOL AIR
HIGH-PRESSURE AIR GREATER THAN 2 1/2" S.P. (HIGH VELOCITY DUCTWORK AND FANS)
FOR INTERIOR
WATER TO AIR HEAT EXCHANGER
COOL AIR
LOW PRESSURE AIR (LOW VELOCITY DUCTWORK AND FANS)
CHILLED WATER (FAN COILS)
HIGH VELOCITY AIR (INDUCTION UNITS)
WATER TO AIR HEAT EXCHANGER
VAV (VARIABLE VOLUME)
DIRECT AIR (CONSTANT VOLUME)
DX UNIT (WINDOW UNIT)
AIR TO SPACE
Combined Heating and Cooling
Cooling Side

Figure 4.9

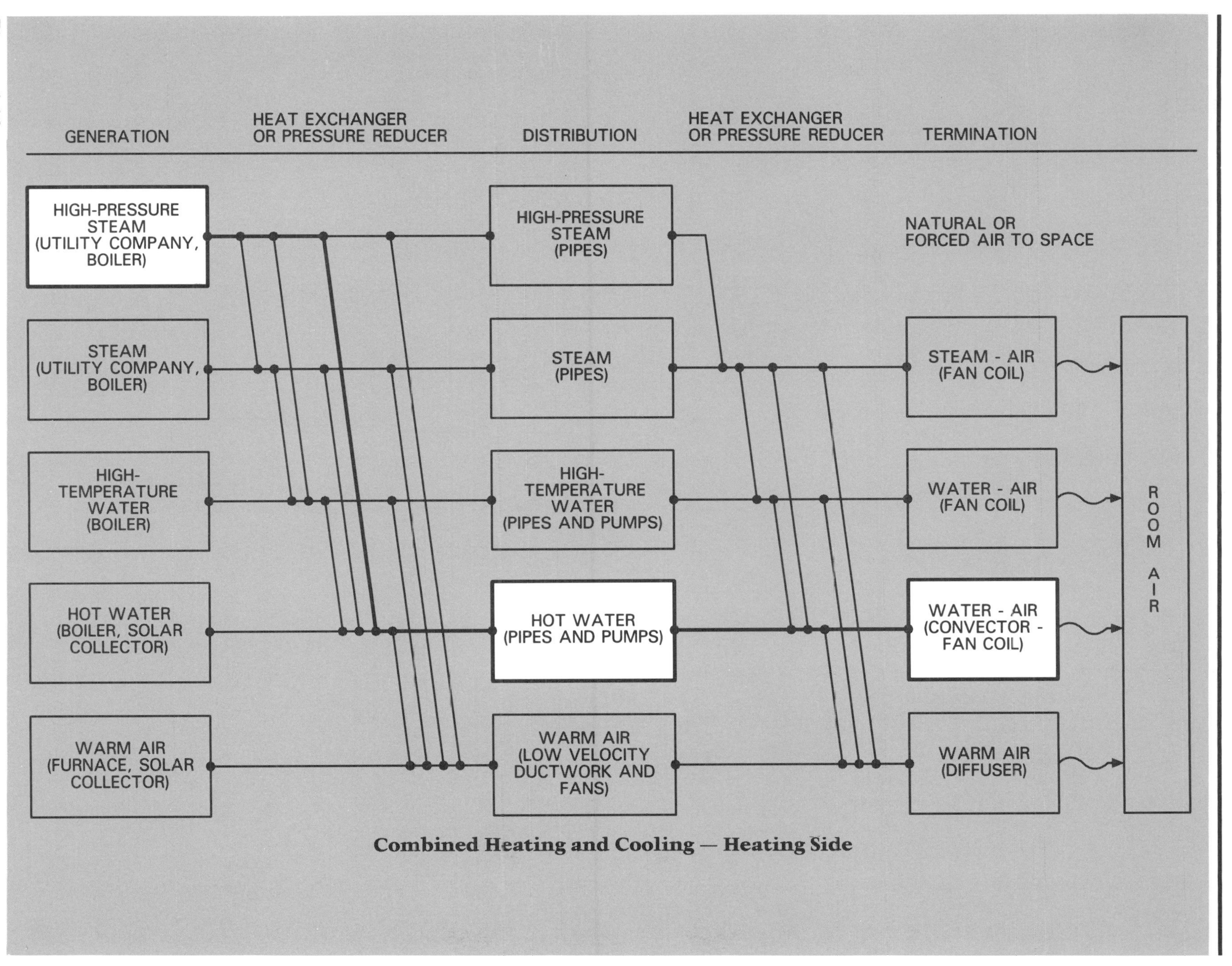

Combined Heating and Cooling — Heating Side

Figure 4.10

PART TWO

EQUIPMENT SELECTION

In Part II, the equipment selection process is discussed, including layouts, sizing instructions, and other special considerations for generation equipment, distribution and driving systems, coils and heat exchangers, terminal units, controls, and accessories. Equipment data sheets are included at the end of each chapter, listing capacities, advantages and disadvantages, and other considerations for each piece of equipment. Means line numbers are included for easy reference to the annual cost guide, *Means Mechanical Cost Data.*

CHAPTER FIVE

GENERATION EQUIPMENT

Generation equipment includes all major pieces of equipment that generate heating or cooling and their accessories, including boilers, furnaces, heat exchangers, expansion tanks, chillers, and cooling towers, as well as associated pumps, fans, and accessories. Generation equipment is divided into systems that provide heating only, systems that provide cooling only, and systems that provide both heating and cooling combined.

Hydronic Systems —Heating Only

Hydronic systems that provide heating only are described in this section, which includes discussions of boilers, burners, storage tanks, and expansion tanks.

Boilers

Heating boilers are designed to produce steam or hot water. The water in the boilers is heated either by coal, oil, gas, wood, electricity, or a combination of these fuels. Boilers are manufactured from cast iron, steel, and copper.

Cast iron sectional boilers may be assembled in place or shipped to the site as a completely assembled package. These boilers can be made larger on site by adding intermediate sections. The boiler sections may be connected by push nipples, tie rods, and gaskets. Cast iron boilers are noted for their durability.

Steel boilers are usually shipped to the site completely assembled. Large steel boilers may be shipped in segments for field assembly. The components of a steel boiler consist of tubes within a shell and a combustion chamber. If the water being heated is inside the tubes, the unit is called a *water tube boiler*. If the water is contained in the shell and the products of combustion pass through tubes surrounded by this water, the unit is called a *fire tube boiler*. Water tube boilers may be manufactured with steel or copper tubes.

Electric boilers have elements immersed in the water and do not fall into either category of tubular boilers. Steel boilers in the larger sizes are often slightly more efficient than cast iron and are generally constructed to be more serviceable, which with proper maintenance adds to their useful life.

Several types of boilers are available to meet the hot water and heating needs of both residential and commercial buildings. Some different boiler types are shown in Figure 5.1.

Most hot water boilers operate at less than 30 psig (low-pressure boilers). Low-pressure steam boilers operate at less than 15 psig. Above 30 psig, high-pressure boilers must conform to stricter requirements. Water is lost to a heating system through minor leaks and evaporation; therefore, a makeup water line to feed the boiler with fresh makeup water must be provided.

Boiler Selection Chart

Boiler Type	Output Capacity Range - MBH Efficiency Range	Fuel Types	Uses
Cast Iron Sectional	80–14,500 80–92%	Oil, Gas, Coal, Wood/Fossil	Steam/Hot Water
Steel	1,200–18,000 80–92%	Oil, Gas, Coal, Wood/Fossil, Electric	Steam/Hot Water
Scotch Marine	3,400–24,000 80–92%	Oil, Gas	Steam/Hot Water
Pulse Condensing	40–150 90–95%	Gas	Hot Water
Residential/Wall Hung	15–60 90–95%	Gas, Electric	Hot Water

Efficiencies shown are averages and will vary with specific manufacturers. For existing equipment, efficiencies may be 60–75%.

Figure 5.1

Heating boilers are rated by their hourly output. The output available at the boiler supply nozzle is referred to as the *gross output*. The gross output in Btu per hour divided by 33,475 indicates the boiler horsepower rating. The net rating of a boiler is the gross output less allowances for the piping tax and the pickup load. The net load should match the actual building heat load. Another term used often in sizing boilers is *square feet of radiation* (also known as equivalent direct radiation (EDR)); one square foot of radiation equals 240 Btu/hour.

Due to the high cost of fuels, efficiency of operation is a prime consideration when selecting a boiler. Recent innovations in the manufacturing field have led to more efficient and compact boiler designs. Figure 5.2 shows a typical boiler installation.

The Department of Energy has established test procedures to compare the "Annual Fuel Utilization Efficiency" (AFUE) of

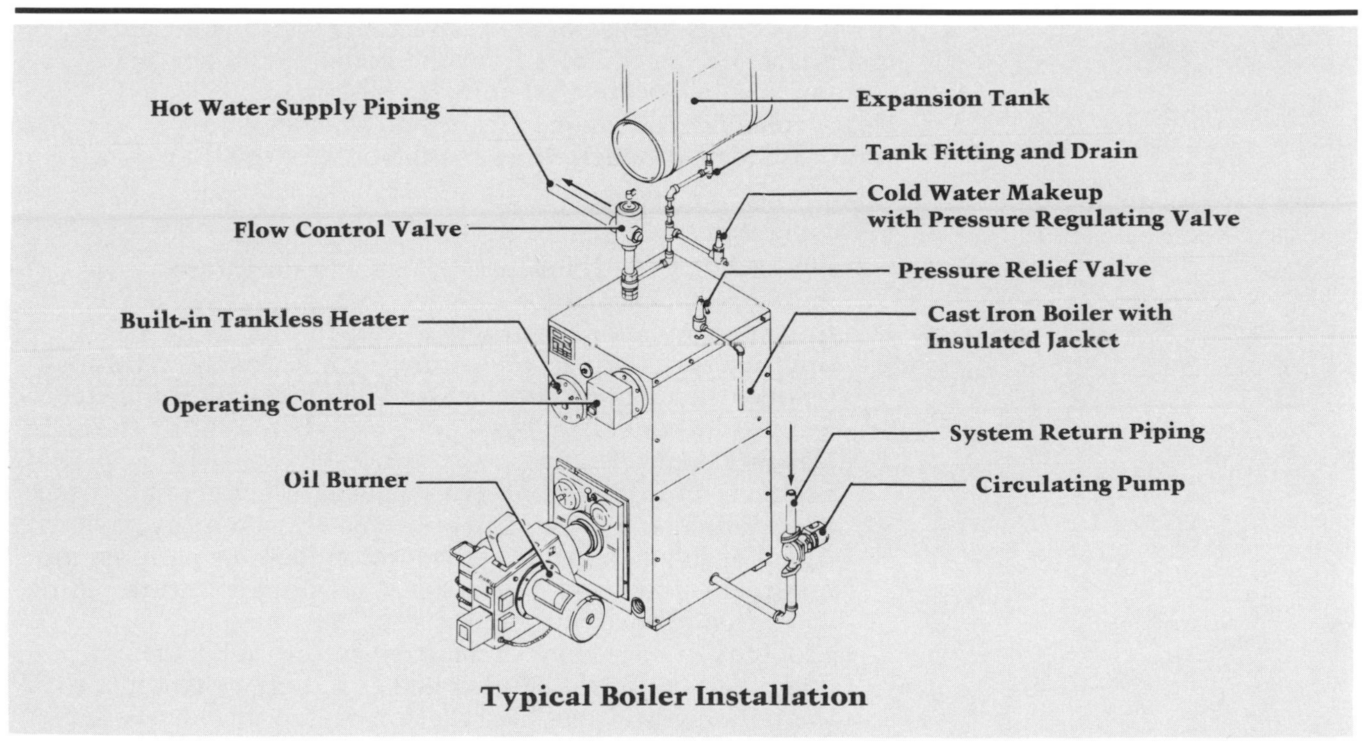

Figure 5.2

comparably sized boilers. Better insulation, heat extractors, intermittent ignition, induced draft, and automatic draft dampers contribute to the near 90 percent efficiencies claimed by manufacturers today.

In the search for higher efficiency, a new concept in gas-fired water boilers has recently been introduced. This innovation is the pulse condensing type boiler, which relies on a sealed combustion system rather than on a conventional gas burner. The AFUE ratings for pulse-type boilers are in the low to mid 90 percent range. Pulse-type boilers cost more initially than conventional types, but savings in other areas help to offset this added cost. Because these units vent through a plastic pipe to a side wall, no chimney is required. The pulse-type boiler also takes up less floor space, and its high efficiency saves on fuel costs.

Another innovation in the boiler field is the introduction from Europe of wall-hung, residential-size boilers. These gas-fired, compact, efficient (up to 80 percent AFUE) boilers may be directly vented through a wall or to a conventional flue. Combustion make-up air is directed to a sealed combustion chamber similar to that of the pulse-type boiler. Storage capacity is not needed in these boilers because the water is heated instantaneously as it flows from the boiler to the heating system. The boiler material consists mostly of steel. Heat-exchanger water tubes are usually made of copper or stainless steel, although some manufacturers use cast iron.

Conditions to consider when selecting a boiler include the following items.

- Accessibility: Both for installation and for future replacement, cast iron boiler sections can be delivered through standard door or window openings. Some steel fire-tube boiler replacement tubes can be installed through strategically located door or window openings. Some steel water-tube boilers are made long and narrow to fit through standard door openings.
- Economy of installation: Truly packaged boilers have been factory-fired and tested and arrive on the job ready to be rigged or manhandled into place. Connections required are minimal—fuel, water, electricity, supply and return piping, and a flue connection to a stack or breeching.
- Economy of operation: In addition to the AFUE ratings, boiler output should be matched as closely as possible to the building heat loss. Installation of two or more boilers (modular), piped and controlled so as to step-fire to match varying load conditions, should be carefully evaluated. Only on maximum design load would all boilers be firing at once. This method of installation not only increases boiler life, but also provides for continued heating capacity in the event that one boiler should fail.

Selecting the Boiler: When selecting boilers, one of the first choices to make is between steam or hot water. For small buildings, hot water is often selected, because the costs of steam generation equipment and distribution piping are higher than for an all-water system. However, steam is an excellent medium because:

1. The high latent heat of steam vaporization permits large quantities of heat to be transmitted from boiler to terminal units with little loss of temperature.
2. Steam promotes its own circulation. It flows from naturally higher pressures (in boiler) to lower pressure (steam lines). Steam does not require pumps or fans.
3. Boiler output is easily modulated by varying steam pressure.

While hot water is the appropriate choice for many situations, steam should be considered when the budget allows and applications are suitable.

When selecting boilers, the general output size and fuel tend to limit the choices. Figure 5.1 shows that the types selected follow a general size relationship. Cast iron, steel, and scotch marine boilers are generally acceptable in the 3,000 to 14,000 MBH range. Other factors, such as cost, availability, and personal preference based on experience, are also considerations. When a boiler type has been selected from Figure 5.1, the generation equipment data sheets at the end of this chapter can be used to make additional determinations and select features to complete the systems.

Boiler Sizing: The main features in sizing a boiler are its type and output. The boiler type is selected from Figure 5.1 and the generation equipment data sheets. Boiler output is designated both as *net* and *gross*: Net output is the actual heat delivered by the boiler to the terminal units; gross or nozzle output is always larger than the net output and represents the heat content of the fuel consumed by the boiler. The boiler efficiency equals the net output, H_N, divided by the gross output, H_G.

$$\text{Boiler efficiency} = \frac{\text{boiler net output } (H_N)}{\text{boiler gross output } (H_G)}$$

To determine the net output desired for the boiler, the heat load, H_T, is used and increased for piping tax and pickup load.

$$\begin{aligned} H_N = H_T &+ 10 - 25\% \text{ contingency reserve (for possible future loads and design uncertainties)} \\ &+ 5 \ - 15\% \text{ pipe tax (5\% if pipes are insulated)} \\ &+ 0 \ - 15\% \text{ pickup load (0 if boiler runs 24 hours)} \end{aligned}$$

Generally, the required net output of the boiler is approximately 1.25 H_T.

Safety Features

Boilers have many incorporated safety features. Pressure relief valves must be used to relieve excess pressure and prevent an explosive buildup of pressure. A low water cut-off senses the water level and shuts the burner off if the water level drops too low, preventing overheating of the metal. Other safety features incorporated into the system include fusible elements, water feeders, alarms, and thermal controls.

Accessories

Accessories, such as breeching, flues, and fuel piping, are also required to operate a boiler and must be considered when selecting a system. (For more information on accessories, see Chapter 12.)

Fuel Consumption

Annual fuel cost is determined using the following formula (see Figure 5.3):

$$F = \frac{24(DD)H_tC}{e(\Delta T)V}$$

F	=	Fuel consumed per year in units (for unit, see Figure 4.3)
DD	=	Degree days per year (see Figure 2.9)
H_t	=	Total heat load (in MBH)
C	=	Correction factor for oversizing and outdoor temperature (see Figure 5.3)
e	=	System efficiency (see Figure 5.1)
ΔT	=	Design indoor minus design outdoor temperature (°F) $(T_i - T_o)$
V	=	Heating value of fuel per unit (MBH/unit) (see Figure 4.3)

As an example, the fuel consumed for a building in Casper, Wyoming, burning #2 oil, using a boiler with an output of 600 MBH for a building heat load of 500 MBH, uses the values listed below.

DD	=	7,400 (see Figure 2.9)
H_t	=	500 MBH (given)
C	=	See Figure 5.3
H_i/H_t	=	600/500 = 1.2
T_o	=	–5°F
C	=	1.05
e	=	.85 (verify w/boiler selected) (see Figure 5.1)
ΔT	=	68 – (–5) = 73°F
V	=	138.5

Thus, for this example, the fuel consumption is calculated as follows:

$$F = \frac{24 \times 7{,}400 \times 500 \times 1.05}{0.85 \times 73 \times 138.5} = 10{,}850 \text{ gal./yr.}$$

Burners

Fuel for boilers is injected and burned in the combustion chamber by the burner. Burners may be fueled by oil, gas, or a combination of the two. The combined burner has the advantage of switching to the more economical of the two. For oil, the burner atomizes the oil into a fine mist, which is then ignited. The finer the mist, the more complete and efficient the combustion. In a gun-type burner, a nozzle is used to mist the oil, which is fed to it under pressure. Rotary burners vaporize the oil as it leaves a rotating cup while mixing with the combustion air.

Burners are controlled to shut down under several conditions for safety. The stack temperature is monitored to ensure that

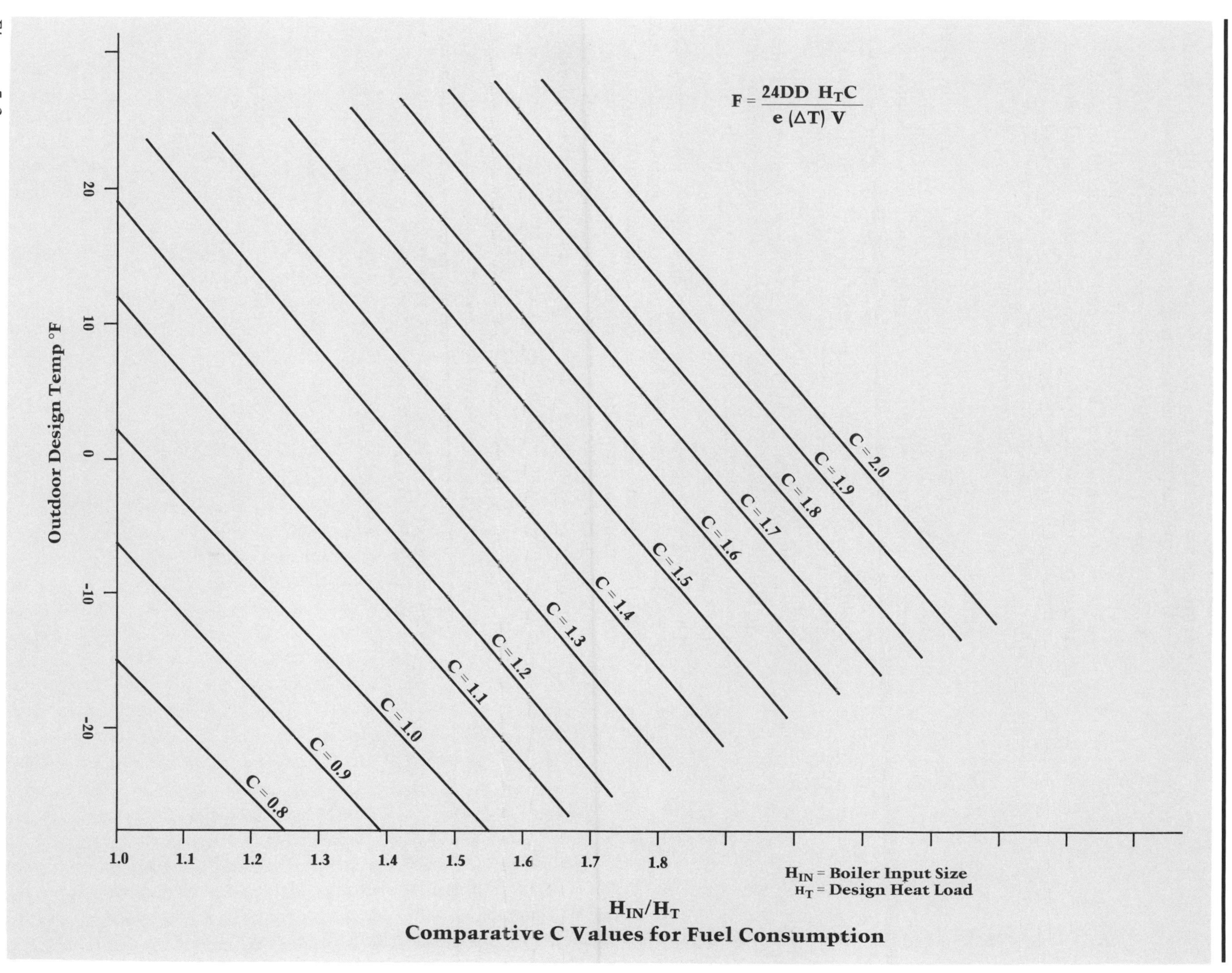

F = 24DD H_T C / e (ΔT) V
C = 2.0
C = 1.9
C = 1.8
C = 1.7
C = 1.6
C = 1.5
C = 1.4
C = 1.3
C = 1.2
C = 1.1
C = 1.0
C = 0.9
C = 0.8
Outdoor Design Temp °F
20
10
0
-10
-20
1.0
1.1
1.2
1.3
1.4
1.5
1.6
1.7
1.8
H_IN = Boiler Input Size
H_T = Design Heat Load
H_IN/H_T
Comparative C Values for Fuel Consumption

Figure 5.3

combustion has taken place. Failure to cause a rise in stack temperature within a few minutes due to loss of fuel or ignition will stop the burner from operating.

Burner flames are ignited using one of two methods, pilot or spark. In pilot ignition, a pilot light burns continually and the oil mist or gas injected over it on call is ignited. With spark ignition (sometimes called electronic ignition), a spark is passed across the oil mist or gas. Some electronic combustion controls have the ability to continuously monitor the combustion air and adjust the quantity for optimum conditions.

Tanks

There are two common tanks used with HVAC equipment: fuel storage tanks and expansion tanks. Both are shown in the generation equipment data sheets at the end of this chapter.

Fuel Storage Tanks: Oil and gas may be stored on site for a building's HVAC system. Oil is the most common fuel typically stored in tanks. Fuel tanks may be above ground or below ground. Above-ground tanks are subject to safety provisions, such as containment areas, to hold a volume equal to the liquid volume of the tank, so that in the event of a spill or major leak, the fuel can be contained. Walls are used to create the containment area.

Underground tanks may be of single or double-wall construction. The double-wall tanks offer better protection against leaks. Leak detection systems are available and required for underground tanks in some communities. Cathodic protection for underground tanks to prevent electrolytic deterioration as well as bituminous coatings are routinely used to extend the tank life.

Buried tanks must be held down because in the spring, when the tanks are likely to be empty, the spring rains flood the surrounding earth. The resulting buoyant force can lift the tank unless it is secured to a concrete pad. The pad is usually sized to weigh 1.5 times the buoyant force of the tank and should be approximately 1′6″ larger than the tank overall. The buoyant force is calculated as follows:

$$B = \frac{62.5 \text{ lbs.}}{\text{ft.}^3} \cdot \frac{0.133 \text{ ft.}^3}{\text{gal.}} = 8.33T$$

B = Buoyant force

T = Tank capacity (gal.)

Concrete used for this purpose should weigh 150 pounds per cubic foot; therefore, the volume of the concrete pad required is calculated as follows:

$$V_p \text{ ft.}^3 = \frac{8.33T \text{ (gal.)} \cdot 1.5}{150 \text{ lbs./ft.}^3} = 0.0833T$$

Example: The concrete pad required to hold down a 5,000-gallon fuel tank 23 feet long and 7.5 feet in diameter is calculated below.

$$V_p \text{ ft.}^3 = 0.0833T$$
$$= 0.0833 \cdot 5{,}000$$
$$= 416.5 \text{ ft.}^3$$

Pad length 23 + 1.5 = 24.5 ft.
width 6 + 3 = 9 ft.
thickness try 2′0″

$$24.5 \cdot 9 \cdot 2 = 441 \text{ ft.}^3 > 416.5 \text{ ft.}^3$$

Therefore, 441 cubic feet of concrete would be adequate to secure the tank.

A more exact analysis that uses the weight of earth above the pad or a low water table can be used to reduce the above requirements, which are generally conservative. Hold down straps or angles and rods must be provided to secure the tank. They should be anchored in the concrete and the straps passed over the top of the tank to hold it down.

Expansion Tanks: All hydronic systems undergo changes in temperature that cause the water to expand and contract. An expansion tank is always provided on each closed loop piping system, because the tank allows the water to expand into it as the water volume increases with the temperature.

There are two formulas for computing the expansion volume required, one for hot water and one for cold water.

For hot water (160–280°F)

$$V_t = \frac{(0.00041T) - 0.0466)\, V_s}{\frac{P_a}{P_f} - \frac{P_a}{P_o}}$$

For chilled water (40–100°F)

$$V_t = \frac{(0.006)\, V_s}{\frac{P_a}{P_f} - \frac{P_a}{P_o}}$$

V_t = expansion tank size (gal.)
V_s = total gallons in system
T = average water temperature (°F)
P_a = atmospheric pressure
P_f = minimum pressure at tank (psig)
P_o = maximum pressure at tank (psig)

Expansion tanks may permit the liquid and air to be in contact or they may be separated by a diaphragm. Properly sized expansion tanks have the ability to maintain a set pressure range at the tank.

Heat Exchangers

If steam or hot water service is available to a building from a remote source of supply, the need for a boiler is eliminated. If the proposed system is forced hot water and the remote source of supply is steam or high-temperature hot water, a shell and tube-type heat exchanger (converter) must be provided. A heat exchanger would also be required if the building is served by low-temperature chilled water or brine for building cooling.

Warm Air Systems — Heating Only

Furnaces

Forced warm air furnaces utilize self-contained fans and combustion chambers. Forced warm air furnaces use burners similar to hydronic boilers. The furnace operates by drawing cool air around the combustion chamber and blowing this heated air into the distribution system utilizing the fan. The net output of the furnace should equal the building design heat loss plus the allowances for losses in a manner similar to boilers.

$$\begin{aligned} H_N = H_T &+ 10 - 25\% \text{ contingency reserve} \\ &+ 2 - 10\% \text{ duct loss} \\ &+ 0 - 5\% \text{ pickup load} \end{aligned}$$

Generally, the required net output, H_N, of the furnace is approximately 1.15 H_T.

Furnaces are specified by the net output and/or by the cfm of air and static pressure the fan delivers. The fan static pressure must be verified to ensure that it has enough pressure to deliver the air through the ductwork. For a supply air temperature of 130°F and a room air temperature of 70°F (a 60°F temperature difference), the quantity of air needed to compensate for heat loss can be found using the following equation.

$$H_T = 1.1 \text{cfm } \Delta T$$

$$\text{cfm} = \frac{H_T}{1.1 \cdot \Delta T} = \frac{H_T}{1.1 \cdot 60}$$

Figure 5.4 illustrates a typical warm air heating system.

Duct Heaters

Duct heaters are frequently used to reheat cool air that needs local tempering before it is delivered to a space. Duct heaters are a convenient device because of the ease with which they can be individually controlled independent of the main heating system.

Many applications require electric duct heaters. A typical example is a fresh air supply intake. Sometimes mounted on the roof, remote from the main distribution system, an electric coil serves easily to heat incoming cold air to 55–60°F without threat of freezing.

Makeup Air Units

When the fresh air requirements are large, special makeup air units are employed. A common example is commercial kitchens where large quantities of air must be introduced to make up for the air exhausted by the range hood. The quantity of air should be ten to twenty percent larger than the exhaust air to keep spaces under positive pressure. The following formula is used to determine the rating (in Btu) of the makeup air system:

$$H = 1.1 \text{cfm } \Delta T$$

$$\text{cfm} = 1.1 \times \text{exhaust air quantity}$$

$$\Delta T = T_i - T_o$$

Hydronic Systems —Cooling Only

The central component of any chilled water air conditioning system is the water chiller. Chillers producc water from 42–55°F, which is then distributed to coils, fan coil units, or central station air handling units for cooling. This is done by passing water through the evaporator section, where refrigerant gas absorbs the heat and chills the water. The chilled water goes out through the distribution system. The refrigerant gas from the evaporator goes to the compressor section. The hot gas is cooled by a condenser and the refrigerant is then returned to the evaporator.

Chillers are sized to meet the total building heat gain in addition to an allowance for piping tax, pickup load, and contingency. Because cooling often is not done twenty-four hours a day, the pickup load can be large or the equipment can be activated two to four hours before being occupied.

Packaged water chillers are available in three basic designs: the reciprocating compressor, direct-expansion type; the centrifugal compressor, direct-expansion type; and the absorption type generator (see Figures 1.16 and 1.17).

Chillers

Chillers and other cooling apparatus are sized by the ton. One ton of cooling equals the melting rate of one ton of ice in a

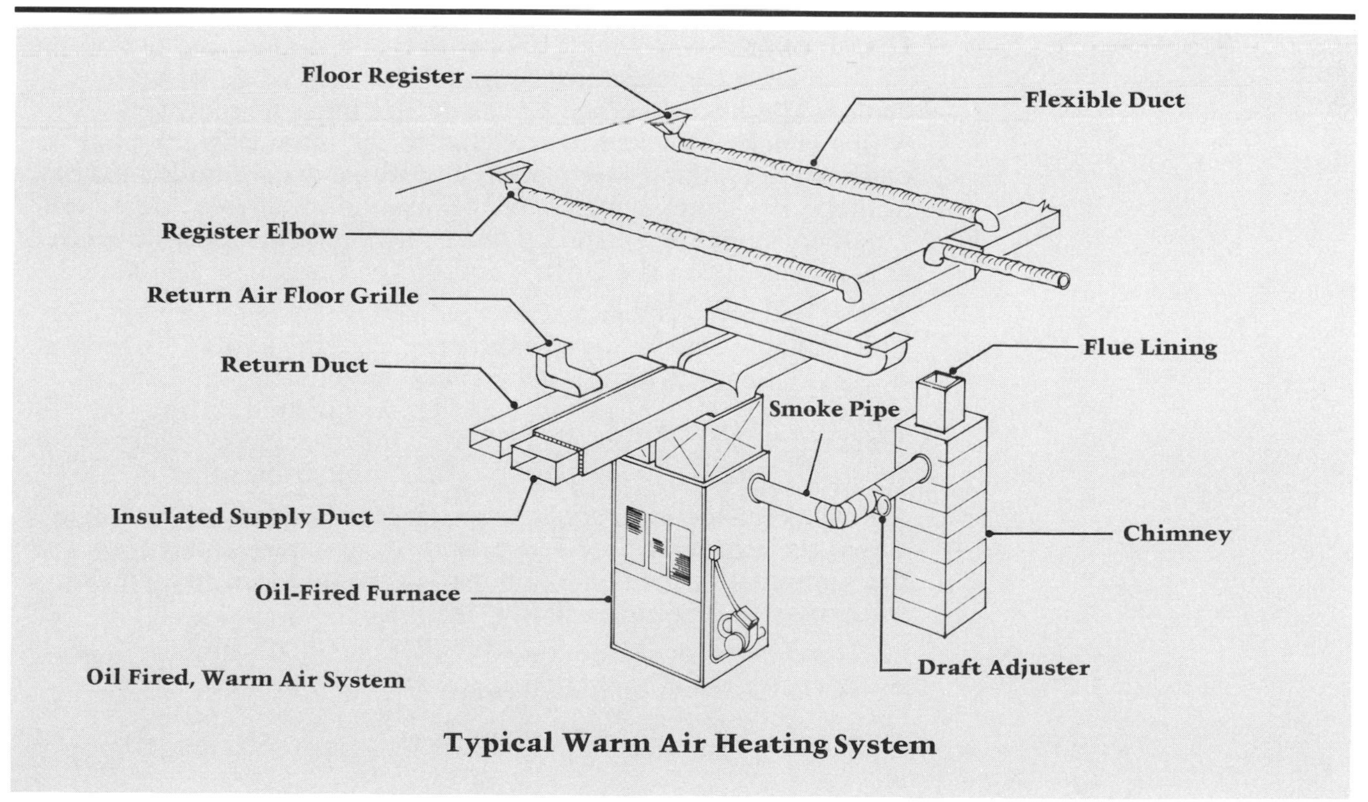

Figure 5.4

twenty-four hour period, or 12,000 Btu's per hour. The three types of chillers vary significantly in their cooling power, as well as in their operation.

The *reciprocating compressor chiller*, which generates cooling capacities in the range of 10 to 200 tons, is usually powered by an electric motor.

The *centrifugal compressor*, which generates cooling capacities ranging from one hundred to several thousand tons, is also commonly powered by an electric motor, but it may be designed for a steam-turbine drive as well. In some instances, both of these types of chillers may be powered by internal combustion engines.

Absorption-type chillers provide cooling capacities ranging from 3–1,600 tons. Because they use water as a refrigerant and lithium bromide or other salts as an absorbent, this system only consumes about 10 percent of the electrical power required to operate the conventional reciprocating and centrifugal direct-expansion chillers. This low consumption advantage is particularly desirable in buildings where an electrical power failure triggers an emergency backup system, such as in hospitals, data processing centers, electronic switching systems locations, and other buildings that must continue to function on auxiliary power. Absorption-type chillers require a heat source to maintain the absorption process, which is economically advantageous in areas where electric power is scarce or costly and where gas rates are low, or where waste or process steam or hot water is available during the cooling season. Solar power may also be used in some areas to generate the heat required for the absorption process. Figure 5.5 shows a typical chiller, condenser, and cooling tower installation.

Condensers

The hot refrigerant gas from the chillers may be air or water cooled. The heat absorbed by the chiller must be removed, which is done by rejecting the heat to the outside. Very small chillers are available with an air cooled condenser built into the package. For larger systems and for systems that may create too much noise for their location, air cooled condensers are installed at a distance from the chiller, and the two units are connected with refrigerant piping.

The basic process for air conditioning systems using refrigerants as the cooling medium is the cooling and condensing back to liquid form of the refrigerant gas that was heated during the evaporation stage of the cycle. This condensation is achieved by cooling the gas with air, water, or a combination of both.

Air cooled condensers cool the refrigerant by blowing air directly across the refrigerant condenser coil; evaporative condensers use the same method of cooling, but with the addition of a spray of water over the coil to expedite the process.

The condenser for water cooled chillers is piped into a remote water source, such as a cooling tower, pond, or river, via the "condenser water system." Completely packaged chiller systems may include built-in chilled water pumps and all interconnecting piping, wiring, and controls. All of the components of the completely packaged unit are factory installed and tested prior to shipment to the installation site for connection to the chilled water and condenser water systems.

When water is used as the condensing medium and is abundant enough that recycling is not required, it may be piped to a drain after performing its cooling function and returned to its source. The source may be a river, a pond, or the ocean. If the water supply is limited, expensive, or regulated by environmental restrictions, a water-conserving or recycling system must be employed. Several types of systems may be installed to conform to these limitations. For example, a water regulating valve, a spray pond, a natural draft cooling tower, or a mechanical draft cooling tower may be used.

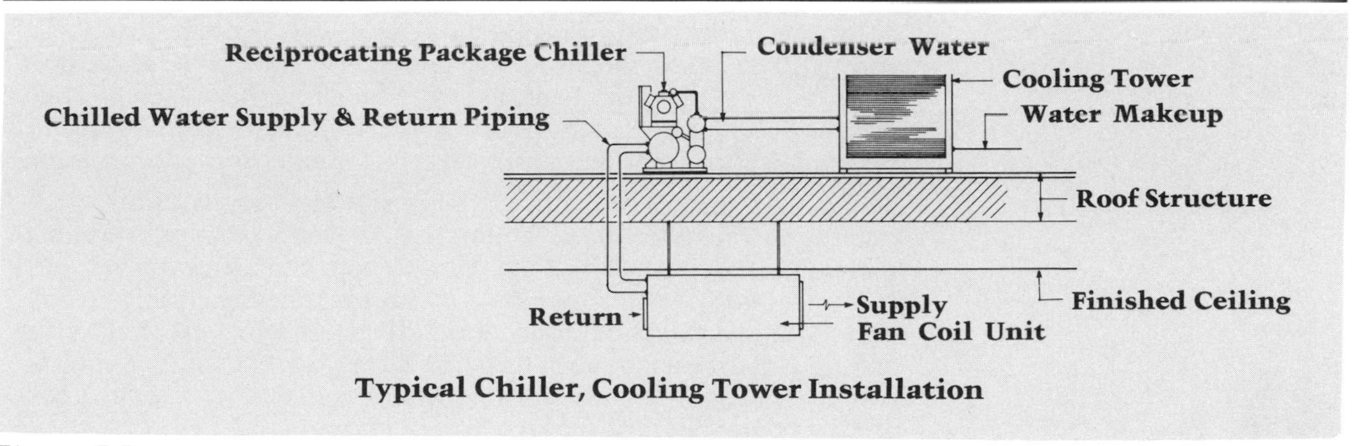

Typical Chiller, Cooling Tower Installation

Figure 5.5

In very small cooling systems, a temperature-controlled, water-regulating valve may be used, provided that such a system is permitted by local environmental and/or building codes. The regulating valve system functions by allowing cooling water to flow when the condenser temperature rises, and conversely, by stopping the flow as the temperature falls. The problem with this system, however, is that the heated condenser water cannot be recycled and is therefore wasted during the flow cycle.

Cooling Towers

Although both are viable and available methods of cooling, *spray pond* and *natural draft* cooling tower systems are not commonly used for building air conditioning. Because of water loss caused by excessive drift and the large amount of space required for their installation and operation, spray pond and natural draft cooling tower systems are less desirable than *mechanical draft* cooling tower systems.

Mechanical draft cooling tower systems are classified in two basic designs: *induced draft* and *forced draft*. In an induced draft tower, a fan positioned at the top of the structure draws air upwards through the tower as the warm condenser water spills down. A cross-flow induced draft tower operates on the same principle, except that the air is drawn horizontally through the spill area from one side. The air is then discharged through a fan located on the opposite side. In a forced draft tower, the fan is located at the bottom or on the side of the structure. Air is forced by the fan into the water spill area, through the water, and then discharged at the top. All designs of mechanical draft towers are rated based on the tons of cooling; three gallons of condenser water per minute per ton is an approximate tower sizing method.

After the water has been cooled in the tower, it cycles back through a heat exchanger, or condenser, in the refrigeration unit. Here, it again picks up heat and is pumped back to the cooling tower. The piping system is called the *condenser water system*. Figure 5.6 shows the layout of a typical condenser water system.

The actual process of cooling within the mechanical draft cooling tower takes place when air is moved across or counter to a stream of water falling through a system of baffles or "fill" to the tower basin. After the cooled water reaches the basin, it is piped back to the condenser. Some of the droplets created by the fill are carried away by the moving air as "drift" and some of the droplets evaporate. This limited loss of water is to be expected as part of the operation of the tower system.

Because of the loss of water by drift, evaporation, and bleed off, replenishment water must be added to the tower basin to maintain a predetermined level and to ensure continuous operation of the system. To prevent scale buildup, algae, bacterial growth, or corrosion, tower water should be treated with chemicals or ozone applications.

The materials used in constructing mechanical draft cooling towers include redwood (which is the most commonly employed material), other treated woods, and various metals, plastics, concrete, or ceramic materials. The fill, which is the most important element in the tower's construction, may be manufactured from the same wide variety of materials used in the tower structure. Factory assembled, prepackaged towers are available and are usually preferable to built-in-place units. Multiple tower installations are now being used for large systems.

The location of cooling towers is an important consideration for both practical and aesthetic reasons. They may be located outside of the building on the roof or on the ground. If space permits, a cooling tower may be installed indoors by substituting centrifugal fans for the conventional noisy propeller type, and adding air intake and exhaust ductwork. The tower discharge should not be directed into the prevailing wind or towards doors, windows, and building fresh air intakes. In general, common sense should be used when determining tower placement so that the noise, heat, and humidity the system creates do not interfere with building operation and comfort. The manufacturer's guidelines for installation should be strictly followed, especially the sections that address clearances for maximum air flow, maintenance, and future unit replacement.

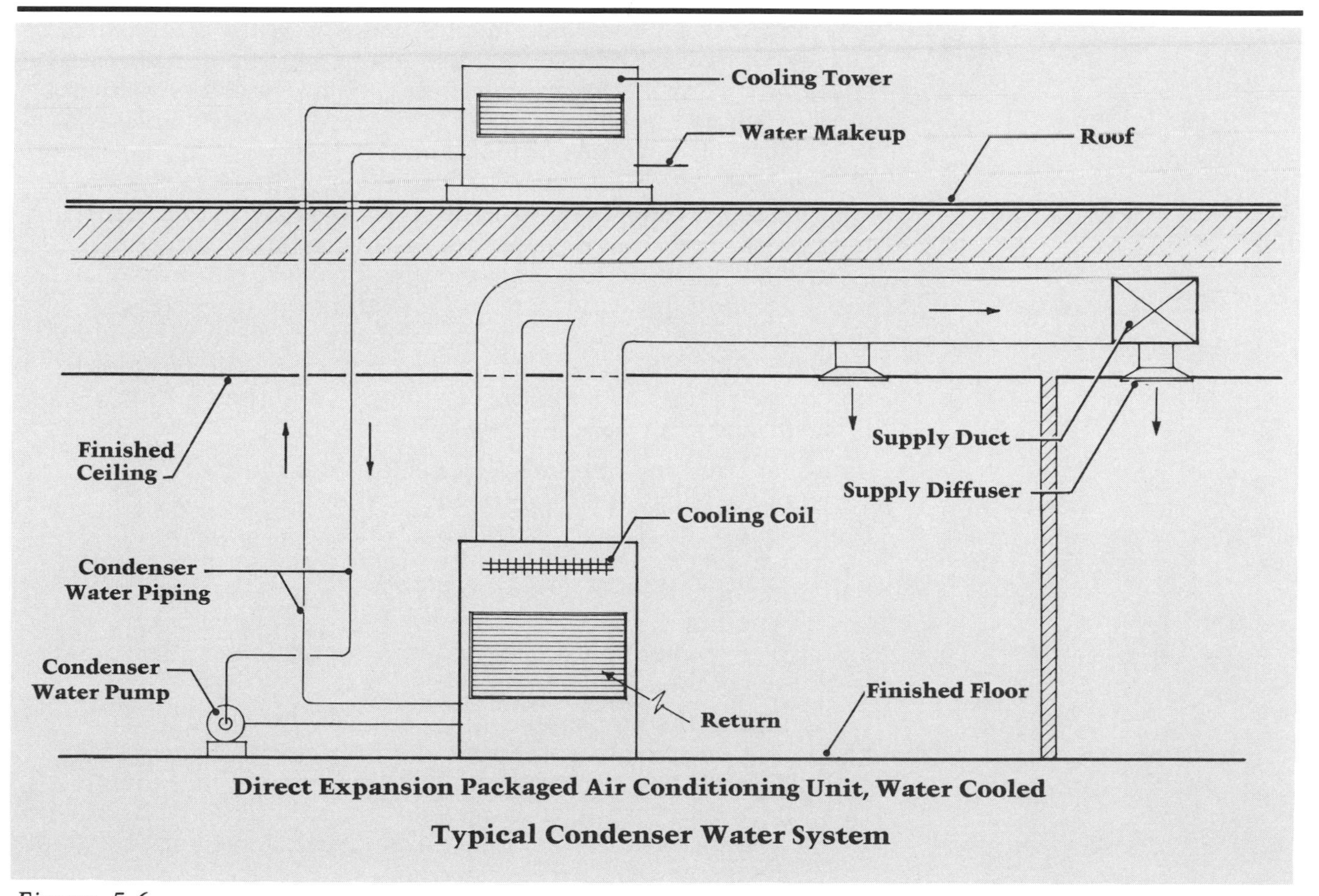

Figure 5.6

Economical operation of a cooling tower system may be achieved through effective control and management of several critical aspects of its operation, including careful monitoring of water treatment, selecting and maintaining the most efficient condensing temperature, and controlling water temperature with fan cycling and two-speed fans.

A recent development in tower water system operation allows the tower to substitute for the water chiller under certain favorable climatic conditions. This method cannot be implemented in all cases, but in situations where it can be employed, substantial savings result in the reduced cost of chiller operation. A plate coil type heat exchanger is used to good advantage in this type of system.

Cooling towers are shown in the generation equipment data sheets at the end of this chapter. Cooling towers are sized to match the load of the chiller (the chiller net output plus the mechanical load of the compressor).

Split Systems—Cooling Only

Split systems are used both for cooling only and for combined heating and cooling systems. Split systems are discussed in the section on combined heating and cooling systems.

Combined Heating and Cooling Systems

Central Station Air Handling Units

Air handling units consist of a filter section, a fan section, and a coil section on a common base. For large installations, each component is selected individually. Smaller units, often called *package units*, contain all components in a single prefabricated enclosure. They are used to distribute clean, cooled or heated air to the occupied building spaces. These units are available in a wide range of capacities, from 200 cubic feet per minute to tens of thousands of cubic feet per minute. The units also vary in complexity of design and versatility of operation. Small units tend to have relatively simple coil and filter arrangements and a modest-sized fan motor. Larger, more sophisticated units usually require remote placement. Because of the need to overcome resistance losses caused by intake and supply ductwork and by complex coil, filter, and damper configurations, the fan motor horsepower must be dramatically increased. Figure 5.7 shows a typical rooftop air handling system.

Small air handling units may be located and mounted in a variety of settings and by different methods. They may be mounted on the floor or hung from walls or ceilings with no discharge ductwork required in the room they service. Small air handling units require supply and return piping for heating and/or cooling. If these units are used for cooling, a drain pan and piping connection is also required to run the condensation from the cooling coil to waste.

Determining the proper size, number, capacity, type, and configuration of coils in the unit is a prime consideration when selecting and/or designing an air handling unit. As a general rule, the amount of air (in cubic feet per minute) to be handled by the unit determines the size and number of the various coils.

Electric or hydronic coils are used for heating; chilled water or direct expansion coils are used for cooling. As the units increase in size and complexity, the coil configurations and arrangements become limitless. A simple heating and cooling unit, for example, may use the same coil for either hot or chilled water. A large unit usually demands different types of coils to perform many separate functions. In humid conditions, the air temperature may be intentionally lowered to remove moisture. In this case, a reheat coil is added to bring the air temperature up to its desired level. Conversely, in dry conditions, a humidifier component is built into the unit. If outside air is introduced to the unit at subfreezing temperature, then a preheat coil is placed in the outside air intake duct. Precautions must be taken to prevent icing or freeze-up of this coil.

Certain precautions should be taken to prevent damage and to ensure the efficiency of the unit's coils and other components. To protect the coil surfaces from accumulating dust and other airborne impurities, a filter section is a necessary addition to the unit. If a unit is designed to cool air, a drain pan must be included beneath the coiling section. This pan is then piped to an indirect drain to dispose of the unwanted condensation.

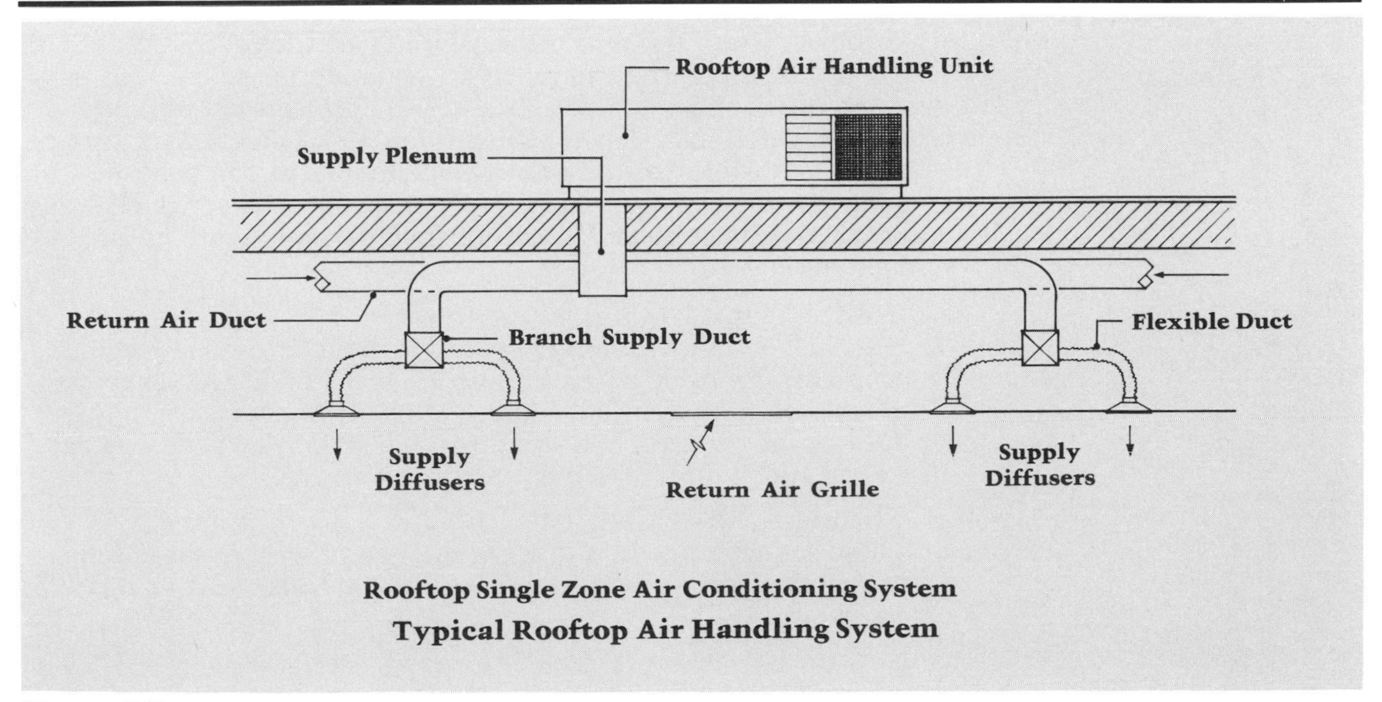

Rooftop Single Zone Air Conditioning System

Typical Rooftop Air Handling System

Figure 5.7

Another protection precaution involves the internal insulation of the fan coil casing. If this precaution is not taken to protect the cooling coil section and all other sections "down-stream," corrosive or rust-causing condensation will damage the casing and discharge ductwork. Insulating this casing also helps deaden the noise of the fan. Noise can be further reduced by installing flexible connections between the unit and its ductwork, as well as by mounting the unit on vibration-absorbing hangers.

Free Cooling

In free cooling, or economizer cycle and enthalpy control, fresh outdoor air is used to cool in spring, fall, and winter, instead of running the chiller whenever the outside temperature is cooler than the inside design temperature and it is not too humid. As mentioned previously, the cooling tower can also be used "in reverse" for producing cool water on mild days for hydronic systems. In air systems, the savings are similar. No compressors need to be run for free cooling. Except for the cost of running the fans, the cooling is "free." As the outside air gets colder than 55°F, it is mixed with the building return air (typically 70°F) to keep the supply air from being too cold. This method automatically provides large quantities of outside air, which also satisfies the ventilation requirements.

As the outside air approaches 32°F, some heat is provided to prevent freezing of the coils. If the free cooling is tied to an automatic control system that measures indoor and outdoor temperatures and controls supply, return, and exhaust air dampers, the building is said to have an *economizer cycle*. If, in addition to the economizer, the control system also measures the humidity of the air and computes the latent and sensible energy of the air in modulating the dampers, the building is said to have *enthalpy control*. Figure 5.8 illustrates an economizer system.

Package Units

For loads under 100 tons, package units are a convenient option. In one assembly, the fan, refrigeration equipment, and heating equipment are assembled and ready to operate with minimal field connections. They are most often air cooled and mounted on a roof or through the wall where outside air can cool the condenser coils. Most manufacturers have a wide range of package units suitable for various installation situations and fuel types.

Rooftop Units

For single-story buildings or spaces with an accessible flat roof, economical rooftop package units are often preferred. They can provide both heating and cooling, supply the necessary fresh outdoor air, have economizer options, and should generally have a local supply of replacement parts and service.

Through-the-Wall Units

These self-contained units are similar to rooftop package units, except they are placed inside a building, usually near or through an exterior wall.

Split Systems

For locations where the space to be cooled is remote from an available wall or roof, split systems are used. In split systems, the condenser and compressor are placed outdoors on a pad or roof, and the evaporator and supply fan with evaporator coil is placed in the space, often in a closet or above the ceiling. The refrigerant gas from the evaporator in the space is piped to the outdoor condensing unit by copper tubing, where it rejects the heat and returns as a liquid to the evaporator. This system removes the noise of the compressor from the conditioned or occupied space, and is, therefore, very desirable. It also allows particular tenants, such as first-floor retail or individual residential units, to have cooling and also pay for it individually. Figure 5.9 illustrates the layout of a split system.

Heat Pumps

A method of saving energy when buildings are simultaneously heated and cooled is with water source heat pumps, illustrated in Figure 5.10. A heat pump is a complete package refrigeration unit that can "pump" heat in or out of the water passing through it. Also available are air-to-air heat pumps for milder climates where only minimal heating is required.

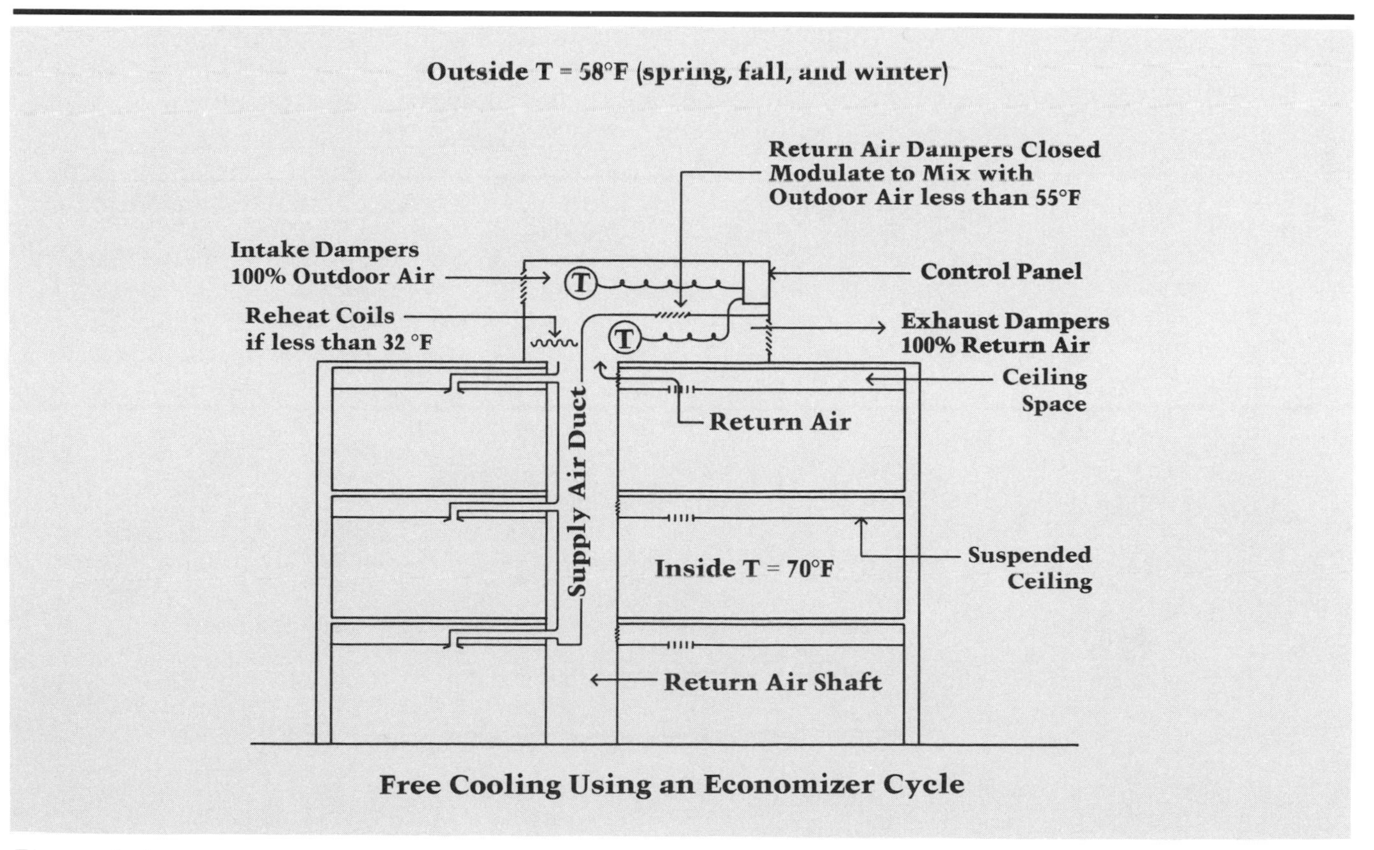

Figure 5.8

In the case of a water source heat pump system, water is continuously pumped in a loop, which connects a boiler, a cooling tower, and a series of heat pumps. The boiler and cooling tower are activated as necessary to maintain the temperature as close to 75°F as possible (always above 60°F and below 90°F). By "pumping" the heat absorbed from the room into the water, the heat pump raises the temperature of the water. This is possible because the heat pump is able to reverse the direction of the refrigerant in its loop. When heating a space, the air stream coil is used as the condenser and the water stream coil acts as the evaporator. When cooling a space, an internal valve in the heat pump reverses the flow of the refrigerant. Now the air stream is used as the evaporator, the air is cooled and the water loop is warmed as it flows over the condenser. By themselves, heat pump systems do not provide ventilation, but they do take advantage of situations where heating and cooling are occurring simultaneously. Because there is considerable heat inertia in the water of the heat pump loop and because some heat pumps are adding heat from the building to the water loop while other heat pumps are removing heat from the water loop, it is possible to move heat from one portion of the building to another without using a boiler or cooling tower. This saves considerable energy on the total fuel bill.

Both heat pumps and economizer systems offer energy savings. Both use fans, but the compressors in the heat pumps consume energy, which can make them slightly more expensive than economizer systems, depending on the specific applications. On the other hand, all of the heat pump energy (fans and compressors) can be individually run, metered, and paid for by the tenant, who can expect to pay only for what is used and to share with other tenants in the benefits of simultaneous heating and cooling during the year.

Heat pumps supply air that can be sent into a constant volume ductwork system or further connected to variable air volume

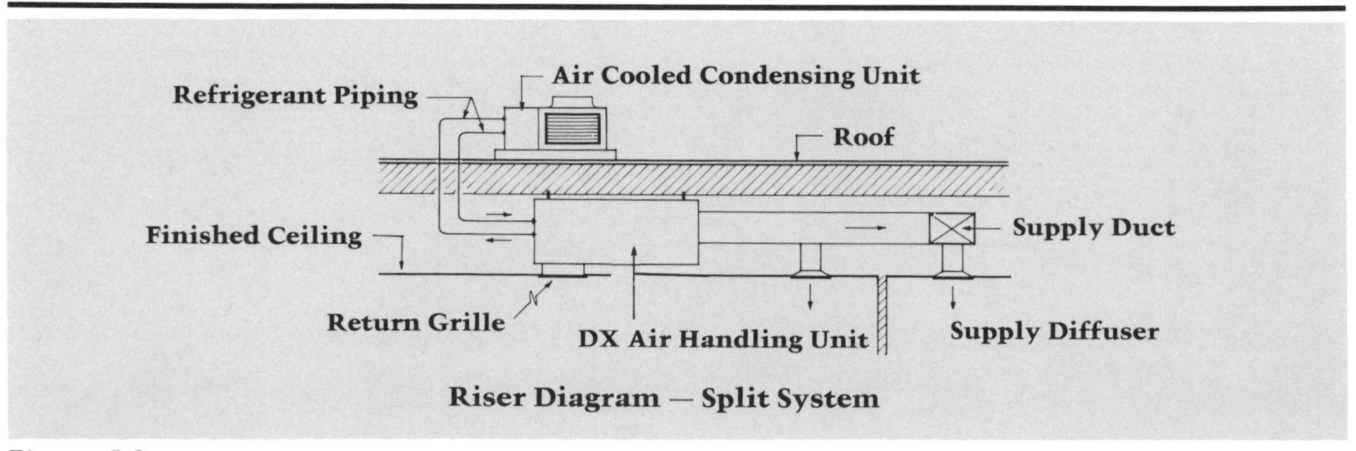

Riser Diagram — Split System

Figure 5.9

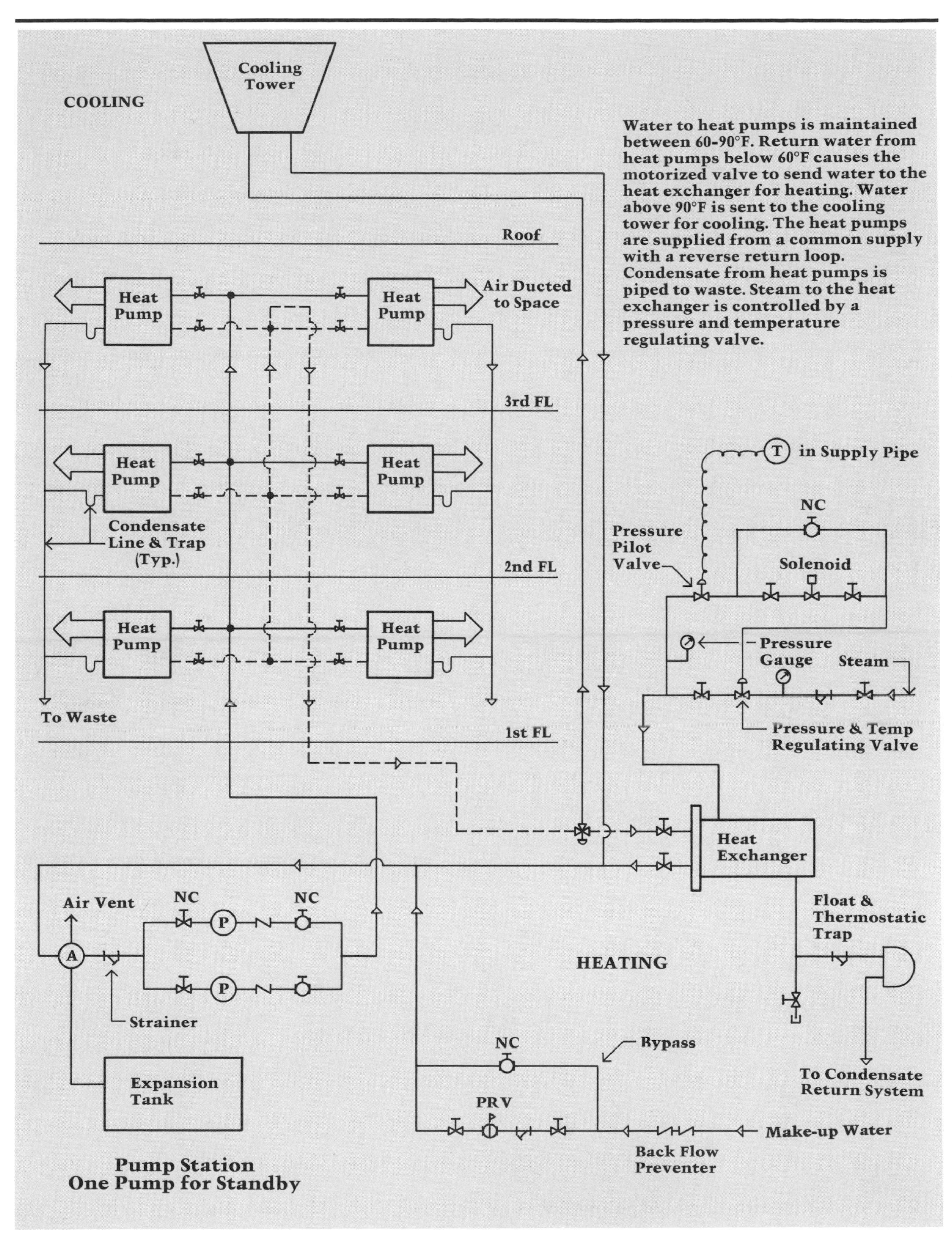

Cooling Tower
COOLING
Water to heat pumps is maintained between 60-90°F. Return water from heat pumps below 60°F causes the motorized valve to send water to the heat exchanger for heating. Water above 90°F is sent to the cooling tower for cooling. The heat pumps are supplied from a common supply with a reverse return loop. Condensate from heat pumps is piped to waste. Steam to the heat exchanger is controlled by a pressure and temperature regulating valve.
Roof
Heat Pump
Heat Pump
Air Ducted to Space
3rd FL
Heat Pump
Heat Pump
T in Supply Pipe
NC
Condensate Line & Trap (Typ.)
Pressure Pilot Valve
Solenoid
2nd FL
Heat Pump
Heat Pump
Pressure Gauge
Steam
To Waste
Pressure & Temp Regulating Valve
1st FL
Heat Exchanger
Air Vent
NC
NC
P
A
P
Float & Thermostatic Trap
HEATING
Strainer
NC
Bypass
To Condensate Return System
Expansion Tank
PRV
Make-up Water
Back Flow Preventer
Pump Station
One Pump for Standby

Figure 5.10

(VAV) boxes. When one heat pump is connected to several spaces or offices, VAV boxes allow a better measure of individual control.

Generation Equipment Data Sheets

The following pages of this chapter contain equipment data sheets for generation equipment. Each of these sheets includes efficiency ratings, useful life, typical uses, capacity ranges, advantages and disadvantages, special considerations, accessories, and additional equipment needed for each piece of generating equipment.

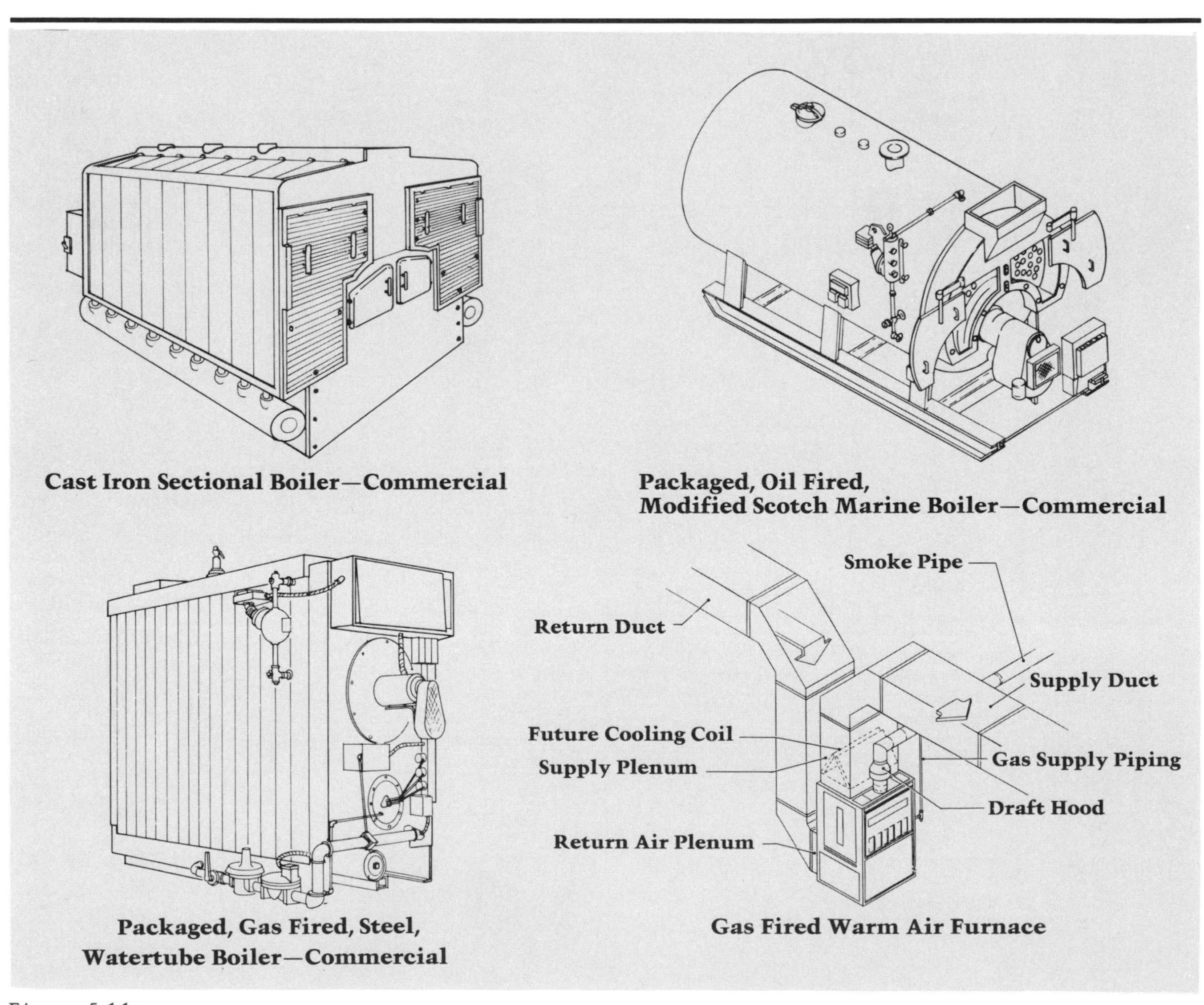

Figure 5.11a

Generation Equipment — Heating

Heating	Type and R.S. Means Co. Line No.	Capa-city (MBH)	Effi-ciency (%)	Advantages	Disadvantages	Related Equipment and Accessories	Considerations
Cast Iron Sectional Boilers	Gas-fired 155-115	80-7,000	80-92	25 years (plus) useful life Long life Durability Expandable Wide range of sizes Ease of replacement Cracked section may be isolated rather than shut down entire boiler	Heavy Hard to clean High standby losses	Insulated jacket Type B vent required for gas Push nipples or gaskets required at sections Flue and breeching Safety devices Two-stage firing available May be shipped as a package or broken down for field assembly	Gas service required
	Oil-fired 155-120	100-7,000	80-92				Fuel oil storage and pumping required
	Gas/oil combination 155-125	720-13,500	80-92				Allows for use of least expensive fuel available
	Solid fuel 155-130	148-4,600	N/A				Allows for use of inexpensive fuel or burning of waste by-products
Steel Boilers	Solid fuel 155-130	1,500-18,000	N/A	15 to 25 years useful life Less expensive to purchase Tubes easily accessed for cleaning or replacement Leaking tubes may be plugged for future replacement	Regular maintenance required to extend boiler life and retain efficiency Space allowance to pull or punch tubes	Insulated jacket Type B vent required for gas Flue and breeching Safety devices Two-stage firing available Must be shipped as one piece. (two-piece available on special order.)	Allows for use of inexpensive fuel or burning of waste by-products
	Oil or gas 155-115/135	144-23,435	80-92				Fuel service and/or storage required
	Electric 155-100	6-2,500 KW	100				No flue or chimney required No fuel storage required Large electric service required
Novel Residential Type Boilers	Pulse/condensing gas-fired 155-115	44-134	90-95	No chimney required Less floor space Ease of installation	Noisy Acid waste	PVC through-wall flue Plastic drain lines	Gas service required
	Wall hung gas-fired 155-115	44-64	85	No chimney required Less floor space Ease of installation	Structural support considerations	Through-wall flue	Gas service required Available in cast iron or steel
Furnaces — Warm Air	Electric 155-420	30-141	100	Quick response No freezeup of system Air cleaning feature Cooling option at approximately 1/3 of heating output 15 to 30 years useful life	Space consideration for ductwork Humidification required Large temperature swings Noisy Air filtering required	Insulated jacket Type B vent required for gas Safety devices Shipped as a package	No flue or chimney required No fuel storage required Large electric service required
	Gas 155-420	42-400	80-92				Gas service required
	Oil 155-420	55-400	80-92				Fuel oil storage and pumping required
	Solid fuel 155-420	112-170	N/A				Allows for use of inexpensive fuel or burning of waste by-products, etc.

Figure 5.11a (cont.)

Generation Equipment — Heating (continued)

Heating	Type and R.S. Means Co. Line No.	Capa-city (MBH)	Effi-ciency (%)	Advantages	Disadvantages	Related Equipment and Accessories	Considerations
Makeup Air Unit	Gas 155-461	168-6,275	80-92	Quick response No freezeup of system Air cleaning feature Cooling option at approximately 1/3 of heating output 15 to 30 years useful life	Space consideration for ductwork Humidification required Large temperature swings Noisy Air filtering required	Insulated jacket Type B vent required for gas Safety devices Shipped as a package	Makeup air for kitchens or other large volume considerations May be roof mounted or indoors
Burners	Gas 155-230	35-5,760	80-92	Clean burning Available fuel	Danger of leakage		Usually furnished with boiler or furnace
	Oil 155-230	.5-12 (GPH)	80-92		Oil spill hard to contain or clean up Oil supply vulnerable		Usually furnished with boiler or furnace
	Gas/Oil combination 155-230	400-6,300	80-92	Choice of available fuel	High original cost		May be furnished as original equipment Takes advantage of least expensive fuel available
	Coal stoker 155-230	1,000-7,300	N/A	Inexpensive fuel	Dirty storage and exhaust Needs coal storage space	Conveyor or manual feed	Uses inexpensive coal or waste by-products
Fuel Oil Storage Tanks	Fiberglass underground 155-671	550-48,000 gal.	N/A	Cannot rust Ease of handling Dielectric material	Leak detection difficult	Fill vent and sound piping to grade or higher Hold down pads and anchors	Double wall available Extreme caution in backfilling
	Steel underground 155-671	500-30,000 gal.	N/A	Inherent strength of steel prevents crushing	Leak detection difficult Subject to electrolytic action		Double wall available Normal caution in backfilling
	Steel above ground	275-5,000 gal.	N/A	Ease of replacement No excavation or backfill required	Takes up valuable space		Visual leakage control
Expansion Tanks	Liquid expansion	15-400 gal.	N/A	Can be atmospherically recharged	Subject to flooding Pressures in system can vary	Diverter fittings Drain valve	20-30 years useful life Used on hot and chilled water systems
	Diaphragm (captive air)	19-528 gal.	N/A	Cannot waterlog Constant pressure possible	Diaphragm can rupture or lose seal		20 years (plus) useful life Used on hot and chilled water systems

Figure 5.11a (cont.)

Generation Equipment — Cooling

Packaged Water Chillers
R. S. Means No. 157-110
157-190

Typical Uses: Produce chilled water for cooling coils, fan coils
Useful Life: 15–30 years
Capacity Range: 2–250 tons

Advantages	Disadvantages	Remarks
Direct Expansion Chilled water is piped throughout the building and available on demand. No environmental concern.	Reciprocating type can be noisy.	Most chillers operate on electricity.
Absorption Very competitive where heat source is inexpensive. Few mechanical parts. Use little electrical power for operating.	High initial cost. Requires frequent maintenance. Disposal of lithium bromide is an environmental concern.	Energy source is steam or hot water. Gas is also used frequently.

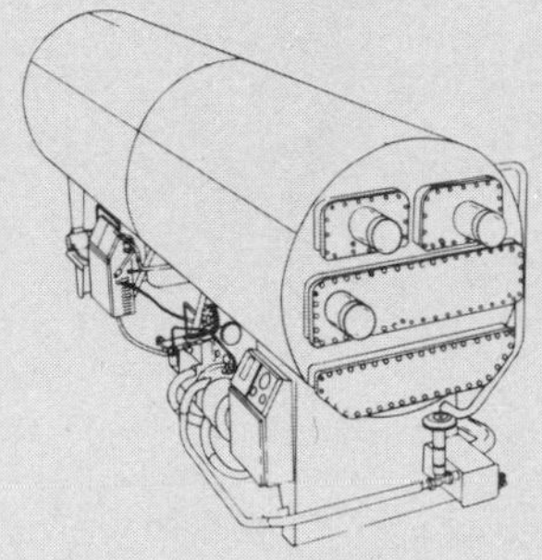

Absorption, Gas-Fired, Air Cooled

Condensers
R. S. Means No. 157-225

Typical Uses: Cools refrigerant gases leaving compressor—air cooled or water cooled
Useful Life: 15–30 years
Capacity Range: 20–100 tons (air cooled)

Advantages	Disadvantages	Remarks
Air Cooled Cools refrigerant gas directly. Fewer overall components to maintain. Can be grouped for larger capacities.	Limited in size but can be grouped as modules. Noisy.	Air cooled preferred when available. Heavy loads—check with structural engineer.
Water Cooled Generally integral with hydronic chiller.	Cooling tower or evaporative condenser required. Freezing of water lines a problem in northern climates if winter operation is needed.	Evaporative condensers are very efficient under certain conditions and can be used for the cooling options. Local codes regulate their use due to heavy water consumption.

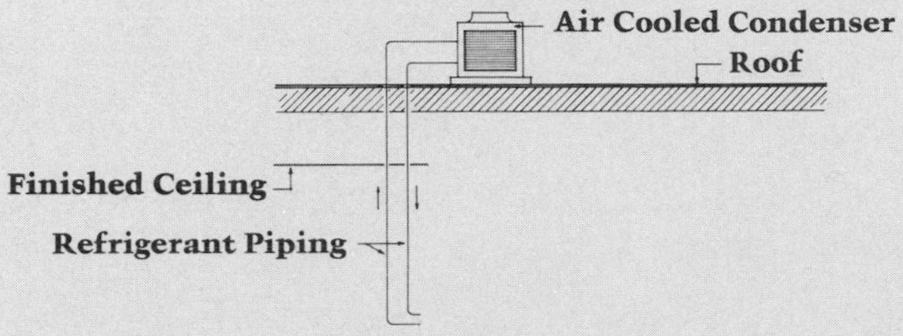

Air Cooled Condenser
Roof
Finished Ceiling
Refrigerant Piping

Air Cooled Condenser

Figure 5.11b

Generation Equipment — Cooling (continued)

Rooftop Air Conditioning Units
R. S. Means No. 157-180

Typical Uses: Commercial applications
Useful Life: 10–20 years
Capacity Range: 3–100 tons

Advantages	Disadvantages	Remarks
Single package, low initial cost. Ease of installation. No flue required. Valuable floor space not taken up for a mechanical room. Very serviceable. Minimum field connections.	Noisy. May have to be screened or shielded for aesthetic purposes.	Electric for cooling. Gas or electric for heating. Multi-zone capability with some units. Economizer option recommended.

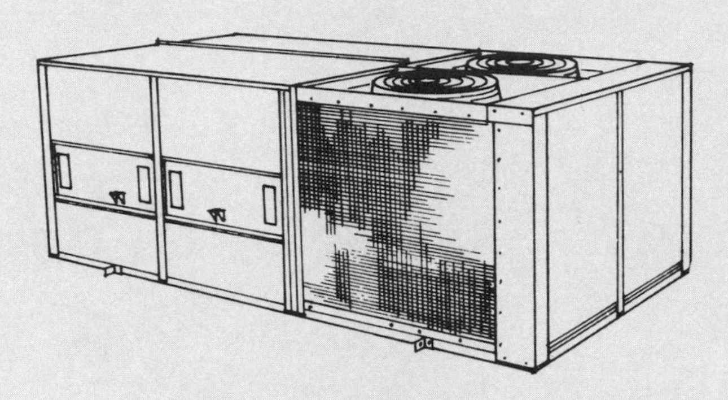

Packaged Rooftop Air Conditioner

Self-Contained Air Conditioning Units
R. S. Means No. 157-185

Typical Uses: Commercial applications
Useful Life: 10–20 years
Capacity Range: 3–60 tons

Advantages	Disadvantages	Remarks
No major remote pieces of generating equipment necessary other than a condenser. Can be free-blow without ducts if located in conditioned space.	Ventilation must be provided separately. Takes up valuable floor space.	Heating coil may be added to these units. Units typically air cooled. Fire regulations may require smoke detector in ductwork to shut down unit upon detection.

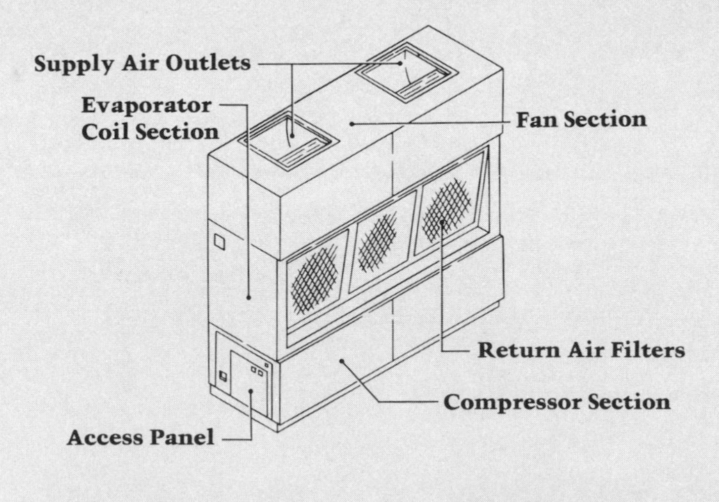

Packaged Vertical Fan Coil Air Handling Unit

Figure 5.11b (cont.)

Generation Equipment — Cooling (continued)

Cooling Towers
R. S. Means No. 157-240

Typical Uses: Cools condenser water from refrigerant condenser
Useful Life: 25–40 years
Capacity Range: 60–1000 tons

Advantages	Disadvantages	Remarks
Natural Draft Low initial cost. Few mechanical parts. Quiet operation.	All cooling towers require anti-bacterial water treatment.	Provide makeup water to replace evaporated water and to wash out (blow down) system as water is fouled.
Mechanical Draft Energy savings possible by modulating fan speed. Acts like natural draft with fans off. High capacity is possible.	Higher initial cost. Noisy.	Mechanical draft towers are available as forced draft or as induced draft. In northern climates, if winter operation is a requirement, closed circuit cooling towers containing antifreeze solution is an alternative.

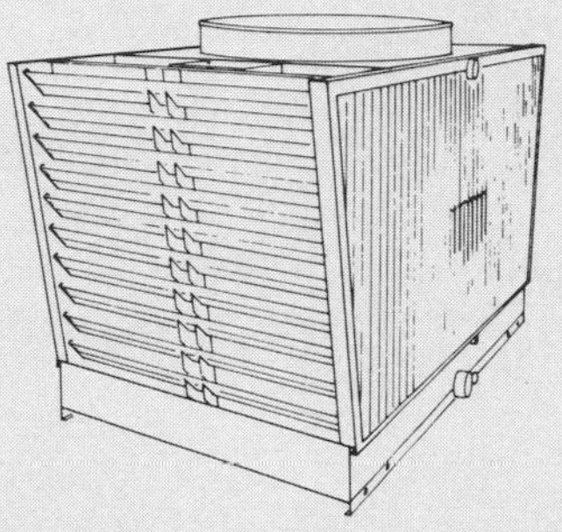

Induced Draft, Double Flow, Cooling Tower

Central Station Air Handling Units
R. S. Means No. 157-125
157-150

Typical Uses: HVAC of building
Useful Life: 15–30 years
Capacity Range: 1300–60,000 cfm

Advantages	Disadvantages	Remarks
Combines all HVAC components into one system. Energy savings via economizers possible. Can be coordinated with VAV systems.	Ducts and shafts require space. Must protect coils from freezing in the event of a power failure.	Heating: steam electricity gas hot water Cooling: electric chilled water direct expansion Check structural loads for floor or roof mounting.

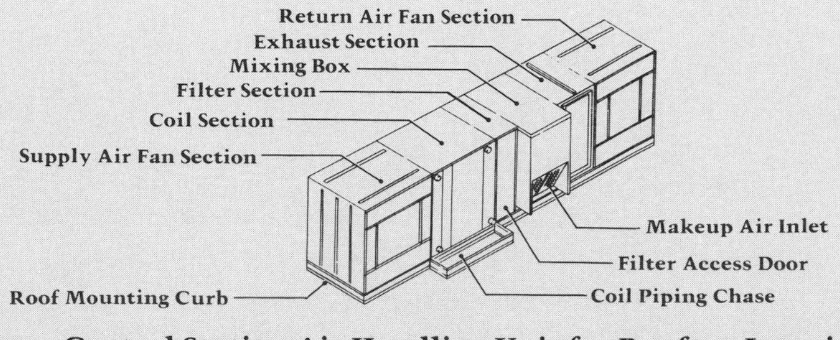

Central Station Air Handling Unit for Rooftop Location

Figure 5.11b (cont.)

Generation Equipment — Cooling (continued)

Split Systems **R. S. Means No. 157-200** **157-270**	Typical Uses: Individual space conditioning remote from mechanical equipment Useful Life: 10–20 years Capacity Range: 1–10 tons	
Advantages	**Disadvantages**	**Remarks**
Individual cooling coil for each tenant. Provides separate cooling for spaces remote from condensing unit location. No noise in conditioned space.	Separate ventilation air required. Long runs of refrigerant piping.	Distance between two sections is limited to approximately 60′. Refrigerant lines between both units must be insulated. Optional heating coil.

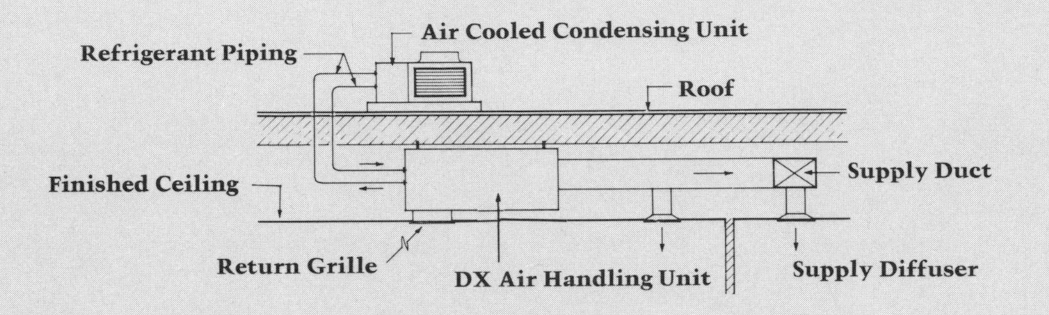

Heat Pumps **R. S. Means No. 157-160**	Typical Uses: Heating and cooling of spaces Useful Life: 10–20 years Capacity Range: 1.5–50 tons	
Advantages	**Disadvantages**	**Remarks**
Individual metering of space possible for electricity. Energy-efficient, as some heat and others cool. Boiler and chiller off for portions of spring and fall.	Compressor noise in space. For winter cooling, a glycol solution in a closed circuit cooling tower or air cooled condenser is necessary.	Boiler must be controlled to limit supply water to 90°F maximum. Low operating temperatures are conducive to plastic piping.

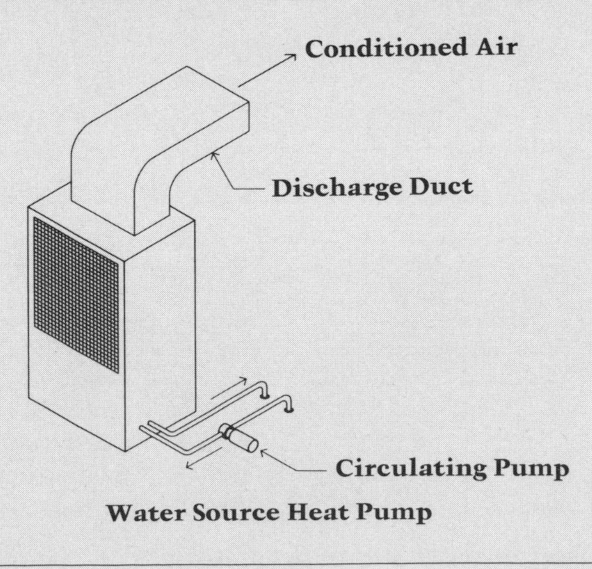

Figure 5.11b (cont.)

CHAPTER SIX

GENERATING EQUIPMENT ASSEMBLIES

The previous chapter describes and illustrates individual generation equipment components. Now those components are put together to create some common generation equipment assemblies. Illustrations and discussion show how generation equipment systems operate, depending on the type of fuel available and the configuration and purpose of the building to be heated or cooled.

Generation equipment for buildings is usually composed of *primary* and *secondary equipment*. The following paragraphs are a review of these basic categories.

Primary Equipment

Primary equipment for heating systems includes boilers and furnaces. Boilers are used with hot water and steam systems. Located in mechanical rooms, boilers require fuel lines, flues, and combustion air. Fire safety regulations also require heat detectors and fire-rated walls. Furnaces are used with air systems and require the same ancillary equipment as boilers, in addition to the necessary space for ductwork.

Primary equipment for cooling systems includes compression cycle and absorption cycle refrigeration equipment.

- *Compression cycle refrigeration equipment*
  - Through-the-wall direct expansion units (common window air conditioners) are noisy, provide poor humidity control, and block windows. These units are routinely used in residences and small offices.
  - Chillers, condensers, and cooling towers are large pieces of equipment and are usually located in a mechanical room or on the roof. Installation of these components requires the attention of a structural engineer who must ensure adequate provisions for loads and vibration. Noise control must also be provided.
  - Split system units are composed of two components installed remotely from one another. The fan and evaporator are usually installed in the ceiling in the conditioned space, and the condenser or condensing unit

placed outdoors. The condensate line from the evaporator coil is then run to a convenient open waste.

- *Absorption cycle refrigeration equipment* is typically cumbersome and heavy. Installation of these components requires the attention of a structural engineer. Also required are vibration isolation and noise control.

Secondary Equipment

Examples of secondary equipment include pumps, fans, and heat exchangers.

- Pumps — circulating pumps for hot or chilled water in the building and pumps between chiller and cooling tower.
- Fans — supply, return, and exhaust fans for buildings.
- Heat exchangers for main equipment — steam to hot water; steam to warm air; hot water to warm air; and chilled water or refrigerant to cool air.

Typical Assemblies

Figures 6.1 through 6.7 illustrate several common generating equipment assemblies. These illustrations can be used as references in the selection and pricing process since they show the major components of various systems and provide Means line numbers for each component.

Figure 6.1 shows a supplied steam assembly. This type of system is commonly used in large urban areas where the local utility has waste steam available, but it may also be used in any large facility that is comprised of a group of buildings serviced by a central steam-generating power plant. The central power source is the *primary* equipment. It is connected to mechanical rooms which house the *secondary* equipment in the serviced buildings. In the supplied steam system, the steam pressure is reduced and can then be used to heat a building, either directly or following conversion to hot water. Among the advantages of this system is the fact that it requires neither a chimney nor fuel storage.

An oil-fired (boiler) steam generation system is shown in Figure 6.2. Where steam is the medium chosen for heating, an oil-fired boiler is typically chosen as the generation equipment. In this case, the system is located in the boiler room of the building it serves. The steam is distributed through convectors, radiators, coils, and air handling units. The oil for this type of steam generation system is stored in a tank in the building or underground.

The system shown in Figure 6.3 differs from the previous model in that it is fired by gas and uses hot water rather than steam. The gas is supplied by a utility in the street. The hot water is distributed by radiators and convectors. This is a very common heating system for both residential and commercial applications. For rural areas not served by a local gas utility, propane gas is an option and can be stored in a tank on the site.

Figure 6.4 is an illustration of an oil-fired warm air furnace system. The primary generation equipment consists of a furnace, rather than a hydronic boiler as shown in the previous systems. The heat is distributed through ductwork to grilles or registers.

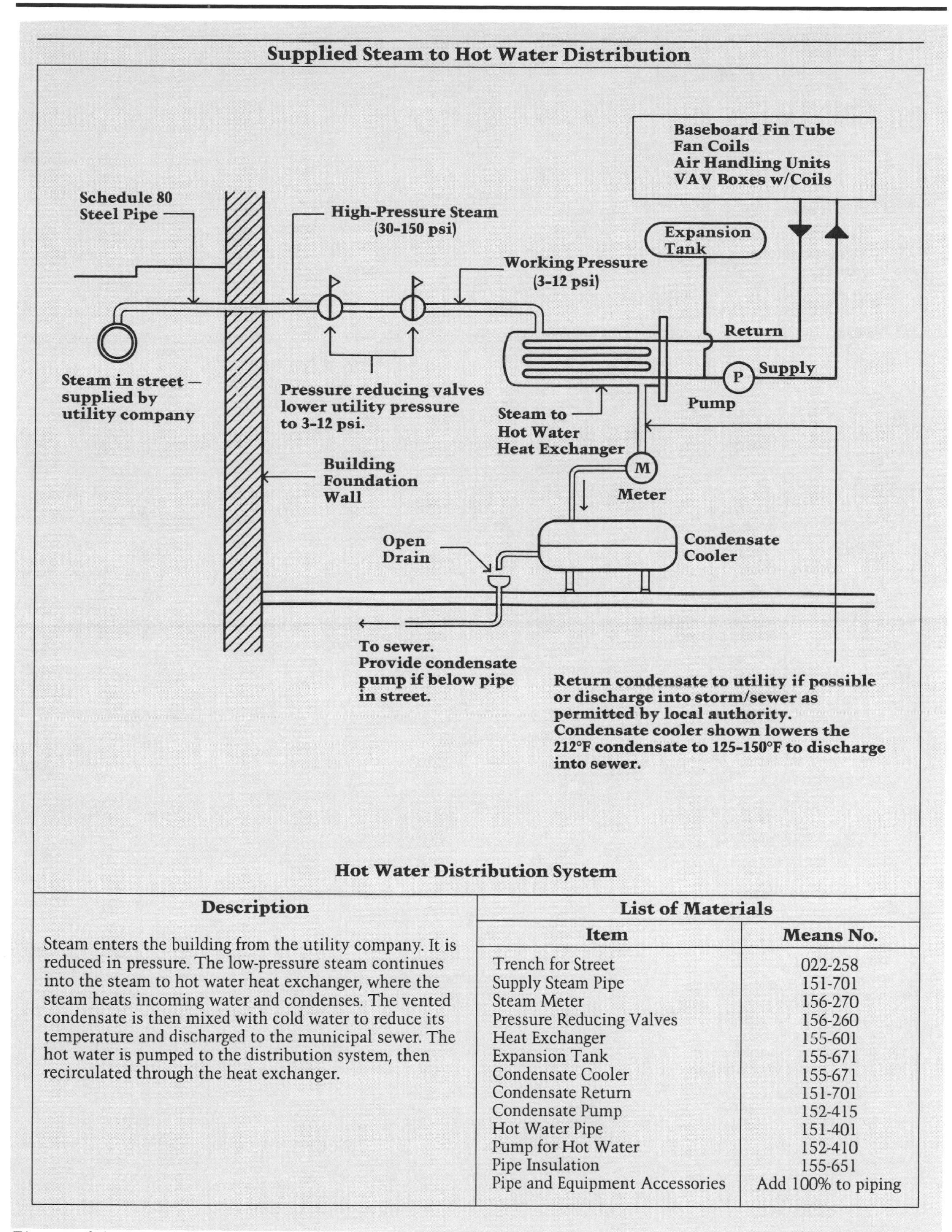

Hot Water Distribution System

Description

Steam enters the building from the utility company. It is reduced in pressure. The low-pressure steam continues into the steam to hot water heat exchanger, where the steam heats incoming water and condenses. The vented condensate is then mixed with cold water to reduce its temperature and discharged to the municipal sewer. The hot water is pumped to the distribution system, then recirculated through the heat exchanger.

List of Materials

Item	Means No.
Trench for Street	022-258
Supply Steam Pipe	151-701
Steam Meter	156-270
Pressure Reducing Valves	156-260
Heat Exchanger	155-601
Expansion Tank	155-671
Condensate Cooler	155-671
Condensate Return	151-701
Condensate Pump	152-415
Hot Water Pipe	151-401
Pump for Hot Water	152-410
Pipe Insulation	155-651
Pipe and Equipment Accessories	Add 100% to piping

Figure 6.1

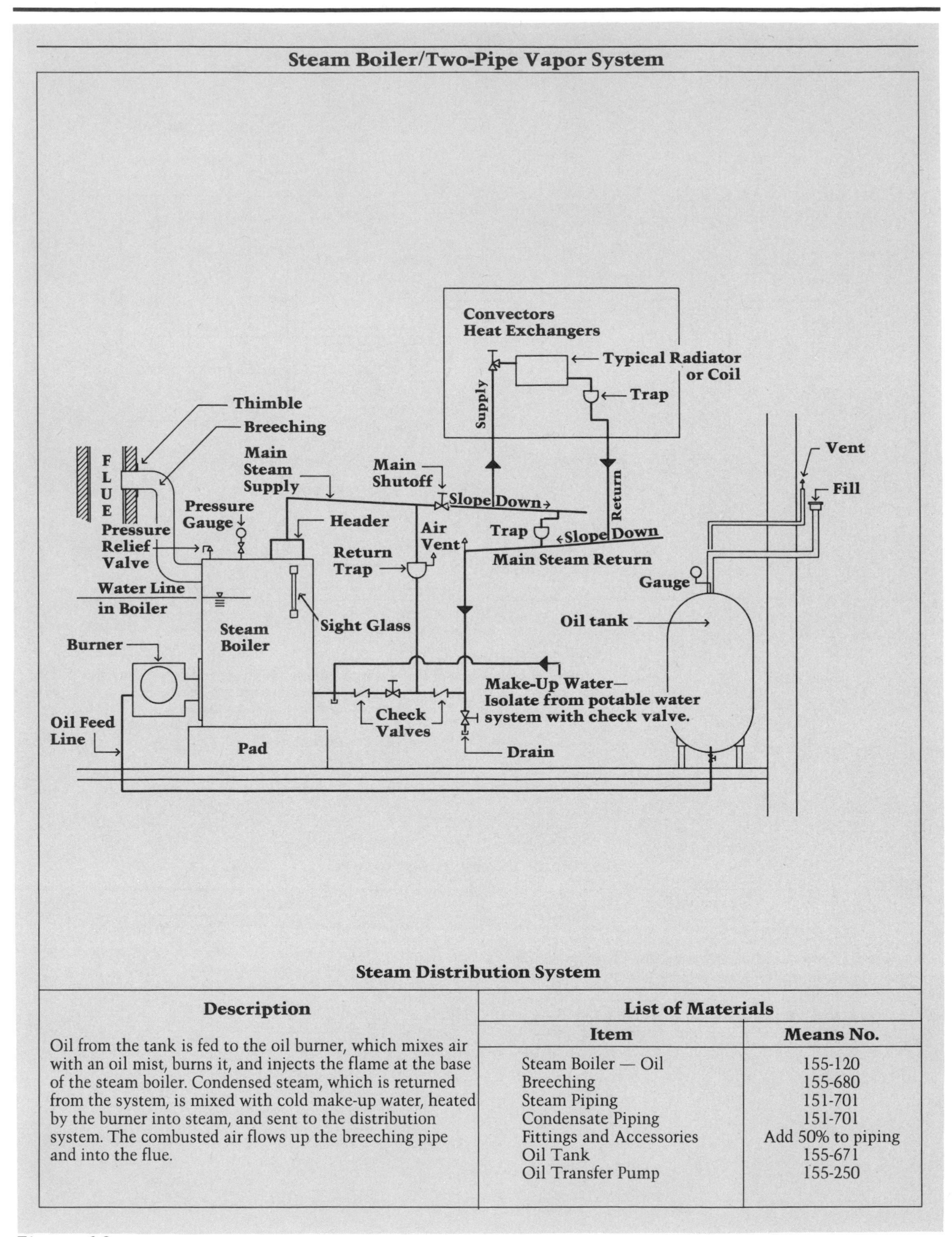

Description

Oil from the tank is fed to the oil burner, which mixes air with an oil mist, burns it, and injects the flame at the base of the steam boiler. Condensed steam, which is returned from the system, is mixed with cold make-up water, heated by the burner into steam, and sent to the distribution system. The combusted air flows up the breeching pipe and into the flue.

List of Materials

Item	Means No.
Steam Boiler — Oil	155-120
Breeching	155-680
Steam Piping	151-701
Condensate Piping	151-701
Fittings and Accessories	Add 50% to piping
Oil Tank	155-671
Oil Transfer Pump	155-250

Figure 6.2

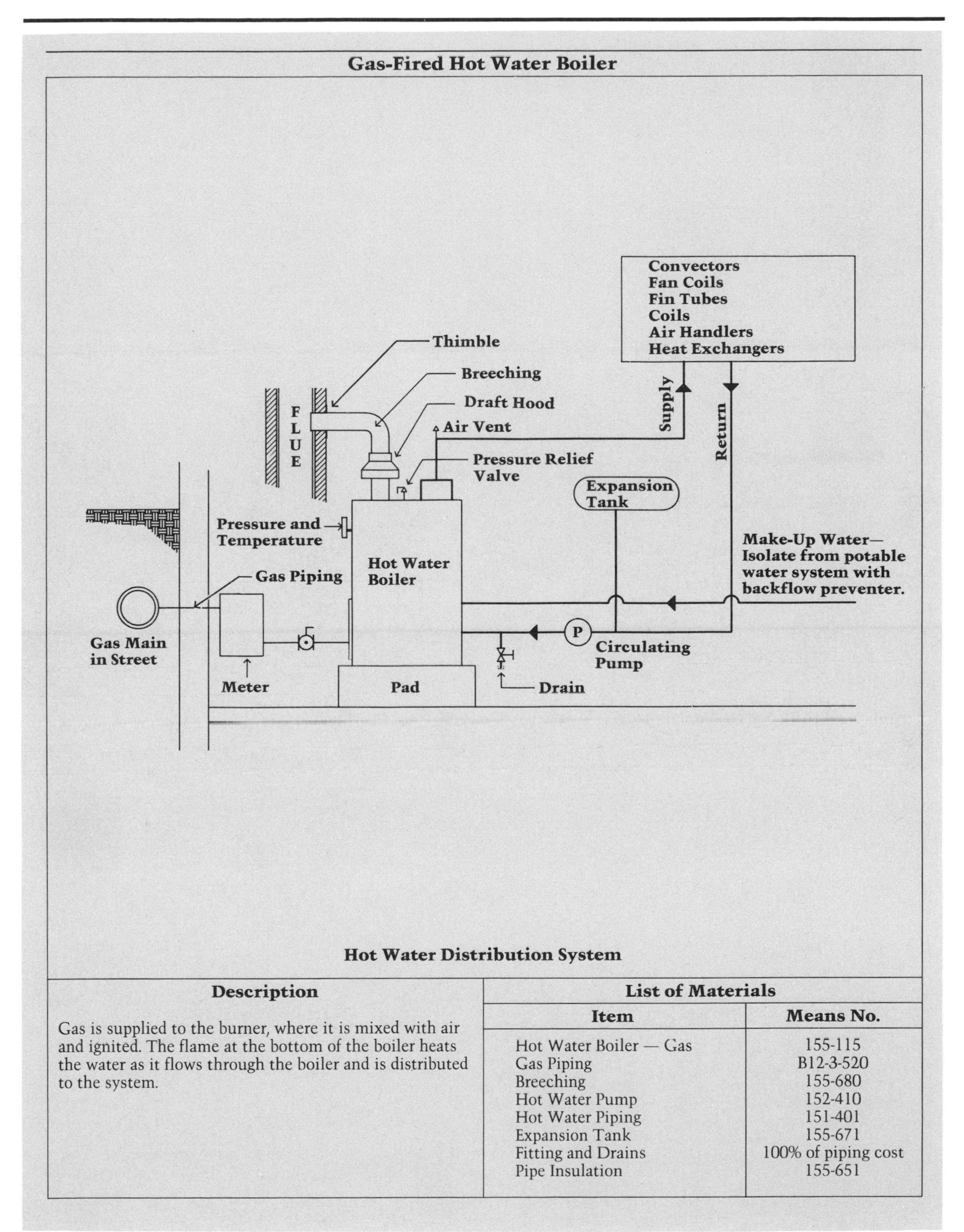

Description

Gas is supplied to the burner, where it is mixed with air and ignited. The flame at the bottom of the boiler heats the water as it flows through the boiler and is distributed to the system.

List of Materials

Item	Means No.
Hot Water Boiler — Gas	155-115
Gas Piping	B12-3-520
Breeching	155-680
Hot Water Pump	152-410
Hot Water Piping	151-401
Expansion Tank	155-671
Fitting and Drains	100% of piping cost
Pipe Insulation	155-651

Figure 6.3

Warm Air Furnace

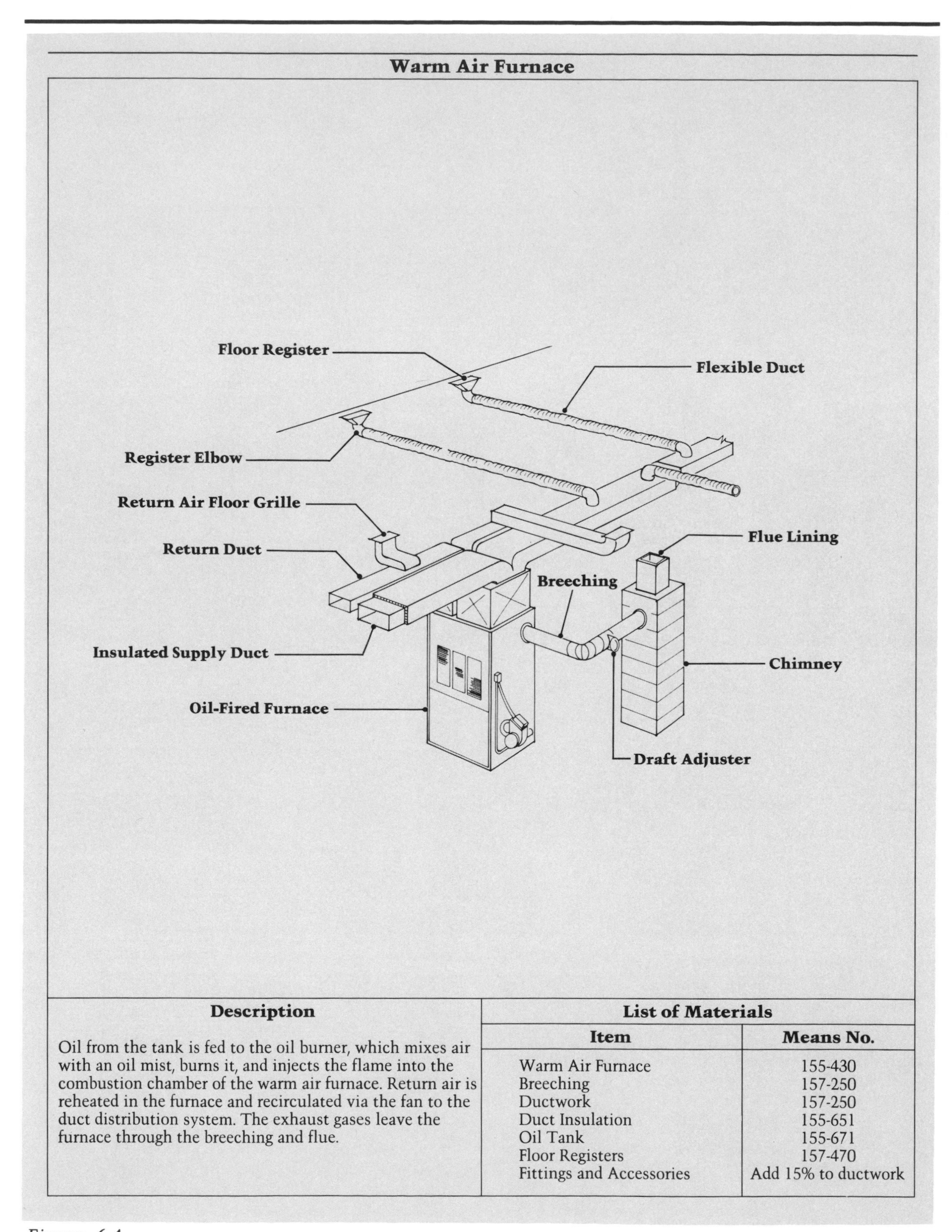

Description

Oil from the tank is fed to the oil burner, which mixes air with an oil mist, burns it, and injects the flame into the combustion chamber of the warm air furnace. Return air is reheated in the furnace and recirculated via the fan to the duct distribution system. The exhaust gases leave the furnace through the breeching and flue.

List of Materials

Item	Means No.
Warm Air Furnace	155-430
Breeching	157-250
Ductwork	157-250
Duct Insulation	155-651
Oil Tank	155-671
Floor Registers	157-470
Fittings and Accessories	Add 15% to ductwork

Figure 6.4

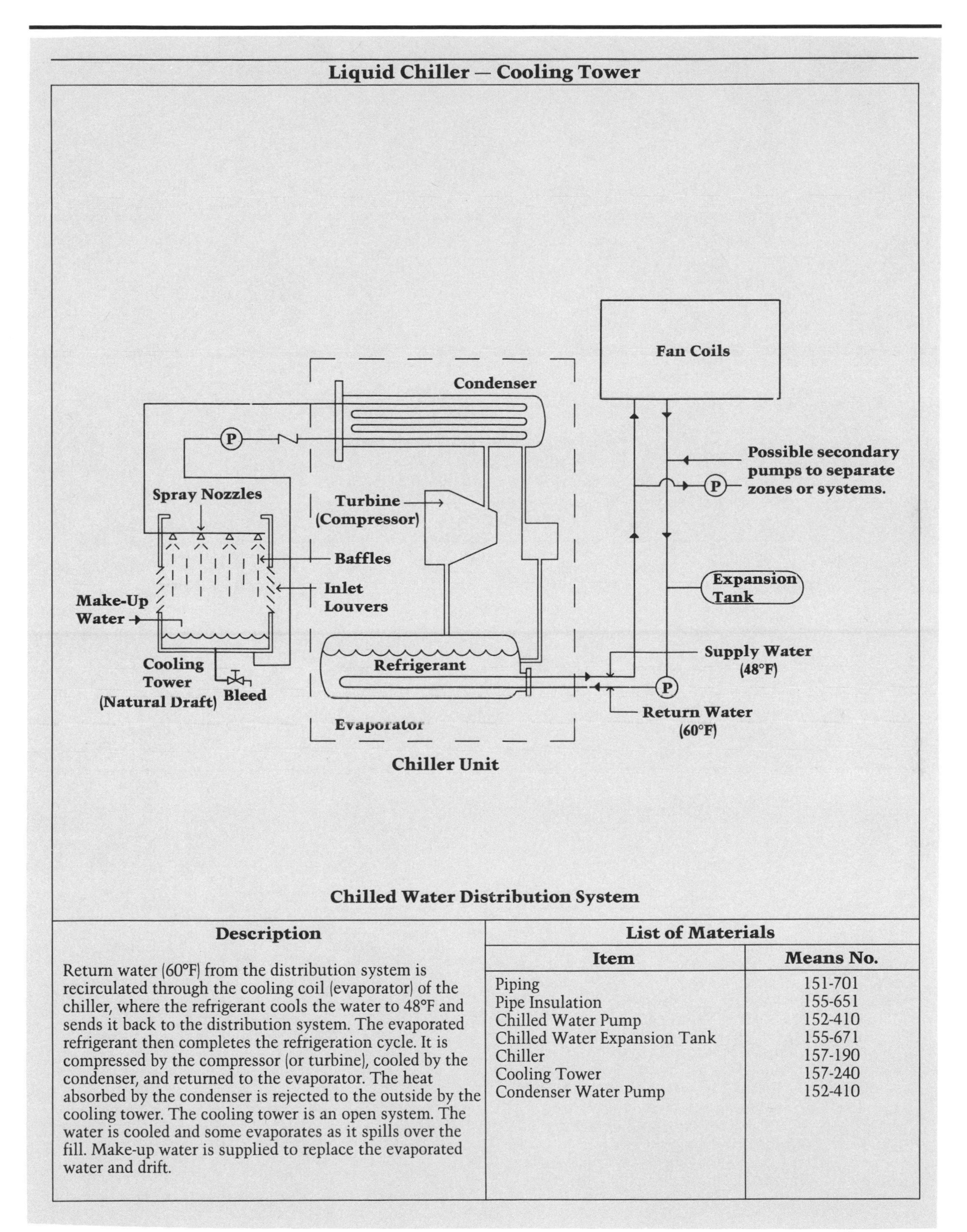

Description

Return water (60°F) from the distribution system is recirculated through the cooling coil (evaporator) of the chiller, where the refrigerant cools the water to 48°F and sends it back to the distribution system. The evaporated refrigerant then completes the refrigeration cycle. It is compressed by the compressor (or turbine), cooled by the condenser, and returned to the evaporator. The heat absorbed by the condenser is rejected to the outside by the cooling tower. The cooling tower is an open system. The water is cooled and some evaporates as it spills over the fill. Make-up water is supplied to replace the evaporated water and drift.

List of Materials

Item	Means No.
Piping	151-701
Pipe Insulation	155-651
Chilled Water Pump	152-410
Chilled Water Expansion Tank	155-671
Chiller	157-190
Cooling Tower	157-240
Condenser Water Pump	152-410

Figure 6.5

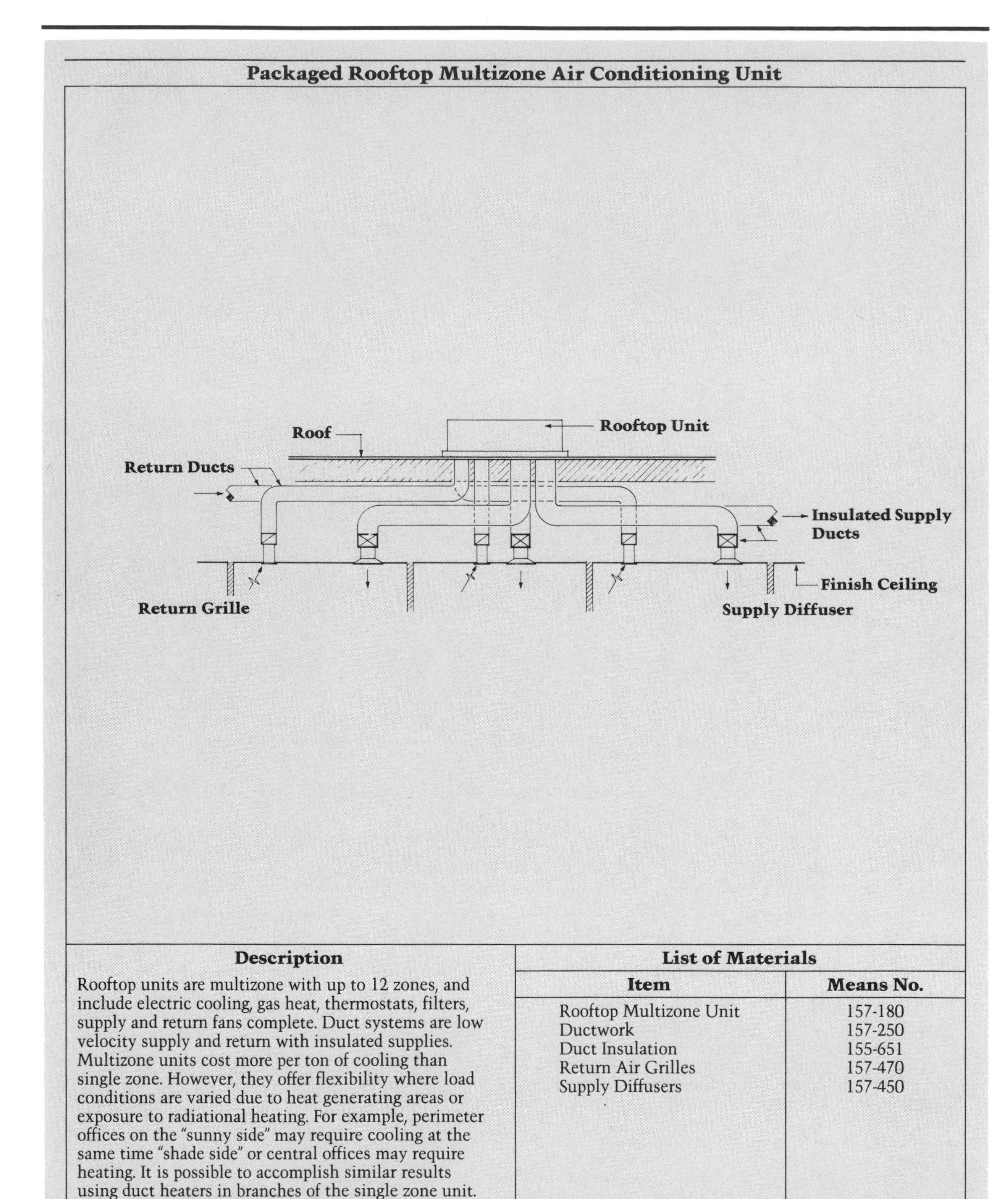

Packaged Rooftop Multizone Air Conditioning Unit

Description

Rooftop units are multizone with up to 12 zones, and include electric cooling, gas heat, thermostats, filters, supply and return fans complete. Duct systems are low velocity supply and return with insulated supplies. Multizone units cost more per ton of cooling than single zone. However, they offer flexibility where load conditions are varied due to heat generating areas or exposure to radiational heating. For example, perimeter offices on the "sunny side" may require cooling at the same time "shade side" or central offices may require heating. It is possible to accomplish similar results using duct heaters in branches of the single zone unit. However, heater location could be a problem and total system operating energy efficiency could be lower.

List of Materials

Item	Means No.
Rooftop Multizone Unit	157-180
Ductwork	157-250
Duct Insulation	155-651
Return Air Grilles	157-470
Supply Diffusers	157-450

Figure 6.6

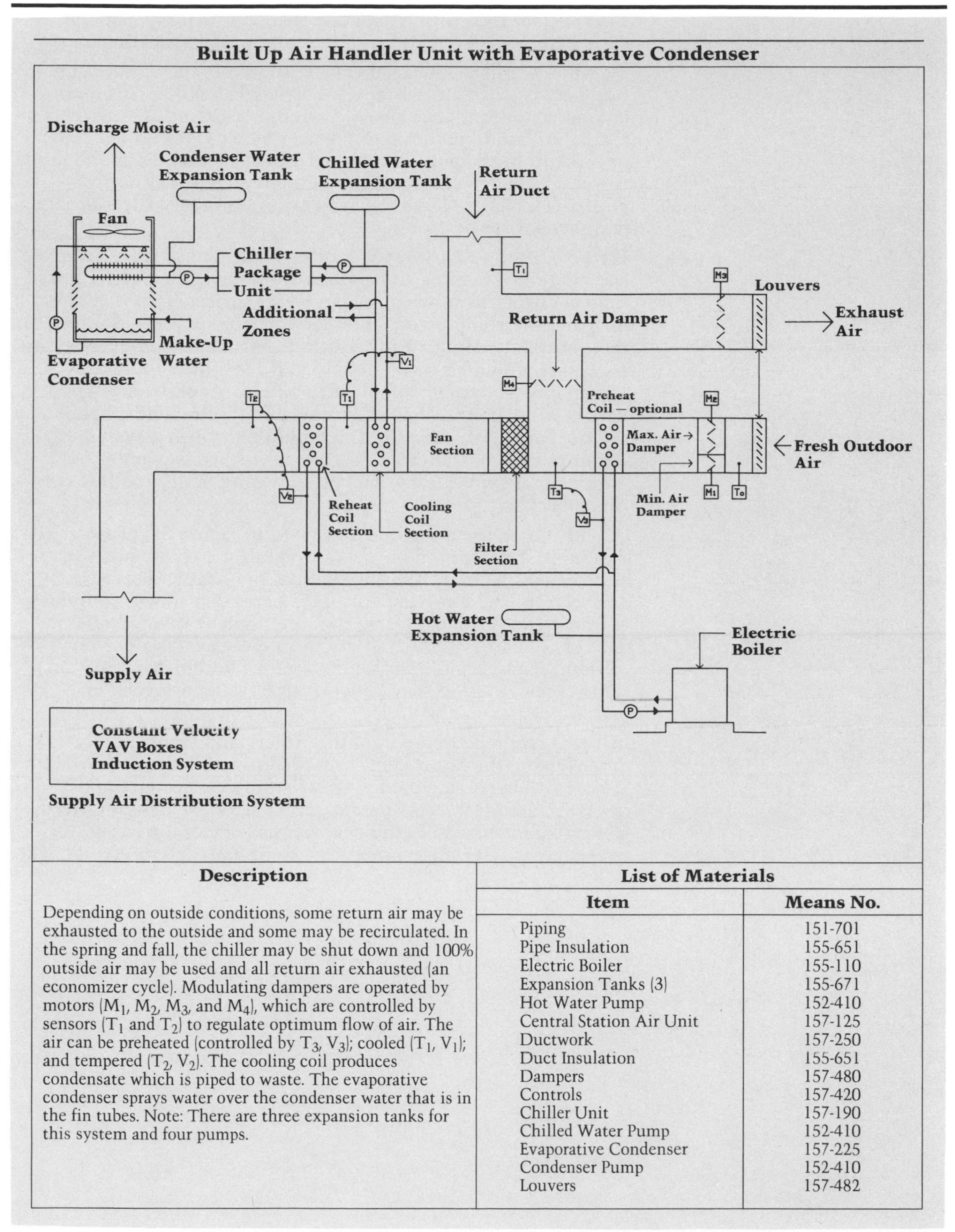

Description

Depending on outside conditions, some return air may be exhausted to the outside and some may be recirculated. In the spring and fall, the chiller may be shut down and 100% outside air may be used and all return air exhausted (an economizer cycle). Modulating dampers are operated by motors (M_1, M_2, M_3, and M_4), which are controlled by sensors (T_1 and T_2) to regulate optimum flow of air. The air can be preheated (controlled by T_3, V_3); cooled (T_1, V_1); and tempered (T_2, V_2). The cooling coil produces condensate which is piped to waste. The evaporative condenser sprays water over the condenser water that is in the fin tubes. Note: There are three expansion tanks for this system and four pumps.

List of Materials

Item	Means No.
Piping	151-701
Pipe Insulation	155-651
Electric Boiler	155-110
Expansion Tanks (3)	155-671
Hot Water Pump	152-410
Central Station Air Unit	157-125
Ductwork	157-250
Duct Insulation	155-651
Dampers	157-480
Controls	157-420
Chiller Unit	157-190
Chilled Water Pump	152-410
Evaporative Condenser	157-225
Condenser Pump	152-410
Louvers	157-482

Figure 6.7

This system requires fuel storage on the site. It is a popular choice for residential and small commercial applications.

A primary chilled water generating source is shown in Figure 6.5. This is a centrifugal-type, water-cooled chiller. It supplies chilled water to a distribution system consisting of duct coils and/or air handling units. Condenser water gives up the heat rejected from the building through the use of an outdoor cooling tower. Cooled condenser water is recirculated back to the chiller. This type of cooling system is commonly used for large commercial applications.

Figure 6.6 shows a packaged rooftop multi-zoned unit. This is a self-contained heating and cooling assembly. The only required field connections for this system are electricity to power the fans and the compressor, and ductwork for supply and return air distribution (cooling). If the heating is also supplied by this unit, a gas line would be required to feed a gas burner or, less commonly, electricity supplied to an electric resistance coil. The multi-zone feature of this unit provides individual zone control for specific areas of the building through a system of dampers and controllers. This type of system is popular for commercial applications, such as shopping malls and low-rise office buildings.

A built-up air handling unit is shown in Figure 6.7. This assembly is comprised of several modules: a cooling coil section, heating coil section, filtering section, pre-heat coil section, fan section, and a mixing box. This particular model combines primary and secondary generation equipment in one location. Hot or cold air is circulated via the supply air distribution system, fed by a chiller assembly and a hydronic source. This type of system may be used for large commercial and industrial applications.

All generation equipment, both primary and secondary, can be priced using the latest edition of *Means Mechanical Cost Data*. This cost reference provides *installed* prices, not just material costs. Using Means cost data together with the design criteria presented in this book, the designer can select the most economical system or component to fit a predetermined budget.

CHAPTER SEVEN

DRIVING SYSTEMS

Driving systems consist primarily of pumps and fans, the equipment used to "drive" water or air through pipes or ducts. These pumps and fans are normally located in the mechanical room and drive air or water across or through the generation equipment. In order to properly select a driving system, the following factors must be determined.

- **Type:** There are several choices of pumps and fans for each specific system.
- **Capacity:** The capacity is measured in gallons of liquid per minute (gpm) for pumps and cubic feet of air per minute (cfm) for fans.
- **Head or pressure:** Head or pressure is measured as feet of head (H) for pumps and inches of water (wg) for fans. One foot of head is the pressure at the bottom of a column of water one foot high. One inch of water is the pressure at the bottom of a column of water one inch high. These pressure designations are illustrated in Figure 7.1.

Pumps

Pumps are generally used on *closed loop systems*, where the fluid is continuously recycled and contained within a piping system. The pump must be sized to provide the proper quantity of water (gpm), and must be rated large enough to provide the required pressure. Pressure from a pump is used to overcome the friction, or drag, that exists between the fluid and the pipe and within itself as it moves through the piping system. (This pressure/friction relationship is discussed in Chapter 9, "Distribution and Driving Systems.")

Some pumps are used in *open systems*. In open systems, the fluid leaves the pipe and is exposed to the atmosphere. Water pumps for evaporative cooling towers are used in open systems, because during the circuit, the water leaves the pipe as it enters "open" air while it spills over the cooling tower fill. Pumps in an open system must have a pump head large enough to compensate for friction losses, and must have the additional

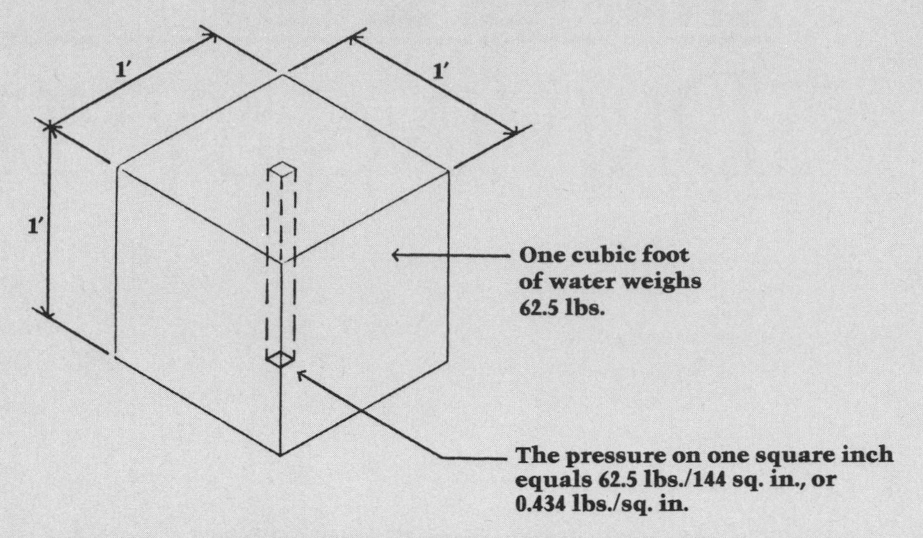

One foot of head — Used for Hydronic Systems

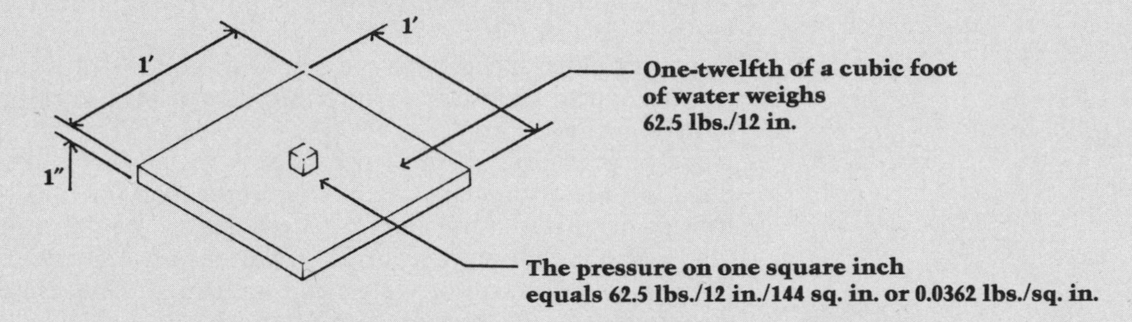

One inch of pressure (Hg)—Used for Gaseous Systems

Pressure and Head

Figure 7.1

capacity to lift the liquid from one elevation to another. This application for the evaporative condenser pump is shown in Figure 6.7.

Centrifugal Pumps

The most commonly used pumps in HVAC systems are centrifugal, driven by an electric motor. Most pump bodies are cast iron. Bronze bodies are available for potable water systems at additional cost.

Centrifugal pumps are comprised of an impeller, which rotates and drives the fluid; the casing, which surrounds the impeller; and the electric motor, which turns the impeller. Pumps over 3/4 horsepower (HP) should be connected with flexible hoses and mounted on vibration isolators or absorbers on concrete inertia pads in order to avoid transmission of noise and vibrations to the piping system.

Figure 7.2 lists the types of pumps that are commonly used in HVAC systems. This chart can be used as a guide for selection of pumps, in conjunction with manufacturers' literature and space considerations. (See also the equipment data sheets at the end of this chapter.)

Pump Curves

Pumps provide less water flow (in gpm) as the head increases. Pump curves are diagrams that plot how a particular pump performs over the range of normal use. Centrifugal pumps, for example, have a predictable pattern, or curve, throughout their range. Typical performance curves for a specific type of pump are shown in Figure 7.3.

Pump curves plot *head* versus *flow* for each pump. Pump *size* on the curves refers to the nominal size of the pump connection. As a pump delivers more gallons per minute (gpm), the head, or pressure, imparted to the water declines. The reverse is also true. A pump connected to a system with a low head requirement will deliver more gpm than when connected to a system with a high head. Generally, centrifugal pump curves drop rapidly and are considered nearly vertical. Examine the "ideal" pump curve shown in Figure 7.4 and compare it with the actual curves from Figure 7.3 shown for 20 gpm at a 6.5 foot head.

Whenever a pump is selected, it is rare that the actual head and flow of the pump exactly match the head and flow requirements of the system design. It is better to slightly oversize the pumps to allow for future connections and site balancing. Figure 7.4 shows the system curve passing through the system requirements of 6.5 feet of head and 20 gpm. It also intersects the two pump curves for 1-1/4" and 1-1/2" pumps. The 1-1/4" pump is too small, since it would, if connected to the system, only deliver 18 gallons per minute, a rate that would "starve" some of the equipment. The 1-1/2" pump, on the other hand, delivers 21 gpm — slightly more than necessary — and it should be selected. A perfect 20 gpm can be achieved by "throttling down" slightly, although this is rarely necessary.

Efficiency

Another consideration in pump selection is efficiency. Usually more than one type of pump is suitable. When this situation occurs, the pump with the highest efficiency in the design range is selected. The precise diameter of the impeller may also be specified.

When selecting a pump, the design engineer should choose a type that will avoid cavitation. Cavitation occurs when the pressure on the suction side of the pump falls below the vapor

Selection Guide for Pump Types

Centrifugal Pump	Means No.	Use	Remarks
Circulator	152-410	Small systems and residential heating and domestic hot water recirculation	In line
Close Coupled	152-410	Hot water circulation (heating and domestic hot water) Cold water booster (domestic water)	High-temperature seals
Base Mounted (single stage double suction)	152-410	Primary and secondary chilled water	
Base Mounted (multi stage)	152-430	Condenser water (cooling tower)	Check suction pressure, strainer on supply side
In Line (vertical)	152-410		
Condensate Pump	152-415	Steam (condensate) return	

Most pumps listed above are single-suction, volute, flexible-coupled with a single impeller except where noted. Multiple impellers are available with most pump types.

Figure 7.2

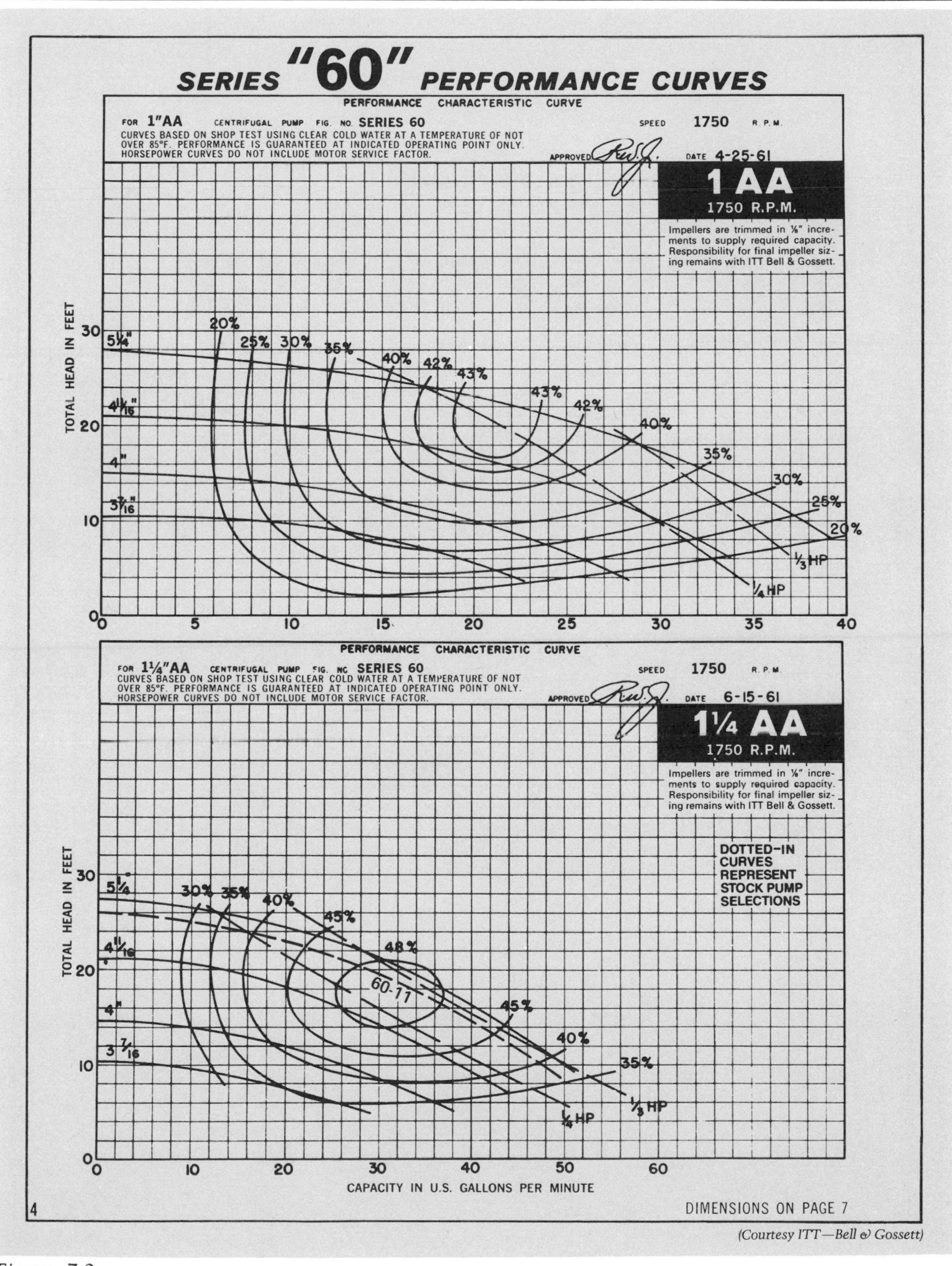

(Courtesy ITT—Bell & Gossett)

Figure 7.3

SERIES "60" CENTRIFUGAL PUMPS

AVAILABLE FROM STOCK

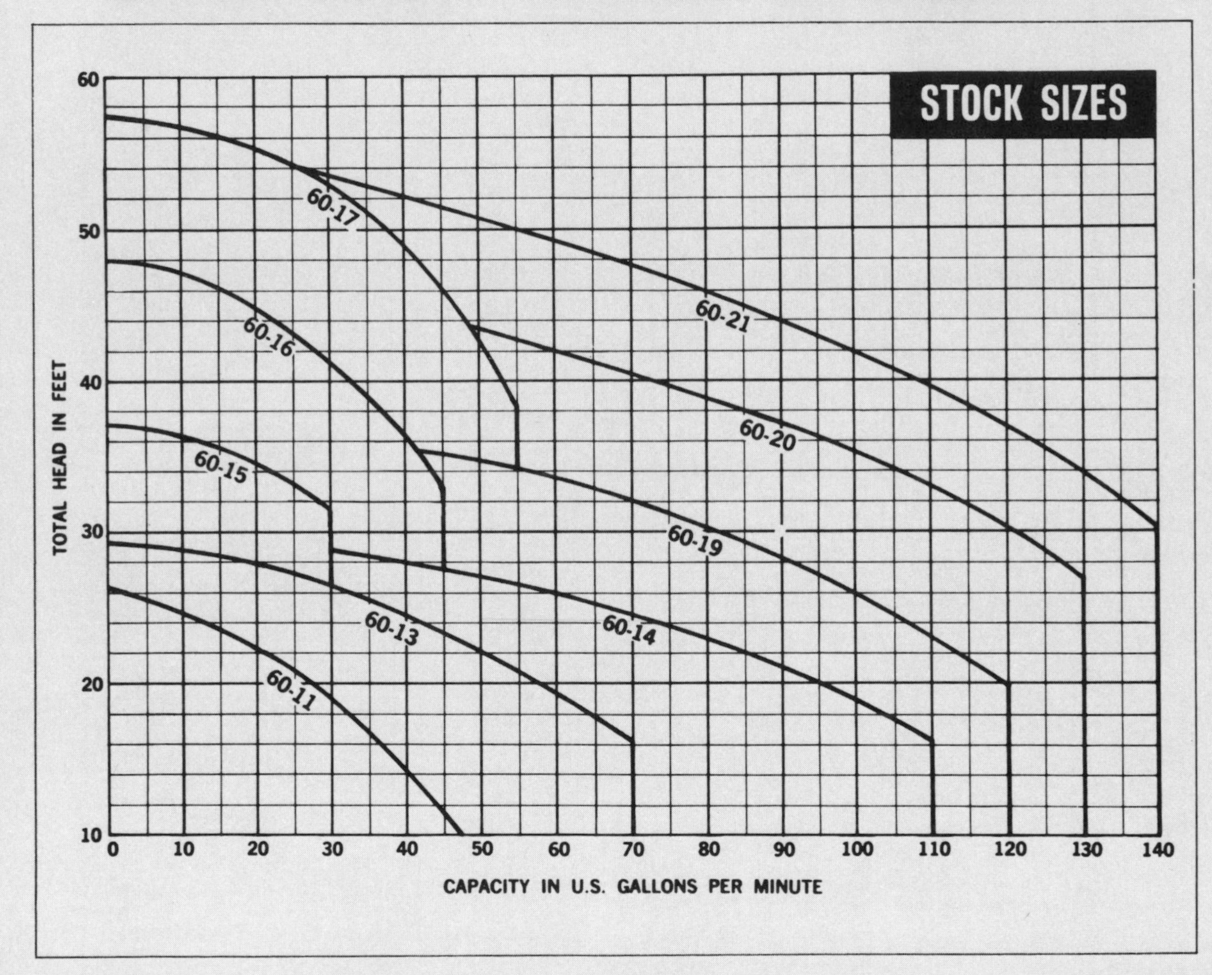

2

(Courtesy ITT—Bell & Gossett)

Figure 7.3 (cont.)

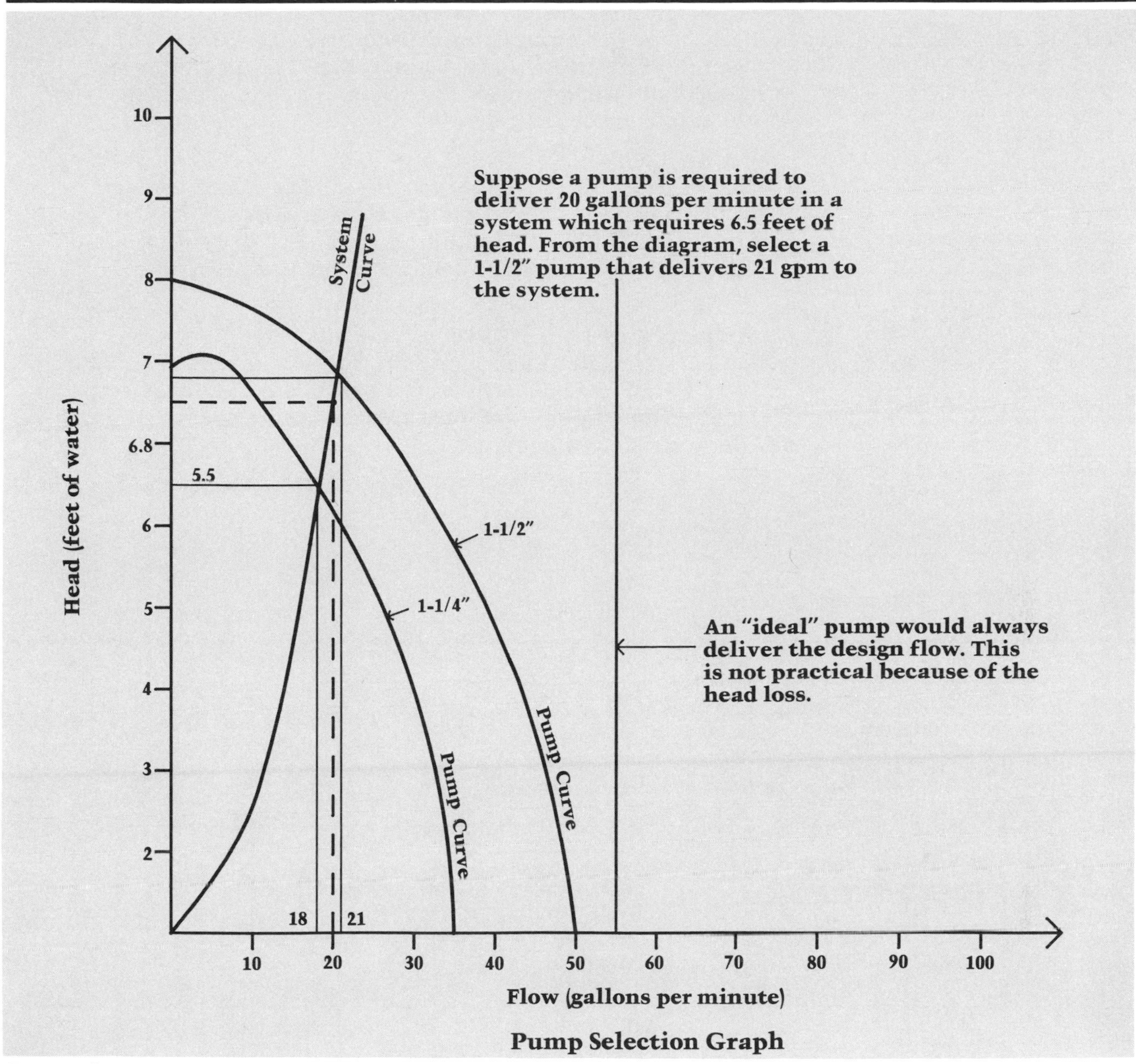
Suppose a pump is required to deliver 20 gallons per minute in a system which requires 6.5 feet of head. From the diagram, select a 1-1/2" pump that delivers 21 gpm to the system.
An "ideal" pump would always deliver the design flow. This is not practical because of the head loss.
System Curve
1-1/2"
1-1/4"
Pump Curve
Pump Curve
5.5
18
21
10
9
8
7
6.8
6
5
4
3
2
Head (feet of water)
10
20
30
40
50
60
70
80
90
100
Flow (gallons per minute)
Pump Selection Graph

Figure 7.4

pressure, thus causing vapor (gas) pockets within the pump. Each pump has a rating for the required *net positive suction head* (NPSH), which must be met. Meeting this rating is particularly important for hot water systems, where the vapor pressure is most critical.

Pump Laws

The relationships between flow, head, speed, horsepower, impeller diameter, and fluid are shown in Figure 7.5. For known conditions, the pump laws are used to predict pump characteristics when one condition, such as pump speed, is changed.

Arrangement of Pumps

For small installations, such as residences or pumps smaller than 3/4 HP, one constant speed pump is normally selected. For larger installations, a variety of pump arrangements are used. These are shown in Figure 7.6.

Pump Laws

Variable	Constant	No.	Law	Formula
Speed of impeller (rpm) (N)	Fluid specific gravity (W)	1	Capacity varies as the speed	$\frac{Q_1}{Q_2} = \frac{N_1}{N_2}$
	Pump impeller diameter (D)	2	Pressure varies as the square of speed	$\frac{P_1}{P_2} = \left\{\frac{N_1}{N_2}\right\}^2$
		3	Horsepower varies as the cube of speed	$\frac{HP_1}{HP_2} = \left\{\frac{N_1}{N_2}\right\}^3$
Pump impeller diameter (D)	Fluid specific gravity (W)	4	Capacity varies as the diameter	$\frac{Q_1}{Q_2} = \frac{D_1}{D_2}$
	Speed of impeller (N)	5	Pressure varies as the square of the diameter	$\frac{P_1}{P_2} = \left\{\frac{D_1}{D_2}\right\}^2$
		6	Horsepower varies as the cube of the diameter	$\frac{HP_1}{HP_2} = \left\{\frac{D_1}{D_2}\right\}^3$
Fluid specific gravity	Speed of impeller (N) Pump impeller diameter (D)	7	Horsepower varies as the specific gravity	$\frac{HP_1}{HP_2} = \frac{W_1}{W_2}$

Q = Flow through pump (gpm)
N = Speed of impeller (rpm)
P = Pressure (feet of head)
HP = Pump horsepower
D = Impeller diameter
W = Fluid specific gravity

Figure 7.5

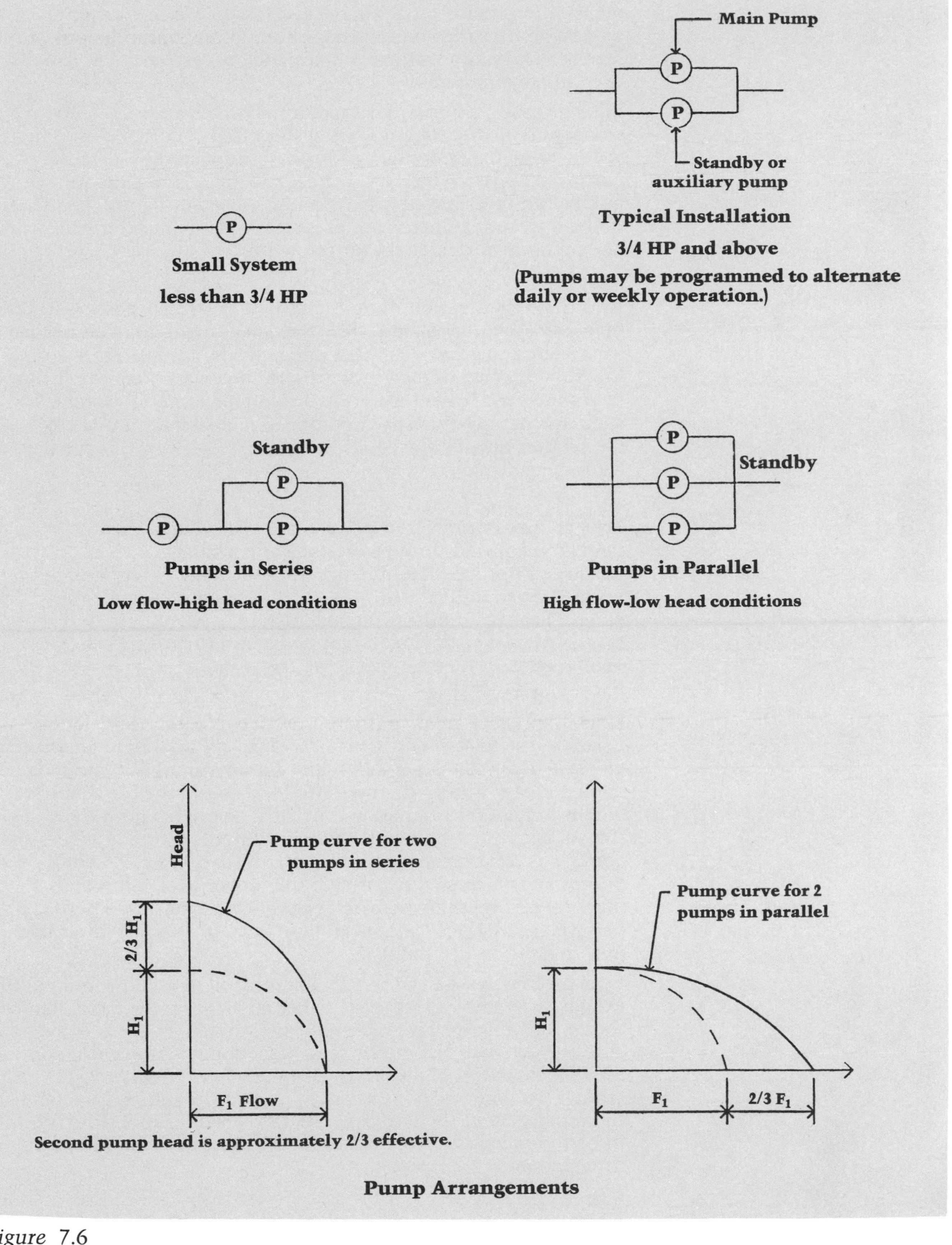
Main Pump
P
P
Standby or
auxiliary pump
Typical Installation
3/4 HP and above
(Pumps may be programmed to alternate daily or weekly operation.)
P
Small System
less than 3/4 HP
Standby
Pumps in Series
Low flow-high head conditions
Standby
Pumps in Parallel
High flow-low head conditions
Head
Pump curve for two pumps in series
2/3 H1
H1
F1 Flow
Second pump head is approximately 2/3 effective.
Pump curve for 2 pumps in parallel
H1
F1
2/3 F1
Pump Arrangements

Figure 7.6

Generally, whenever a pump is greater than 3/4 HP, it is good practice to provide an auxiliary or back-up pump so that there will be no interruption in service should one pump be out of service. Thus, two pumps, a main and an auxiliary, are usually used in large systems.

In some cases, a particular system may have an odd characteristic system curve. A system that requires a high head at low flow, for example, can be accommodated by placing pumps in series. Where a low head but high flow is required, the pumps can be arranged in parallel as shown in Figure 7.6. These two pump arrangements are provided when it is determined that one pump is not practical for the required flow and head.

Staging

Another consideration in pumping is that two speeds (or stages), high and low, may need to be provided. Staging may be needed for systems that have variable demands on pumps. For example, low flow may be needed in winter for hot water, and high flow in summer for chilled water on a two-pipe fan coil system. In such a situation, two-speed pumps or additional pumps are brought on line, or staged, as necessary to pick up the greater cooling load.

Fans

Fans are used most often in open systems. The air from the fan is transmitted through ducts and discharged into or exhausted from the building spaces. When air is recirculated within a system, it is still not a tight enough system to be called closed. Leakage, infiltration, exhaust air, makeup air, and room losses mean that the air is supplied and returned by separate open systems.

Size and Location

Fans must be of an appropriate size to overcome the friction losses in the ducts and fittings. Because air is so light relative to water (0.075 pounds per cubic foot for air versus 62.5 pounds per cubic foot for water), the pressure head to lift the air from one height to another is small and usually ignored. The weight of the water can not be overlooked in pumping systems where the density of water is so much more than that of air. As with pumps, fans are sized to provide the proper flow (cfm) and pressure (in wg) that the particular system requires. There are performance curves for fans and fan laws to regulate their use, just as there are for pumps.

Fans may be exposed in the area served, as in a bathroom ceiling exhaust, or remotely located, either on a roof or in a mechanical room. Fans may be direct drive or belt driven. A direct drive fan is less expensive, but might send objectionable noise directly into the ductwork. Belt drives contain a belt which wraps around two pulleys, or sheaves — one on the motor drive and the other around the fan drive. Belt drives offer more flexibility in performance since the diameter of the sheaves can be changed to regulate the fan speed.

Fan Types

There are three broad categories of fans commonly used in driving systems: centrifugal fans, axial flow fans, and special purpose fans (see Figures 7.7 and 7.8). Centrifugal fans take in air at the side (circular opening) and discharge it radially into the supply stream (rectangular opening). Axial flow fans are mounted in the air stream and boost the pressure of the air. Figures 7.7 and 7.8 may be used to determine the basic fan type; Means line numbers are indicated for use in determining costs in the current edition of *Means Mechanical Cost Data*.

Fans supply air from as low as 50 cfm to more than 500,000 cfm and at pressures from 0.1″ wg to 6″ wg. Fan speeds are adjustable by using two-speed motors, variable pitch sheaves or blades, or variable frequency motors. Noise control is a particular concern in fan design and noise ratings are taken into account in fan selection. When fans are near combustible fumes, nonsparking materials are usually specified for blades and housing.

Fan Curves

Each type of fan has a characteristic performance curve. This information is supplied by the manufacturer. As with pumps, the fan curves plot pressure versus flow and indicate other characteristics, such as efficiencies and horsepower over the range of operation. Typical fan performance characteristics are shown in Figure 7.9 for forward, backward, and radial pitched centrifugal fans.

As with pumps, it is important to select a fan large enough to handle the pressure and flow required for the system. Oversizing slightly (by 10–15 percent) is common practice. Comparisons between various manufacturers and models should be made to select for overall durability, ease of maintenance, cost, and efficiency.

Fan Laws

As with pumps, fan laws may be used to predict fan performance under conditions different from the conditions listed in a fan curve. The fan laws, shown in Figure 7.10, are used to determine the proper fan speed for a system. Slowing the fan to a lower speed is always less expensive than running it at high speed and partially closing dampers (throttling).

Fan Arrangement

Fans may be placed in series or parallel in a manner similar to pumps. In practice, however, this approach is rarely taken because fans come in a wide range of capacities suitable for most installation conditions. Also, the compressibility of air (as compared to water) makes it more difficult to properly pair fans in series or parallel.

When design conditions in a system vary, it is more common to select a two-stage fan or, in larger installations, to utilize variable pitched blades. Advancements have recently been made in varying the frequency of power to motors from the normal 60 hertz. Further developments are likely in this area.

Selection Guide for Fan Types		
Fan Type	**Use**	**Remarks**
Centrifugal Fans		
Backward inclined airfoil	Large systems where additional cost of airfoil is justified by energy efficiency savings	High efficiency Low noise
Backward curved	Most commonly used in high speed applications	High efficiency
Radial	Industrial applications, medium speed	Least efficient Easily repaired
Forward curved	Smaller systems — residential and package units, low speed	Fans in series
Axial Fans		
Propeller	Large volume at low pressure (e.g., room circulation, wall exhaust, and systems with no ducts)	Low efficiency Noisy
Tube axial	Used where space limitations prohibit centrifugal fan	Heavy duty propeller fans
Vane axial	Used where space limitations and costs permit	Tube axial and vanes
Special Purpose Fans		
Roof exhaust	General building exhaust Direct or belt drive Copper, aluminum, fiberglass	Curb-bird screen Self-acting shutter
Variable pitched blade	Variable air volume systems—activated on pressure	Efficient
Variable speed	VAV systems—Frequency of current to motor is varied from 60 Hz (cycles per second) to adjust speed of blades	Economical

Figure 7.7

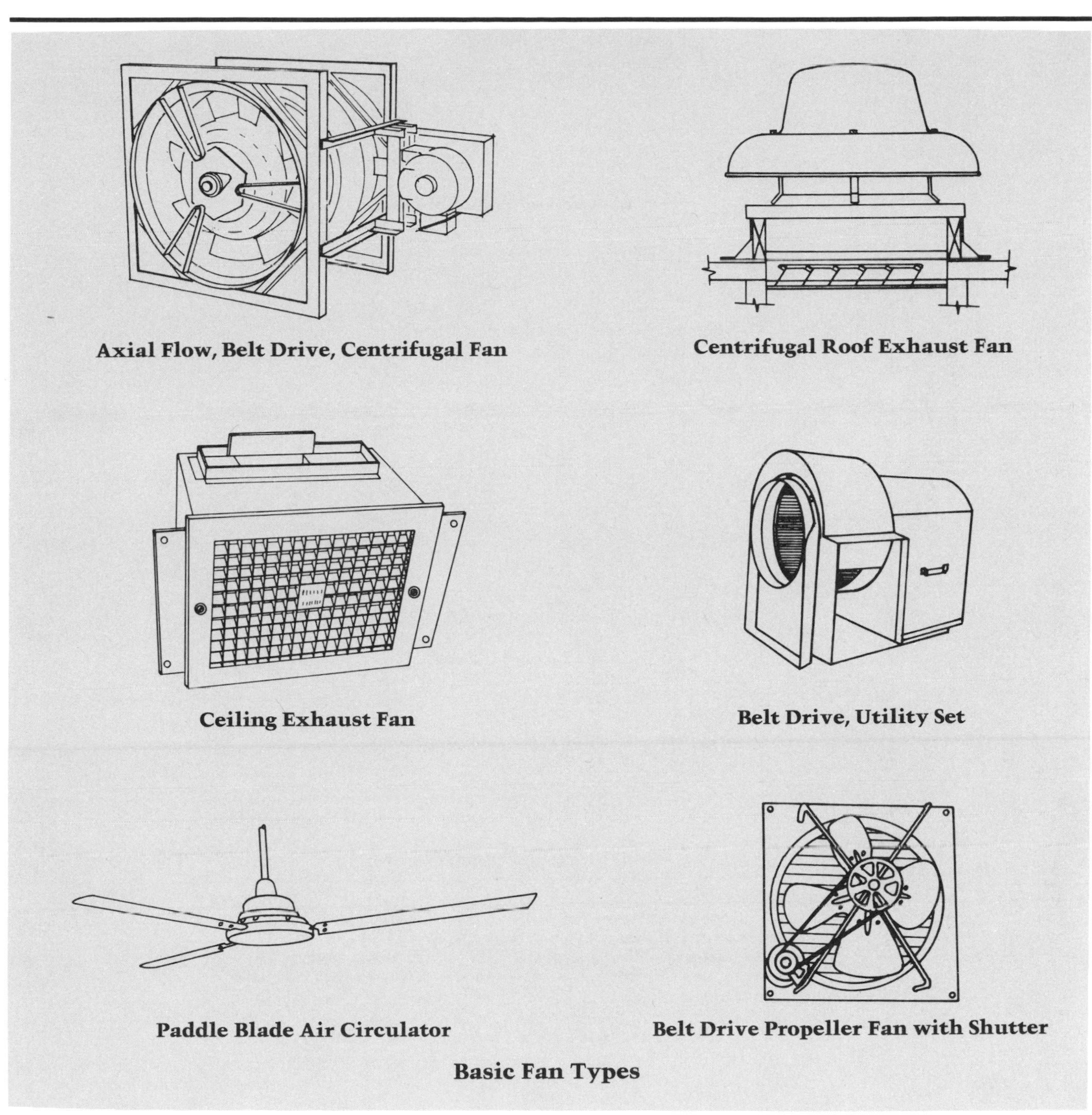

Basic Fan Types

Figure 7.8

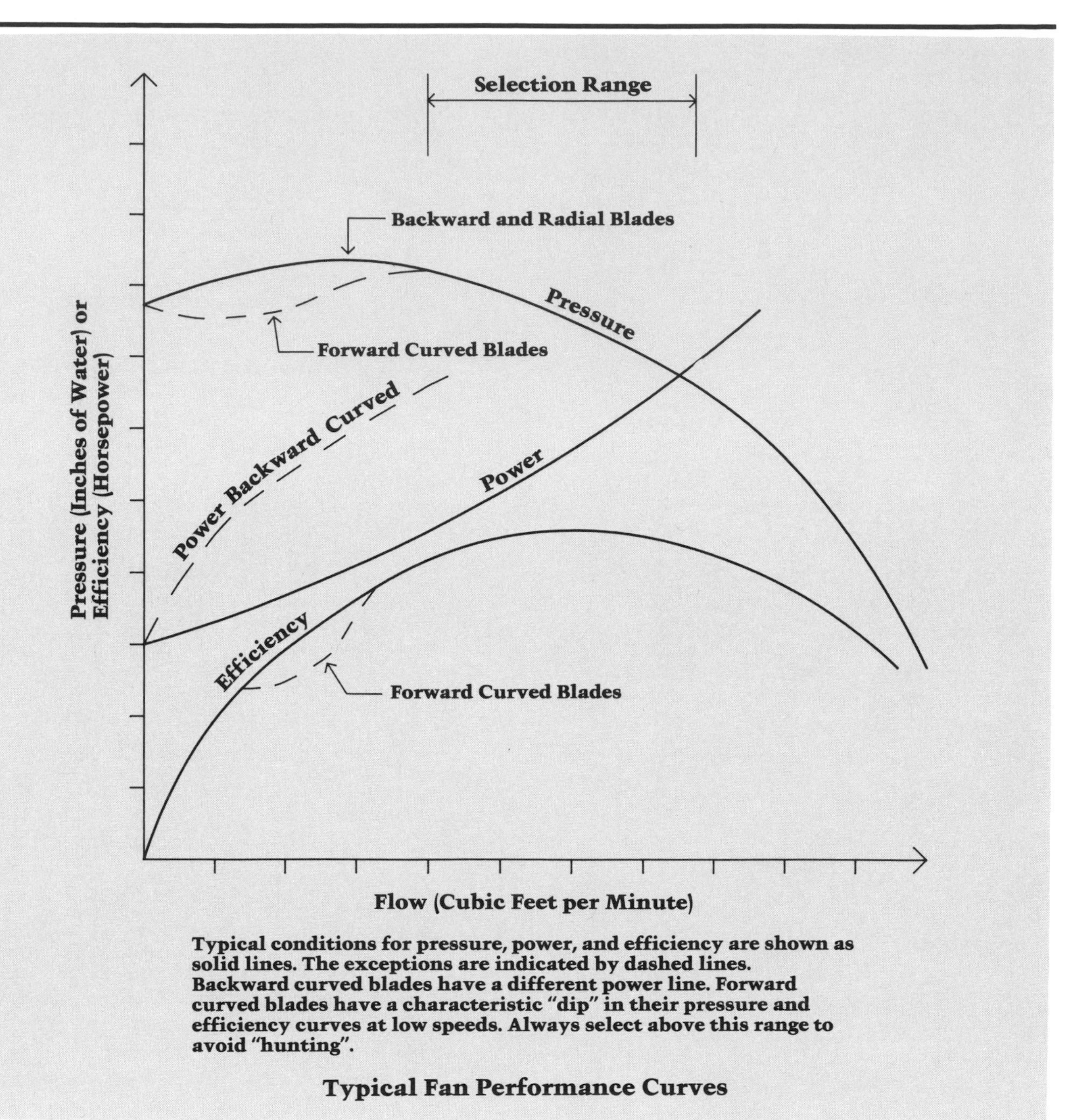

Typical conditions for pressure, power, and efficiency are shown as solid lines. The exceptions are indicated by dashed lines. Backward curved blades have a different power line. Forward curved blades have a characteristic "dip" in their pressure and efficiency curves at low speeds. Always select above this range to avoid "hunting".

Typical Fan Performance Curves

Figure 7.9

Equipment Data Sheets

The following pages of this chapter contain equipment data sheets for fans and pumps. These sheets describe the efficiency, useful life, typical uses, capacity range, special considerations, accessories, and additional equipment needed for driving system components.

These sheets can be used in the selection of each piece of equipment. A Means line number, where applicable, is included. Costs can be determined by looking up the line numbers in the current edition of *Means Mechanical Cost Data.*

LAWS OF FAN PERFORMANCE

Fan laws are used to predict fan performance under changing operating conditions or fan size. They are applicable to all types of fans.

The fan laws are stated in *Table 5.* The symbols used in the formulas represent the following quantities:

Q – Volume rate of flow thru the fan.

N – Rotational speed of the impeller.

P – Pressure developed by the fan, either static or total.

Hp – Horsepower input to the fan.

D – Fan wheel diameter. The fan size number may be used if it is proportional to the wheel diameter.

W – Air density, varying directly as the barometric pressure and inversely as the absolute temperature.

TABLE 5—FAN LAWS

VARIABLE	CONSTANT	NO.	LAW	FORMULA
SPEED	Air Density Fan Size Distribution System	1	Capacity varies as the Speed.	$\frac{Q_1}{Q_2} = \frac{N_1}{N_2}$
		2	Pressure varies as the square of the Speed.	$\frac{P_1}{P_2} = \left(\frac{N_1}{N_2}\right)^2$
		3	Horsepower varies as the cube of the Speed.	$\frac{Hp_1}{Hp_2} = \left(\frac{N_1}{N_2}\right)^3$
FAN SIZE	Air Density Tip Speed	4	Capacity and Horsepower vary as the square of the Fan Size.	$\frac{Q_1}{Q_2} = \frac{Hp_1}{Hp_2} = \left(\frac{D_1}{D_2}\right)^2$
		5	Speed varies inversely as the Fan Size.	$\frac{N_1}{N_2} = \frac{D_2}{D_1}$
		6	Pressure remains constant.	$P_1 = P_2$
	Air Density Speed	7	Capacity varies as the cube of the Size.	$\frac{Q_1}{Q_2} = \left(\frac{D_1}{D_2}\right)^3$
		8	Pressure varies as the square of the Size.	$\frac{P_1}{P_2} = \left(\frac{D_1}{D_2}\right)^2$
		9	Horsepower varies as the fifth power of the Size.	$\frac{Hp_1}{Hp_2} = \left(\frac{D_1}{D_2}\right)^5$
AIR DENSITY	Pressure Fan Size Distribution System	10	Speed, Capacity and Horsepower vary inversely as the square root of Density.	$\frac{N_1}{N_2} = \frac{Q_1}{Q_2} = \frac{Hp_1}{Hp_2} = \left(\frac{W_2}{W_1}\right)^{1/2}$
	Capacity Fan Size Distribution System	11	Pressure and Horsepower vary as the Density.	$\frac{P_1}{P_2} = \frac{Hp_1}{Hp_2} = \frac{W_1}{W_2}$
		12	Speed remains constant.	$N_1 = N_2$

(Courtesy Carrier Corporation, McGraw–Hill Book Company)

Figure 7.10

Driving Equipment Data Sheets

Driving Equipment		
In-Line Centrifugal Pump	**In-Line Centrifugal Pumps**	**Means No. 152-410**
	Typical Uses: Residential, Light Commercial Hydronic Heating	**Capacity Range: 1/40-1-1/2 HP**
	Comments: Common close coupled pump for small, residential applications. No base required. Pump often supported directly by piping. Easily repaired or replaced without disturbing the piping system. Non-ferrous models available for domestic water recirculation systems.	
Close Coupled Centrifugal Pump	**Close Coupled Centrifugal Pumps**	**Means No. 152-410**
	Typical Uses: Commercial Hot and Chilled Water Systems	**Capacity Range: 1-1/2-25 HP/40-1,500 gpm**
	Comments: Factory assembled for longer life but difficult to repair. Entire unit can be readily replaced. Vibration connections advisable.	
Base Mounted Centrifugal Pump	**Base Mounted Pumps—Single Stage**	**Means No. 152-410**
	Typical Uses: Hot and Chilled Water Systems	**Capacity Range: 1-1/2-25 HP/40-1,500 gpm**
	Comments: Most common for intermediate and large commercial installations. Many configurations available for higher capacities. Vibration bases and/or couplings usually required. Chill water pumps may require a drip ledge base and drain tapping. Units available as factory assembled or with motor furnished by others for field assembly.	
Condensate Return Pump	**Condensate Return Pumps**	**Means No. 152-415**
	Typical Uses: Return Condensate to Boiler or Utility	**Capacity Range: Modular—No Limit**
	Comments: Condensate return pumps may be single or duplex and are available with cast iron or steel receivers. A vent line from the receiver relieves any pressure buildup and prevents steam from entering the pump itself. On larger systems, with the addition of sophisticated controls, a boiler feed system capable of handling several boilers can be evolved. A duplex alternator adds to the pump life expectancy.	

Figure 7.11

Driving Equipment Data Sheets (continued)

Driving Equipment	Axial Flow Fans	Means No. 157-410
Axial Flow, Belt Drive, Centrifugal Fan	Typical Uses:	Capacity Range: 500-3,480 CFM
	Comments: Provides pressure increase with minimum dimensions. More expensive to purchase and difficult to access for maintenance. In-line configuration adds installation costs.	
	Belt Drive Centrifugal Fans	**Means No. 157-290**
Belt Drive, Utility Set	Typical Uses:	Capacity Range: 800-20,000 CFM
	Comments: Belt drive allows speed to be varied by changing sheaves. 4-8 blades per wheel. Low through intermediate pressures obtained. Vibration bases and duct connections required.	
	Vane Axial Fans	**Means No. 157-290**
Vane Axial Fan	Typical Uses:	Capacity Range: 2,000-16,000 CFM
	Comments: Blades shaped for higher performance. Overall better flow and performance and highest pressures for propeller-type fans. Contains the most blades per wheel.	
	Direct Drive Blowers	**Means No. 157-290**
	Typical Uses:	Capacity Range: 1,000-1,720 CFM
	Comments: Isolated heating/cooling or ventilation. Connect to duct heaters or duct coils.	

Figure 7.11 (cont.)

Driving Equipment Data Sheets (continued)

Driving Equipment		
Ceiling Exhaust Fan	**Right Angle Ceiling Fans**	**Means No. 157-290**
	Typical Uses: Exhaust	**Capacity Range: 95-2,960 CFM**
	Comments: Typical exhaust from offices, toilets, and conference rooms where no other return or exhaust air system is used.	
Paddle Blade Air Circulator	**Paddle Blade Circulator Fans**	**Means No. 157-290**
	Typical Uses: High Ceilings	**Capacity Range: 3,000-7,000 CFM**
	Comments: Weak pressure and general circulation within space.. Good for high ceiling spaces to prevent stratification. Fan rotation may be reversed for heating or cooling applications.	
Belt Drive, Utility Set	**Centrifugal Air Foil Fans (Backward Inclined Air Foil)**	**Means No. 157-290**
	Typical Uses: Large systems with high pressure requirements	**Capacity Range: 1,000-12,000 CFM**
	Comments: High efficiency fans that require large cfm to offset cost of air foil blades. Vibration mounts and flexible duct connections.	
Centrifugal Roof Exhaust Fan	**Centrifugal Rooftop Fans**	**Means No. 157-290**
	Typical Uses: Roof Exhaust	**Capacity Range: 300-21,600 CFM**
	Comments: Common exhaust for commercial buildings at top of toilet, kitchen, dryer, or building exhaust systems. Sometimes coordinated with economizer and smoke exhaust systems. Fans available direct or belt drive. Self-acting shutters prevent air backflow.	

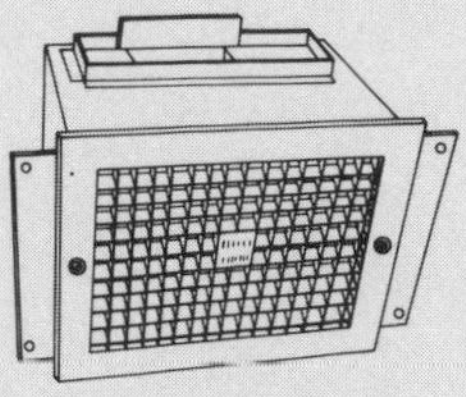

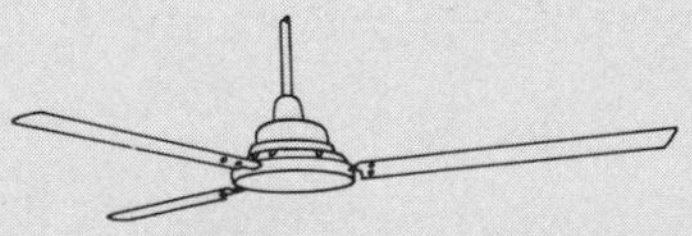

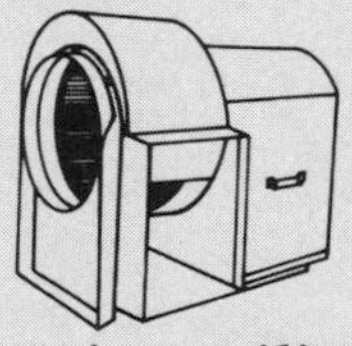

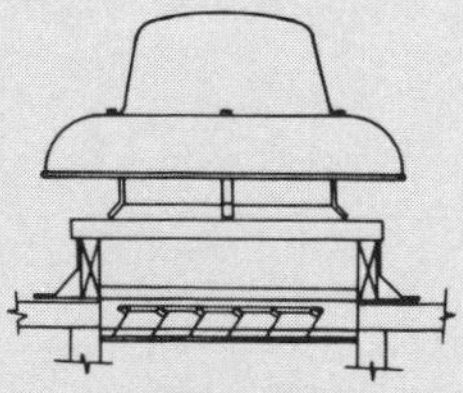

Figure 7.11 (cont.)

Driving Equipment Data Sheets (continued)

Driving Equipment	Propeller Fans	Means No. 157-290
Belt Drive Propeller Fan with Shutter	Typical Uses:	Capacity Range: 375-51,500 CFM
	Areas where large volumes must be handled and noise is not a consideration. Commonly used through walls for mechanical, electrical, and elevator rooms to prevent overheating. Sometimes required to be connected to emergency systems.	
	Residential Whole House Fans	**Means No. 157-290**
Belt Drive Propeller Fan with Shutter	Typical Uses:	Capacity Range: 300-16,000 CFM
	Cooling and ventilation in summer. Installed in the attic or crawl space, this type of fan can draw cool air from the basement or from outside through the building and discharge the hot stale air through an attic window or louver.	
	Bathroom Exhaust Fans	**Means No. 157-290**
Ceiling Exhaust Fan	Typical Uses:	Capacity Range: 50-110 CFM
	Bathroom exhaust, toilet exhaust in residential buildings. Low pressure limit ducts to 10–20'. Fan motor is usually wired to the light switch in the bathroom.	

Figure 7.11 (cont.)

CHAPTER EIGHT

HEAT EXCHANGERS

A heat exchanger is required whenever there is a change in the type of heating or cooling medium. In a boiler room, for example, a shell and tube heat exchanger converts steam to hot water. This hot water is eventually converted to warm air through a fan coil heat exchanger. Figures 1.6, 1.11, and 1.23 in Chapter 1 show the locations of heat exchangers within a system. In these figures, a diagonal line designates a change in medium and a transfer of heat, requiring a heat exchanger.

There are two basic types of heat exchangers. *Shell and tube* heat exchangers are used when converting from a liquid to a gas, and vice versa. *Fin and tube* heat exchangers (*coils*) are used to warm or cool air. This is accomplished with steam, hot water, or chilled water contained within the tubes, and air passing through the fins. Types of heat exchangers are illustrated in Figurc 8.1.

Shell and Tube Exchangers

The shell of a shell and tube heat exchanger surrounds the tubes and carries the primary heating or cooling medium (steam or water). The tubes contain the secondary heated or cooled liquid. There is typically one shell and several tubes.

The important selection criteria for shell and tube heat exchangers are:

Transfer load: the MBH of heating or cooling that is to be delivered to the system by the exchanger. All heat exchangers are rated in MBH.

Primary fluid rate: the flow rate of the primary fluid through the shell, expressed in gpm or pounds of steam per hour.

Entering primary fluid condition: the temperature and pressure of the fluid supplied to the shell side of the heat exchanger.

Leaving primary fluid condition: the temperature and pressure of the fluid leaving the shell, typically 20°F lower than the primary entering fluid temperature for hot water systems.

Entering secondary fluid condition: the temperature and pressure of the fluid entering the tubes of the heat exchanger.

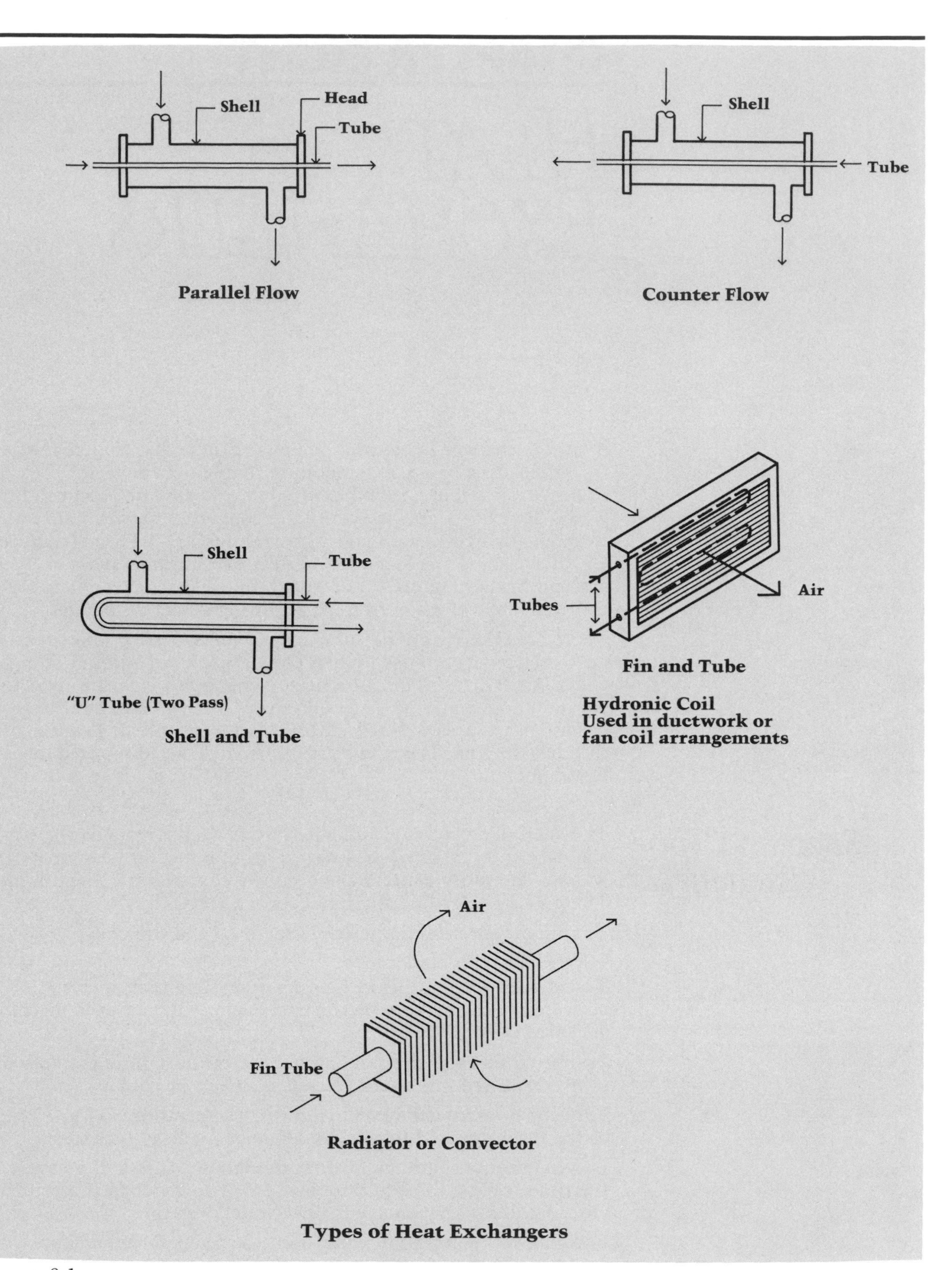
Shell
Head
Tube
Parallel Flow
Shell
Tube
Counter Flow
Shell
Tube
"U" Tube (Two Pass)
Shell and Tube
Tubes
Air
Fin and Tube
Hydronic Coil
Used in ductwork or
fan coil arrangements
Air
Fin Tube
Radiator or Convector
Types of Heat Exchangers

Figure 8.1

Leaving secondary fluid condition: the temperature and pressure of the fluid leaving the tubes of the heat exchanger. It is usually 15°F to 20°F above the entering secondary fluid temperature for hot water systems.

Approach temperature: the entering primary temperature minus the leaving secondary fluid temperature. An approach temperature of zero is impossible. For practical considerations, an approach temperature of 5°F to 20°F is usually selected.

Secondary fluid rate: the flow (in gpm or pounds) of steam per hour through the secondary (tube) side of the heat exchanger.

Each of the above factors are generally determined by a trained engineer in consultation with an equipment manufacturer. The primary temperatures are made as high as possible to improve overall efficiency. The secondary temperatures are usually determined by room or terminal unit performance criteria recommendations (130°F warm air supply, 200°F hot water supply).

To specify a heat exchanger, the MBH required, primary and secondary fluid conditions, and flow rates must be determined. Reduction in the pressures of the fluids as they proceed through the shell and tube must be compensated for in calculating the total head for the primary and secondary pumps.

Heat Exchanger Performance

A perfect heat exchanger would make the temperature of the water *leaving* the tubes equal to the temperature of the water *entering* the shell. However, because this is theoretically impossible, a 5°F to 20°F difference is usually selected between the entering temperature of the primary medium and the exiting temperature of the secondary medium. This difference is called the *approach temperature* (see Figure 8.2).

Figure 8.2 illustrates a heat exchanger sized for a 15°F approach temperature. A lower approach temperature requires a larger, more efficient heat exchanger. The size of the heat exchanger is influenced by the following factors:

- total Btu/hour to be transferred;
- gallons per minute flow in the primary and secondary loops;
- fluids used in each loop (i.e., water, glycol, steam, or refrigerant);
- whether the heat exchanger is parallel, counterflow, or multipass;
- long-term cleanliness (fouling factor) of the coils.

The theoretical equations for heat exchangers are complex. Manufacturers' tables and charts are typically used in the selection of shell and tube heat exchangers.

Fin and Tube Heat Exchangers

Fin and tube heat exchangers exist in many forms. The most common types used are *fin tube baseboards* for heating and *direct expansion coils* (DX) in window air conditioners. In both cases, the primary fluid is circulated in the tube and air moves around the tube (or coil), typically between parallel, closely-spaced metal fins, for increased efficiency. Fin and tube coils, in one form or another, are required somewhere in most HVAC systems.

Fin and tube exchangers are selected using a similar method as that used for shell and tube exchangers. The pressure required to drive the air through the coil must be added to the fan head.

Equipment Data Sheets

The following equipment data sheets for coils and heat exchangers summarize the useful life, capacity range, typical uses, and special considerations for heat exchangers.

A Means line number, where applicable, is included in order to determine costs in the current edition of *Means Mechanical Cost Data*.

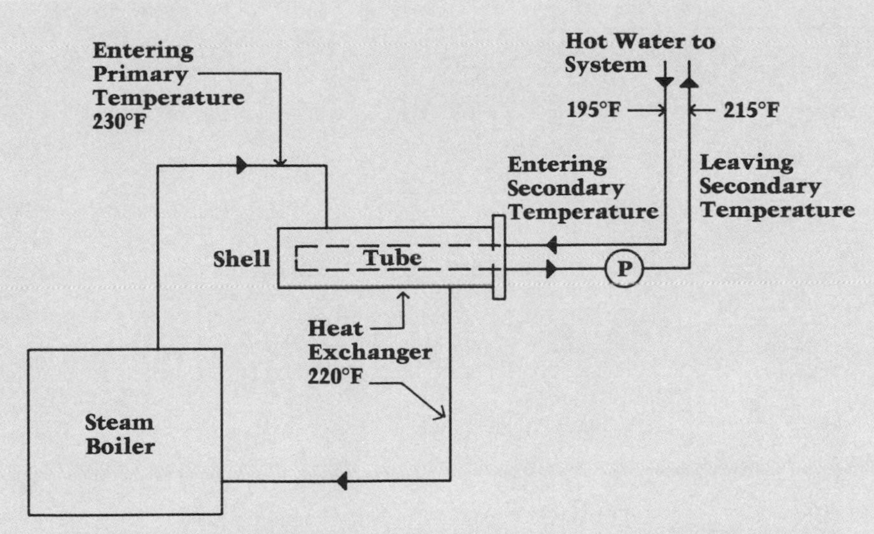

Approach Temp. = Entering Primary Temp. - Leaving Secondary Temp.
= 230°F - 215°F = 15°F Temperature Drop

Heat Exchanger Performance

Shell Diameter (in.)	Length (ft.)	Temperature Rise		
		40°-140°F	40°-180°F	140°-180°F
4	3	100	30	100
4	4	150	45	150
4	5	215	60	205
4	6	275	85	260
6	3	230	60	250
6	4	360	95	365
6	5	490	145	475
6	6	625	200	590
8	4	745	190	720
8	5	1,020	300	960
8	6	1,270	390	1,200
8	7	1,500	480	1,440
10	4	1,285	325	1,200
10	5	1,755	480	1,560
10	6	2,100	635	1,920
10	7	2,460	765	2,260
12	4	1,770	455	1,440
12	5	2,430	645	2,040
12	6	3,000	840	2,640
12	7	3,510	1,020	3,240

Preliminary Heat Exchanger Selection

GPH of Water inside Tubes (supply water at 200°F)

Figure 8.2

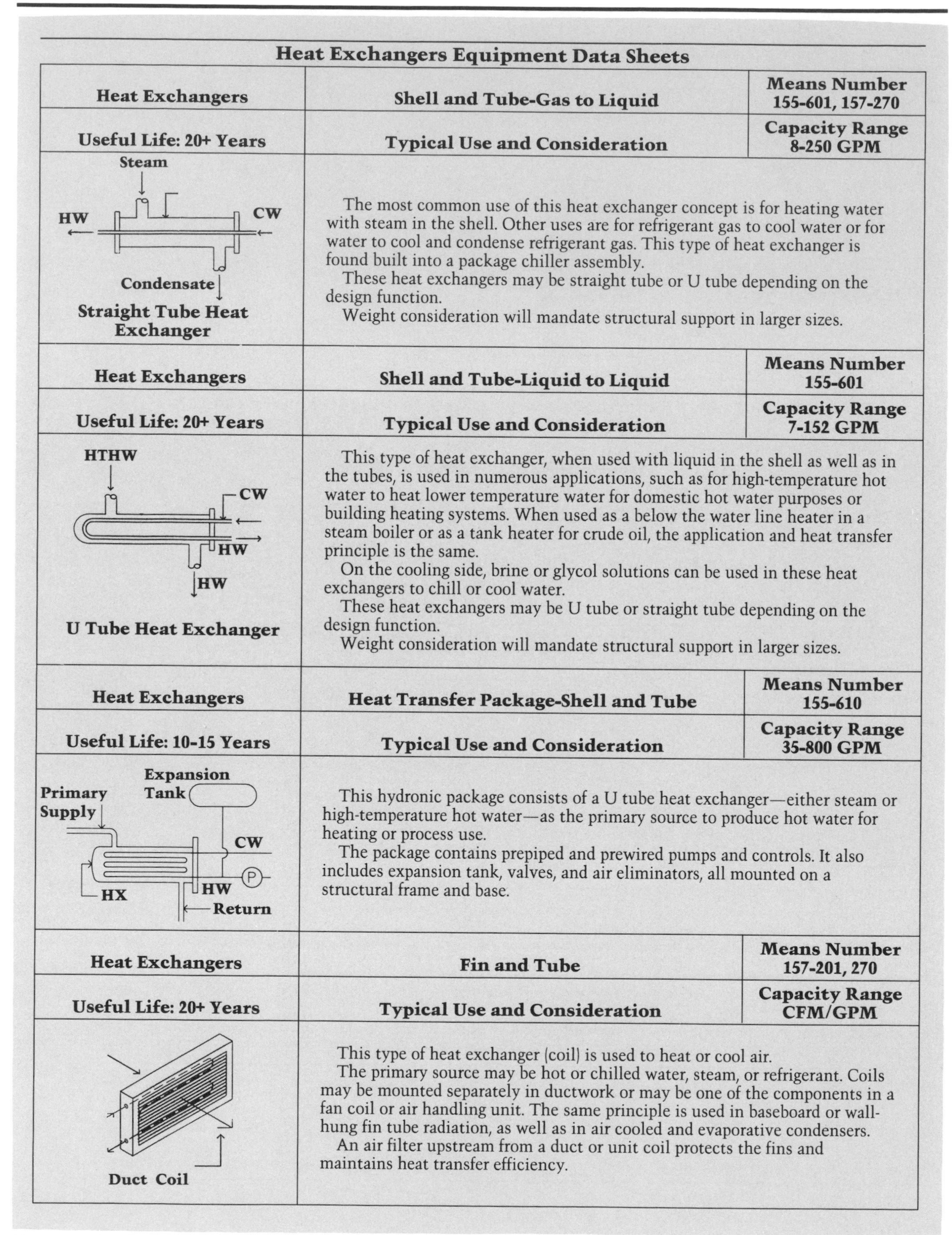

Heat Exchangers Equipment Data Sheets

Heat Exchangers	Shell and Tube-Gas to Liquid	Means Number 155-601, 157-270
Useful Life: 20+ Years	Typical Use and Consideration	Capacity Range 8-250 GPM
Steam HW CW Condensate **Straight Tube Heat Exchanger**	The most common use of this heat exchanger concept is for heating water with steam in the shell. Other uses are for refrigerant gas to cool water or for water to cool and condense refrigerant gas. This type of heat exchanger is found built into a package chiller assembly. These heat exchangers may be straight tube or U tube depending on the design function. Weight consideration will mandate structural support in larger sizes.	

Heat Exchangers	Shell and Tube-Liquid to Liquid	Means Number 155-601
Useful Life: 20+ Years	Typical Use and Consideration	Capacity Range 7-152 GPM
HTHW CW HW HW **U Tube Heat Exchanger**	This type of heat exchanger, when used with liquid in the shell as well as in the tubes, is used in numerous applications, such as for high-temperature hot water to heat lower temperature water for domestic hot water purposes or building heating systems. When used as a below the water line heater in a steam boiler or as a tank heater for crude oil, the application and heat transfer principle is the same. On the cooling side, brine or glycol solutions can be used in these heat exchangers to chill or cool water. These heat exchangers may be U tube or straight tube depending on the design function. Weight consideration will mandate structural support in larger sizes.	

Heat Exchangers	Heat Transfer Package-Shell and Tube	Means Number 155-610
Useful Life: 10-15 Years	Typical Use and Consideration	Capacity Range 35-800 GPM
Expansion Tank Primary Supply CW P HX HW Return	This hydronic package consists of a U tube heat exchanger—either steam or high-temperature hot water—as the primary source to produce hot water for heating or process use. The package contains prepiped and prewired pumps and controls. It also includes expansion tank, valves, and air eliminators, all mounted on a structural frame and base.	

Heat Exchangers	Fin and Tube	Means Number 157-201, 270
Useful Life: 20+ Years	Typical Use and Consideration	Capacity Range CFM/GPM
Duct Coil	This type of heat exchanger (coil) is used to heat or cool air. The primary source may be hot or chilled water, steam, or refrigerant. Coils may be mounted separately in ductwork or may be one of the components in a fan coil or air handling unit. The same principle is used in baseboard or wall-hung fin tube radiation, as well as in air cooled and evaporative condensers. An air filter upstream from a duct or unit coil protects the fins and maintains heat transfer efficiency.	

Figure 8.3

CHAPTER NINE

DISTRIBUTION AND DRIVING SYSTEMS

Distribution is the movement of heated or conditioned air to desired locations. Distribution equipment includes any equipment that connects generating equipment to the terminal units, such as ductwork and piping, and also the pumps, fans, and accessories that are separate from the generating equipment and conduct the steam, water, or air through the distribution system. (For information on valves and dampers, refer to Chapter 10, "Terminal Units.") This chapter addresses the basic principles involved in the selection, laying out, and sizing of distribution equipment. Cost guides such as *Means Mechanical Cost Data* and manufacturers' price lists should be consulted for the specific costs of the components that make up distribution and driving systems.

The distribution equipment is part of either an *open* or *closed* loop system. In an open system, the fluid leaves the pipe or duct and is "open" to the atmosphere. In a closed system, the fluid remains in the containing pipes and pressure vessels, usually completing a recirculating loop. In general, most systems involving liquids (water) are closed, most systems involving air are open. Open and closed systems are illustrated in Figure 9.1.

The basic types of distribution systems and their ranges are listed below.

• **Steam**	High-pressure steam	above 15 psig
	Low-pressure steam	0 - 15 psig
• **Water**	High-temperature water	above 250°F
	Hot water	100–250°F
	Chilled water	35–100°F
• **Air**	High velocity air	above 2,000 fpm and above 2.5″ wg
	Low velocity air	0–2,000 fpm and above 2.5″ wg
• **Gases**	Refrigerants	varies with gas
	Natural gas	
	Propane	

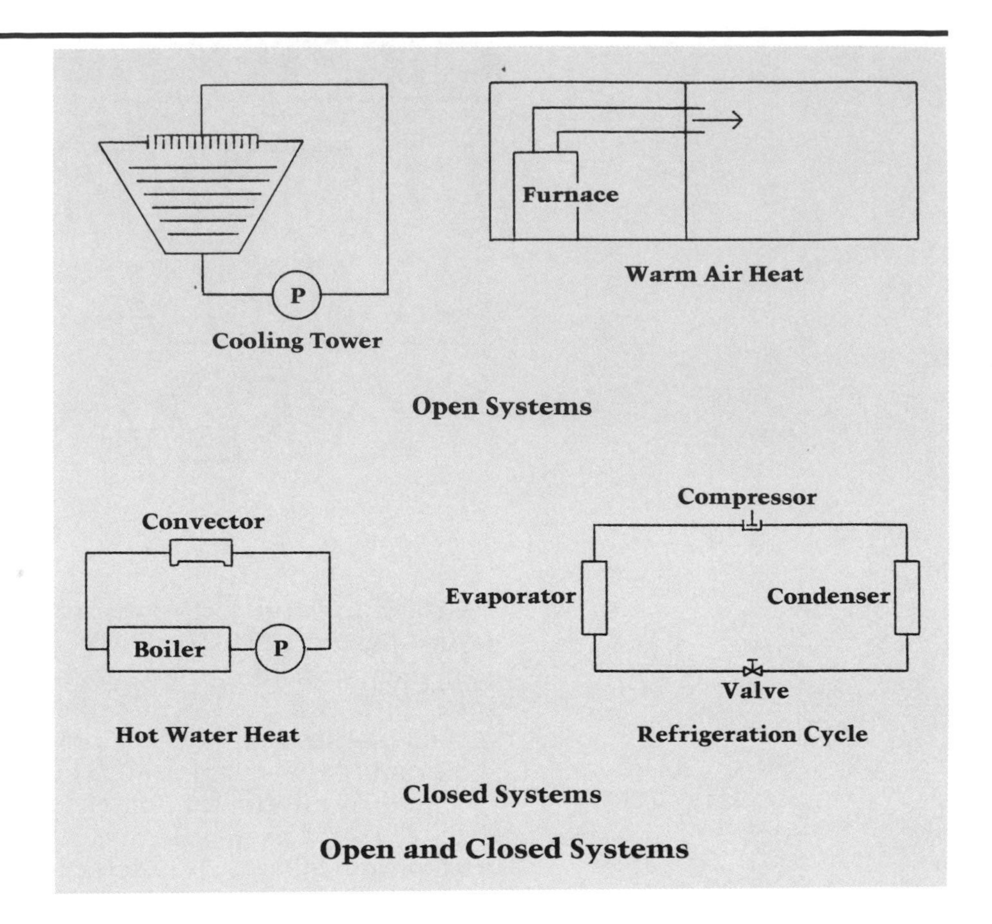
Furnace
Warm Air Heat
P
Cooling Tower
Open Systems
Compressor
Convector
Evaporator
Condenser
Boiler
P
Valve
Hot Water Heat
Refrigeration Cycle
Closed Systems
Open and Closed Systems

Figure 9.1

Distribution equipment conveys the heating or cooling medium (steam, water, air, or gas) from the generation equipment to the terminal units. When the medium used in the generating system is different from that of the distribution system, a heat exchanger and additional pumps or fans are required. The following sections describe distribution system layouts for steam, hot water, and air. Figure 9.2 is a selection guide for choosing which distribution system is appropriate for a particular building.

Steam Distribution

Distribution by steam transfers the most heat per pound of water. At 212°F, one pound of steam releases 970 Btu's. Accounting for the additional heat from the condensate and using a slightly higher operating temperature, the customary value of 1,000 Btu's per pound of steam is used for most applications. The term *square foot of radiation* (SFR) is usually used to size radiation for steam (also equivalent direct radiation (EDR)). One square foot of radiation equals 240 Btu/hour, which is the amount of heat delivered by one square foot of radiator surface with a one pound steam output.

The force to drive steam is created by the boiler, which increases steam volume and pressure, and by the condensing process, which draws more steam into the system. Thus, steam systems are largely self-propelled, requiring no pumps. With proper maintenance, they are also long lasting, due to the absence of motor-driven equipment.

A particular advantage of the use of steam in high-rise buildings is that it does not present the pressure problems associated with water distribution systems. Steam heating is often used for industrial applications, hospitals, and large facilities, as the steam is already necessary for cleaning, sterilization, food preparation, and generation of hot water.

Layout

The condensed steam (condensate in the form of hot water) is returned to the boiler for re-use through either a one-pipe or a two-pipe system. In a one-pipe system, the condensate returns to the boiler in the same pipe that supplies steam to the terminal unit. A two-pipe system uses the supply pipe for steam and a separate return pipe for the condensate. Two-pipe systems are more expensive, because they contain roughly twice as much piping. However, they do offer better overall control of steam systems.

Because steam is lighter than air, it displaces the heavier air and rises upward through risers to the upper floors. As steam condenses and becomes water, the water is returned to the boiler through pipes. All pipes in both one- and two-pipe steam systems are pitched toward the boiler or mechanical room to return the condensate by gravity directly to the boiler or a condensate return pump.

Most units are *upfed*, meaning that the terminal unit is located above the steam main. When steam flows down to a terminal unit below the main, the unit is said to be *downfed*. Vertical

Selection Guide for Steam, Water, and Air Distribution Systems		
Distribution System	**Common Usage**	**Type of Piping Used**
Steam 0–15 psig—low-pressure, 212°F to 250°F Over 15 psig—high-pressure	Hospitals, industrial plants, laundries, restaurants. Supply to units for heating coils of central systems for dry cleaning, commercial buildings. Campus heating systems between buildings, district heating. Insulation required.	Carbon steel
Hot Water 30 psig—low-temperature, 250°F 150 psig—medium-temperature, 350°F 300 psig—high-temperature, 450°F	Residences, offices, commercial, institutional. Widest use for loads under 5,000 MBH. Campus type heating systems. Supply to convectors, cast iron radiators, baseboard fin tubes, VAV coils, fan coils, air handling units. Insulation required.	Carbon steel or copper tubing
Chilled Water 125 psig—40–55°F	Offices, commercial, institutional. Supplies coils, fan coil units, VAV coils, air handling units. Also used in campus type cooling systems. Vapor barrier insulation mandatory.	Carbon steel or copper tubing
Warm Air 130°F	Residences, offices, commercial, institutional. Supplies warm air from furnaces, air handling units, and discharges through grilles, registers or diffusers. Insulation and sound lining optional.	Aluminum, galvanized steel, or fiberglass duct
Cool Air 55°F	Residences, offices, commercial, institutional. Supplies cool air from cooling units such as package units and discharges through grilles, registers or diffusers. Sound lining is optional, vapor barrier insulation is mandatory in unconditioned spaces.	Aluminum, galvanized steel, or fiberglass duct
Mixed Air 55°F to 130°F	Residences, offices, commercial, institutional. Supplies cool or heated air through common ductwork. Ductwork must be insulated for cooling.	Aluminum, galvanized steel, or fiberglass duct

Figure 9.2

pipes that feed the system are called *risers*. Horizontal pipes are called *mains*. Sloped pipes between the mains and risers or radiators are called *runouts*.

Types of Steam Systems

There are three types of steam heating systems, characterized by the way that the steam is propelled and/or the condensate is returned. The three types of steam heating systems are *gravity*, *vapor*, and *vacuum* systems.

Gravity Systems: Gravity systems are usually single-pipe, low-pressure systems. Steam is propelled by the initial boiler pressure, and by the fact that it is lighter than air, to radiators in upper floors. The steam and condensate are carried in the same pipe, or main. The main is pitched and gravity carries the condensate downward, sometimes in the opposite direction of steam flow, back to the boiler.

Two-pipe gravity systems are used on larger systems where a separate return pipe from the terminal units enters a return main and flows by gravity back to the boiler. Two-pipe systems generally require smaller size pipe (smaller diameter), but require twice as much pipe as a one-pipe system.

Vapor Systems: A vapor system is a two-pipe, low-pressure steam heating system utilizing thermostatic traps and combination float and thermostatic traps at terminal units and at drip connections between supply and return mains and risers. Traps let water and air pass through them; the thermostatic element prevents steam from getting by. This arrangement allows the unwanted water and air to flow through the return piping back to the boiler, while retaining the steam in the system. A condensate return unit is often utilized in this type of system.

Vacuum Systems: A vacuum added to a vapor system reduces pressure in the return pipe and improves performance and control. Because the pressure in a vacuum system is below atmospheric pressure, performance and control are improved.

Pressure

A particular advantage of steam systems is that the pressure can be adjusted to increase the heat content. By varying the pressure to meet the demand, a smooth and efficient delivery of energy can be achieved. The ranges of pressure for steam systems are shown below.

above 15 psig	High-pressure system (most efficient; strict safety controls)
0–15 psig	Low-pressure steam (most common)
Less than 0 psig	Vacuum system (vacuum pumps required to maintain pressure below atmospheric pressure)

Note that 15 psig means 15 pounds per square inch (psi) gauge pressure. Atmospheric pressure is 14.7 psi at sea level. However, this is not generally read by the gauge, as atmospheric pressure has no effect on the pressure of a steam system. A gauge pressure of 15 psig combined with atmospheric pressure equals 29.7 psia, and is known as the *absolute pressure* of the 15 psig reading.

The higher the pressure, the more efficient the system. Utility companies routinely supply steam at a rate of between 80 and 200 psig. Large boilers generating steam above 15 psig usually require a licensed attendant at all times. Power plants, hospitals, or other industrial buildings usually contain such boilers. High-pressure steam from a boiler room or utility must go through a pressure-reducing station—a system of pressure reducing valves—in order to meet such standards. In commercial and residential buildings, pressures above 15 psig are considered dangerous; these buildings are required to use low-pressure or vacuum systems.

Venting Steam Heating Systems

With steam systems, air must be removed from the pipe prior to boiler start-up and must be continually removed from the piping, the boiler, and the heated water to ensure that the steam flows evenly. Venting must be performed for gravity, vacuum, and vapor systems.

In gravity systems, air is vented throughout the piping network at the high points to allow gravity to drain the condensate back to the boiler. Air vents are located at the high point of each terminal unit, on the high point of the main, at the top of each riser, and at the boiler. Air vents are specifically designed for radiators, mains, and other pieces of equipment.

Air in vapor and vacuum systems is vented centrally at a condensate receiver, usually at the end of a system. No air vents are provided at the terminal units. In addition, steam traps are provided at each terminal unit and at the drip connections between the supply and return main. The traps allow the steam condensate to return to the boiler, but trap the steam on the supply side. Systems with traps are sometimes called "mechanical," "trapped," or "vapor" systems.

The most frequently used types of systems are illustrated in Figure 9.3.

Steam Pipe Sizing

The size of steam pipe must be determined for each branch or run of pipe. Generally, the larger the pipe, the more pounds of steam per hour it will carry and the more heat will be delivered. The variables that affect pipe sizing include the pipe material, pipe length, type of system, system pressure, condensate flow, size of risers, and heat capacity.

- **Pipe material:** The surface roughness of the pipe and the actual inside pipe diameter affect flow. Schedule 40 steel pipe is the most frequently used pipe material for low-pressure steam and is the basis of most steam pipe tables used.
- **Pipe length:** The total equivalent pipe length equals the length of the straight pipe plus the equivalent length of all fittings (see Figure 9.4). When fittings are not known, total equivalent length is assumed to be twice the straight length.

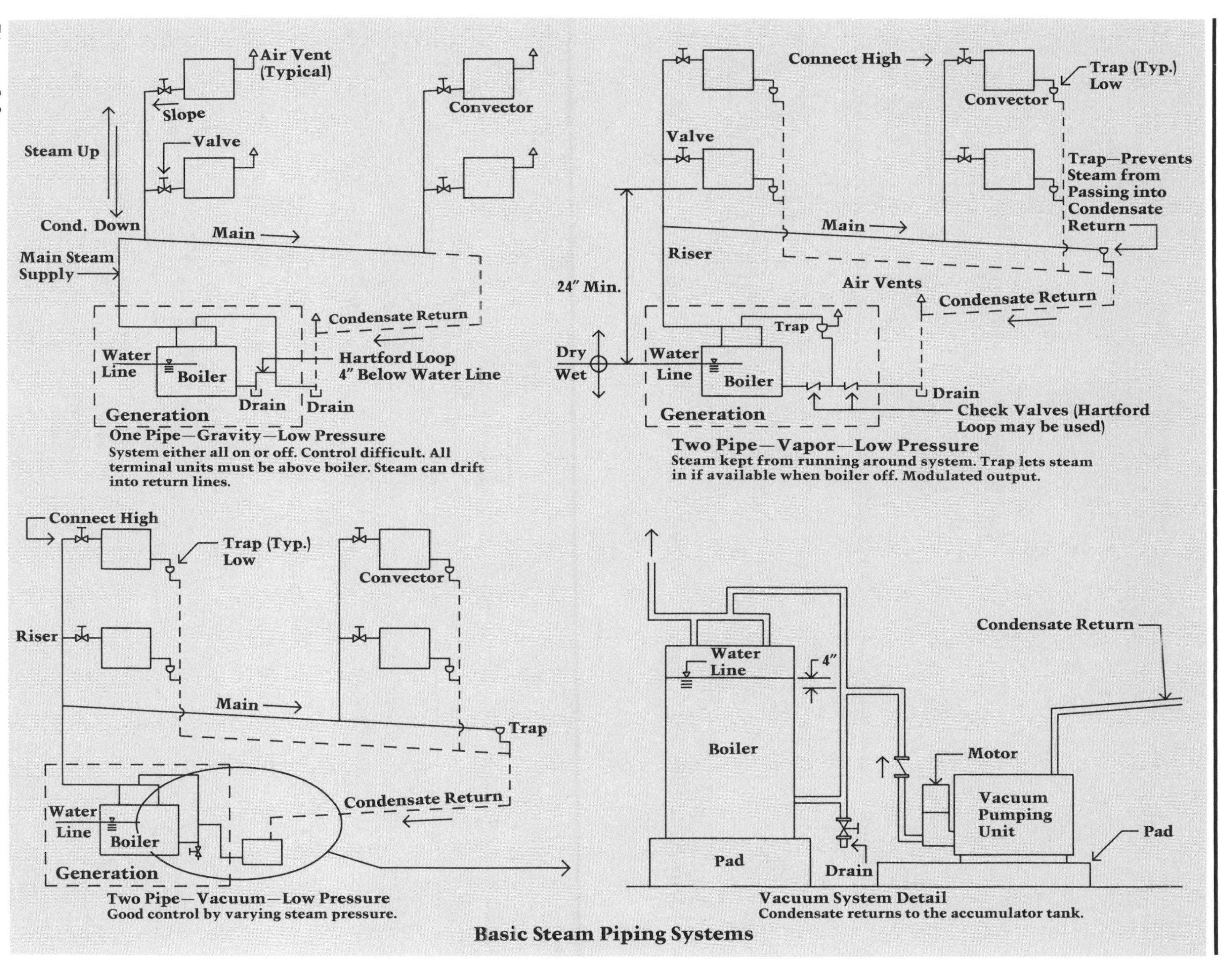

Air Vent (Typical)
Slope
Valve
Convector
Steam Up
Cond. Down
Main Steam Supply
Main
Condensate Return
Hartford Loop 4" Below Water Line
Drain
Drain
Water Line
Boiler
Generation
One Pipe—Gravity—Low Pressure
System either all on or off. Control difficult. All terminal units must be above boiler. Steam can drift into return lines.
24" Min.
Dry
Wet
Connect High
Valve
Riser
Main
Convector
Trap (Typ.) Low
Trap—Prevents Steam from Passing into Condensate Return
Condensate Return
Air Vents
Trap
Water Line
Boiler
Generation
Drain
Check Valves (Hartford Loop may be used)
Two Pipe—Vapor—Low Pressure
Steam kept from running around system. Trap lets steam in if available when boiler off. Modulated output.
Connect High
Riser
Trap (Typ.) Low
Convector
Main
Condensate Return
Trap
Water Line
Boiler
Generation
Two Pipe—Vacuum—Low Pressure
Good control by varying steam pressure.
Condensate Return
Pad
Motor
Vacuum Pumping Unit
4"
Water Line
Boiler
Pad
Drain
Vacuum System Detail
Condensate returns to the accumulator tank.
Basic Steam Piping Systems

Figure 9.3

Equivalent Pipe Lengths for Steam Fittings

Length in Feet of Pipe to be Added to Actual Length of Run — Owing to Fittings — To Obtain Equivalent Length

Size of Pipe (Inches)	Length in Feet to be Added to Run				
	Standard Elbow	Side Outlet Tee[b]	Gate Valve[a]	Globe Valve[a]	Angle Valve[a]
1/2	1.3	3	0.3	14	7
3/4	1.8	4	0.4	18	10
1	2.2	5	0.5	23	12
1¼	3.0	6	0.6	29	15
1½	3.5	7	0.8	34	18
2	4.3	8	1.0	46	22
2½	5.0	11	1.1	54	27
3	6.5	13	1.4	66	34
3½	8	15	1.6	80	40
4	9	18	1.9	92	45
5	11	22	2.2	112	56
6	13	27	2.8	136	67
8	17	35	3.7	180	92
10	21	45	4.6	230	112
12	27	53	5.5	270	132
14	30	63	6.4	310	152

[a]Valve in full open position.
[b]Values given apply only to a tee used to divert the flow in the main to the last riser.
Example: Determine the length in feet of pipe to be added to actual length of run illustrated.

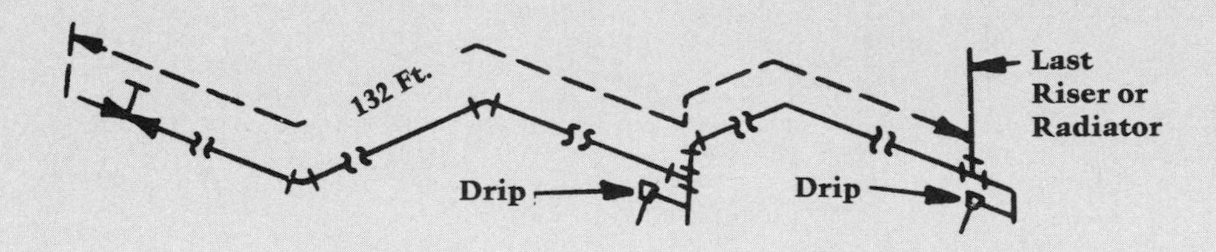

Measured Length	=	132.0 ft.
4 in. Gate Valve	=	1.9 ft.
4-4 in. Elbows	=	36.0 ft.
2-4 in. Tees	=	36.0 ft.
Equivalent	=	205.9 ft.

(Courtesy ASHRAE, 1985 Fundamentals)

Figure 9.4

- **Type of system:** See the sizing data for one- and two-pipe systems for low-pressure steam in Figures 9.5 through 9.7. More piping is required for a two-pipe system.
- **Pressure of system:** Vacuum, vapor, low-pressure, and high-pressure systems deliver more heat by increasing steam pressure, which compresses the steam. The data shown in Figures 9.5 through 9.7 is for 3.5 psig average pressure, the most common low-pressure systems. (Any system under 15 psig, or pounds, is considered to be a low-pressure system.)
- **System pressure:** The difference in pressure between the supply and return pipe is called the *pressure drop*. The difference in pressure conveys the steam through the system. Most systems operate below 1 psi pressure drop (1 psi = 144 pounds per square foot, or a column of water 28" high). Most radiators are placed at least 28" above the boiler

Flow Rate of Steam in Schedule 40 Pipe[a] at Initial Saturation Pressure of 3.5 and 12 Psig[b]
(Flow Rate Expressed in Pounds per Hour)

Pressure Drop — Psi Per 100 Ft. in Length

Nom. Pipe Size (Inches)	1/16 Psi (1 oz.) Sat. press. psig		1/8 Psi (2 oz.) Sat. press. psig		1/4 Psi (4 oz.) Sat. press. psig		1/2 Psi (8 oz.) Sat. press. psig		3/4 Psi (12 oz.) Sat. press. psig		1 Psi Sat. press. psig		2 Psi Sat. press. psig	
	3.5	12	3.5	12	3.5	12	3.5	12	3.5	12	3.5	12	3.5	12
3/4	9	11	14	16	20	24	29	35	36	43	42	50	60	73
1	17	21	26	31	37	46	54	66	68	82	81	95	114	137
1¼	36	45	53	66	78	96	111	138	140	170	162	200	232	280
1½	56	70	84	100	120	147	174	210	218	260	246	304	360	430
2	108	134	162	194	234	285	336	410	420	510	480	590	710	850
2½	174	215	258	310	378	460	540	660	680	820	780	950	1,150	1,370
3	318	380	465	550	660	810	960	1,160	1,190	1,430	1,380	1,670	1,950	2,400
3½	462	550	670	800	990	1,218	1,410	1,700	1,740	2,100	2,000	2,420	2,950	3,450
4	640	800	950	1,160	1,410	1,690	1,980	2,400	2,450	3,000	2,880	3,460	4,200	4,900
5	1,200	1,430	1,680	2,100	2,440	3,000	3,570	4,250	4,380	5,250	5,100	6,100	7,500	8,600
6	1,920	2,300	2,820	3,350	3,960	4,850	5,700	5,700	7,000	8,600	8,400	10,000	11,900	14,200
8	3,900	4,800	5,570	7,000	8,100	10,000	11,400	14,300	14,500	17,700	16,500	20,500	24,000	29,500
10	7,200	8,800	10,200	12,600	15,000	18,200	21,000	26,000	26,200	32,000	30,000	37,000	42,700	52,000
12	11,400	13,700	16,500	19,500	23,400	28,400	33,000	40,000	41,000	49,500	48,000	57,500	67,800	81,000

(Courtesy ASHRAE, 1985 Fundamentals)

[a]Based on Moody Friction Factor, where flow of condensate does not inhibit the flow of steam.
[b]The flow rates at 3.5 psig can be used to cover sat. press. from 1 to 6 psig, and the rates at 12 psig can be used to cover sat. press. from 8 to 16 psig with an error not exceeding 8%.

Figure 9.5

outlet as a safety factor to avoid condensate flooding the system. Most low-pressure and vacuum systems use a 1/16 psi (1 ounce) pressure drop for systems under 200 feet total equivalent length (pipe and fittings).

Assume the pressure drop to be one-third to one-half of the average operating gauge pressure. This figure is converted to pressure drop per 100 feet. Thus, for a boiler operating at 5 psig on a system with 300 total equivalent feet of length, the pressure drop would be calculated as follows:

$$1/3 \times 5 \times 100/300 = 5/9 \text{ psi}$$

The 5/9 could be reduced, or rounded, to 1/2 psi, thus producing the figure of 1/2 psi pressure drop per 100 feet of pipe.

Steam Pipe Capacities for Low-Pressure Systems
(For Use on One-Pipe Systems or Two-Pipe Systems in which Condensate Flows Against the Steam Flow)

Nominal Pipe Size, (Inches)	Capacity in Pounds per Hour				
	Two-Pipe Systems		One-Pipe Systems		
	Condensate Flowing Against Steam		Supply Risers Up-Feed	Radiator Valves and Vertical Connections	Radiator and Riser Runouts
	Vertical	Horizontal			
A	B[a]	C[c]	D[b]	E	F[c]
3/4	8	7	6	—	7
1	14	14	11	7	7
1¼	31	27	20	16	16
1½	48	42	38	23	16
2	97	93	72	42	23
2½	159	132	116	—	42
3	282	200	200	—	65
3½	387	288	286	—	119
4	51	425	380	—	186
5	1,050	788	—	—	278
6	1,800	1,400	—	—	545
8	3,750	3,000	—	—	—
10	7,000	5,700	—	—	—
12	11,500	9,500	—	—	—
16	22,000	19,000	—	—	—

(Courtesy ASHRAE, 1985 Fundamentals)

Note: Steam at an average pressure of 1 psig is used as a basis of calculating capacities.

[a]Do not use Column B for pressure drops of less than 1/16 psi per 100 ft. of equivalent run.
[b]Do not use Column D for pressure drops of less than 1/24 psi per 100 ft. of equivalent run except on sizes 3 in. and over.
[c]Pitch of horizontal runouts to risers and radiators should be not less than 1/2 in. per ft. Where this pitch cannot be obtained, runouts over 8 ft. in length should be one pipe size larger than called for in this table.

Figure 9.6

Figure 9.7

Return Main and Riser Capacities for Low-Pressure Systems — Pounds per Hour*

	Pipe Size (Inches)	1/32 Psi or 1/2 Oz. Drop per 100 Ft.			1/24 Psi or 2/3 Oz. Drop per 100 Ft.			1/16 Psi or 1 Oz. Drop per 100 Ft.			1/8 Psi or 2 Oz. Drop per 100 Ft.			1/4 Psi or 4 Oz. Drop per 100 Ft.			1/2 Psi or 8 Oz. Drop per 100 Ft.		
		Wet	Dry	Vac.	Wet	Dry	Vac.	Wet	Dry	Vac.	Wet	Dry	Vac.	Wet	Dry	Vac.	Wet	Dry	Vac.
	G	H	I	J	K	L	M	N	O	P	Q	R	S	T	U	V	W	X	Y
Return Main	¾	—	—	—	—	—	42	—	—	100	—	—	142	—	—	200	—	—	283
	1	125	62	—	145	71	143	175	80	175	250	103	249	350	115	350	—	—	494
	1¼	213	130	—	248	149	244	300	168	300	425	217	426	600	241	600	—	—	848
	1½	338	206	—	393	236	388	475	265	475	675	340	674	950	378	950	—	—	1,340
	2	700	470	—	810	535	815	1,000	575	1,000	1,400	740	1,420	2,000	825	2,000	—	—	2,830
	2½	1,180	760	—	1,580	868	1,360	1,680	950	1,680	2,350	1,230	2,380	3,350	1,360	3,350	—	—	4,730
	3	1,880	1,460	—	2,130	1,560	2,180	2,680	1,750	2,680	3,750	2,250	3,800	5,350	2,500	5,350	—	—	7,560
	3½	2,750	1,970	—	3,300	2,200	3,250	4,000	2,500	4,000	5,500	3,230	5,680	8,000	3,580	8,000	—	—	11,300
	4	3,880	2,930	—	4,580	3,350	4,500	5,500	3,750	5,500	7,750	4,830	7,810	11,000	5,380	11,000	—	—	15,500
	5	—	—	—	—	—	7,880	—	—	9,680	—	—	13,700	—	—	19,400	—	—	27,300
	6	—	—	—	—	—	12,600	—	—	15,500	—	—	22,000	—	—	31,000	—	—	43,800
Riser	¾	—	48	—	—	48	143	—	48	175	—	48	249	—	48	350	—	—	494
	1	—	113	—	—	113	244	—	113	300	—	113	426	—	113	600	—	—	848
	1¼	—	248	—	—	248	388	—	248	475	—	248	674	—	248	950	—	—	1,340
	1½	—	375	—	—	375	815	—	375	1,000	—	375	1,420	—	375	2,000	—	—	2,830
	2	—	750	—	—	750	1,360	—	750	1,680	—	750	2,380	—	750	3,350	—	—	4,730
	2½	—	—	—	—	—	2,180	—	—	2,680	—	—	3,800	—	—	5,350	—	—	7,560
	3	—	—	—	—	—	3,250	—	—	4,000	—	—	5,680	—	—	8,000	—	—	11,300
	3½	—	—	—	—	—	4,480	—	—	5,500	—	—	7,810	—	—	11,000	—	—	15,500
	4	—	—	—	—	—	7,880	—	—	9,680	—	—	13,700	—	—	19,400	—	—	27,300
	5	—	—	—	—	—	12,600	—	—	15,500	—	—	22,000	—	—	31,000	—	—	43,800

(Courtesy ASHRAE, 1985 Fundamentals)

*Reference to this table is made by column letter G through Y.

- **Condensate Flow (with or against steam):** When the condensate flows in the same direction as the steam, the pipe has a greater capacity. If the condensate flows against the steam, the capacity is reduced and larger diameter pipe is required.
- **Risers:** Riser pipe sizes have a greater steam capacity because of the effect of gravity, which removes the interference from the condensate flow. (In an upfed riser, the steam is going up and the condensate is flowing down.)
- **Capacity:** Pipe size varies with the number of Btu's per hour that must be delivered. Generally, one pound of steam equals 1,000 Btu's, or 1 MBH, for both low-pressure and vacuum systems.

Adequate steam pipe size is the most critical during system start-up. Air must be removed from the piping system, which may be overloaded with additional condensate and produce excessive noise. To prevent this, it is good practice to design the horizontal steam mains larger than two inches in diameter (at least 2-1/2 inches). In addition, all pipe should be pitched for drainage. The recommended minimum pitch for mains is 1/4" per 10 feet and 1/2" per foot for branches (runouts) to risers and radiators. Pipe that cannot be adequately pitched must be increased in size to compensate for any hindrance to the steam and condensate flow.

Example: Size the main riser in an upfeed two-pipe low-pressure (4.0 psig) system for 40 MBH where the actual length of the pipe is 175 feet. What pressure drop should be used in sizing the runs of pipe?

In this example, the steam and condensate flow in opposite directions; and therefore, Figure 9.6 is used to size the main riser. A 40 MBH load requires 40 pounds of steam per hour. In a two-pipe system, 1-1/2" vertical pipe is suitable, because it can provide 48 pounds of steam per hour (see Column B in Figure 9.6).

The pressure drop for this example is calculated using the tables in Figures 9.5 and 9.7. The calculations for a 4.0 psig system would be:

$$1/3 \times 4.0 = 1\text{-}1/3 \text{ psi (20 oz.)}$$

The pipe length with fittings is estimated initially at twice the straight length, or 350 feet total equivalent pipe length.

$$\frac{20 \text{ oz.} \times 100}{350}$$

Therefore, the pressure drop per 100 feet is approximately 6 oz.

Safety and Operational Features

There are several safety and operational features in steam systems, including provisions for monitoring both the water level in the boiler and temperature changes, properly supporting the pipe, and maintaining uniform flow and traps.

The *Hartford Loop* is a safety feature designed to prevent the water level in the boiler from dropping too low (see Figure 9.3). Whenever the water level falls below the Hartford loop connection tie-in (approximately two to four inches below water level in the boiler), the return line is no longer isolated from the supply line by the water in the boiler. As a result, the pressures equalize and the abnormal siphon that caused the boiler to drop is broken.

Because steam pipes undergo a wide range of temperature changes, *expansion loops*, or joints, and *swing joints* are provided for expansion and contraction throughout a system. Expansion loops and joints are used at connectors to mains at terminal units and on straight runs of more than 40 feet.

Expansion loops and joints also require that the piping be properly anchored and guided. Steam pipes should be *insulated* and *supported* on properly designed hangers at suitable spacings.

In order to maintain continuous flow of condensate and avoid water hammer (water flowing through the pipes at too high a speed), *eccentric pipe reducers* are required to maintain uniform flow at pipe inverts whenever pipe size reduces in the direction of the condensate flow. Eccentric reducers reduce the pipe off-center, maintaining a flat invert. The small end of a *concentric reducer* is positioned centrally at the end of the fitting. Eccentric and concentric reducers are shown in Figure 9.8.

All units must be properly trapped, and dirt pockets, or drain cocks, should be provided at the base of every riser and at main drips to remove sludge build-up.

Water Distribution

Most hot and chilled water piping networks are closed-loop water distribution systems. The four basic types of water distribution systems—series, monoflow, direct return, and reverse return—are illustrated in Figure 9.9. When separate heating and cooling are provided to the same terminal unit, three- or four-pipe systems are generally used.

Series/Monoflow

The series and monoflow loops are similar configurations routinely used for systems and subsystems under 50 MBH. They require the least amount of piping and have the lowest cost, but offer the least control. A series circuit connects all convectors/radiators by a single pipe in one continuous loop. A monoflow circuit has an uninterrupted loop and the convectors/radiators are tapped off the loop.

The series circuit has no control; shutting off one terminal shuts off the entire circuit. Each terminal unit in a monoflow system may be individually controlled. In either system, the radiators at the end of the run should be sized based on a lower supply water temperature; the water temperature is lower as it reaches the end of the loop.

Direct/Reverse Return

In a direct return water distribution system, the return pipe takes the shortest, most direct route back to the pump. The distance that the water travels from the pump to the first terminal unit and back is shorter than the loop distance to the last terminal unit (see Figure 9.9). In a reverse return system, the return piping leaving a terminal unit follows the supply piping to the next terminal unit. The result is that the loop distance to all terminal units is equal. Some reverse return systems have short supply pipe and long return pipe runs, while others have long supply and short return pipe runs, but the total loop length to all terminal units is equal. In short, in a direct return system, the first terminal unit fed is also the first one returned to the boiler, whereas in a reverse return system, the first one fed is the last one returned and the last one fed is the first returned.

Reverse return piping systems are sometimes called *three-pipe systems*, because of the apparent extra return pipe. This situation can be seen by comparing the elevations of the direct and reverse piping systems shown in Figure 9.9.

The reverse return piping layout provides the best overall control and reliability. It supplies the same temperature water to each terminal unit (minus negligible line losses) and distributes the proper quantities of water in each pipe, because all loops are the same length and no short circuiting can occur.

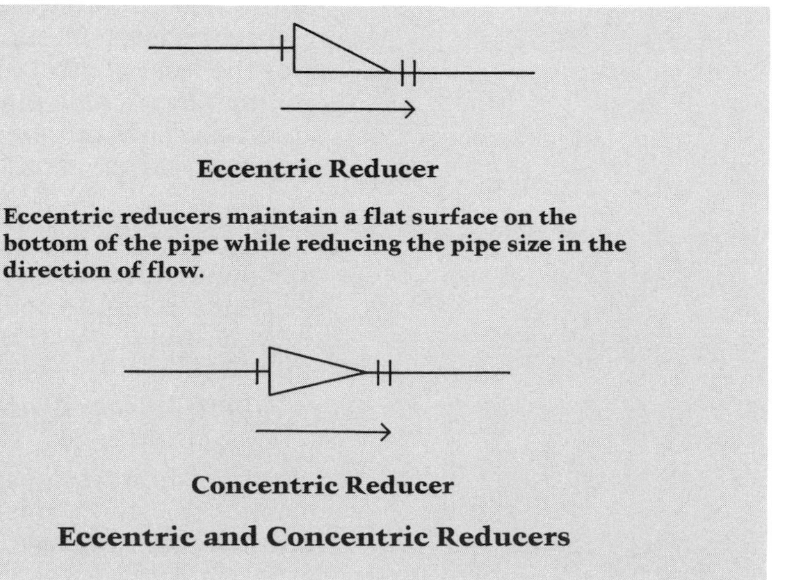

Eccentric and Concentric Reducers

Figure 9.8

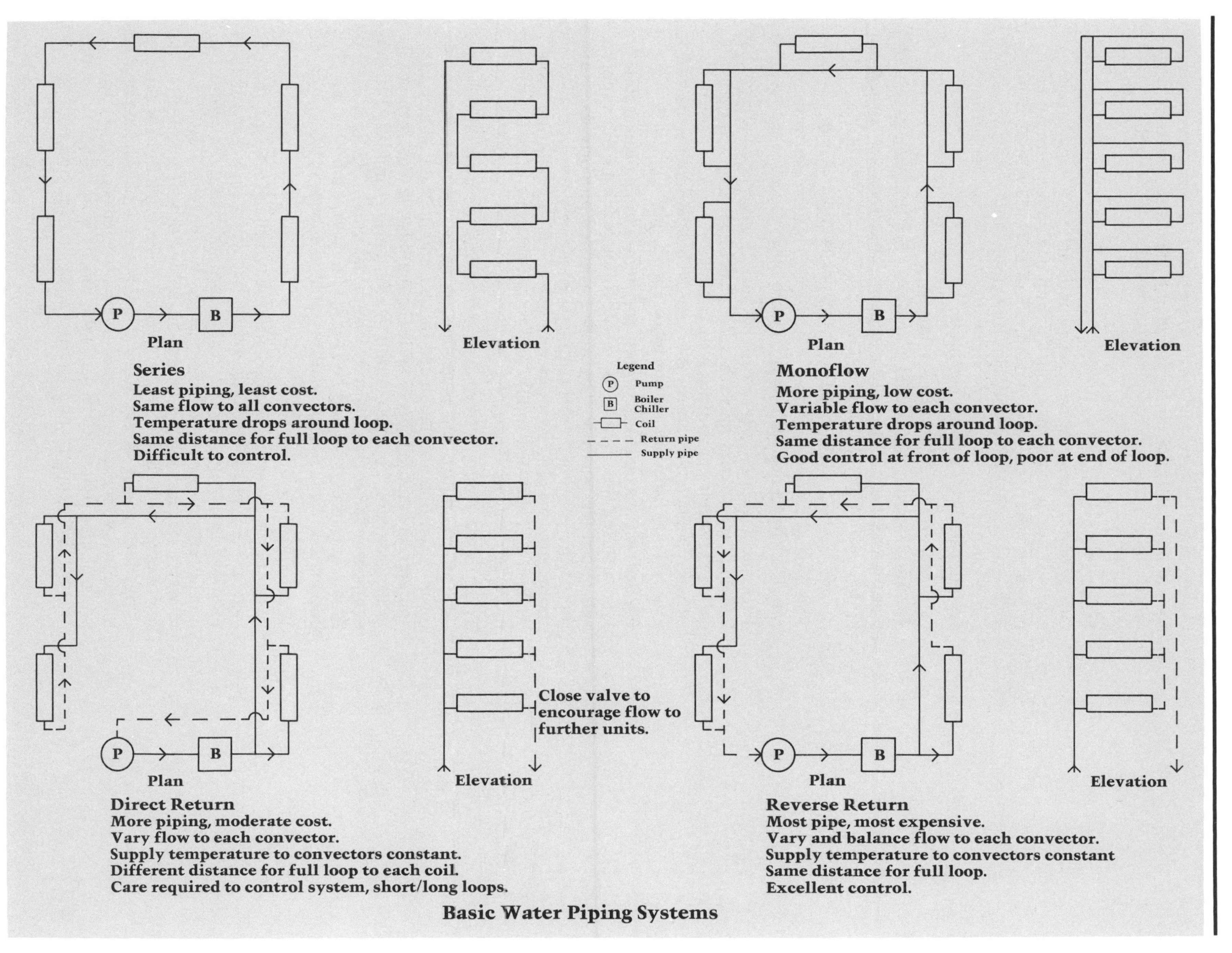
P
B
Plan
Elevation
Series
Least piping, least cost.
Same flow to all convectors.
Temperature drops around loop.
Same distance for full loop to each convector.
Difficult to control.
Legend
P Pump
B Boiler Chiller
Coil
Return pipe
Supply pipe
Monoflow
More piping, low cost.
Variable flow to each convector.
Temperature drops around loop.
Same distance for full loop to each convector.
Good control at front of loop, poor at end of loop.
Close valve to encourage flow to further units.
Direct Return
More piping, moderate cost.
Vary flow to each convector.
Supply temperature to convectors constant.
Different distance for full loop to each coil.
Care required to control system, short/long loops.
Reverse Return
Most pipe, most expensive.
Vary and balance flow to each convector.
Supply temperature to convectors constant
Same distance for full loop.
Excellent control.
Basic Water Piping Systems

Figure 9.9

The direct return system is usually used in loops or sub-loops under 50 MBH. Terminal units closest to the pump typically should have partially closed balancing valves to constrict flow on the short runs and to encourage water to flow to the farther units.

Heat Per GPM

One gallon of water weighs 8.1 pounds at 180°F. One Btu is released into a room for each 1°F drop in temperature for each pound of water at a terminal unit. For heating systems, using a 20°F temperature drop between the supply and return water temperatures is an industry standard. A 1 gallon per minute system equals 60 gallons per hour.

1 gpm = 60 × 8.1 = 486 lbs./hr.

The result is rounded to 500 pounds per hour. If each pound of water drops 20°F, it loses 20 Btu/hour. Therefore, the heat given up by 1 gallon per minute equals:

500 lbs./hr. × 20°F = 10,000 Btu/hr., or 10 MBH

The general rule for heating systems is: 1 gallon per minute equals 10 MBH at a 20°F temperature drop.

This calculation is easy to use, because it directly converts heat loss to flow. A building with a 100 MBH heat load requires 10 gallons of water per minute with a 20°F drop in temperature between supply and return. A room in that building with a heat loss of 3,500 Btu/hour would require 0.35 gpm.

Chilled water, condenser water, and high-temperature water systems use other standards for the difference between supply and return temperature. The rule is adjusted proportionally for these other systems. For example, a condenser pump to a cooling tower is likely to operate at a lower temperature drop, such as 8°F, between supply and return, and therefore requires more water flow (20/8).

To lay out a piping system, first determine the load (MBH) for the entire building, and then for each room. At least one terminal unit should be placed in each room. Next, select an appropriate combination of series, monoflow, direct, or reverse return systems. Then the pump, boiler, chiller, and other major equipment are connected to the terminal units. The load and flow requirements should be recorded next to each pipe length designation. This process is illustrated in Figure 9.10.

Water Piping

Water is the most frequently used medium in HVAC piping systems. A typical low-temperature hot water piping network is used as a base guide, but the rules used for sizing the piping for hot water apply to all water systems. As the temperature varies in water systems from 40°F for chilled water to 250°F for low-temperature hot water systems, the only significant change is in the viscosity of the water. Dual systems use the same pipes for chilled water in summer and hot water in winter. Many types of pipe material are used; the most common are Schedule 40 steel pipe, copper tubing (types K, L, and M), and Schedule 80

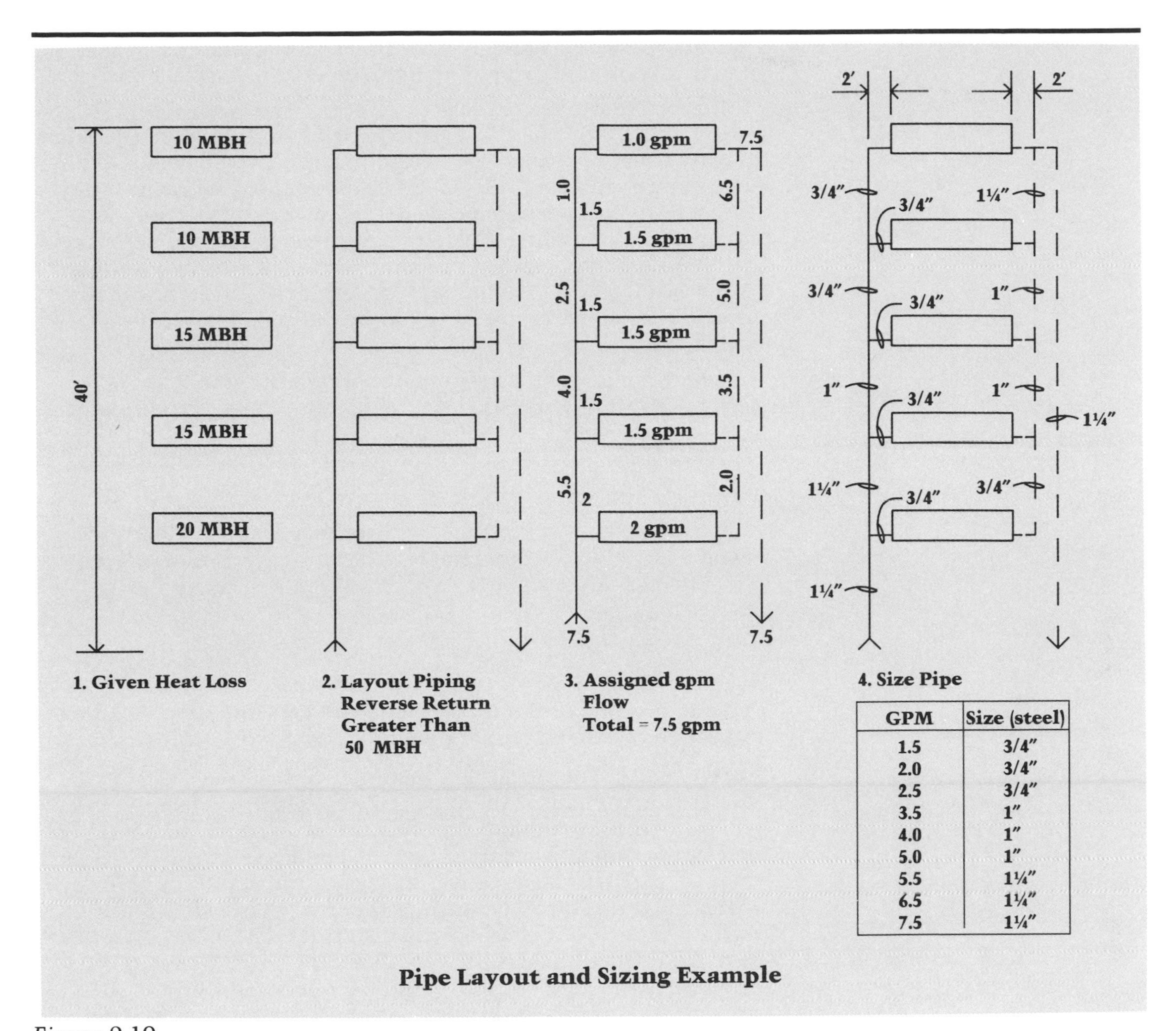

GPM	Size (steel)
1.5	3/4″
2.0	3/4″
2.5	3/4″
3.5	1″
4.0	1″
5.0	1″
5.5	1¼″
6.5	1¼″
7.5	1¼″

Figure 9.10

PVC and CPVC (plastic) pipe. Figure 9.10 illustrates an example of pipe layout and sizing. The method used is explained in the following paragraphs.

Pipe size is based on either *velocity* or *head loss*, which provide roughly equal results. Good design practice provides pipe sizes large enough to avoid the velocity noise caused by rapid water flow. Large pipe sizes also reduce the amount of friction that pumps must overcome. Velocities of three to six feet per second are considered acceptable. Pressure head losses of two to four feet of head per 100 feet of pipe are also acceptable. Generally, a rating of 2.5 feet of head per 100 feet of pipe is used as design criteria.

To select a pipe size, determine the design flow for the particular run of pipe based on the recommended head loss per 100 feet (see Figure 9.11). For example, select a pipe for a flow of 30 gallons per minute at 2.5 feet of head per 100 feet in steel, copper type L, and Schedule 80 PVC.

Type	Interior Diameter	Nominal Size to Use	Velocity in Selected Pipe
Steel (Sched. 40)	2.067″	2″	2.9 fps
Type L Copper	1.985″	2″	3.0 fps
PVC (Sched. 80)	1.913″	2″	3.5 fps

For this example, a two-inch diameter pipe is selected, the water velocity is approximately 3 fps, and the pressure drop is 2.5 feet of head for each 100 feet of pipe.

Flow Temperature and Viscosity

The charts in Figure 9.11 are based on a water temperature of 60°F, which means that for chilled water systems (40 – 55°F), they are reasonably accurate; for hot water systems (200 – 250°F), they are slightly conservative. However, the actual head losses will be slightly lower, because of the lower viscosity of the warmer water — the warmer the water, the less force is required to make it move. When friction losses for hot water are calculated, the pipe section should be oversized by 15 to 20 percent. PVC pipe should not be used in systems with water temperatures greater than 110°F.

Many hydronic systems use a mixture of glycol and water, which has a low freezing point, to prevent freezing in pipes. Water in outside air intake coils, makeup air units, solar systems, and pipes that pass through unheated spaces are sometimes filled with this mixture to minimize the problems associated with freezing. A fifty-fifty water/glycol mixture has a freezing point of –34°F, which is low enough for most applications.

The selection of pipe size is affected when glycol is used, because glycol has a higher viscosity and requires more force to pump. Correction factors for glycol mixtures are used to account for friction losses (see Figure 9.12).

Pipe Length/Pump Size

Like steam systems, hot and chilled water systems are also affected by friction losses for fittings such as valves, elbows,

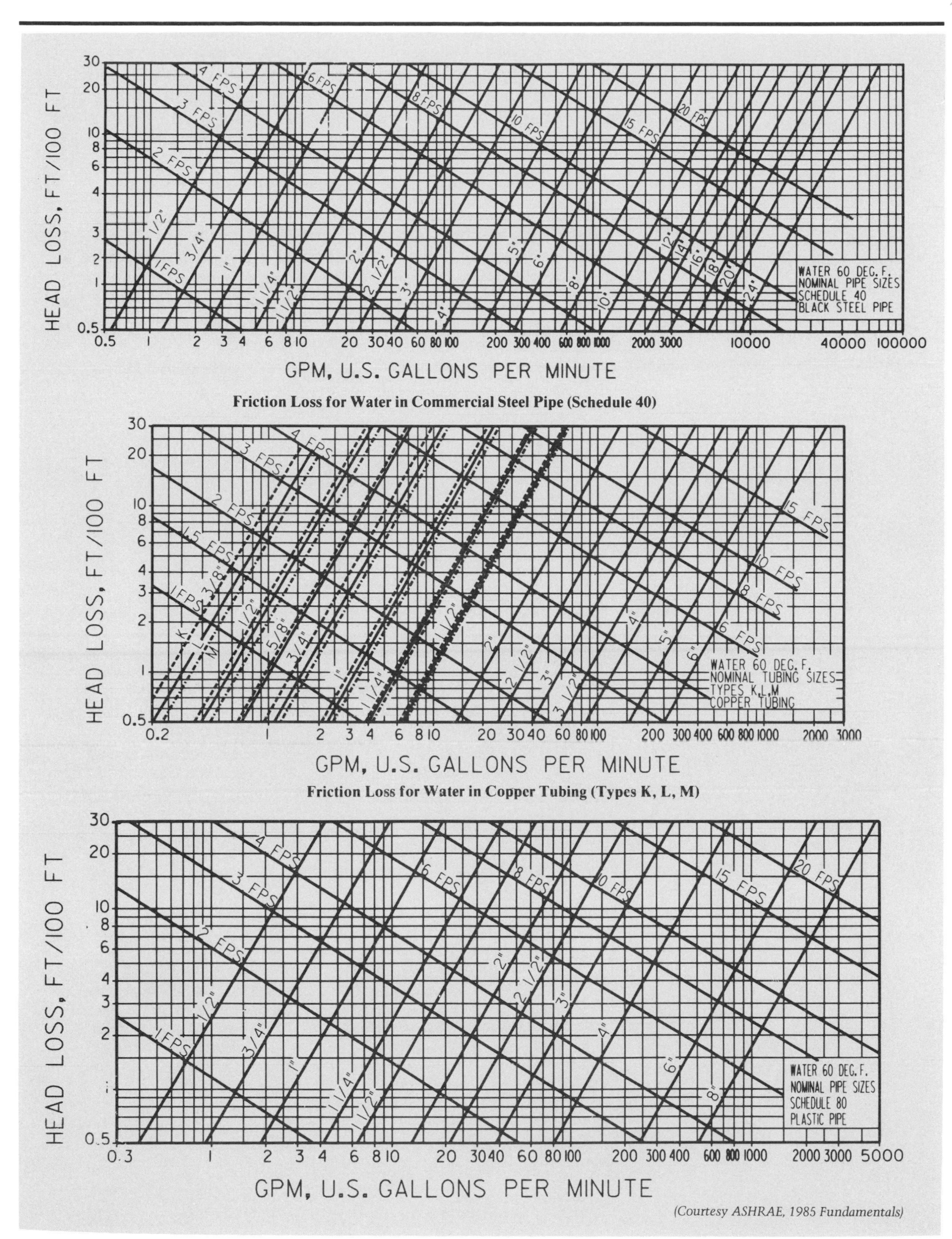
HEAD LOSS, FT/100 FT
GPM, U.S. GALLONS PER MINUTE
WATER 60 DEG. F.
NOMINAL PIPE SIZES
SCHEDULE 40
BLACK STEEL PIPE
Friction Loss for Water in Commercial Steel Pipe (Schedule 40)
HEAD LOSS, FT/100 FT
GPM, U.S. GALLONS PER MINUTE
WATER 60 DEG. F.
NOMINAL TUBING SIZES
TYPES K,L,M
COPPER TUBING
Friction Loss for Water in Copper Tubing (Types K, L, M)
HEAD LOSS, FT/100 FT
GPM, U.S. GALLONS PER MINUTE
WATER 60 DEG. F.
NOMINAL PIPE SIZES
SCHEDULE 80
PLASTIC PIPE
(Courtesy ASHRAE, 1985 Fundamentals)

Figure 9.11

tees, and specialties. The pump must be sized for the total flow (in gpm) of the total equivalent run of pipe and the total head. The total flow is calculated from the heat load and temperature drop of the water. The total equivalent run of pipe equals the measured straight run of pipe plus the straight run equivalent for the fittings. The total head is determined by adding up the friction head loss for both the pipes and the fittings and adding (for open systems only) the lift head from one elevation to another.

Figure 9.13 lists the equivalent length for elbows over a range of velocities and pipe sizes. For example, a two-inch elbow in a system with water flowing at 3 feet per second has an equivalent straight run of 5.4 feet. An open gate valve for steel pipe has an elbow equivalent of 0.5. Therefore, the equivalent length of the valve is

$$0.5 \times 5.4 = 2.7 \text{ ft.}$$

Correction Factors for Glycol Mixtures

Increased Flow Requirement for Same Heat Conveyance 50% Glycol as Compared with Water	
Fluid Temperature (°F)	**Flow Increase Needed for 50% Glycol as Compared with Water**
40	1.22
100	1.16
140	1.15
180	1.14
220	1.14

Pressure Drop Correction Factors; 50% Glycol Solution Compared with Water		
Fluid Temperature (°F)	**Pressure Drop Correction Flow Rates Equal**	**Combined Pressure Drop Correction; 50% Glycol Flow Increased as Shown Above**
40	1.45	2.14
100	1.1	1.49
140	1.0	1.32
180	.94	1.23
220	.9	1.18

(Courtesy ITT—Bell & Gossett)

Figure 9.12

34.2 CHAPTER 34 1985 Fundamentals Handbook

Table 2 K Values—Flanged Welded Pipe Fittings[1]

Nominal Pipe Dia.	90° Ell Reg.	90° Ell Long	45° Ell Long	Ret. Bend Reg.	Ret. Bend Long	Tee-Line	Tee-Branch	Globe Valve	Gate Valve	Angle Valve	Swing Check Valve
1	0.43	0.41	0.22	0.43	0.43	0.26	1.0	13	—	4.8	2.0
1-1/4	0.41	0.37	0.22	0.41	0.38	0.25	0.95	12	—	3.7	2.0
1-1/2	0.40	0.35	0.21	0.40	0.35	0.23	0.90	10	—	3.0	2.0
2	0.38	0.30	0.20	0.38	0.30	0.20	0.84	9	0.34	2.5	2.0
2-1/2	0.35	0.28	0.19	0.35	0.27	0.18	0.79	8	0.27	2.3	2.0
3	0.34	0.25	0.18	0.34	0.25	0.17	0.76	7	0.22	2.2	2.0
4	0.31	0.22	0.18	0.31	0.22	0.15	0.70	6.5	0.16	2.1	2.0
6	0.29	0.18	0.17	0.29	0.18	0.12	0.62	6	0.10	2.1	2.0
8	0.27	0.16	0.17	0.27	0.15	0.10	0.58	5.7	0.08	2.1	2.0
10	0.25	0.14	0.16	0.25	0.14	0.09	0.53	5.7	0.06	2.1	2.0
12	0.24	0.13	0.16	0.24	0.13	0.08	0.50	5.7	0.05	2.1	2.0

Table 3 Approximate Range of Variation for K Factors[1]

90° Elbow	Regular Screwed	±20% above 2 inch
		±40% below 2 inch
	Long Radius Screwed	±25%
	Regular Flanged	±35%
	Long Radius Flanged	±30%
45° Elbow	Regular Screwed	±10%
	Long Radius Flanged	±10%
Return Bend (180°)	Regular Screwed	±25%
	Regular Flanged	±35%
	Long Radius Flanged	±30%
Tee	Screwed, Line or Branch	±25%
	Flanged, Line or Branch	±35%
Globe Valve	Screwed	±25%
	Flanged	±25%
Gate Valve	Screwed	±25%
	Flanged	±50%
Angle Valve	Screwed	±20%
	Flanged	±50%
Check Valve	Screwed	±50%
	Flanged	+200%
		−80%

Example 1: Determine the pressure drop for 60 F water flowing at 4 ft/s through a nominal 1 in. 90° screwed ell.

Solution: From Table 1, the k for a 1 in. 90° screwed ell is 1.5. $\Delta p = 1.5 \cdot 62.4/32.2 \cdot 4^2/2 = 23.3\ lb_f/ft^2$ or 0.16 psi.

The loss coefficient for valves appears in another form as C_v, a dimensional coefficient expressing the flow through a valve at a specified pressure drop.

$$Q = C_v \sqrt{\Delta p}$$

where

Q = volume flow, gal/min for a 1 $lb_f/in.^2$ pressure drop

Example 2: Determine the volume flow through a valve with C_v = 10 (gpm, psi units) for an allowable pressure drop of 5 psi.

Solution: $Q = 10\sqrt{5}$ = 22.4 gpm.

Alternative formulations express fitting losses in terms of equivalent lengths of straight pipe (Tables 4 and 5, Fig. 4).

Calculating Pressure Losses

The most common engineering design flow loss calculation selects a pipe size for the desired total flow rate and available or allowable pressure drop.

Since either formulation of fitting losses requires a known diameter, pipe size must be selected before calculating the detailed influence of fittings. A frequently used rule-of-thumb assumes that the design length of pipe is 50% to 100% longer than actual to account for fitting losses. After a pipe diameter has been selected on this basis, the influence of each fitting can be evaluated.

"Water Hammer"

When any moving fluid (not just water) is abruptly stopped as when a valve closes suddenly, large pressures can develop. While detailed analysis requires knowledge of the elastic properties of the pipe and the flow-time history, the limiting case of rigid pipe and instantaneous closure is simple to calculate. Under these conditions,

$$p_h = \varrho c_s V/g_c$$

where

p_h = pressure rise caused by water hammer, lb_f/ft^2
ϱ = fluid density, lb_m/ft^3
c_s = velocity of sound in the fluid, ft/s
V = fluid flow velocity, ft/s

c_s for water is 4720 ft/s, although elasticity of the pipe reduces the effective value.

Example 3: What is the maximum pressure rise if water flowing at 10 ft/s is stopped instantaneously?

Solution: $p_h = 62.4 \cdot 4720 \cdot 10/32.2 = 91468\ lb_f/ft^2$
$= 635$ psi

Other Considerations

Not discussed in detail in this chapter, but of potentially great importance are a number of physical and chemical considerations: pipe and fitting design, materials and joining methods must be appropriate for working pressures and temperatures encountered, as well as being suitably resistant to chemical attack by the fluid.

Table 4 Equivalent Length in Feet of Pipe for 90-Deg Elbows

Vel. Fps	Pipe Size ½	¾	1	1¼	1½	2	2½	3	3½	4	5	6	8	10	12
1	1.2	1.7	2.2	3.0	3.5	4.5	5.4	6.7	7.7	8.6	10.5	12.2	15.4	18.7	22.2
2	1.4	1.9	2.5	3.3	3.9	5.1	6.0	7.5	8.6	9.5	11.7	13.7	17.3	20.8	24.8
3	1.5	2.0	2.7	3.6	4.2	5.4	6.4	8.0	9.2	10.2	12.5	14.6	18.4	22.3	26.5
4	1.5	2.1	2.8	3.7	4.4	5.6	6.7	8.3	9.6	10.6	13.1	15.2	19.2	23.2	27.6
5	1.6	2.2	2.9	3.9	4.5	5.9	7.0	8.7	10.0	11.1	13.6	15.8	19.8	24.2	28.8
6	1.7	2.3	3.0	4.0	4.7	6.0	7.2	8.9	10.3	11.4	14.0	16.3	20.5	24.9	29.6
7	1.7	2.3	3.0	4.1	4.8	6.2	7.4	9.1	10.5	11.7	14.3	16.7	21.0	25.5	30.3
8	1.7	2.4	3.1	4.2	4.9	6.3	7.5	9.3	10.8	11.9	14.6	17.1	21.5	26.1	31.0
9	1.8	2.4	3.2	4.3	5.0	6.4	7.7	9.5	11.0	12.2	14.9	17.4	21.9	26.6	31.6
10	1.8	2.5	3.2	4.3	5.1	6.5	7.8	9.7	11.2	12.4	15.2	17.7	22.2	27.0	32.0

(Courtesy ASHRAE, 1985 Fundamentals)

Figure 9.13

Fitting	Iron Pipe	Copper Tubing
Elbow, 90-deg	1.0	1.0
Elbow, 45-deg	0.7	0.7
Elbow, 90-deg long turn. . .	0.5	0.5
Elbow, welded, 90-deg. . . .	0.5	0.5
Reduced coupling	0.4	0.4
Open return bend.	1.0	1.0
Angle radiator valve.	2.0	3.0
Radiator or convector	3.0	4.0
Boiler or heater	3.0	4.0
Open gate valve	0.5	0.7
Open globe valve	12.0	17.0

Table 5 Iron and Copper Elbow Equivalents[a, 3-5]

[a]See Table 4 for equivalent length of one elbow.

Other Piping Materials and Fluids

For fluids not included in this chapter or for piping materials of different dimensions, manufacturer's literature frequently supplies pressure drop charts. The Darcy-Weisbach equation and the Moody chart or the Colebrook equation can be used as an alternative to pressure drop charts or tables.

HOT AND CHILLED WATER PIPE SIZING

The theoretical basis for calculating pressure drop in hot and chilled water piping is discussed in the previous section.

The Darcy-Weisbach equation with friction factors from the Moody chart or Colebrook equation (or alternatively, the Hazen-Williams equation) is fundamental to these calculations; however, charts calculated from these equations (such as Fig. 1, 2 and 3) provide easy determination of pressure drops for specific fluids and pipe standards. In addition, tables of pressure drops can be found in Ref. 1 and 2.

Most tables and charts for water are calculated for properties at 60 F. Using these for hot water introduces some error, although the answers are conservative; i.e., cold water calculations overstate the pressure drop for hot water. Using 60 F water charts for 200 F water should not result in errors in Δp exceeding 20%.

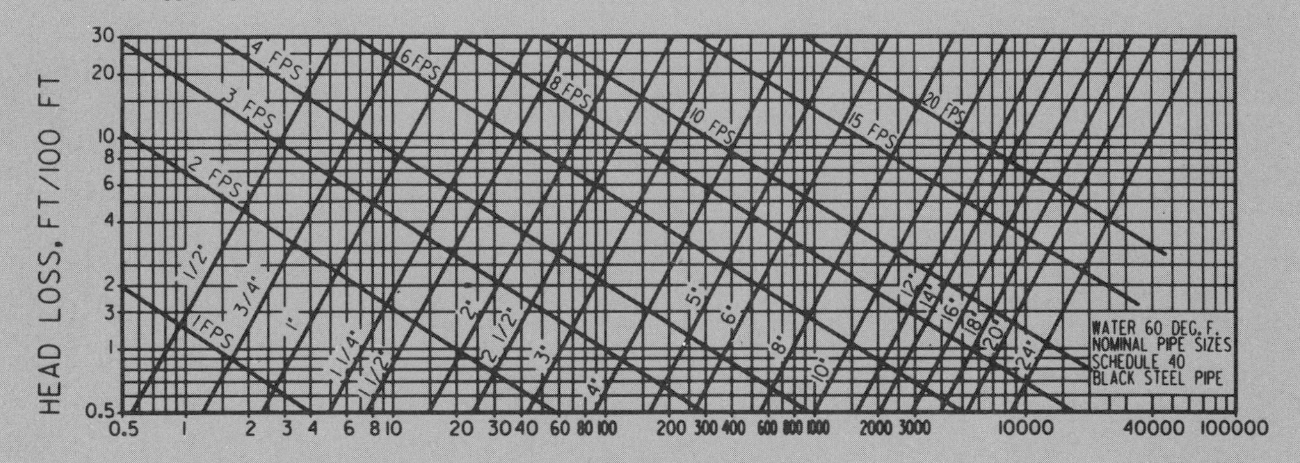

Fig. 1 Friction Loss for Water in Commercial Steel Pipe (Schedule 40)

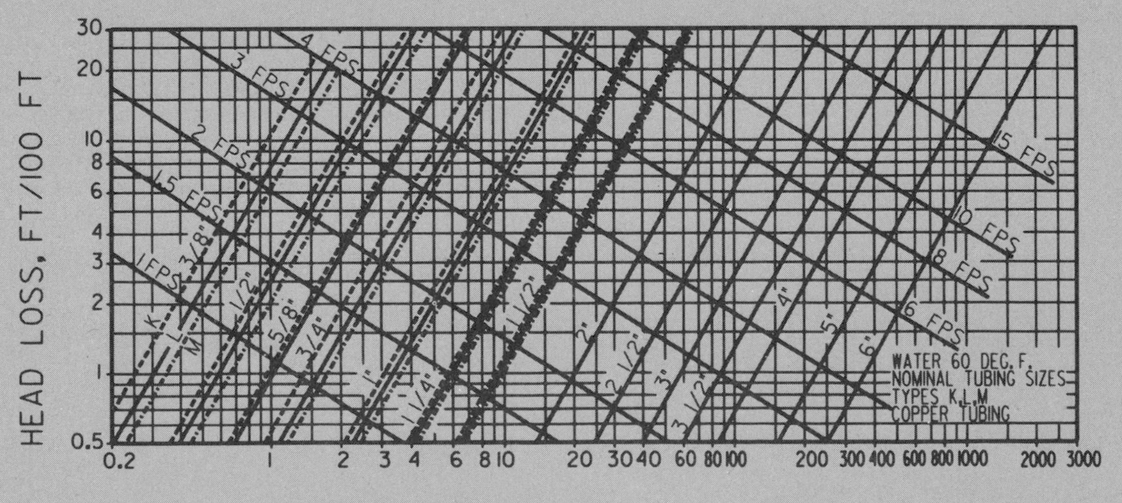

Fig. 2 Friction Loss for Water in Copper Tubing (Types K, L, M) *(Courtesy ASHRAE, 1985 Fundamentals)*

Figure 9.13 (cont.)

For initial trial designs, it is commonly assumed that the friction of all fittings around a loop will equal that of the straight run of pipe, meaning that the total equivalent run of pipe equals twice the measured run of straight pipe. This assumption should be confirmed after the design layout is completed.

For the system shown in Figure 9.10, the pump head required is based on the friction head loss around one radiator loop. A one-inch pipe in the system has water flowing at about 1.5 feet per second; the elbow equivalent is 2.3 feet (see Figure 9.13).

Straight runs of pipe			
3 Risers at 40 ft.		=	120 ft.
Radiator runouts	(2 × 2 ft.)	=	4
			124
Fittings on one loop			
15 Elbows (estimated)	(15 × 1 elbow equiv. × 2.3)	=	34.5
1 Boiler	(1 × 3 elbow equiv. × 2.3)	=	6.9
1 Radiator	(1 × 3 elbow equiv. × 2.3)	=	6.9
2 Tees	(2 × 0.5 elbow equiv. × 2.3)	=	2.3
	Total equivalent straight pipe	**=**	**174.6 ft.**

Because all pipe was sized not to exceed 2.5 feet of head per 100 feet, the pump head equals

$$2.5/100 \times 174.6 = 4.4 \text{ feet of head}$$

Therefore, a pump that delivers 7.5 gpm with a head of 4.4 feet is required. These requirements, combined with the information from the pump curves (see Figure 7.3), result in the choice of a 3/4" pump. Figure 9.14 illustrates the basic piping layouts for various low-pressure systems.

Hydronic Specialties

Hot water systems must remove accumulated air. Air vents or bleed valves at terminal units and high points allow air to be mechanically "bled." Air scoops are also used in the mechanical room to remove air.

Pipe runs over 40 feet long should have provisions for expansion. In addition to anchors, guides, loops, or expansion joints, additional elbows (swing joints) are normally used at branch takeoffs to allow this kind of flexibility. All piping runs should be pitched to draw off valves for drainage.

Temperature/Pressure Considerations

Medium- and high-temperature water systems are designed using principles similar to those applied to chilled water and low-temperature hot water systems. An important difference is the pressure in the piping. At 212°F, the atmospheric pressure of 14.7 psi is the minimum pressure that is needed to keep water from boiling or "flashing" into the system. At 250°F, 29.8 psi is needed; at 350°F, 134.6 psi; and at 450°F, 422.6 psi is the pressure exerted by the hot water against the piping. To avoid future problems, care must be taken in selecting safety features and materials and ensuring good construction techniques for hot water systems.

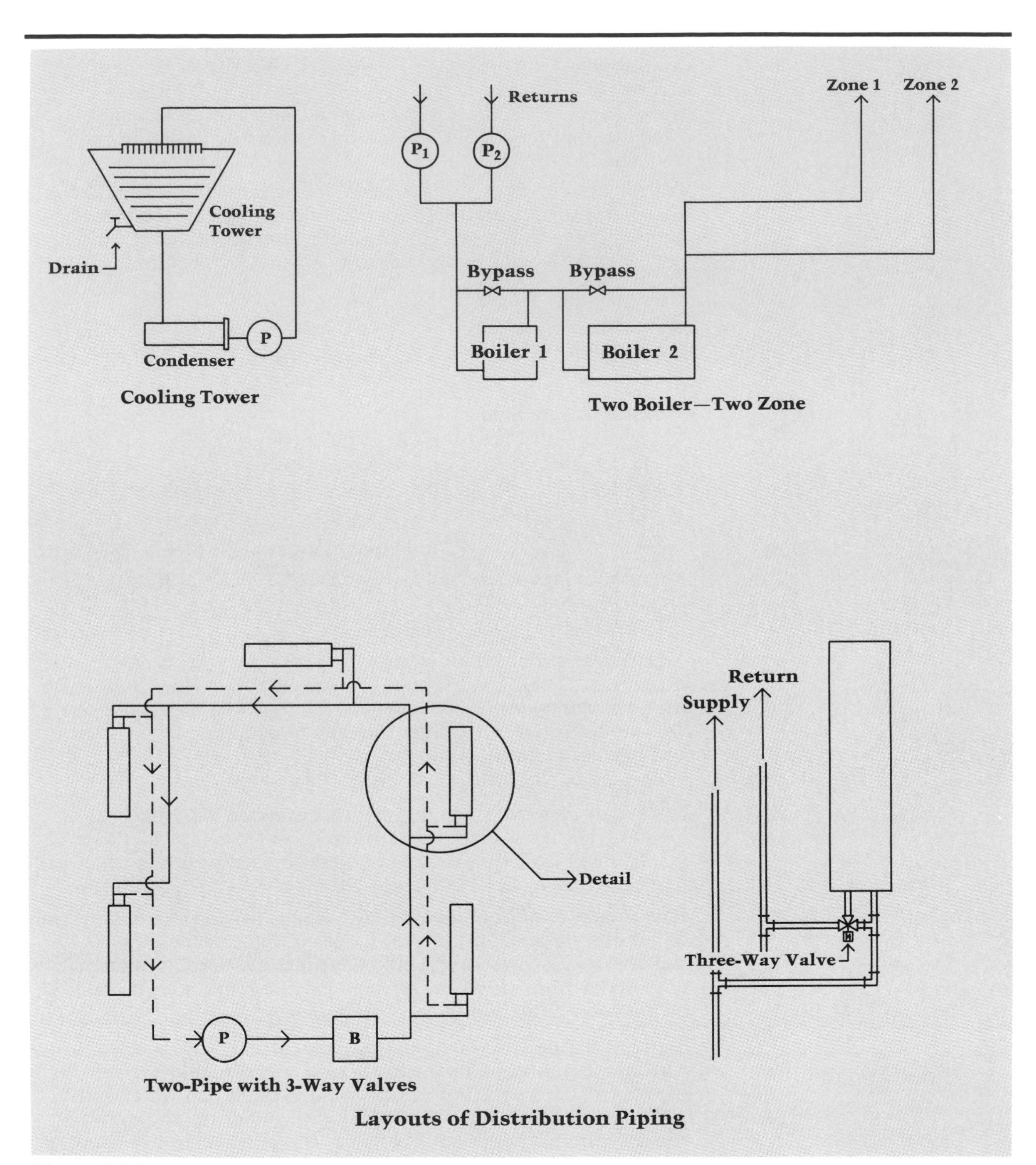
Returns
Zone 1
Zone 2
P1
P2
Cooling
Tower
Drain
Bypass
Bypass
Boiler 1
Boiler 2
P
Condenser
Cooling Tower
Two Boiler—Two Zone
Return
Supply
Detail
Three-Way Valve
P
B
Two-Pipe with 3-Way Valves
Layouts of Distribution Piping

Figure 9.14

Air Distribution Systems

Air systems are widely used for both heating and cooling. Air systems offer the advantages of being free of freeze-up problems; providing full temperature control treatment of a space, including filtration, humidification, and cleaning; and being easily adapted for alterations. In addition, air systems have few mechanical or electrical devices to repair outside of the main equipment areas. Some disadvantages of air systems include large ducts, which can be difficult to conceal, and balancing air quantities for different seasons. Rapid temperature changes, duct leakage, and temperature losses in transmission of the air are also concerns. Overall, air is the most flexible of any type of system and is used in office, commercial, and industrial buildings, as well as laboratories, universities, and hotels.

Heat Per Cubic Foot

Air transfers the least amount of heat per cubic foot. At room conditions, cool air occupies approximately 13.3 cubic feet per pound of air and one cubic foot of air equals 1/13.3, or 0.075, pounds per cubic foot. 0.244 Btu's are required to heat or cool one pound of air 1°F. To cool one cubic foot of air 1°F,

$$0.075 \times 0.244 \text{ (lbs./ft.}^3 \times \text{Btu/lb.)} = 0.0183 \text{ Btu/min. or}$$

$$0.0183 \times 60 = 1.1 \text{ Btu/hr.}$$

Therefore,

$$H = 1.1 \text{ cfm } \Delta T \quad \text{Cooling}$$

$$= 1.08 \text{ cfm } \Delta T \quad \text{Heating (the air is even less dense)}$$

The factor of 1.1 will be used throughout the rest of the book.

Layout

Duct systems are generally classified as single or dual duct systems. Single duct systems, the most common type, have one supply duct to the terminal units and are limited to either heating or cooling at a given time. Dual duct systems have two supply air ducts, one with warm air and the second with cool air, which go to a mixing box above the space. The mixing box proportions the mixture of air to the space to achieve the design temperature. Both single and dual duct systems may be constant or variable volume, single or multi-zone, high or low velocity (see Figure 9.15).

High Velocity Systems

High velocity systems (the velocity of air in the ductwork exceeds 2,000 fpm or the pressure exceeds 2.5") are used for special applications. High velocity systems allow the use of smaller duct sizes, but require stronger ducts (or pipe) to transmit the high-pressure/high velocity air. These systems also require specialized high-pressure fans and sound attenuating capabilities. Induction systems in high-rise buildings represent a common application for high velocity systems. High velocity systems are designed and sized in the same manner as low velocity systems.

VAV Systems

Variable air volume systems (VAV) are economical and energy efficient. In a VAV system, the amount of air sent to a space is varied to meet the load. At the peak hour, the dampers are fully open and the maximum air is delivered to a space. At other

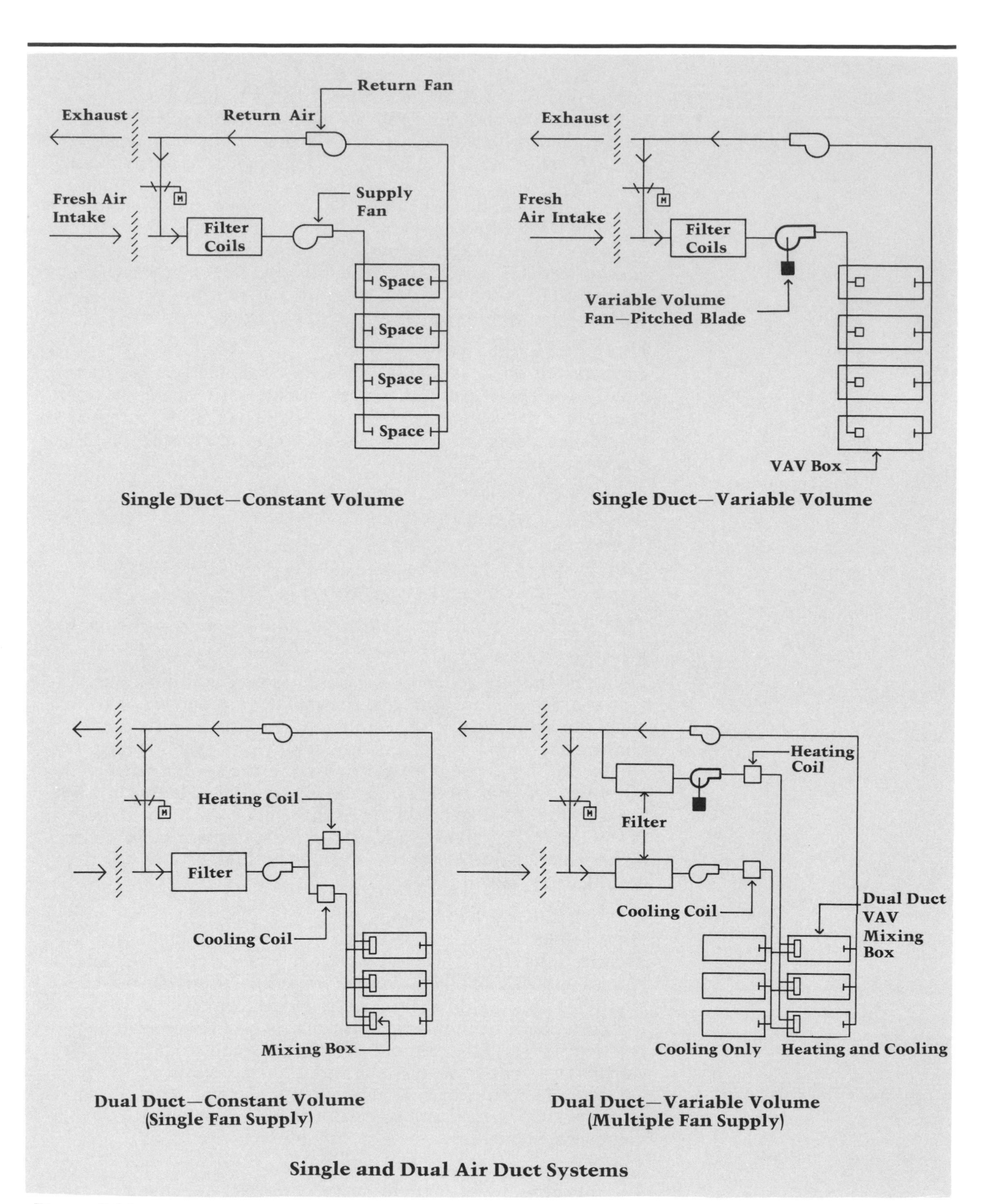

Return Fan
Exhaust
Return Air
Fresh Air
Intake
Supply
Fan
Filter
Coils
Space
Space
Space
Space
Single Duct—Constant Volume
Exhaust
Fresh
Air Intake
Filter
Coils
Variable Volume
Fan—Pitched Blade
VAV Box
Single Duct—Variable Volume
Heating Coil
Filter
Cooling Coil
Mixing Box
Dual Duct—Constant Volume
(Single Fan Supply)
Heating
Coil
Filter
Cooling Coil
Dual Duct
VAV
Mixing
Box
Cooling Only
Heating and Cooling
Dual Duct—Variable Volume
(Multiple Fan Supply)
Single and Dual Air Duct Systems

Figure 9.15

times, the dampers in the VAV box close partially, and less air is delivered to the space to match the decreased load. The ducts for VAV systems are sized using the same methods as for low velocity systems. However, special care must be taken to select each part of the variable system based on a worst case scenario. In any system, a minimum amount of air must be circulated for health and comfort and to meet ventilation requirements. The main advantage of the VAV system is that the equipment is sized for the maximum simultaneous peak in each item, rather than the sum of all individual peaks, as for constant volume systems.

Reheat

Many single duct systems in larger buildings have reheat coils incorporated into the terminal units. These reheat and booster coils may be hydronic or electric. Reheat coils are necessary in systems where cool air must be tempered for a particular zone. For example, the north side of an office building's interior zone in winter may need cooling, but at 60°F, not 55°F. The reheat coils sense the room temperature and "fine tune" the temperature to the particular zone requirements.

Reheat coils are not energy efficient, because they heat up air that has previously been cooled, thereby consuming energy for both processes. Generally, reheat coils are used only where special conditions are required, such as in laboratories and hospitals.

Grilles, Registers, and Diffusers

In Chapter 1, Figure 1.21 illustrates the general principle for locating supply diffusers. Basically, for heating, supply low at the perimeter; and for cooling, supply high near the perimeter.

A basic principle in locating air supply registers is the *throw*. Throw is the distance travelled by air from the diffuser to the point at which its velocity falls below 50 feet per minute. Diffusers should be located so that the space is fully covered. The "dead" band (between diffuser throws) should not exceed 25 percent of the throw. This principle is illustrated in Figure 9.16.

Ideally, the return grilles should be diagonally opposite the supply diffusers. In laying out a system, the number and location of supply air outlets must be considered in terms of the effect on room comfort. Supply air outlets should be located so as to avoid drafts and convection currents, while providing even, overall distribution. Because most air systems are located above ceilings, they are ideally suited for cooling. Some form of perimeter heat (low at the exterior windows) must often be added to supplement the air distribution system in order to cover winter conditions.

Once the location of the diffusers has been determined, the distribution ductwork can be laid out and sized.

Duct Sizing

The following four methods are used to size ducts (see Figure 9.17):

1. Velocity method
2. Constant pressure drop method

3. Balanced pressure drop method
4. Static regain

Each method builds on the techniques of the previous method. The methods become progressively more involved and are used for more complicated layouts.

The velocity method sizes ducts by targeting an appropriate velocity. This method is basically a rough guess and is used to size simple, short air systems. The second and third methods, constant and balanced pressure drop, are used by designers to accurately determine appropriate fan pressures. The last method, static regain, is the most difficult to design. It is also the most economical to build, although it requires advanced techniques.

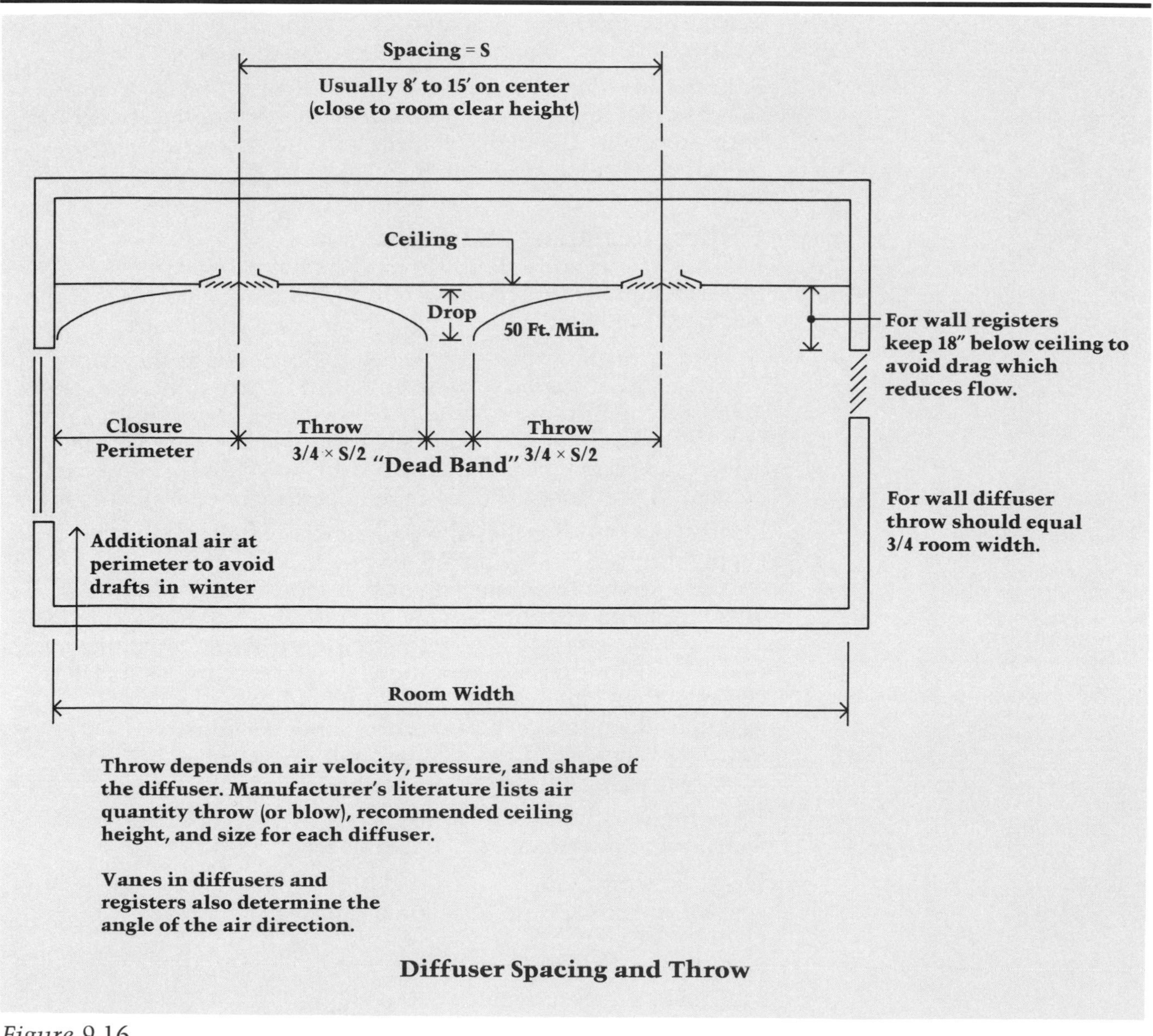

Figure 9.16

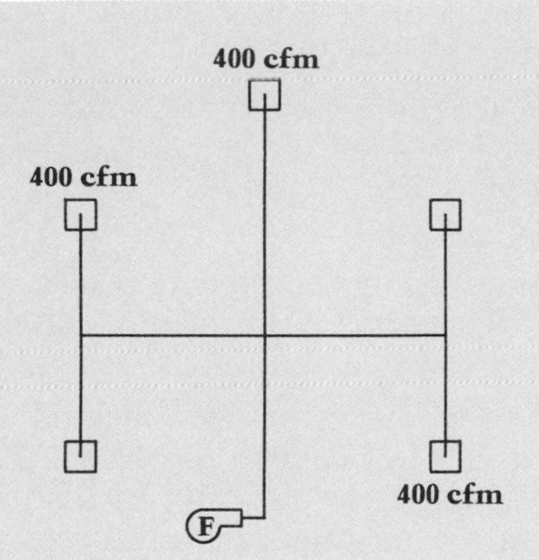

Velocity Method
Equal distance from fan to all registers.
Registers distribute same air quantity.
Easy to size ducts.
Volume dampers and field balance critical.
Small systems.

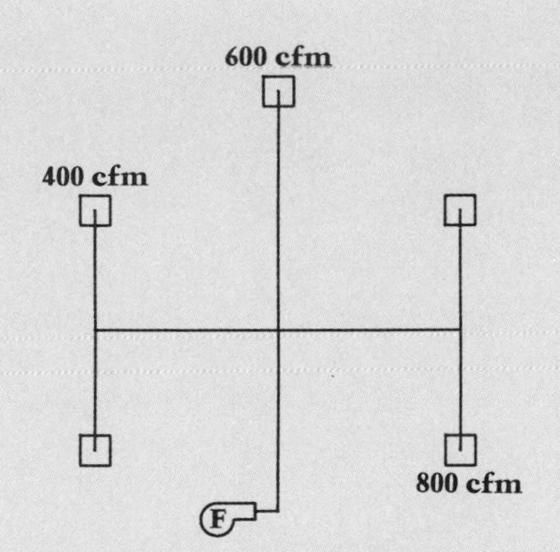

Constant Pressure Drop Method
Equal distance from fan to all registers.
Varied air quantity to each register.
Size duct from equivalent round.

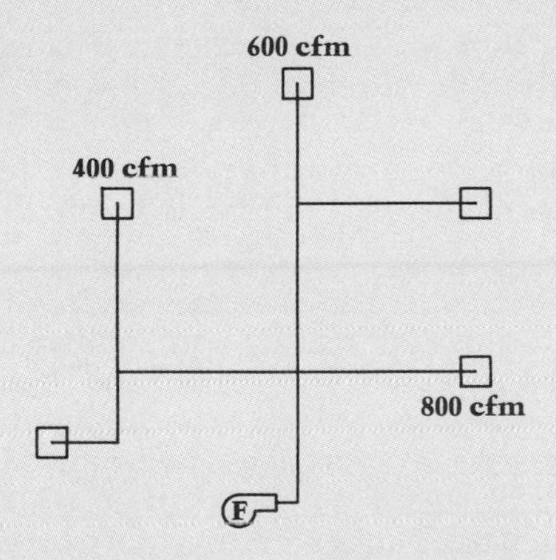

Balanced Pressure Drop Method
Varied distance from fan to registers.
Varied air quantity to each register.
Size duct by balancing pressures.

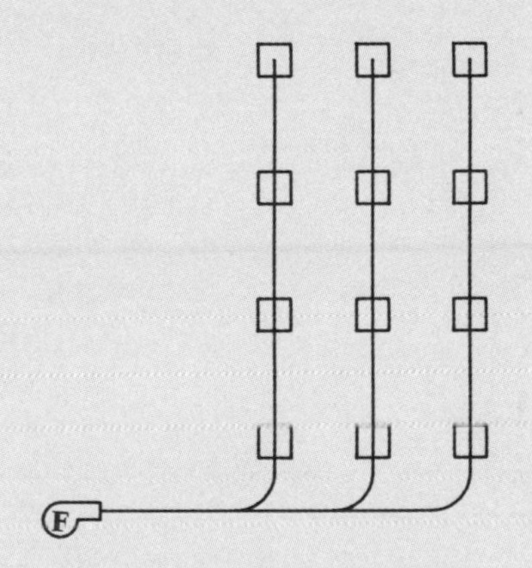

Static Regain Method
Registers take off along duct.
Varied air quantity to each register.
Most difficult to design.
Lowest fan and duct cost.

Legend

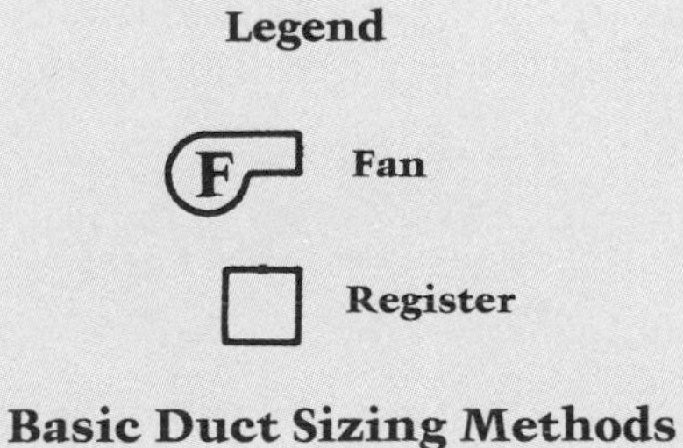

Basic Duct Sizing Methods

Figure 9.17

The Velocity Method: The following formula is used to determine duct size using the velocity method.

$Q = VA$

Q = the quantity of air in a duct (cfm)

V = the velocity of the air in the duct (fpm)

A = the duct cross-section area (s.f.)

As the quantity of air to be moved in the duct increases, the required duct size increases. The velocity is selected based on the criteria shown in Figure 9.18.

When using the velocity method, duct sizes are selected independent of shape, layout, or duct length; therefore, the pressures of the fan to drive the system can only be assumed. The inability to more precisely determine pressure drops in the system is the main disadvantage of the velocity method.

The quantity of air required is proportional to the load. For warm air,

$H_T = 1.1\ Q\ (\Delta T)$

H_T = net gain or heat loss of the space

1.1 = constant for cool air (1.08 for warm air)

Q = quantity of air to be delivered (cfm)

ΔT = The difference between the room design temperature and the temperature of the air in the duct. For cooling, ΔT = 15 – 20°F maximum. For heating, ΔT = 40 – 70°F.

Using these formulas, the recommended velocity, the required air quantity, and the area of the duct can be determined (see Figure 9.19).

The velocity method is best suited to calculations involving small systems, such as residential heating and cooling. The fans for these systems are usually slightly oversized and dampers can be used to control flow. When the velocity method is used, the duct should be as nearly square or circular as possible. The aspect ratio of 1:1 would be a square duct, with an equal proportion of width to height. The aspect ratio for the velocity method should not exceed 3:1. High aspect ratios cause additional friction, which affects overall balancing. This problem must be corrected using one of the other sizing methods. In duct design, a 4″ round or 6″ × 6″ square are generally the smallest sizes used.

Example: Size a cooling duct to carry 4,000 Btu/hr. in a main supply duct (for a store). Referring to Figures 9.18 and 9.19,

$$Q = \frac{H_T}{1.1 \times \Delta T} = \frac{4{,}000}{1.1 \times 15} = 242 \text{ cfm (rounded to 250 cfm)}$$

$$A = \frac{Q}{V}$$

A = duct area

Q = required air quantity

V = 1,800 fpm (recommended velocity for a main supply duct in an average store (see Figure 9.18)

RECOMMENDED MAXIMUM DUCT VELOCITIES FOR LOW VELOCITY SYSTEMS (FPM)

APPLICATION	CONTROLLING FACTOR NOISE GENERATION Main Ducts	CONTROLLING FACTOR—DUCT FRICTION			
		Main Ducts		Branch Ducts	
		Supply	Return	Supply	Return
Residences	600	1000	800	600	600
Apartments Hotel Bedrooms Hospital Bedrooms	1000	1500	1300	1200	1000
Private Offices Directors Rooms Libraries	1200	2000	1500	1600	1200
Theatres Auditoriums	800	1300	1100	1000	800
General Offices High Class Restaurants High Class Stores Banks	1500	2000	1500	1600	1200
Average Stores Cafeterias	1800	2000	1500	1600	1200
Industrial	2500	3000	1800	2200	1500

(Courtesy Carrier Corporation, McGraw–Hill Book Company)

Figure 9.18

1. **For each run of duct, determine the air quantity (Q).**

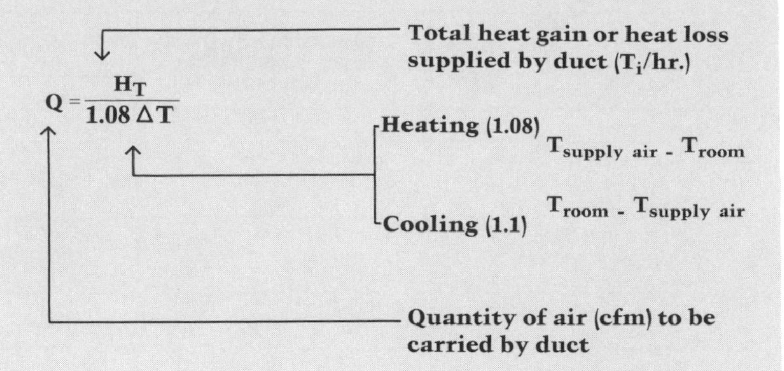

2. **Determine the recommended velocity from Figure 9.18.**

3. **Determine duct area (A) required.**

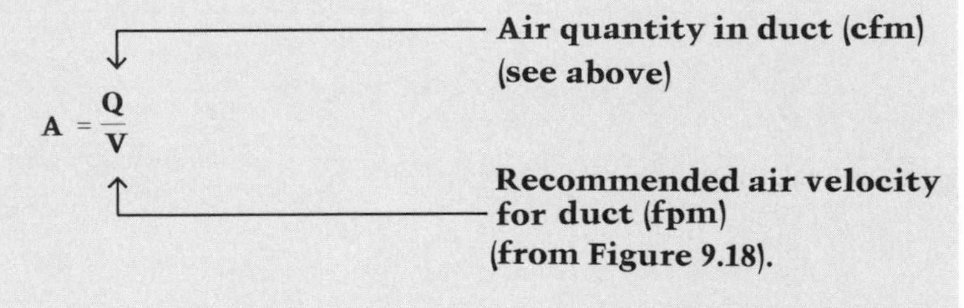

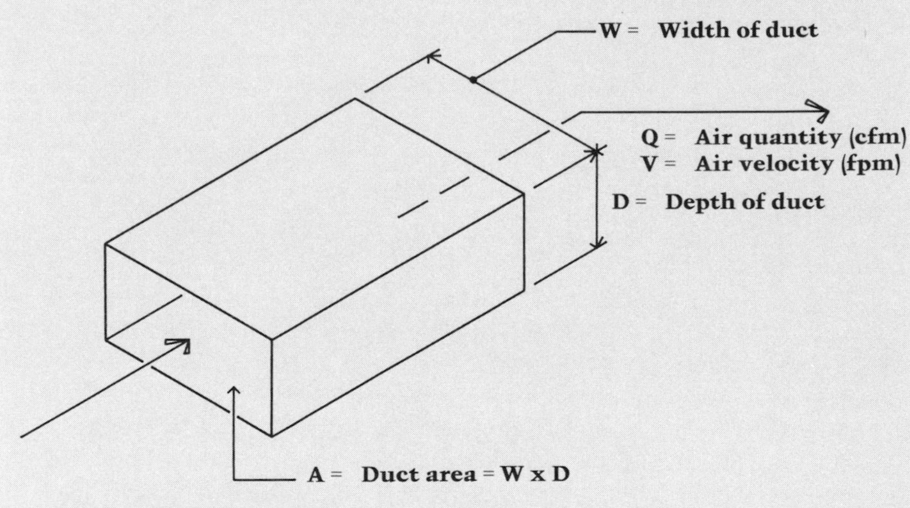

4. **Determine duct width and depth.**

A = W x D—Make D as deep as possible to fit construction.

(1 : 1) Ideal
(3 : 1) Maximum

The Velocity Method

Figure 9.19

$$A = \frac{250 \text{ cfm}}{1{,}800 \text{ fpm}}$$

$$= 0.14 \text{ s.f.}$$

$$= 0.14 \text{ s.f.} \times 144 \text{ sq. in. per s.f.}$$

$$A = 20 \text{ sq. in.}$$

Therefore, use a 6″ × 6″ duct, because six inches is the smallest rectangular duct size recommended.

Constant Pressure Drop Method: Using the constant pressure drop method, the maximum limit for pressure drop per 100 feet of duct is set and the duct is made large enough to maintain this criteria. Familiarity with the terms pressure drop per 100 feet and equivalent round is necessary in order to understand the constant pressure drop method.

Pressure drop per 100 feet: In a low velocity system, it is common to select 0.1 inches of water per 100 feet as a design pressure loss rate. This means that, when selected, the duct will require (or use up) 0.1 inches of water (0.0036 lbs. per square inch—the pressure under a column of water 0.1″ high) for every 100 feet of straight duct as it transports the required air quantity. Consider a duct that is 12 inches in diameter, 100 feet long, and open at both ends. Without a fan to push it, the air will not move. Now suppose one end has a pressure of 0.1 inches (0.0036 psi) higher than that of the other end. The added pressure will force air through the duct. Figure 9.20 illustrates the relationship between duct diameter, friction loss per 100 feet, and air quantity. For the 12-inch diameter duct at 0.1 inches per 100 feet, it can be seen that 700 cfm will move in the duct at 1,000 fpm velocity.

Equivalent round: Because ducts are sized to account for friction, they can be interchanged with no substantial effect in performance, provided they have the same equivalent round diameter. Figure 9.20 can be used to determine the diameter of a circular duct. Round and spiral ductwork are gaining popularity. However, most buildings have traditionally used rectangular or square ducts. Figure 9.21 is a chart that can be used to convert a round duct to its equivalent size rectangular duct. Using Figure 9.21, a 12-inch diameter equivalent would be a square 11″ × 11″, close to the 12 inch, but less to account for the corners.

An advantage to the constant pressure drop method is that it allows the designer to properly select from a wide range of ducts. Based on Figures 9.20 and 9.21, the following ducts will all conduct 700 cfm of air at 0.1 inches per 100 feet:

12″ diameter	113 sq. in.
11″ x 11″	121 sq. in.
10″ x 12″	120 sq. in.
9″ x 14″	126 sq. in.
8″ x 16″	128 sq. in.
7″ x 18″	126 sq. in.
6″ x 22″	132 sq. in.

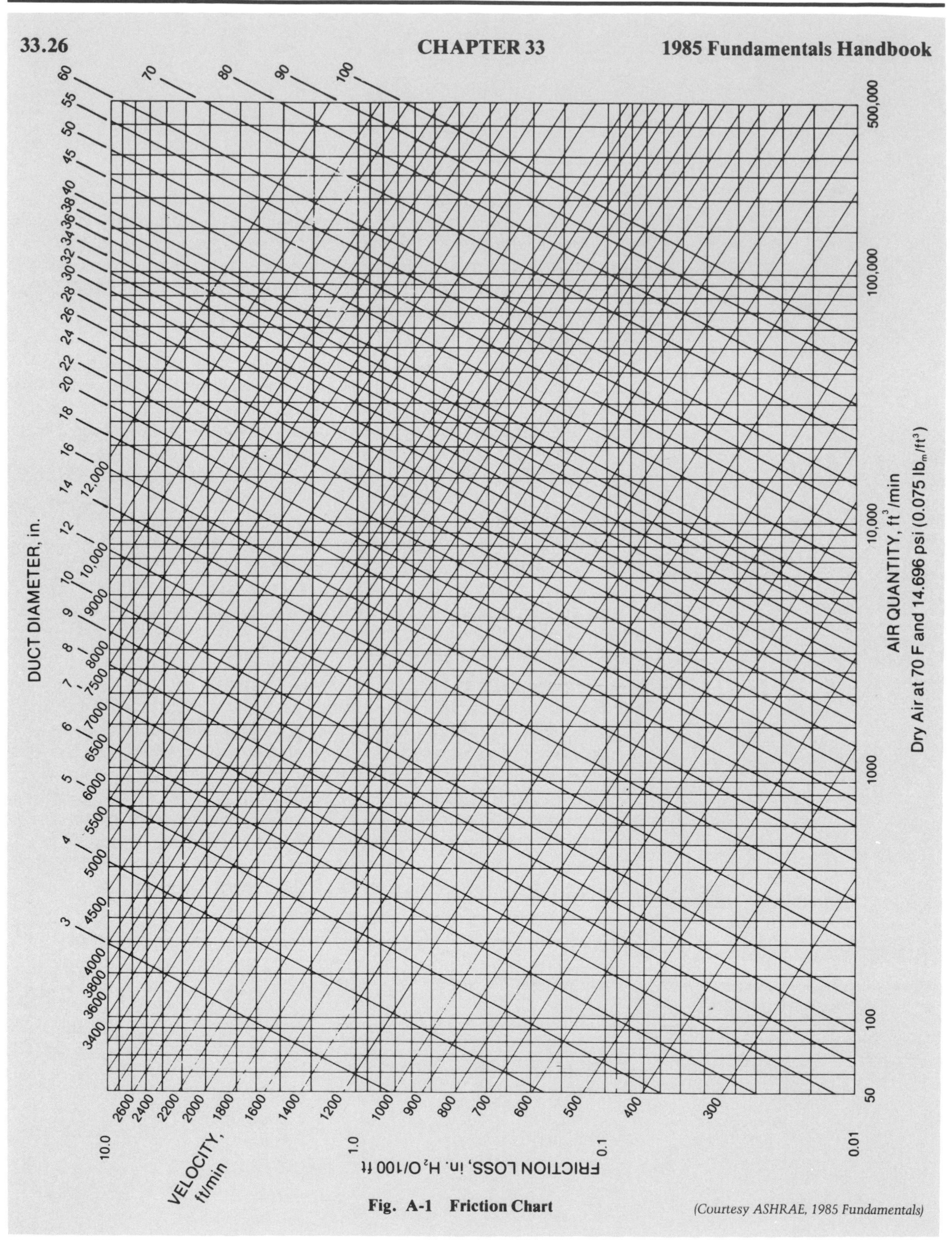

Fig. A-1 Friction Chart

(Courtesy ASHRAE, 1985 Fundamentals)

Figure 9.20

Table A-2 Circular Equivalents of Rectangular Ducts for Equal Friction and Capacity[a]

Lgth Adj.[b]	Length of One Side of Rectangular Duct (a), in.																
	4.0	4.5	5.0	5.5	6.0	6.5	7.0	7.5	8.0	9.0	10.0	11.0	12.0	13.0	14.0	15.0	16.0
3.0	3.8	4.0	4.2	4.4	4.6	4.7	4.9	5.1	5.2	5.5	5.7	6.0	6.2	6.4	6.6	6.8	7.0
3.5	4.1	4.3	4.6	4.8	5.0	5.2	5.3	5.5	5.7	6.0	6.3	6.5	6.8	7.0	7.2	7.5	7.7
4.0	4.4	4.6	4.9	5.1	5.3	5.5	5.7	5.9	6.1	6.4	6.7	7.0	7.3	7.6	7.8	8.0	8.3
4.5	4.6	4.9	5.2	5.4	5.7	5.9	6.1	6.3	6.5	6.9	7.2	7.5	7.8	8.1	8.4	8.6	8.8
5.0	4.9	5.2	5.5	5.7	6.0	6.2	6.4	6.7	6.9	7.3	7.6	8.0	8.3	8.6	8.9	9.1	9.4
5.5	5.1	5.4	5.7	6.0	6.3	6.5	6.8	7.0	7.2	7.6	8.0	8.4	8.7	9.0	9.3	9.6	9.9

Lgth Adj.[b]	Length of One Side of Rectangular Duct (a), in.																				Lgth Adj.[b]
	6	7	8	9	10	11	12	13	14	15	16	17	18	19	20	22	24	26	28	30	
6	6.6																				6
7	7.1	7.7																			7
8	7.6	8.2	8.7																		8
9	8.0	8.7	9.3	9.8																	9
10	8.4	9.1	9.8	10.4	10.9																10
11	8.8	9.5	10.2	10.9	11.5	12.0															11
12	9.1	9.9	10.7	11.3	12.0	12.6	13.1														12
13	9.5	10.3	11.1	11.8	12.4	13.1	13.7	14.2													13
14	9.8	10.7	11.5	12.2	12.9	13.5	14.2	14.7	15.3												14
15	10.1	11.0	11.8	12.6	13.3	14.0	14.6	15.3	15.8	16.4											15
16	10.4	11.3	12.2	13.0	13.7	14.4	15.1	15.7	16.4	16.9	17.5										16
17	10.7	11.6	12.5	13.4	14.1	14.9	15.6	16.2	16.8	17.4	18.0	18.6									17
18	11.0	11.9	12.9	13.7	14.5	15.3	16.0	16.7	17.3	17.9	18.5	19.1	19.7								18
19	11.2	12.2	13.2	14.1	14.9	15.7	16.4	17.1	17.8	18.4	19.0	19.6	20.2	20.8							19
20	11.5	12.5	13.5	14.4	15.2	16.0	16.8	17.5	18.2	18.9	19.5	20.1	20.7	21.3	21.9						20
22	12.0	13.0	14.1	15.0	15.9	16.8	17.6	18.3	19.1	19.8	20.4	21.1	21.7	22.3	22.9	24.0					22
24	12.4	13.5	14.6	15.6	16.5	17.4	18.3	19.1	19.9	20.6	21.3	22.0	22.7	23.3	23.9	25.1	26.2				24
26	12.8	14.0	15.1	16.2	17.1	18.1	19.0	19.8	20.6	21.4	22.1	22.9	23.5	24.2	24.9	26.1	27.3	28.4			26
28	13.2	14.5	15.6	16.7	17.7	18.7	19.6	20.5	21.3	22.1	22.9	23.7	24.4	25.1	25.8	27.1	28.3	29.5	30.6		28
30	13.6	14.9	16.1	17.2	18.3	19.3	20.2	21.1	22.0	22.9	23.7	24.4	25.2	25.9	26.6	28.0	29.3	30.5	31.7	32.8	30
32	14.0	15.3	16.5	17.7	18.8	19.8	20.8	21.8	22.7	23.5	24.4	25.2	26.0	26.7	27.5	28.9	30.2	31.5	32.7	33.9	32
34	14.4	15.7	17.0	18.2	19.3	20.4	21.4	22.4	23.3	24.2	25.1	25.9	26.7	27.5	28.3	29.7	31.0	32.4	33.7	34.9	34
36	14.7	16.1	17.4	18.6	19.8	20.9	21.9	22.9	23.9	24.8	25.7	26.6	27.4	28.2	29.0	30.5	32.0	33.3	34.6	35.9	36
38	15.0	16.5	17.8	19.0	20.2	21.4	22.4	23.5	24.5	25.4	26.4	27.2	28.1	28.9	29.8	31.3	32.8	34.2	35.6	36.8	38
40	15.3	16.8	18.2	19.5	20.7	21.8	22.9	24.0	25.0	26.0	27.0	27.9	28.8	29.6	30.5	32.1	33.6	35.1	36.4	37.8	40
42	15.6	17.1	18.5	19.9	21.1	22.3	23.4	24.5	25.6	26.6	27.6	28.5	29.4	30.3	31.2	32.8	34.4	35.9	37.3	38.7	42
44	15.9	17.5	18.9	20.3	21.5	22.7	23.9	25.0	26.1	27.1	28.1	29.1	30.0	30.9	31.8	33.5	35.1	36.7	38.1	39.5	44
46	16.2	17.8	19.3	20.6	21.9	23.2	24.4	25.5	26.6	27.7	28.7	29.7	30.6	31.6	32.5	34.2	35.9	37.4	38.9	40.4	46
48	16.5	18.1	19.6	21.0	22.3	23.6	24.8	26.0	27.1	28.2	29.2	30.2	31.2	32.2	33.1	34.9	36.6	38.2	39.7	41.2	48
50	16.8	18.4	19.9	21.4	22.7	24.0	25.2	26.4	27.6	28.7	29.8	30.8	31.8	32.8	33.7	35.5	37.2	38.9	40.5	42.0	50
52	17.1	18.7	20.2	21.7	23.1	24.4	25.7	26.9	28.0	29.2	30.3	31.3	32.3	33.3	34.3	36.2	37.9	39.6	41.2	42.8	52
54	17.3	19.0	20.6	22.0	23.5	24.8	26.1	27.3	28.5	29.7	30.8	31.8	32.9	33.9	34.9	36.8	38.6	40.3	41.9	43.5	54
56	17.6	19.3	20.9	22.4	23.8	25.2	26.5	27.7	28.9	30.1	31.2	32.3	33.4	34.4	35.4	37.4	39.2	41.0	42.7	44.3	56
58	17.8	19.5	21.2	22.7	24.2	25.5	26.9	28.2	29.4	30.6	31.7	32.8	33.9	35.0	36.0	38.0	39.8	41.6	43.3	45.0	58
60	18.1	19.8	21.5	23.0	24.5	25.9	27.3	28.6	29.8	31.0	32.2	33.3	34.4	35.5	36.5	38.5	40.4	42.3	44.0	45.7	60
62		20.1	21.7	23.3	24.8	26.3	27.6	28.9	30.2	31.5	32.6	33.8	34.9	36.0	37.1	39.1	41.0	42.9	44.7	46.4	62
64		20.3	22.0	23.6	25.1	26.6	28.0	29.3	30.6	31.9	33.1	34.3	35.4	36.5	37.6	39.6	41.6	43.5	45.3	47.1	64
66		20.6	22.3	23.9	25.5	26.9	28.4	29.7	31.0	32.3	33.5	34.7	35.9	37.0	38.1	40.2	42.2	44.1	46.0	47.7	66
68		20.8	22.6	24.2	25.8	27.3	28.7	30.1	31.4	32.7	33.9	35.2	36.3	37.5	38.6	40.7	42.8	44.7	46.6	48.4	68
70		21.1	22.8	24.5	26.1	27.6	29.1	30.4	31.8	33.1	34.4	35.6	36.8	37.9	39.1	41.2	43.3	45.3	47.2	49.0	70
72			23.1	24.8	26.4	27.9	29.4	30.8	32.2	33.5	34.8	36.0	37.2	38.4	39.5	41.7	43.8	45.8	47.8	49.6	72
74			23.3	25.1	26.7	28.2	29.7	31.2	32.5	33.9	35.2	36.4	37.7	38.8	40.0	42.2	44.4	46.4	48.4	50.3	74
76			23.6	25.3	27.0	28.5	30.0	31.5	32.9	34.3	35.6	36.8	38.1	39.3	40.5	42.7	44.9	47.0	48.9	50.9	76
78			23.8	25.6	27.3	28.8	30.4	31.8	33.3	34.6	36.0	37.2	38.5	39.7	40.9	43.2	45.4	47.5	49.5	51.4	78
80			24.1	25.8	27.5	29.1	30.7	32.2	33.6	35.0	36.3	37.6	38.9	40.2	41.4	43.7	45.9	48.0	50.1	52.0	80
82				26.1	27.8	29.4	31.0	32.5	34.0	35.4	36.7	38.0	39.3	40.6	41.8	44.1	46.4	48.5	50.6	52.6	82
84				26.4	28.1	29.7	31.3	32.8	34.3	35.7	37.1	38.4	39.7	41.0	42.2	44.6	46.9	49.0	51.1	53.2	84
86				26.6	28.3	30.0	31.6	33.1	34.6	36.1	37.4	38.8	40.1	41.4	42.6	45.0	47.3	49.6	51.7	53.7	86
88				26.9	28.6	30.3	31.9	33.4	34.9	36.4	37.8	39.2	40.5	41.8	43.1	45.5	47.8	50.0	52.2	54.3	88
90				27.1	28.9	30.6	32.2	33.8	35.3	36.7	38.2	39.5	40.9	42.2	43.5	45.9	48.3	50.5	52.7	54.8	90
92					29.1	30.8	32.5	34.1	35.6	37.1	38.5	39.9	41.3	42.6	43.9	46.4	48.7	51.0	53.2	55.3	92
96					29.6	31.4	33.0	34.7	36.2	37.7	39.2	40.6	42.0	43.3	44.7	47.2	49.6	52.0	54.2	56.4	96

(Courtesy ASHRAE, 1985 Fundamentals)

Figure 9.21

33.28 CHAPTER 33 1985 Fundamentals Handbook

Table A-2 Circular Equivalents of Rectangular Ducts for Equal Friction and Capacity[a] *(Concluded)*

Lgth Adj.[b]	Length of One Side Rectangular Duct (a), in.																				Lgth Adj.[b]
	32	34	36	38	40	42	44	46	48	50	52	56	60	64	68	72	76	80	84	88	
32	35.0																				32
34	36.1	37.2																			34
36	37.1	38.2	39.4																		36
38	38.1	39.3	40.4	41.5																	38
40	39.0	40.3	41.5	42.6	43.7																40
42	40.0	41.3	42.5	43.7	44.8	45.9															42
44	40.9	42.2	43.5	44.7	45.8	47.0	48.1														44
46	41.8	43.1	44.4	45.7	46.9	48.0	49.2	50.3													46
48	42.6	44.0	45.3	46.6	47.9	49.1	50.2	51.4	52.5												48
50	43.6	44.9	46.2	47.5	48.8	50.0	51.2	52.4	53.6	54.7											50
52	44.3	45.7	47.1	48.4	49.7	51.0	52.2	53.4	54.6	55.7	56.8										52
54	45.1	46.5	48.0	49.3	50.7	52.0	53.2	54.4	55.6	56.8	57.9										54
56	45.8	47.3	48.8	50.2	51.6	52.9	54.2	55.4	56.6	57.8	59.0	61.2									56
58	46.6	48.1	49.6	51.0	52.4	53.8	55.1	56.4	57.6	58.8	60.0	62.3									58
60	47.3	48.9	50.4	51.9	53.3	54.7	60.0	57.3	58.6	59.8	61.0	63.4	65.6								60
62	48.0	49.6	51.2	52.7	54.1	55.5	56.9	58.2	59.5	60.8	62.0	64.4	66.7								62
64	48.7	50.4	51.9	53.5	54.9	56.4	57.8	59.1	60.4	61.7	63.0	65.4	67.7	70.0							64
66	49.4	51.1	52.7	54.2	55.7	57.2	58.6	60.0	61.3	62.6	63.9	66.4	68.8	71.0							66
68	50.1	51.8	53.4	55.0	56.5	58.0	59.4	60.8	62.2	63.6	64.9	67.4	69.8	72.1	74.3						68
70	50.8	52.5	54.1	55.7	57.3	58.8	60.3	61.7	63.1	64.4	65.8	68.3	70.8	73.2	75.4						70
72	51.4	53.2	54.8	56.5	58.0	59.6	61.1	62.5	63.9	65.3	66.7	69.3	71.8	74.2	76.5	78.7					72
74	52.1	53.8	55.5	57.2	58.8	60.3	61.9	63.3	64.8	66.2	67.5	70.2	72.7	75.2	77.5	79.8					74
76	52.7	54.5	56.2	57.9	59.5	61.1	62.6	64.1	65.6	67.0	68.4	71.1	73.7	76.2	78.6	80.9	83.1				76
78	53.3	55.1	56.9	58.6	60.2	61.8	63.4	64.9	66.4	67.9	69.3	72.0	74.6	77.1	79.6	81.9	84.2				78
80	53.9	55.8	57.5	59.3	60.9	62.6	64.1	65.7	67.2	68.7	70.1	72.9	75.4	78.1	80.6	82.9	85.2	87.5			80
82	54.5	56.4	58.2	59.9	61.6	63.3	64.9	66.5	68.0	69.5	70.9	73.7	76.4	79.0	81.5	84.0	86.3	88.5			82
84	55.1	57.0	58.8	60.6	62.3	64.0	65.6	67.2	68.7	70.3	71.7	74.6	77.3	80.0	82.5	85.0	87.3	89.6	91.8		84
86	55.7	57.6	59.4	61.2	63.0	64.7	66.3	67.9	69.5	71.0	72.5	75.4	78.2	80.9	83.5	85.9	88.3	90.7	92.9		86
88	56.3	58.2	60.1	61.9	63.6	65.4	67.0	68.7	70.2	71.8	73.3	76.3	79.1	81.8	84.4	86.9	89.3	91.7	94.0	96.2	88
90	56.8	58.8	60.7	62.5	64.3	66.0	67.7	69.4	71.0	72.6	74.1	77.1	79.9	82.7	85.3	87.9	90.3	92.7	95.0	97.3	90
92	57.4	59.3	61.3	63.1	64.9	66.7	68.4	70.1	71.7	73.3	74.9	77.9	80.8	83.5	86.2	88.8	91.3	93.7	96.1	98.4	92
94	57.9	59.9	61.9	63.7	65.6	67.3	69.1	70.8	72.4	74.0	75.6	78.7	81.6	84.4	87.1	89.7	92.3	94.7	97.1	99.4	94
96	58.4	60.5	62.4	64.3	66.2	68.0	69.7	71.5	73.1	74.8	76.3	79.4	82.4	85.3	88.0	90.7	93.2	95.7	98.1	100.5	96

[a]Table based on Eq. (31)
[b]Length of adjacent side of rectangular duct (b), in.

(Courtesy ASHRAE, 1985 Fundamentals)

Figure 9.21 (cont.)

When using the velocity method, it is important to keep the aspect ratio close to 1:1, thereby minimizing friction errors. Using the constant pressure drop method, a wide range of duct sizes can be selected to clear obstructions and yet meet the flow requirements.

To size a system by the constant pressure drop method, select the circular duct required from Figure 9.20 for each branch according to the cfm in the duct. Then use Figure 9.21 to determine an equivalent rectangular duct. This procedure is illustrated in Figure 9.22. All duct runs, long and short, are then sized by this system, which is similar to the procedures for sizing water piping described in the water distribution section.

Fan Size: Fans for constant pressure drop (and balanced pressure drop) design methods are sized to supply the proper quantity of air and total pressure head to drive the air through the system. The total quantity of air equals the total air simultaneously supplied to all spaces in the building, plus an allowance for leakage. Leakage losses vary from a low of about 2 percent for tight, caulked and taped systems, to about 10 percent for those that have loose, plain joints. The pressure head is the total pressure necessary to drive the air from the fan to the

---

1. **For each run of duct determine the air quantity (Q).**

$$Q = \frac{H_T}{1.08\Delta T}$$

(Refer to velocity method for additional information.)

2. **Determine the recommended pressure drop per 100′.**
   Low-pressure systems—0.1″ supply, 0.07″ return
   High-pressure systems—1.7″ supply, 1.4″ return
3. **Determine the circular duct diameter (see Figure 10.19) using the air quantity (Q) and the recommended pressure drop.**

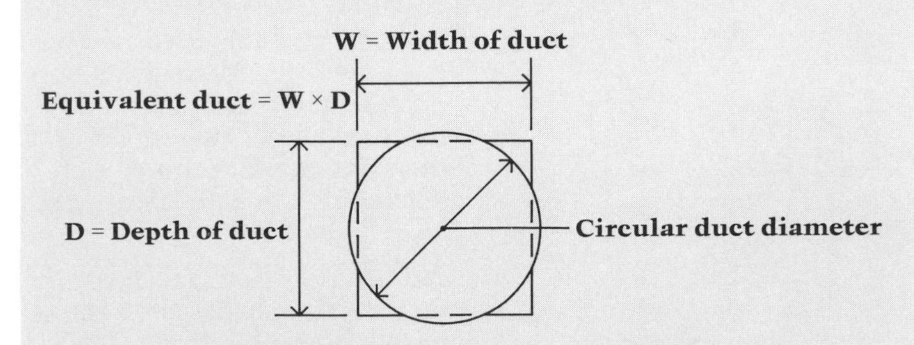

4. **Determine the duct width and depth using Figure 10.20.**

Constant Pressure Drop Method

Figure 9.22

room supply. The required total fan head is the sum of the following items.

Item		Total Pressure in Water
Change the velocity of air from 0 to 1,000 feet per minute at the fan		0.07
Internal equivalent losses— coils 0.15 filters 0.10		0.25
Duct (say 200 equivalent feet at 0.1″/100′ for longest run)		0.20
Pressure to drive across registers to room		0.15
Additional reserve for dampers and balancing (ten percent of above)		0.07
Required total fan head	=	0.74, use 3/4″
Required external fan pressure	=	0.20 + 0.15 + 0.07
	=	0.42, use 1/2″

The procedure for sizing fans is similar to that used to size water pumps. The external fan pressure is often listed by the manufacturer. This figure represents the fan pressure still remaining after overcoming the resistance encountered in the fan equipment. This is the pressure available to drive air through the distribution system. Most systems in small buildings require between 3/8″ and 1/2″ fan pressure.

Equivalent straight run of duct: The pressure drop through grilles, fittings, and dampers must be determined for the longest run. Each fitting equivalent is shown in Figure 9.23.

Using these tables, the equivalent length of a fitting can be determined. The equivalent length of a fitting is added to the straight length to arrive at the overall straight run of duct. This is similar to the calculation for equivalent pipe lengths.

To size the fan, select the longest total equivalent run. Then, add up all of the friction losses from elbows, fittings, filters, and registers that the fan must overcome.

Elbow equivalents can be obtained using Figure 9.23. For example, for a duct carrying 2,000 cubic feet of air per minute, based on an average velocity of 1,000 feet per minute, the approximate duct area will be two square feet. The elbow equivalent length depends on the duct depth, which must be approximated in order to size the duct. A 1′ × 2′ duct would have an area of two square feet. Referring to Figure 9.23, the duct has an L/D ratio of 9 on a 90-degree smooth elbow. Therefore, the equivalent length of the duct elbow would be 9′ × 2′, or 18 feet.

Balanced Pressure Drop Method: The balanced pressure drop method is based on the constant pressure drop method, adapted for all duct runs regardless of length or air quantity. The longest duct run is sized using a constant pressure drop. All other takeoffs from the longest duct are sized in such a way that their friction loss equals that of the longest run. This procedure is illustrated in Figure 9.24.

TABLE 9—FRICTION OF ROUND DUCT SYSTEM ELEMENTS

ELEMENT	CONDITION	L/D RATIO*
90° Smooth Elbow 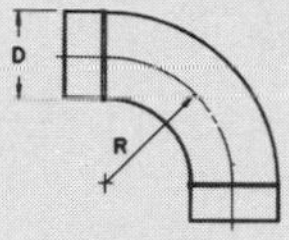	R/D = 1.5	9
90° 3-Piece Elbow 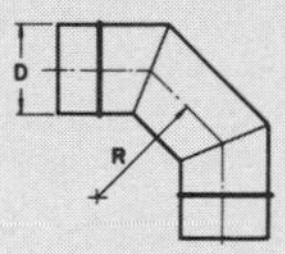	R/D = 1.5	24
90° 5-Piece Elbow 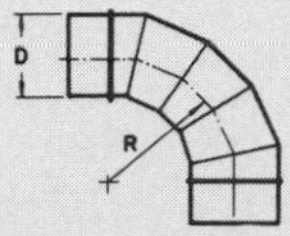	R/D = 1.5	12
45° 3-Piece Elbow 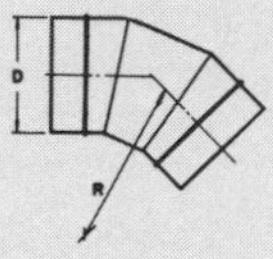	R/D = 1.5	6
45° Smooth Elbow 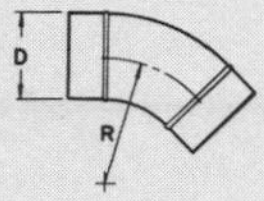	R/D = 1.5	4.5
90° Miter Elbow 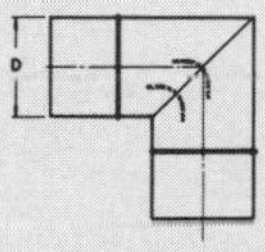	Vaned Not Vaned	22 65

ELEMENT	CONDITION	VALUE OF n†
90° Tee‡ and 90°, 135° & 180° Cross‡ 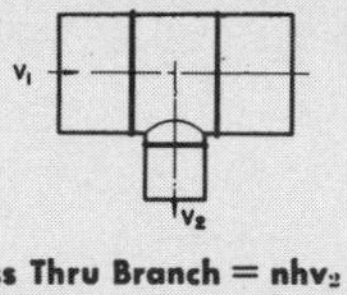 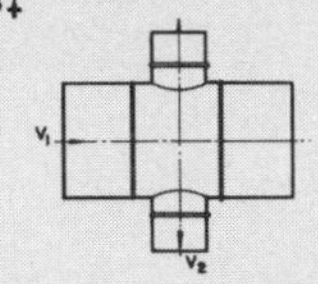 Pressure Loss Thru Branch = nhv_2	$\frac{V_2}{V_1} =$ 0.2 0.5 1.0 5.0	4.0 2.0 1.75 1.6
45° Tee‡ 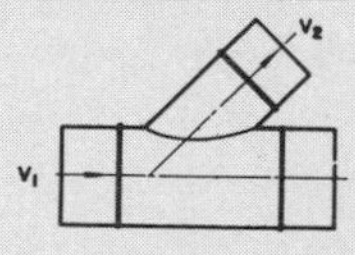 Pressure Loss Thru Branch = nhv_2	$\frac{V_2}{V_1} =$ 0.8 1.0 2.0 3.0	.10 . 44 1.21 1.47
90° Conical Tee and 180° Conical Cross 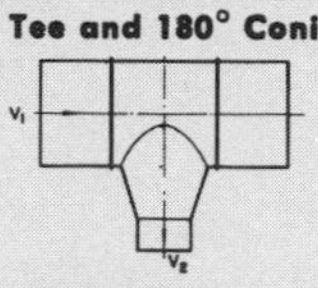 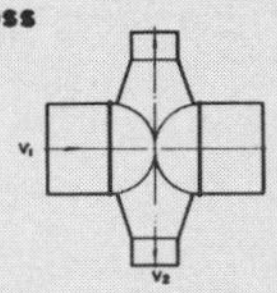 Pressure Loss Thru Branch = nhv_2	$\frac{V_2}{V_1} =$ 0.5 1.0 2.0 5.0	0.2 0.5 1.0 1.2

Notes on page 42.

(Courtesy Carrier Corporation, McGraw–Hill Book Company)

Figure 9.23

TABLE 10—FRICTION OF RECTANGULAR DUCT SYSTEM ELEMENTS

ELEMENT	CONDITIONS	L/D RATIO †
Rectangular Radius Elbow	(see table below)	
Rectangular Vaned Radius Elbow	(see table below)	
X° Elbow 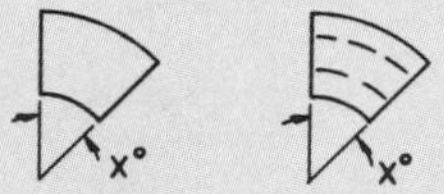	Vaned or Unvaned Radius Elbow	X/90 times value for similar 90° elbow
Rectangular Square Elbow 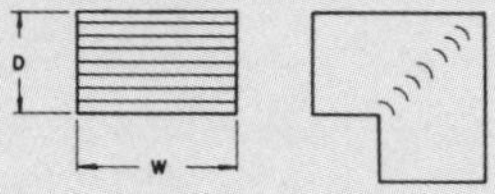	No Vanes	60
	Single Thickness Turning Vanes	15
	Double Thickness Turning Vanes	10
Double Elbow W/D = 1, R/D = 1.25*	S = O	15
	S = D	10
Double Elbow W/D = 1, R/D = 1.25*	S = O	20
	S = D	22
Double Elbow W/D = 1, R/D = 1.25* For Both	S = O	15
	S = D	16
Double Elbow W/D = 2, R_1/D = 1.25*, R_2/D = .5	Direction of Arrow	45
	Reverse Direction	40
Double Elbow W/D = 4, R/D = 1.25* for both elbows	Direction of Arrow	17
	Reverse Direction	18

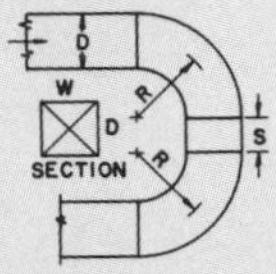

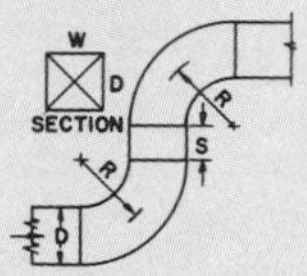

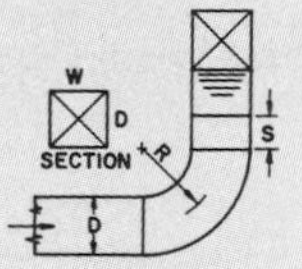

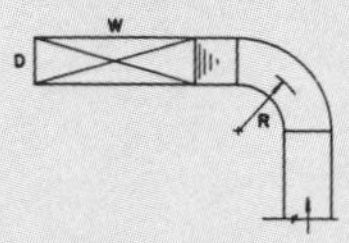

Rectangular Radius Elbow:

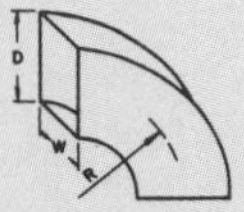

W/D	R/D .5	.75	1.00	1.25*	1.50
	L/D Ratio				
.5	33	14	9	5	4
1	45	18	11	7	4
3	80	30	14	8	5
6	125	40	18	12	7

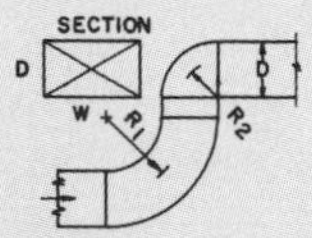

Rectangular Vaned Radius Elbow:

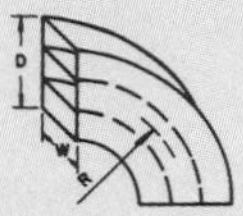

Number of Vanes	R/D .50	.75	1.00	1.50
	L/D Ratio			
1	18	10	8	7
2	12	8	7	7
3	10	7	7	6

(Courtesy Carrier Corporation, McGraw-Hill Book Company)

Figure 9.23 (cont.)

CHAPTER 2. AIR DUCT DESIGN 2–41

TABLE 10—FRICTION OF RECTANGULAR DUCT SYSTEM ELEMENTS (Contd)

ELEMENT	CONDITIONS	VALUE OF n‡
Transformer 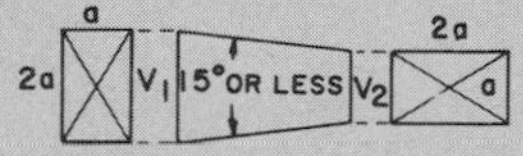 (2a, a, V_1, 15° OR LESS, V_2)	$V_2 = V_1$ S.P. Loss = nhv_1	.15
Abrupt Entrance (V_1)	S.P. Loss = nhv_1	.35
Bellmouth Entrance (V_1)	S.P. Loss = nhv_1	.03
Abrupt Exit (V_1)	S.P. Loss or Regain Considered Zero	
Bellmouth Exit (V_1)	S.P. Loss or Regain Considered Zero	
Re-Entrant Entrance (V_1)	S.P. Loss = nhv_1	.85

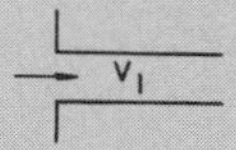

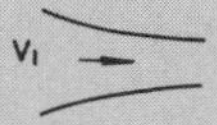

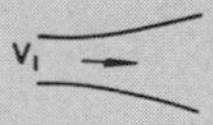

Expansion (V_1, a, V_2)

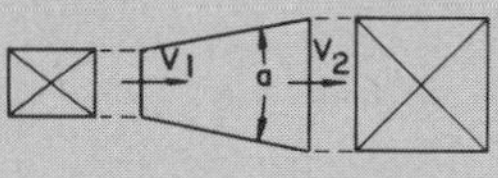

"n"

V_2/V_1	Angle "a" 5°	10°	15°	20°	30°	40°
.20	.83	.74	.68	.62	.52	.45
.40	.89	.83	.78	.74	.68	.64
.60	.93	.87	.84	.82	.79	.77

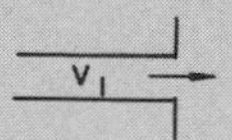

S.P. Regain = $n(hv_1 - hv_2)$

Contraction (V_1, a, V_2)

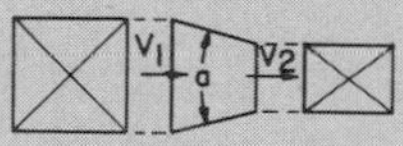

a	30°	45°	60°
n	1.02††	1.04	1.07

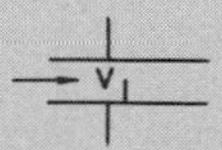

S.P. Loss = $n(hv_2 - hv_1)$ ††Slope 1″ in 4″

Sharp Edge Round Orifice (A_1, A_2)

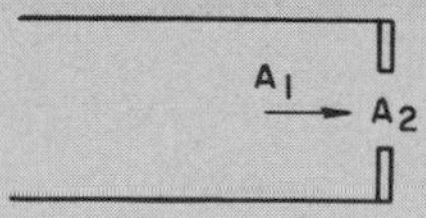

A_2/A_1	0	.25	.50	.75	1.00
n	2.5	2.3	1.9	1.1	0

S.P. Loss = nhv_2

Abrupt Contraction (V_1, V_2)

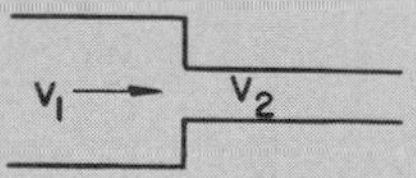

V_1/V_2	0	.25	.50	.75
n	1.34	1.24	.96	.52

S.P. Loss = nhv_2

Abrupt Expansion (V_1, V_2)

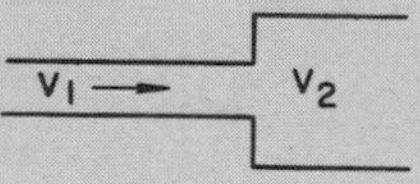

V_2/V_1	.20	.40	.60	.80
n	.32	.48	.48	.32

S.P. Regain = nhv_1

Pipe Running Thru Duct (V_1, E, D)

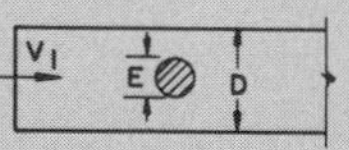

E/D	.10	.25	.50
n	.20	.55	2.00

S.P. Loss = nhv_1

Bar Running Thru Duct (V_1, E, D)

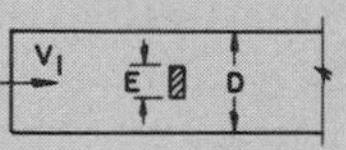

E/D	.10	.25	.50
n	.7	1.4	4.00

S.P. Loss = nhv_1

Easement Over Obstruction (V_1, E, D)

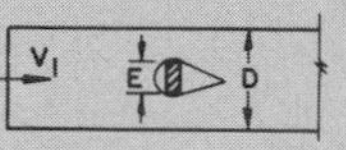

E/D	.10	.25	.50
n	.07	.23	.90

S.P. Loss = nhv_1

Notes on page 42.

(Courtesy Carrier Corporation, McGraw-Hill Book Company)

Figure 9.23 (cont.)

NOTES FOR TABLE 9

*L and D are in feet. D is the elbow diameter. L is the additional equivalent length of duct added to the measured length. The equivalent length L equals D in feet times the ratio listed.

†The value of n is the loss in velocity heads and may be converted to additional equivalent length of duct by the following equation.

$$L = n \times \frac{h_v \times 100}{h_f}$$

where: L = additional equivalent length, ft

h_v = velocity pressure at V_2, in. wg (conversion line on *Chart 7* or *Table* 8).

h_f = friction loss/100 ft, duct diameter at V_2, in. wg (*Chart 7*).

n = value for tee or cross

‡Tee or cross may be either reduced or the same size in the straight thru portion

NOTES FOR TABLE 10:

*1.25 is standard for an unvaned full radius elbow.

†L and D are in feet. D is the duct dimension illustrated in the drawing. L is the additional equivalent length of duct added to the measured duct. The equivalent length L equals D in feet times the ratio listed.

‡The value n is the number of velocity heads or differences in velocity heads lost or gained at a fitting, and may be converted to additional equivalent length of duct by the following equation:

$$L = n \times \frac{h_v \times 100}{h_f}$$

where: L = additional equivalent length, ft.

h_v = velocity pressure for V_1 or V_2, in. wg (conversion line on *Chart 7* or *Table* 8).

h_f = friction loss/100 ft, duct cross section at h_v, in. wg (*Chart 7*).

n = value for particular fitting.

(Courtesy Carrier Corporation, McGraw–Hill Book Company)

Figure 9.23 (cont.)

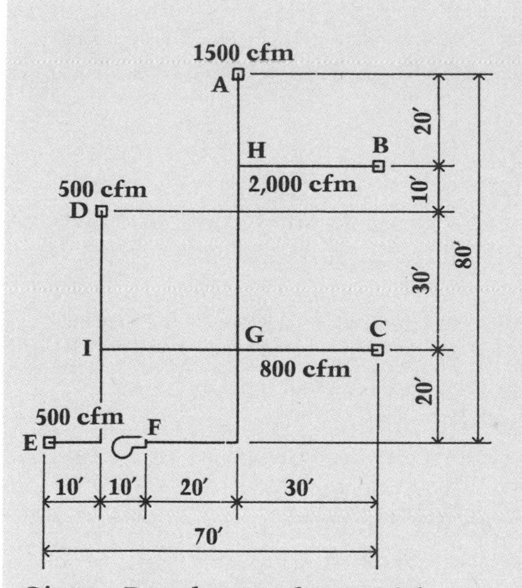

Given—Duct layout shown with dimensions and air quantities shown. Keep duct depth below 12".

Find—Size ducts and fan by balanced pressure drop method.

For 90° round elbows L/D = 9.
HB 2,000 cfm = 12 x 24 duct approx. at 1,000 fpm.
2' wide x L/D of 9' = 18' for elbows.

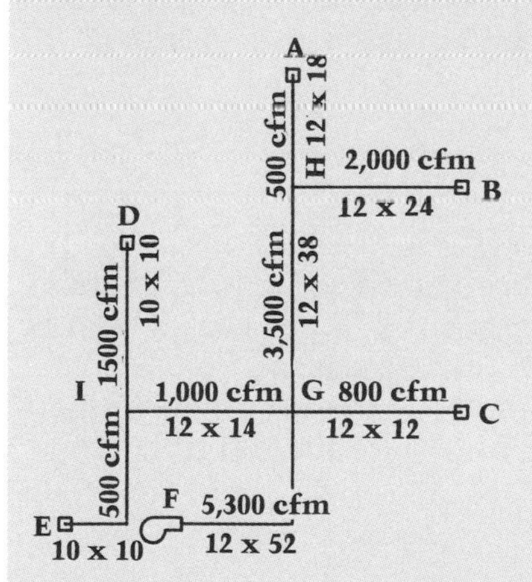

Solution

1. **Select "longest run" from fan to furthest register.**
   F-A at 20' + 80' = 100' is the longest.
2. **Determine air quantities in each run of duct. These are shown in the lower diagram.**
3. **Size each length of duct on longest run by the constant pressure drop method. (0.1"/100').**

Run	Air Quantity	(Fig. 9.20) Equivalent Round	(Fig. 9.21) Rect. Size
FG	5,300 cfm	26"	12 x 52"
GH	3,500 cfm	22"	12 x 38"
HA	1,500 cfm	16"	12 x 18"

4. **For each take-off from longest run, determine the design pressure loss per 100' by proportion to match losses on longest run.**

Run	Lengths Straight	Elbows	Total
HA	20'	0	20'
HB	30'	18	48'
GA	60'	0	60'
GC	30'	9	39'
GID	60'	9	69'
IE	30'	5	35'

Run	Ratio	Pressure	cfm	Eq. Round	Rect.
HA	$\frac{20}{48}$ x .1	—	—	—	—
HB		0.04	2,000	21½"	12 x 24
GA	$\frac{60}{39}$ x .1	—	—	—	—
GC		0.15	800	12"	12 x 12
GID	$\frac{60}{69}$ x .1	0.09	1,000	14"	12 x 14
		.10	500	11"	10 x 10
IE	$\frac{35}{30}$ x .09	1.1	500	11"	10 x 10

Fan—Longest Run F-A $\times$ **100' + 9** $\times \frac{52}{12}$ **= 140'**

Fan Pressure Required:

1. Velocity (0 to 1,000 fpm)	0.07
2. Internal coils	0.10
3. Filters	0.10
4. Duct 140' at 0.1/100	0.14
5. Residual pressure grilles	0.15
6. Reserve for balancing	0.05
	0.61 use 0.75"

Use 3/4" pressure at 5,300 cfm fan

Balanced Pressure Drop Duct Sizing Example

Figure 9.24

The balanced pressure drop method enables the designer to size a system with a wide variety of duct runs, and "balances" the system so that the fan will provide the proper quantities of air in each run.

Static Regain Method: Static regain refers to the increase (or regain) of static pressure in ductwork when the air velocity decreases. A main trunk duct, for example, may have diffusers at constant intervals for air takeoff (see the static regain example in Figure 9.17). If the main duct size remains constant, the air velocity in the duct will decrease after each takeoff. As the air slows down, it "bunches up" and its pressure is increased; the fast-moving air before the takeoff is converted to higher pressure/lower velocity air after the takeoff. By carefully sizing the ductwork, a designer uses the increased pressure from one takeoff to overcome pressure friction losses on the run to the next takeoff. If the air starts at about 1,600 feet per minute and loses velocity with each succeeding takeoff, it is possible to put about five takeoffs on a run before the supply velocity reaches the low cut-off value of about 800 feet per minute. The following formula is used to determine the increase in static pressure:

$$\Delta P = K[(V_1/4005)^2 - (V_2/4005)^2]$$

ΔP = increase in static pressure after the takeoff in inches of water

K = between 0.5 and 0.75 for decrease in velocity, −0.7 and −1.1 for pressure loss

V_1 = velocity of air (fpm) before the takeoff

V_2 = velocity of air (fpm) after the takeoff

The pressure head is measured in feet of air. As the movement of the air slows down, the pressure head increases as follows:

$$P = 1/2 \times V_s^2/g$$

P = pressure in feet of air

V_s = velocity of air (ft./sec.)

g = gravitational constant 32.2 ft./sec.^{2}

The pressure in feet of air must be converted to head in inches of water. A cubic foot of 68°F air weighs 0.075 pounds, which is equivalent to a column of water one foot square and ((0.075/62.3) = 0.0012 feet) approximately 0.01445 inches high. Velocity in feet per second must be multiplied by 60 to convert it to the standard feet per minute.

$$P \text{ (feet of air)} = P \text{ (in. wg)} \frac{1}{.01445}$$

$$P \text{ (in. wg)} = \frac{1}{2}\left(\frac{V}{60}\right)^2 \left(\frac{1}{32.2}\right)\left(.01445\right)$$

$$= \left(\frac{.01445\ V^2}{3{,}600 \times 32.2}\right)$$

$$= \frac{V^2}{16{,}044{,}000} = \left(\frac{V}{4{,}005}\right)^2$$

P = velocity pressure (in. wg)
V = fluid mean velocity (fpm)
Air = 0.075 lbs./ft.^{3} at standard conditions

Actual details of duct construction must be taken into account. For example, smooth connections result in ease of flow, whereas abrupt connections cause turbulence in the air stream, thereby diminishing the maximum theoretical pressure head. Thus, for the decrease in velocity after a takeoff (where V_1 is greater than V_2 and pressure is regained), a factor between 0.5 and 0.75 is generally used. When conditions cause the post-takeoff velocity to increase, V_2 is greater than V_1 and a factor between –0.7 and –1.1 is used. In this situation, there is a theoretical decrease in pressure, with K values between –0.7 and –1.1. (The (–) sign means that the pressure drops, which is the opposite of regain).

Using the static regain method requires more care and experienced judgment than the other duct sizing methods. Solutions may sometimes be based on "trial and error" and may be sensitive to any changes in air balancing and variable air flow. Figure 9.25 contains charts used to determine static regain.

Miscellaneous Fluids: In addition to steam, water, and air, other fluids are transported in an HVAC system, such as the fuels and refrigerants listed below.

- **Fuels**
  oil
  propane
  natural gas
- **Refrigerants**
  ammonia
  brine
  fluorocarbons
  other refrigerants

The piping design for these systems is regulated by a wide range of codes. The fuel piping in particular generally follows local fire, plumbing, and gas fitting codes. Requirements for refrigerants are often established by equipment manufacturers. Basic characteristics of typical refrigerants are shown in Figure 9.26. (For further information, refer to Chapter 3, "Codes, Regulations, and Standards.")

Distribution Equipment Data Sheets

The data sheets in Figure 9.27 summarize the efficiency, useful life, typical uses, capacity range, advantages and disadvantages, special considerations, accessories, and additional equipment needed for distribution system components. A Means line number, where applicable, is included in order to determine costs in the current edition of *Means Mechanical Cost Data*.

CHART 10—L/Q RATIO

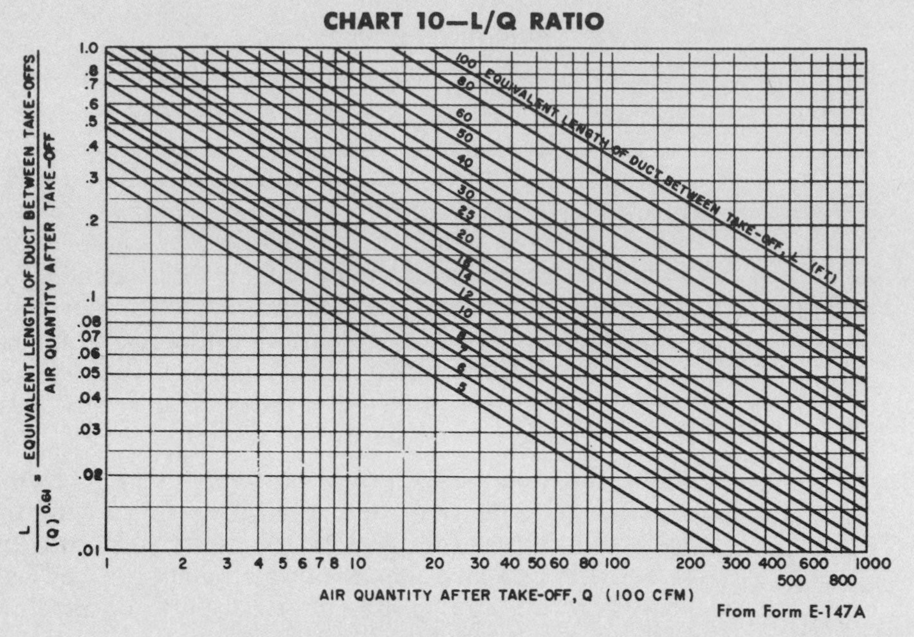

From Form E-147A

CHART 11—LOW VELOCITY STATIC REGAIN

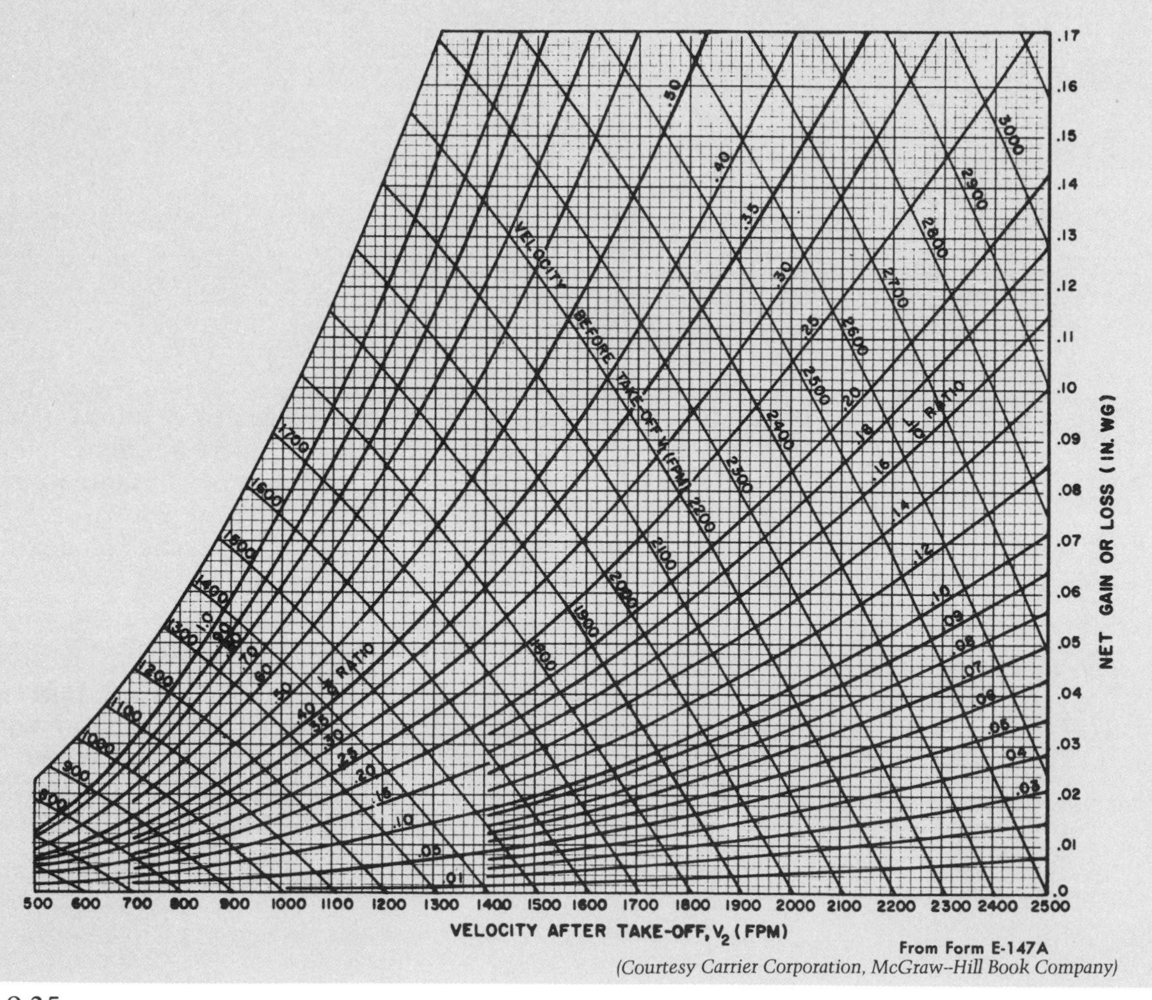

From Form E-147A

(Courtesy Carrier Corporation, McGraw-Hill Book Company)

Figure 9.25

REFRIGERANT TABLES

TABLE 1—COMPARATIVE DATA OF REFRIGERANTS

REFRIGERANT NUMBER (ARI DESIGNATION)	11	12	22	113	114	500
Chemical Name	Trichloromono-fluoromethane	Dichlorodi-fluoromethane	Monochlorodi-fluoromethane	Trichlorotri-fluoroethane	Dichlorotetra-fluoroethane	Azeotrope of Dichlorodi-fluoromethane and Difluoroethane
Chemical Formula	CCl_3F	CCl_2F_2	$CHClF_2$	$CCl_2F\text{-}CClF_2$	$C_2Cl_2F_4$	73.8% CCl_2F_2 26.2% CH_3CHF_2
Molecular Wt	137.38	120.93	86.48	187.39	170.93	99.29
Gas Constant, R (ft-lb/lb-R)	11.25	12.78	17.87	8.25	9.04	15.57
Boiling Point at 1 atm (F)	74.7	−21.62	−41.4	117.6	38.4	−28.0
Freezing Point at 1 atm (F)	−168	−252	−256	−31	−137	−254
Critical Temperature (F)	388.0	233.6	204.8	417.4	294.3	221.1
Critical Pressure (psia)	635.0	597.0	716.0	495.0	474.0	631.0
Specific Heat of Liquid, 86 F	.220	.235	.335	.218	.238	.300
Specific Heat of Vapor, C_p 60 F at 1 atm	*	.146	.149	*	.156	.171
Specific Heat of Vapor, C_v 60 F at 1 atm	*	.130	.127	*	.145	.151
Ratio $\frac{C_p}{C_v}$ = K (86 F at 1 atm)	1.11	1.14	1.18	1.12	1.09	1.13
Ratio of Specific Heats $\frac{\text{Liquid, 105 F}}{\text{Vapor, } C_p\text{, 40 F sat. press.}}$	2.04	1.55	2.14	1.47	1.59	1.77
Liquid Head (ft), 1 psi at 105 F	1.61	1.84	2.04	1.51	1.65	2.10
Saturation Pressure (psia) at: −50 F	0.52	7.12	11.74	*	1.35	*
0 F	2.55	23.85	38.79	0.84	5.96	27.96
40 F	7.03	51.67	83.72	2.66	15.22	60.94
105 F	25.7	141.25	227.65	11.58	50.29	167.85
Net Refrigerating Effect (Btu/lb) 40 F-105 F (no subcooling)	67.56	49.13	66.44	54.54	43.46	59.82
Cycle Efficiency (% Carnot Cycle) 40 F-105 F	90.5	83.2	81.8	87.5	84.9	82.0
Solubility of Water in Refrigerant	Negligible	Negligible	Negligible	Negligible	Negligible	Negligible
Miscibility with Oil	Miscible	Miscible	Limited	Miscible	Miscible	Miscible
Toxic Concentration (% by vol)	Above 10%	Above 20%	*	*	Above 20%	Above 20%
Odor	Ethereal, odorless when mixed with air	Same as R 11	Same as R 11	Same as R 11	Same as R 11	Same as R 11
Warning Properties	None	None	None	None	None	None
Explosive Range (% by vol)	None	None	None	None	None	None
Safety Group, U.L.	5	6	5A	4-5	6	5A
Safety Group, ASA B9.1	1	1	1	1	1	1
Toxic Decomposition Products	Yes	Yes	Yes	Yes	Yes	Yes
Viscosity (centipoises) Saturated Liquid 95 F	.3893	.2463	.2253	.5845	.3420	.2150
105 F	.3723	.2395	.2207	.5472	.3272	.2100
Vapor at 1 atm 30 F	.0101	.0118	.0120	.0097	.0108	*
40 F	.0103	.0119	.0122	.0098	.0109	*
50 F	.0105	.0121	.0124	.0100	.0111	*
Thermal Conductivity (k) Saturated Liquid 95 F	.0596	.0481	.0573	.0512	.0435	*
105 F	.0581	.0469	.0553	.0500	.0421	*
Vapor at 1 atm 30 F	.0045	.0047	.0060	.0037	.0056	*
40 F	.0046	.0049	.0061	.0039	.0057	*
50 F	.0046	.0051	.0063	.0040	.0059	*
Liquid Circulated, 40 F-105 F (lb/min/ton)	2.96	4.07	3.02	3.66	4.62	3.35
Theoretical Displacement, 40 F-105 F (cu ft/min/ton)	16.1	3.14	1.98	39.5	9.16	2.69
Theoretical Horsepower Per Ton 40 F-105 F	0.676	C 736	0.75	0.70	0.722	0.747
Coefficient of Performance 40 F-105 F (4.71/hp per ton)	6.95	6.39	6.29	6.74	6.52	6.31
Cost Compared With R 11	1.00	1.57	2.77	2.15	2.97	2.00

*Data not available or not applicable.

(Courtesy Carrier Corporation, McGraw–Hill Book Company)

Figure 9.26

Distribution Equipment Data Sheets

Designation	Carbon Steel Pipe	Means Number 151-701
Piping	**Typical Use and Consideration**	**Size Range (Diameter)**
Black Steel Schedule 40	Schedule 40 indicates the wall thickness and is also known as the standard weight. Used for low- and medium-pressure hydronic service as well as chilled water, condenser water and gas service and distribution.	1/4″ through 12″
Galvanized Steel Schedule 40	Galvanized pipe is used in corrosive atmospheres or where exposed to the elements. This piping is usually found in drainage, waste, vent and condenser water applications. Galvanized (zinc-coated) pipe should not be used for glycol solutions nor should it be welded after galvanizing.	1/4″ through 12″
Black Steel Schedule 80	Schedule 80, or extra heavy steel pipe, is used for high pressure service and for underground installations.	1/4″ through 12″
Black Steel Schedule 10	This thin wall lightweight pipe is used in many HVAC applications. Special groove type fittings and valves are used to join the piping in low-pressure systems.	**151-801** 2″ through 10″

Steel pipe, regardless of wall thickness, retains its designated outside diameter. For example, 3″ pipe, regardless of wall thickness or schedule number, is always 3-1/2″ in diameter. This is necessary for threading, socket welding, and other fitting and mating purposes. The inside diameter decreases or increases depending on the wall thickness specified. The larger the schedule number, the heavier the wall.

Piping	Copper Tubing	Means Number 151-401
Type K	This heavy wall tubing, available in hard-drawn straight lengths or in soft-drawn coils, is found in severe applications such as underground water piping and higher pressures. K is also used in some refrigeration work.	1/4″ through 8″
Type L	This medium wall tubing, available in hard-drawn straight lengths or in soft-drawn coils, is used predominantly in indoor hot or cold water piping throughout HVAC systems.	1/4″ through 8″
Type M	This thin wall tubing, available in straight lengths only, is used in non-critical (low-pressure) hydronic heating and cooling systems.	1/4″ through 8″

Figure 9.27

Distribution Equipment Data Sheets (continued)

Designation	Copper Tubing	Means Number 151-401
Piping	**Typical Use and Consideration**	**Size Range (Diameter)**
Type ACR	This tubing, available in hard-drawn straight lengths or soft-drawn coils, is actually type L tubing that has been cleaned and capped prior to shipping. As the ACR designation implies, it is used for air conditioning and refrigeration work. This tubing is called by its actual outside diameter rather than the usual nominal sizing.	3/8″ OD through 4-1/8″ OD
Type DWV	This tubing, available in straight lengths only, is produced for drainage waste and vent applications.	1-1/4″ through 8″
Copper tubing, regardless of type, has the same outside diameter per size. Only one class of fittings for copper tubing is made, whether they are soldered, brazed, flared or compression type.		
Piping	**Plastic Pipe and Tubing**	**Means Number 151-551**
Type PVC (Polyvinyl Chloride)	This rigid pipe is used in many HVAC applications, both above and below ground, such as chilled water, drains, wastes and vents, oil and gas. Available in Schedule 40 and 80 with the same outside diameters as steel pipe, it may be cut and threaded using standard steel pipe dies. Flanged, victaulic and socket weld fittings and valves are available. Hot gas or solvent welds are used with the socket fittings.	1/2″ through 16″
Type CPVC (Chlorinated Polyvinyl Chloride)	CPVC piping has all the features and dimensions of PVC but has the added feature of higher operating temperatures in the 120°F to 180°F range. This piping is very difficult to weld by the hot gas method and only highly qualified welders should attempt this procedure. This piping can be used in low-temperature hot water and heat pump applications.	1/2″ through 8″
Polybutylene SDR-11 (Standard Dimension Ratio)	This tubing is available in both straight lengths and coils. Having an operating temperature of 180°F under pressure, and being available in 1,000 foot coils, makes it a popular selection for low-temperature hot water radiant panels and earth-coupled heat pump systems. Available in both copper tubing size (CTS) and iron pipe size (IPS), it is joined by insert type fittings and bands.	1/2″ through 3″

Figure 9.27 (cont.)

Distribution Equipment Data Sheets (continued)

Designation	Rigid Rectangular	Means Number 151-250
Ductwork	**Typical Use and Consideration**	**Size Range**
Aluminum Alloy	Used in low-pressure low velocity air conditioning systems, especially in residential work. Aluminum expands and contracts to a greater extent than steel and allowances must be made to control this movement. Additional bracing and support is required with aluminum to make up for its lack of rigidity compared to steel.	Infinite (priced by weight)
Galvanized Steel	Used in both low- and high-pressure velocity systems, particularly in commercial and institutional HVAC applications. This is the most popular material used for sheet metal ductwork in the HVAC industry.	Infinite (priced by weight)
Stainless Steel Type 304	Used in lieu of aluminum or galvanized ductwork in corrosion-resistant applications.	Infinite (priced by weight)
Fiberglass Rigid Board	Used extensively in residential and light commercial applications where thermal insulation and sound isolation are considerations. Restricted to areas where physical abuse is not a factor.	Infinite (priced by the square foot)
Ductwork	**Rigid Round**	**Size Range (Diameter)**
Spiral	This type of duct is available in galvanized steel, aluminum and stainless steel. It is found in low and high velocity systems. Its ease of installation and fabrication is an economic consideration compared to shop-fabricated rectangular duct.	3″ through 48″
PVC (Polyvinyl Chloride)	Used in special corrosive exhaust applications such as laboratory hoods.	6″ through 48″
	Flexible Round	
Various Materials	This form of ductwork is available in many forms and materials, from uninsulated flexible metal and uninsulated helical wound wire with aluminum or poly foil wrap, to insulated adaptations of the same. Flexible round PVC duct is also available. Flexible duct is a labor-saving device in HVAC systems in space limited areas where many special fabricated fittings can be eliminated. The use of flexible duct in lengths over fifteen feet may be restricted by local labor agreements.	3″ through 20″

Figure 9.27 (cont.)

CHAPTER TEN

TERMINAL UNITS

A terminal unit is a unit at the end of a duct that transfers heating or cooling from the distribution system to the conditioned space. The end product of terminal units is conditioned (warm or cool) air. With the exception of air distribution systems, all terminal units are localized heat exchangers; they take in heat (in the form of steam or hot water) or cooling (from the direct expansion or chilled water system) and transfer it to the air in the conditioned space. Hydronic systems use two-, three-, and four-pipe systems with two- or three-way control valves. Air systems are connected directly to terminal units (grilles, registers, and diffusers) or through mixing boxes or VAV (variable air volume) units.

Steam Systems

The terminal units used for steam and hot water (hydronic) systems are outlined below.

Terminal Unit	Advantages	Disadvantages	Common Use
Cast iron convector or radiator	Rugged	Most expensive and cumbersome	Residential
Fin tube commercial	Inexpensive; blankets exterior wall	Lose wall space	Commercial
Fin tube baseboard	Inexpensive	Covers come loose; interference by drapes, etc.	Residential
Fan coils	Heating and cooling uses	Space considerations	Office and institutional
Unit heaters (hydronic coils with fans)	Inexpensive	Noisy	Garages and factories

The size of the unit depends on the heating or cooling load and the temperature of the steam or hot water. Figure 10.1 lists the sizes of typical hydronic terminal units. Accessories for steam terminal units are described in the following sections.

Valves

Steam radiator valves control the flow of steam to the heating element. Radiator valves are furnished in straight or angle pattern and have a union tailpiece connection. Automatic radiator valves, either self-contained or controlled by a remote thermostat, are popular in most commercial installations.

Air Vents

Air vents eliminate air from steam systems by using a float mechanism that rises when buoyed by the condensate to prevent the discharge of steam. Air vents are used on individual radiators and at the high points of risers and mains in gravity systems.

Traps

Traps are used to keep steam in the terminal units and to pass only condensate and air to the return system. Two types of steam traps that are commonly used are *thermostatic* (radiator) and *float and thermostatic* (combination) traps. Thermostatic traps are activated by temperature, and expand an internal bellows. Float and thermostatic traps contain a float that rises to let condensate pass and closes when there is no condensate, as well as a thermostatic element to pass the air. Float and thermostatic traps are used for main and riser drips and for heavy equipment condensate loads. The accessories for steam systems are illustrated in Figure 10.2.

Water Systems

Hydronic systems are among the most frequently used types of heating systems. Hydronic equipment should be sized properly to match the loads required to heat or cool the space. Common terminal units for water systems include the following:

- Cast iron radiators/convectors (heating only)
- Fin tube baseboard (heating only)
- Unit heaters (heating only)
- Fan coil units (heating and cooling)
- Induction units (heating and cooling)
- Heat pumps — water source (heating and cooling)

There are many manufacturers of terminal units, which differ in appearance, color, enclosure, mounting, and controls. Typical performance characteristics for these units are shown in Figures 10.1 and 10.3. To select the proper unit, determine the load that the unit must deliver to the space (refer to Chapter 2 for heating and cooling loads); determine the entering water temperature, which is usually done by the designer; and read the size of the unit required (from manufacturer's data).

Valves

There are many types of valves used in hydronic systems. Valves are usually sized to match the piping in the system. Automatic

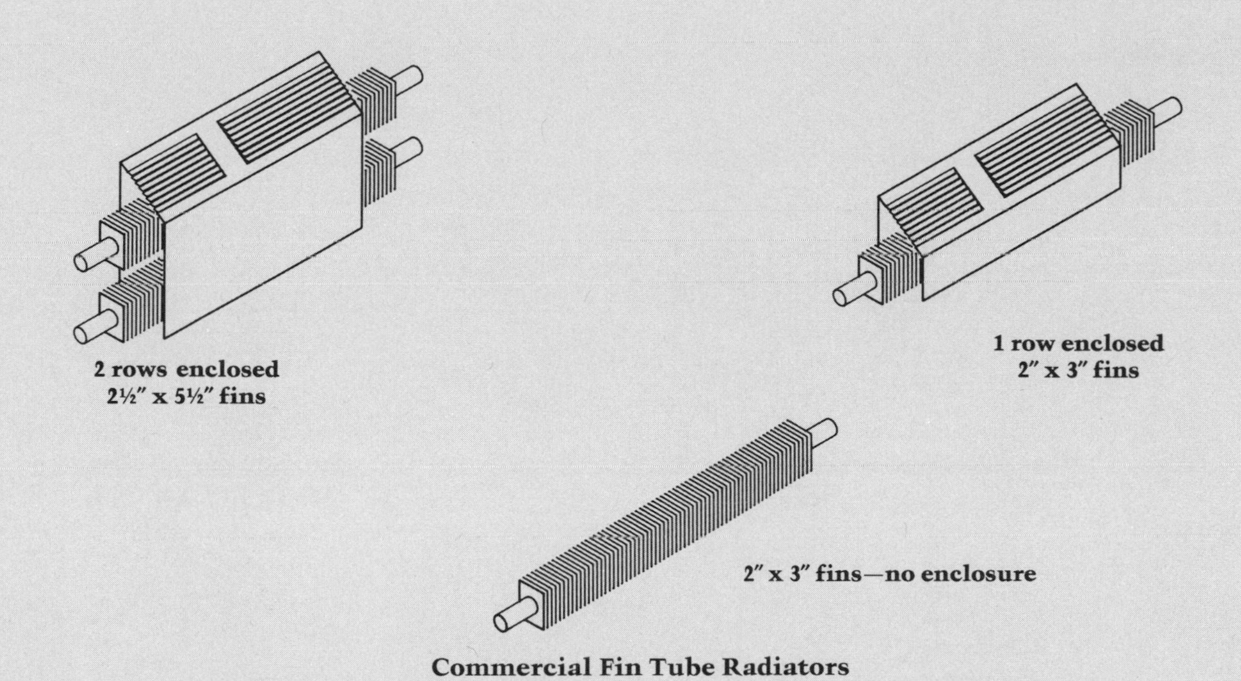

Commercial Fin Tube Radiators

Sizes for Typical Hydronic Terminal Heating Units

Fin Tube Radiation (Output in Btu/hr./l.f.)

Temp. (°F)	170°	180°	190°	200°	210°	220°
2 Rows	1,000	1,150	1,300	1,425	1,575	1,725
1 Row	700	800	875	975	1,075	1,175
No enclosure	575	650	750	825	900	1,000

Unit Heaters (Output in MBH)

Temp. (°F)	180°	200°	220°	240°	260°	300°
cfm = 500	14	19	24	28	33	43
cfm = 1,000	31	40	48	56	65	80
cfm = 1,500	50	62	74	85		
cfm = 2,000	85	100	118	135		

For steam, see manufacturers' data.

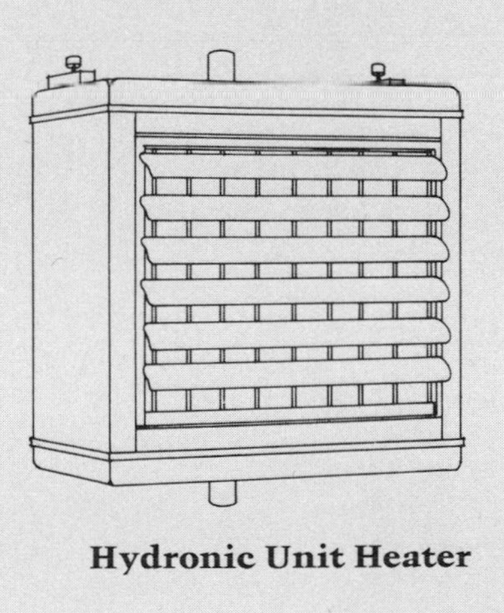

Hydronic Unit Heater

Figure 10.1

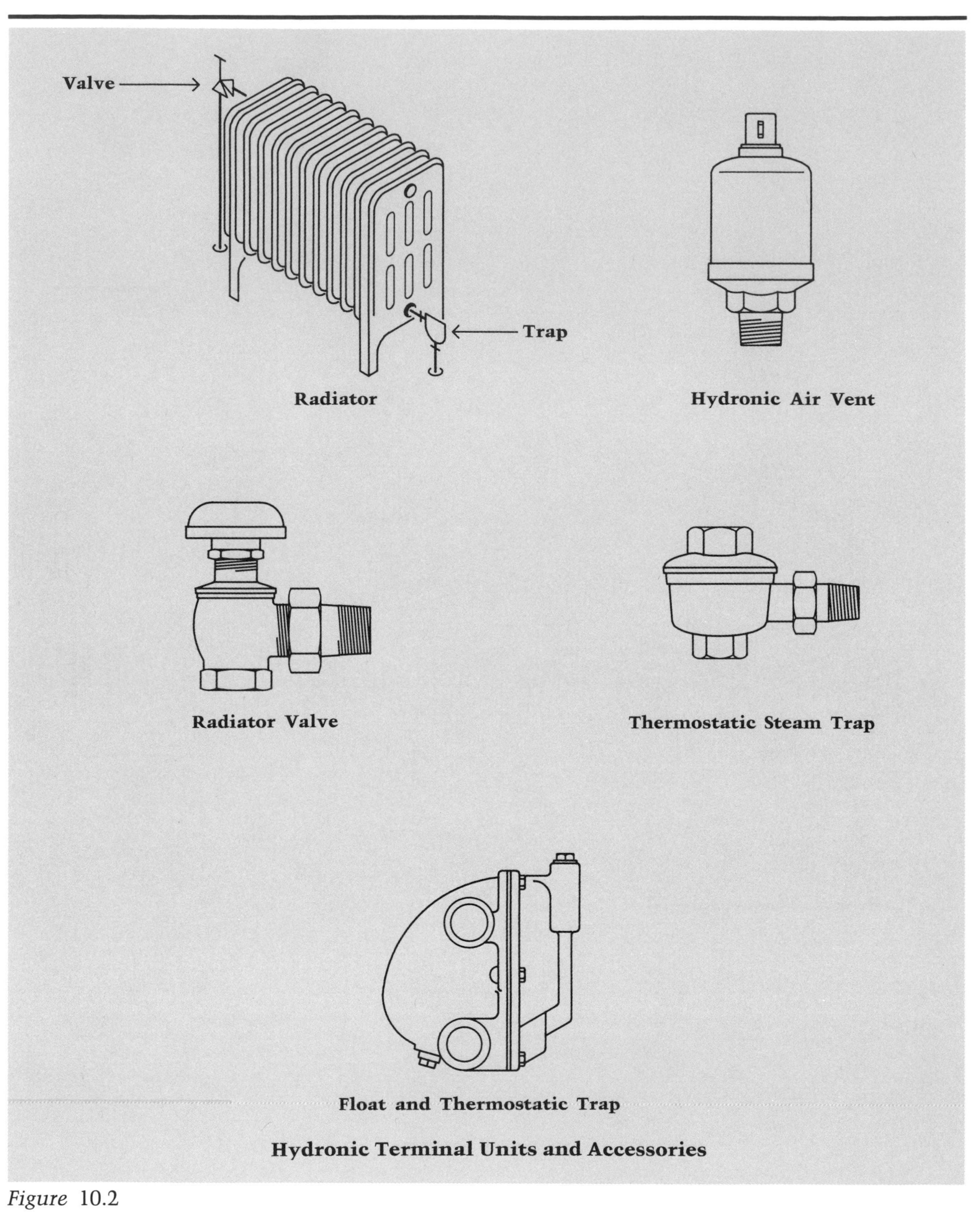
Valve
Trap
Radiator
Hydronic Air Vent
Radiator Valve
Thermostatic Steam Trap
Float and Thermostatic Trap
Hydronic Terminal Units and Accessories

Figure 10.2

Performance Characteristics of Hydronic Unit Heating and Cooling Coils

Size	32 x 25 x 9	40 x 25 x 9	44 x 25 x 9	56 x 25 x 9	60 x 28 x 11	72 x 28 x 11
cfm	200	300	400	600	800	1,000
MBH Heating	20	25	32	40	58	68

Fan Coils—Rated at 180°F water

Size	25 x 44 x 11			25 x 55 x 12	
cfm	230	270	325	380	530
MBH Cooling	7,000	8,700	11,500	15,000	18,000

Heat Pumps

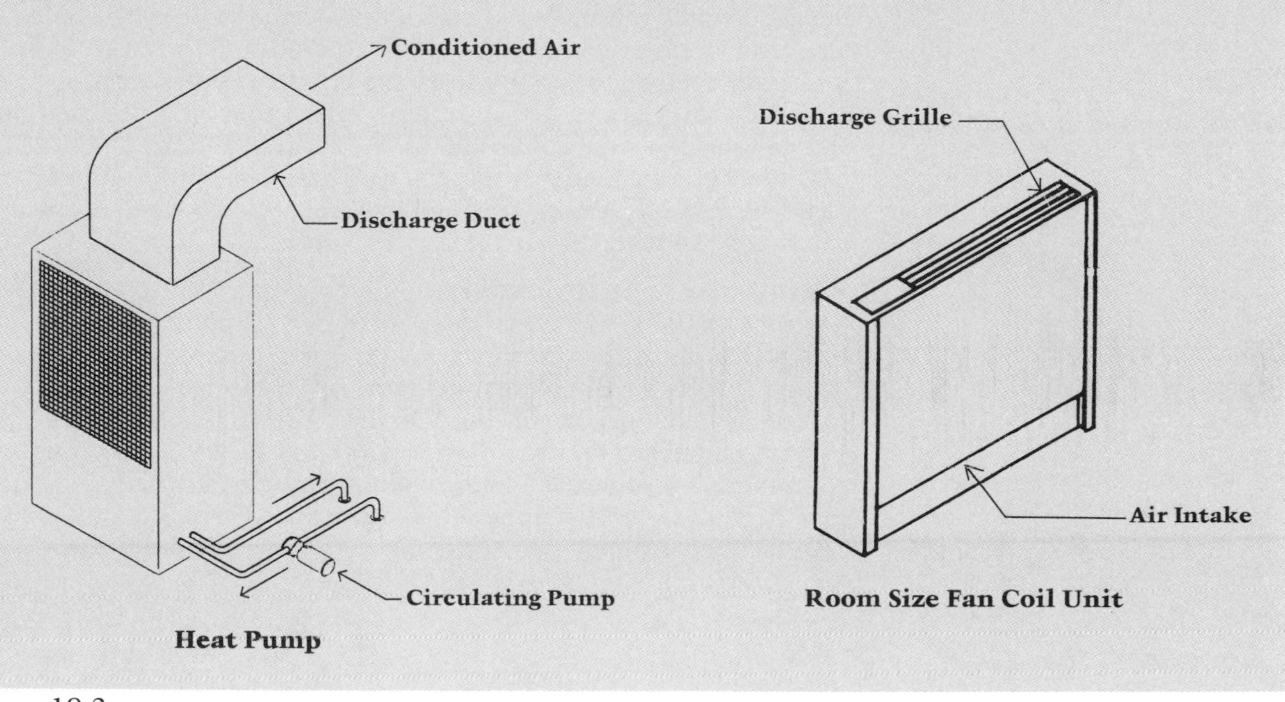

Figure 10.3

temperature control valves and pressure regulators may be one or two sizes smaller than line size due to precise body sizing for actual capacity.

Air Vents

Manual air valves and automatic air vents are used to purge air from the system. Located at or near the high point of each radiator riser and main, they permit air to escape the system. Even though hydronic systems are typically closed, fresh makeup water and minor reactants contain air bubbles, which must be released from the piping and coils via air vents.

Expansion Tank

Because water expands and contracts as it is heated or cooled, an air cushion must be provided in all hydronic systems to maintain system pressure . In small residential systems, a 30-gallon tank is usually sufficient, but actual size depends on the gallons of water contained in the entire system. Manufacturers usually provide sizing data based on overall system capacity. Tanks are also available with a diaphragm that separates the water from the air cushion.

Automatic Control Valves

Two-way valves, the most frequently used type of valve, have two positions, open or closed. When open, two-way valves allow flow (typically water) to a coil or convector. When closed, no flow occurs on the loop. As a result, the overall system balancing and the distribution and flow rates throughout the system are affected. These valves may be controlled from a built-in (self-contained thermostatic element) or a remote thermostat.

Three-way valves act as bypass valves. On *call*, they send water to the unit. On *no call*, they direct the flow past the unit, but maintain overall flow conditions. Three-way valves are also used as mixing or tempering valves. These valves may be self-contained, but usually are remote controlled. Figure 10.5 illustrates typical uses of two- and three-way valves.

Two-, Three-, and Four-Pipe Systems

The number of pipes feeding a piece of equipment varies according to the type of system. If the equipment provides heating only or cooling only, it is fed by a two-pipe system, one supply and one return. This is the most common arrangement.

Equipment that can heat and cool independently of each other are fed by a two-, three-, or four-pipe system (see Figure 10.4). The three- and four- pipe systems permit either heating or cooling to occur as required in a particular space.

Air Systems

Air is distributed through ductwork in a heating or cooling system using single duct, dual duct, constant volume, and variable volume systems (see Figure 10.6).

Single duct systems are the most effective and most often used type of system for common applications. **Dual duct systems** (one heating and one cooling) with mixing boxes are used for

The flow of fluids in a piping system is controlled or regulated by the use of valves. Valves are used to start, stop, divert, relieve, or regulate the flow, pressure, or temperature in a piping system. Valves are manufactured in several configurations according to use. Some of the types include: gate, globe, angle, check, ball, butterfly, and plug. Valves are further classified by their piping connections, stem position, pressure and temperature limits, as well as by the materials from which they are made.

Stainless steel, or steel alloy, valves can be used effectively in most instances for corrosion protection.

Bronze is one of the oldest materials used to make valves. It is most commonly used in steam, hot- and cold-water systems and other non-corrosive services. Bronze is often used as a seating surface in larger iron-body valves to ensure tight closure. Pressure ratings of 300 psi and temperatures up to 150 degrees F are typical.

Iron valves are normally used in medium to large pipe lines to control non-corrosive fluids and gases. Pressures for these valves should not exceed 250 psi at 450 degrees F, or 500 psi cold working pressures for water, oil, or gas.

Carbon steel is a high-strength material, and the valves made from this metal are therefore used in higher-pressure services, such as steam lines up to 600 psi at 850 degrees F. Many steel valves are available with butt-weld ends for economy and are generally used in high-pressure steam service, as well as other higher-pressure, non-corrosive services.

Forged steel valves are made of tough carbon steel. They are used at pressures up to 2,000 psi and temperatures up to 1,000 degrees F.

Plastic is used for a great variety of valves, generally in high-corrosive service, at low temperatures and low pressures. Plastic lining of metal valves for corrosive service and high-purity applications and temperatues are also available.

Gate valves provide full flow, minute pressure drop, and minimum turbulence. They are normally used where operation is infrequent, such as for equipment isolation.

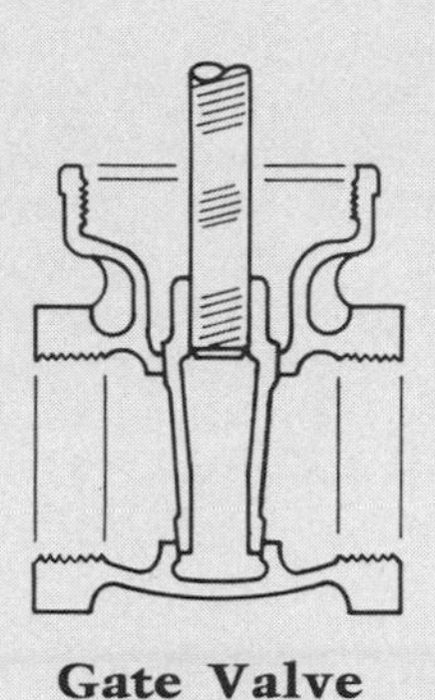

Gate Valve

Globe valves are designed for throttling and/or frequent operation with positive shutoff. Particular attention must be paid to the several types of seating material available to avoid unnecessary wear. The seats must be compatible with the fluid in service and may be composition or metal in construction. The configuration of the globe valve opening causes turbulence, which results in increased flow resistance.

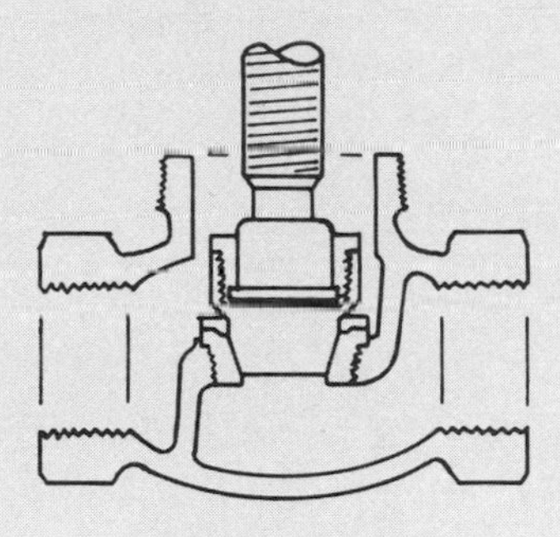

Globe Valve

Check valves are one-way valves and are designed to prevent backflow by automatically seating when the direction of fluid is reversed. Swing check valves are generally installed with gate valves, as they provide comparable full flow, and are usually recommended for lines where flow velocities are low. They should not be used on lines with pulsating flow. They are also recommended for horizontal installation, or in vertical lines where flow is only upward.

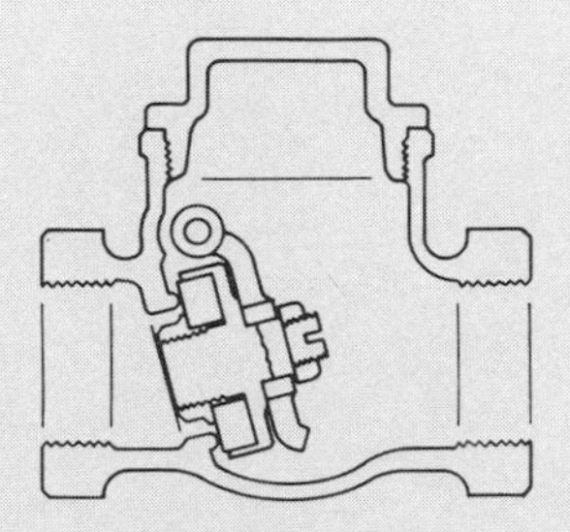

Check Valve

Hydronic Terminal Units—Valves and Fittings

Figure 10.4

Ball valves are light and easily installed, yet because of modern elastomeric seats, they provide tight closure. Flow is controlled by rotating 90 degrees a drilled ball that fits tightly against resilient seals. This ball seals with flow in either direction, and the valve handle indicates the degree of opening. This type of valve is recommended for frequent operation, such as for tank filling, and is readily adaptable to automation. Ball valves are ideal for installation where space is limited.

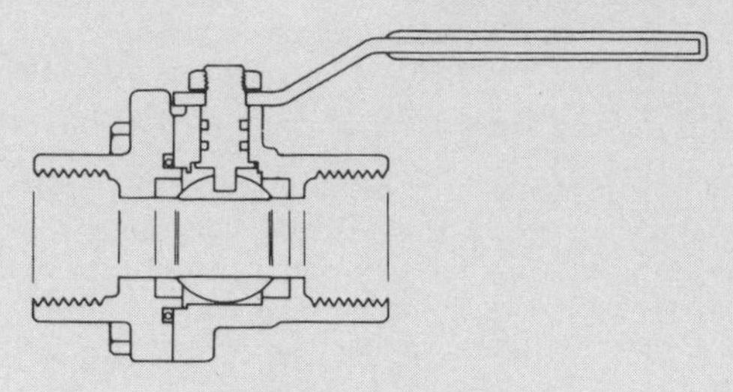

Ball Valve

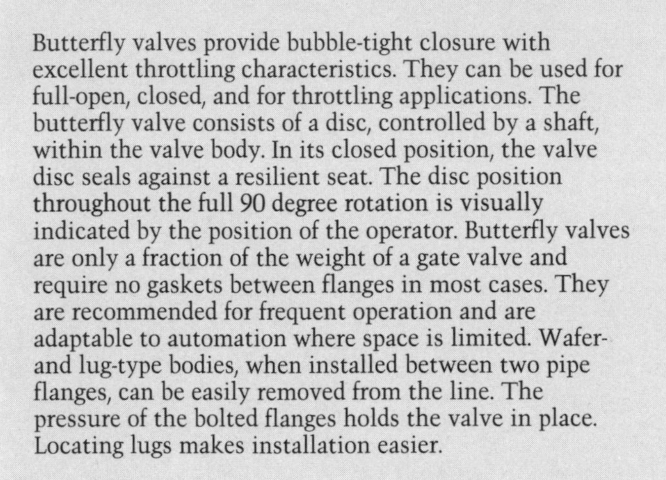

Butterfly valves provide bubble-tight closure with excellent throttling characteristics. They can be used for full-open, closed, and for throttling applications. The butterfly valve consists of a disc, controlled by a shaft, within the valve body. In its closed position, the valve disc seals against a resilient seat. The disc position throughout the full 90 degree rotation is visually indicated by the position of the operator. Butterfly valves are only a fraction of the weight of a gate valve and require no gaskets between flanges in most cases. They are recommended for frequent operation and are adaptable to automation where space is limited. Wafer- and lug-type bodies, when installed between two pipe flanges, can be easily removed from the line. The pressure of the bolted flanges holds the valve in place. Locating lugs makes installation easier.

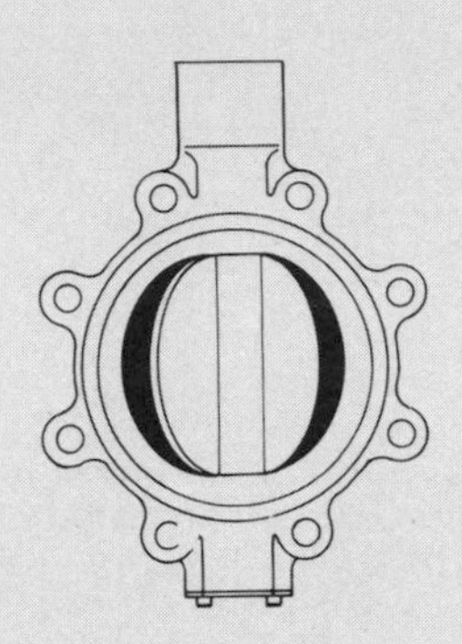

Butterfly Valve

Backflow preventers are mechanical devices installed to stop contaminated fluids from entering the potable water system. This reversal of flow may be caused by back pressure or back syphonage. The assembly incorporates double check valves, a relief valve and vacuum breaker all contained in one housing.

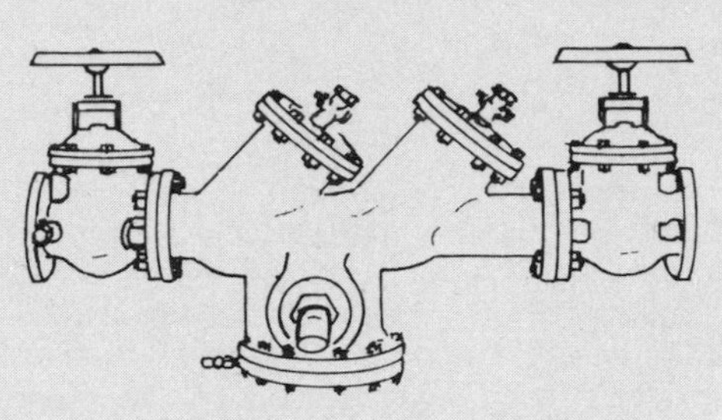

Backflow Preventer

Hydronic Terminal Units—Valves and Fittings (cont.)

Figure 10.4 (cont.)

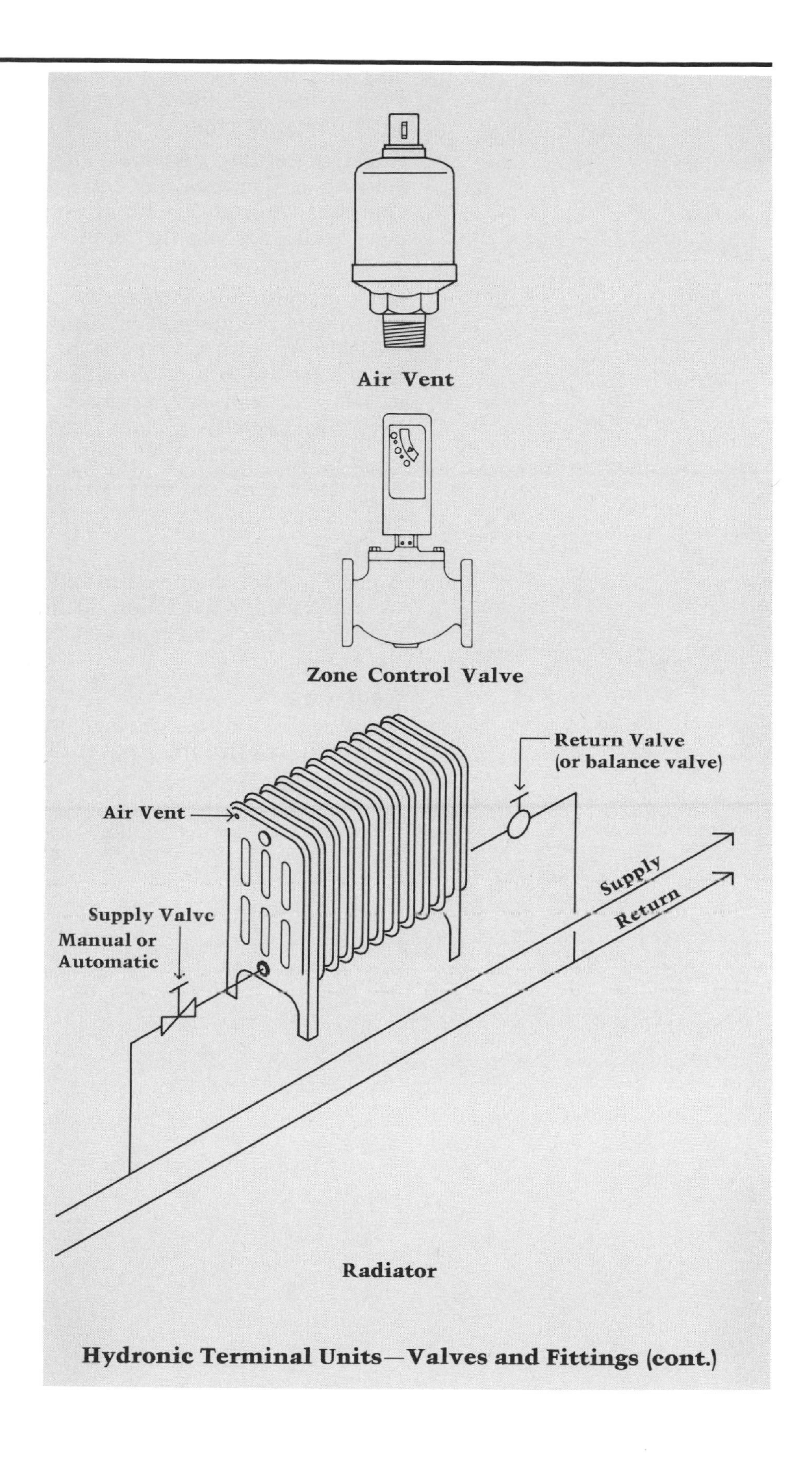

Figure 10.4 (cont.)

hospitals, laboratories, and special control environments. Dual duct systems are more expensive to build and operate than single duct systems.

Constant volume systems contain ducts with a constant velocity at all times. They may be single or dual duct systems. Constant volume systems work best for spaces with relatively even loads or where the temperature of the air must be adjusted to meet varying loads.

Variable volume systems use variable air volume (VAV) boxes, which vary the amount of air to a space to meet the load. Variable air volume systems are efficient, because they supply the precise amount of air needed. A VAV box is a device with a variable position damper that controls flow of air to a space. The VAV box may also contain a reverse-acting thermostatic control for cooling, a reheat coil, and a plenum recirculating feature.

Air system terminal units consist of grilles, registers, diffusers, and louvers, which supply or return air from spaces.

Grilles

A grille is a screened or perforated connection to ductwork, usually used for return air. Grilles are nondirectional and typically act as a cover to visually screen the ductwork from the room.

Registers

A register is a grille with a manually adjusted damper behind it, used to control the flow of the conditioned air into a conditioned space.

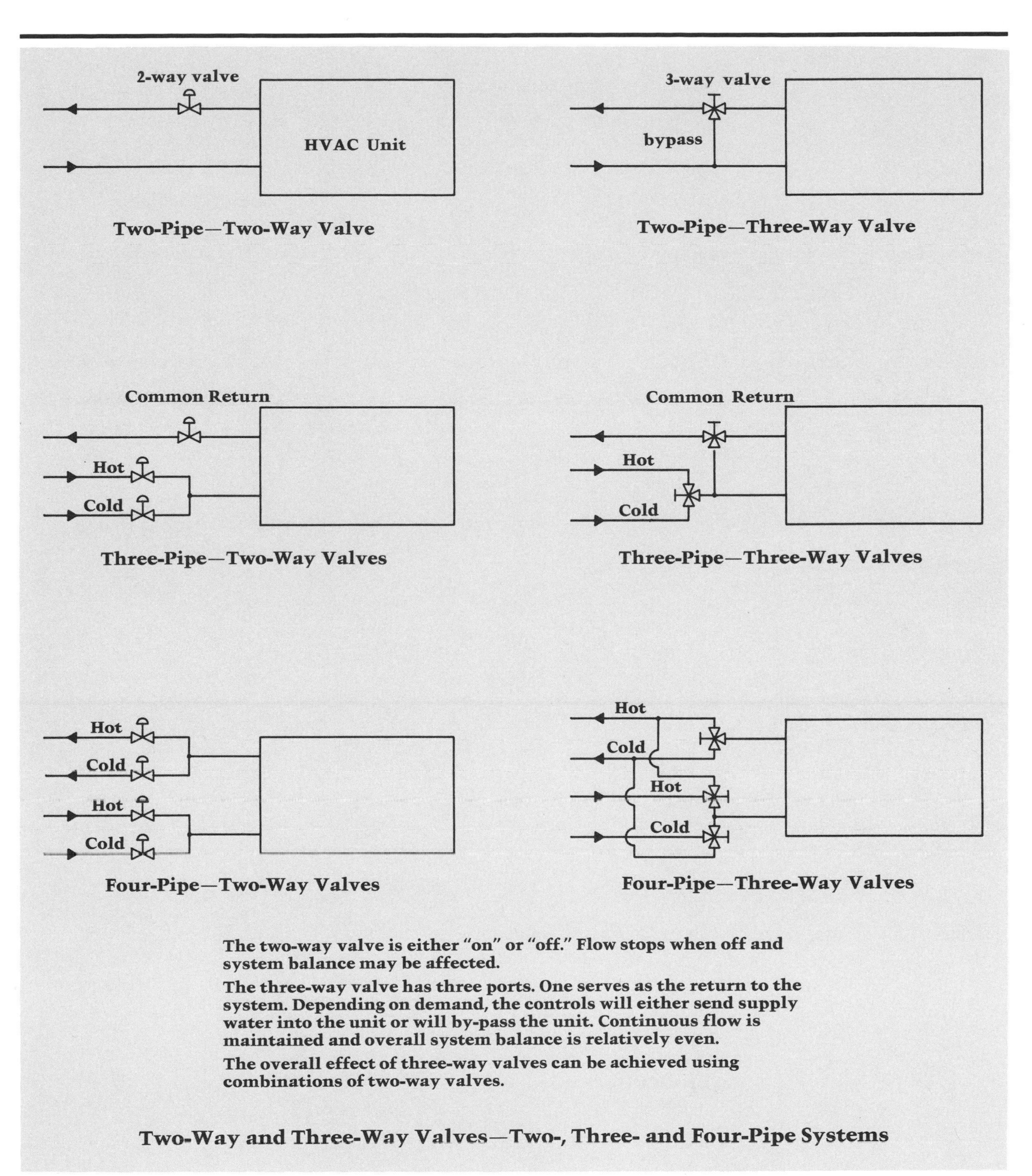

Two-Way and Three-Way Valves—Two-, Three- and Four-Pipe Systems

Figure 10.5

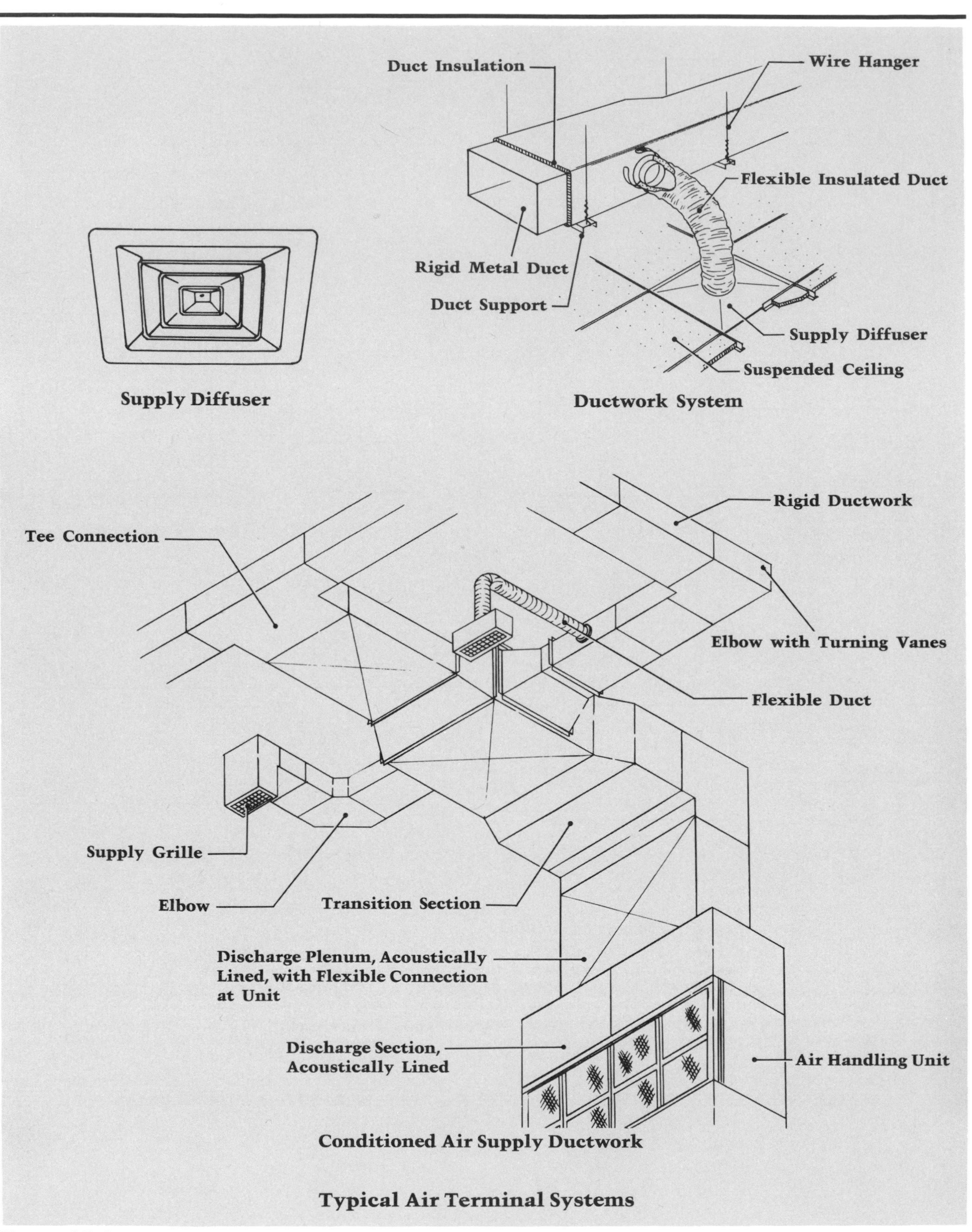

Typical Air Terminal Systems

Figure 10.6

Diffusers

Diffusers direct the conditioned air stream into a space through fixed or adjustable blades and may also have an integral damper to control flow. The most frequently used types of diffusers are ceiling, wall, and slot or linear diffusers.

Louvers

Louvers are openings in exterior walls to intake or exhaust air. Louvers typically have blades to shed rain, insect, or bird screens, and many have dampers.

Dampers

Dampers are devices within a terminal unit that provide manual or automatic control of air flow. Dampers operate in a manner similar to the valves in hydronic systems.

Figure 10.7 illustrates typical air system terminal units.

Electric Systems

Several types of electric terminal units are used to provide electric heat:

- Baseboard radiators
- Unit heaters
- Wall heaters
- Radiant heaters
- In-line duct heaters

Electric terminal units are used for air heating (see Figure 10.8). (Many of the previously mentioned pieces of hydronic equipment use electric power, but are not considered electric terminal units.)

Equipment Data Sheets

The equipment data sheets for terminal units (Figure 10.9) can be used in the selection of each piece of equipment. These sheets provide the efficiency, useful life, typical uses, capacity range, advantages and disadvantages, accessories, and additional equipment for each component listed. A Means line number, where applicable, is provided. Costs can be determined by looking up the line number in the current edition of *Means Mechanical Cost Data.*

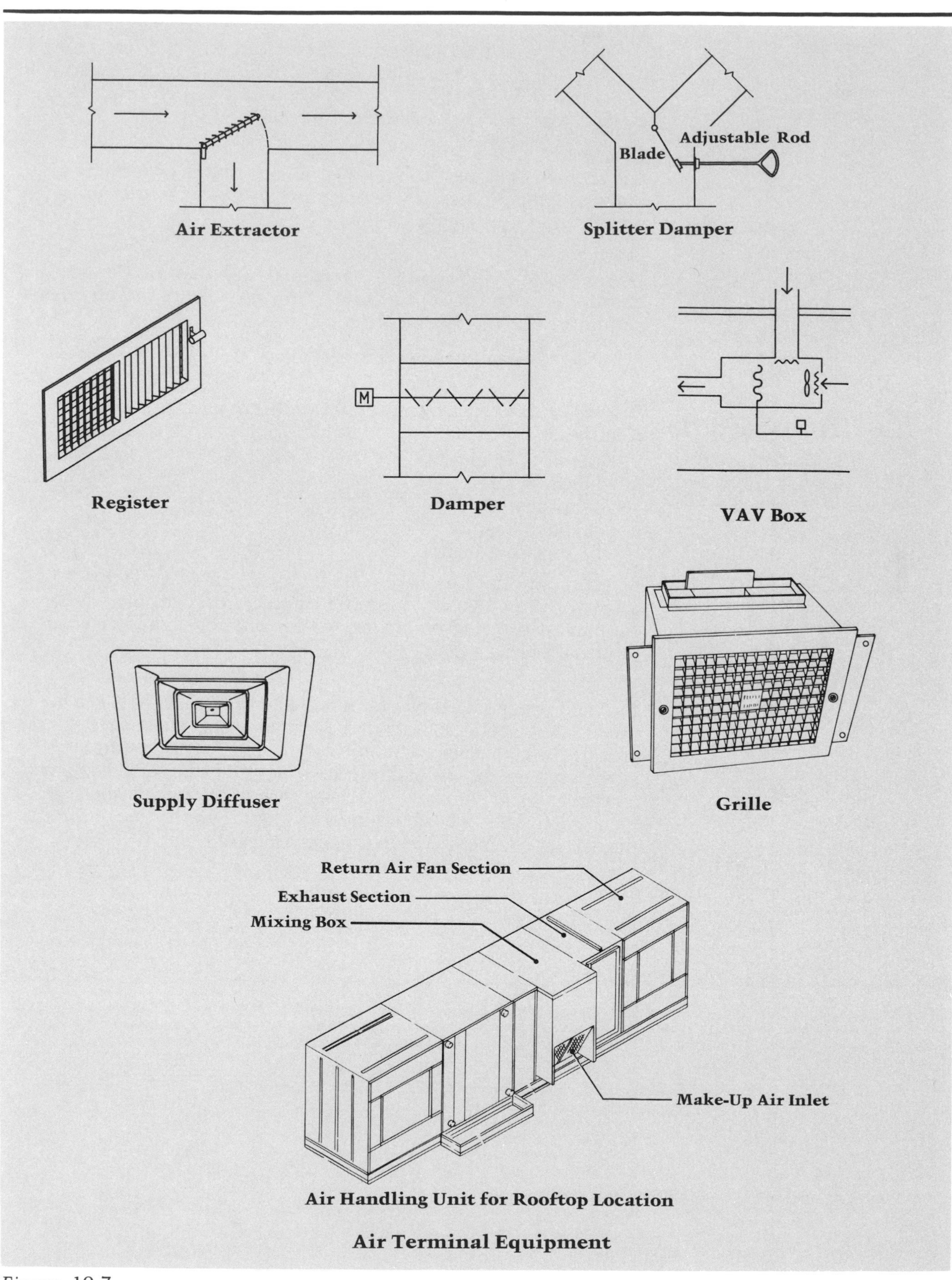

Air Terminal Equipment

Figure 10.7

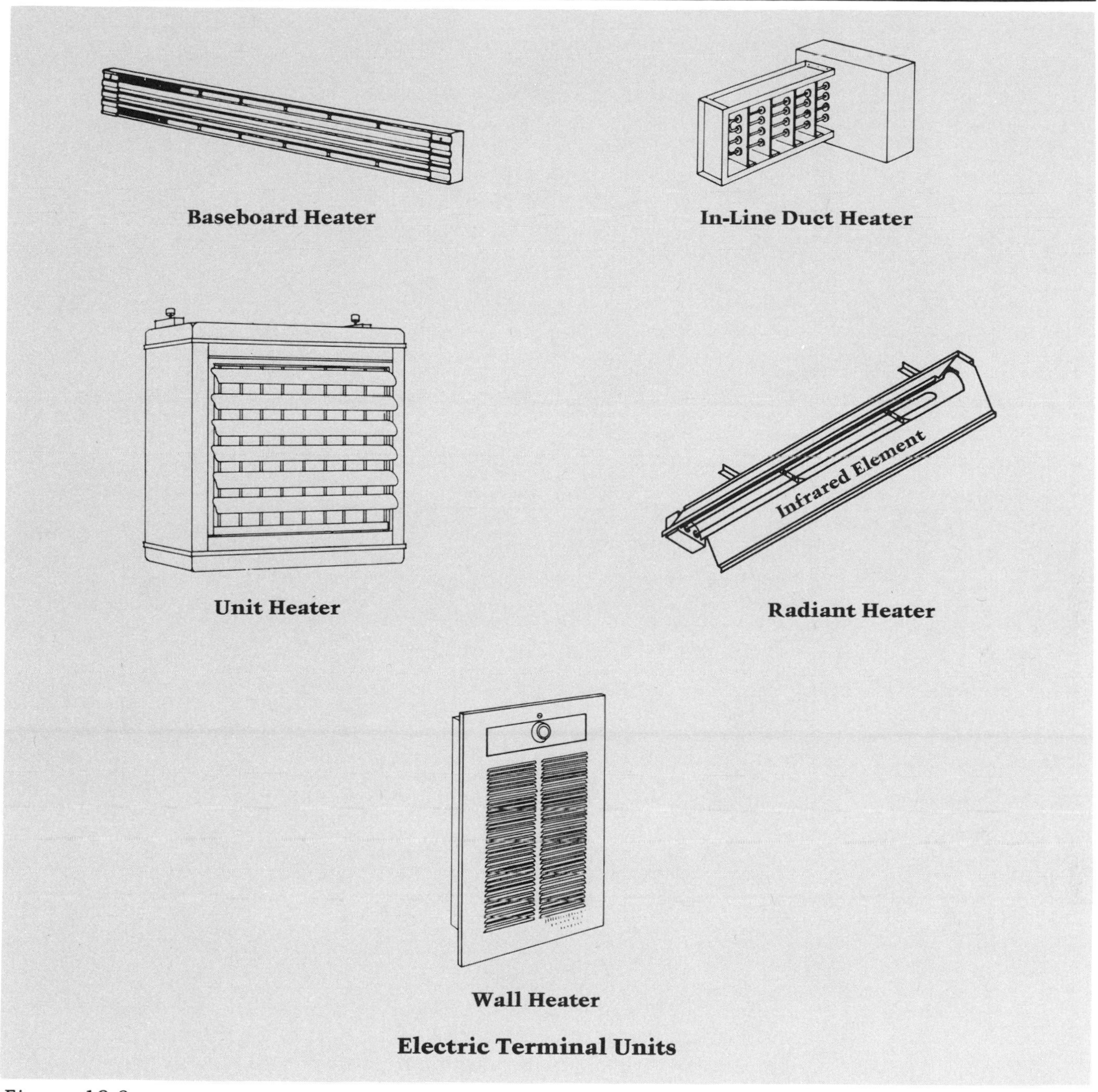
Baseboard Heater
In-Line Duct Heater
Unit Heater
Infrared Element
Radiant Heater
Wall Heater
Electric Terminal Units

Figure 10.8

Terminal Units Equipment Data Sheets

Terminal Units	Hydronic Automatic Air Vents	Means No. 156-201
Useful Life: 10-20 years	**Typical Use and Consideration**	**Size Range**
Float-Type Air Vent	Air vents are used on radiators, unit heaters, and high points of mains and risers in all gravity steam systems. Air vents are used on radiators, coils, and high points of mains and risers in hot water systems. Air vents are used on coils and high points of mains and risers in chilled water systems. Air must be vented from the piping systems, as it may impede or even stop the flow of the heating/cooling medium. Manual air vents are sometimes used. Air vents in hard to reach places often have the discharge piped to an accessible location.	1/8″–2″
Useful Life: 7-15 years	**Steam Specialties**	**Means No. 156-272, 156-240**
Angle Radiator Valve	Manual radiator valves, both angle and straightway, are used to control the steam or hot water supply to radiators and convectors. In many instances these manual valves are replaced by thermostatically controlled valves, either self-contained or with remote thermostats.	1/2″–2″
Angle Thermostatic (Radiator) Trap	Radiator traps are installed on the return end of the two-pipe steam radiators and convectors. They do not have sufficient capacity to remove the condensate from unit heaters or coils. A thermostatic bellows stops the flow of steam and reopens upon temperature drop to allow the flow of condensate. In rare instances under light loads, a thermostatic trap can be used as a drip trap.	1/2″–1″
Combination Float and Thermostatic Trap	Combination float and thermostatic (F & T) traps are designed to handle large volumes of air and condensate. A built-in float mechanism controls the flow of condensate and a thermostatic bellows vents the accumulated air. It is good practice to precede an F & T trap with a sediment strainer and to connect it with a three-valve by-pass. F & T traps handle the large condensate loads generated by piping drips, heating coils and other low-pressure steam points of condensation. For high-pressure steam and industrial applications, bucket traps are used rather than F & T.	3/4″–2″

Figure 10.9

Terminal Units Equipment Data Sheets (continued)		
Terminal Units	**Hydronic Cast Iron Convectors and Radiators**	**Means No. 155-630**
Useful Life: 20-40 years	**Typical Use and Consideration**	**Capacity Range**
Cast Iron Tubular Radiator	Cast iron has an advantage not only in durability but in its capability of transferring heat evenly due to its mass. This heat is given up both by radiation and convection. The type of radiator pictured here can be free standing or wall hung. It consists of sectional elements which may be replaced or even added to in the field in the event of leakage or undersizing. A smaller cast iron element enclosed in a steel compartment with inlet and outlet grilles allowing air to pass over the element increases the capacity by convection. Cast iron, like other hydronic convectors or radiators, can be efficiently used for steam or hot water.	Infinite — priced by the section
Useful Life: 15-25 years	**Hydronic Baseboard Radiation Cast Iron and Fin Tube**	**Means No. 155-630**
Baseboard Radiation	Baseboard radiation is extensively used in residential applications, and in some light commercial. Due to cost considerations, fin tube is more popular than the more solid and substantial cast iron. Baseboard radiation is almost exclusively used in forced hot water heating systems, as the ports are too small for gravity flow systems.	Infinite — priced by the linear foot
Useful Life: 10-25 years	**Hydronic Fan Coil Units**	**Means No. 157-150**
Concealed Fan Coil Element	Most commonly used in commercial applications where space consideration precludes the use of direct radiation. With the addition of a drain pan, a forced hot water unit may be used for cooling applications. Steam coils are a heating option and direct expansion coils are a cooling option. Built-in fan speed control is a popular option also. As fan coil capacities increase in size, separate coils for heating, cooling, preheat and/or reheat modules may be added.	1/2-50 tons
Useful Life: 7-20 years	**Hydronic Unit Heaters**	**Means No. 157-630**
Unit Heater	A unit heater is simply a hydronic coil within the same enclosure as an air moving fan. These units may discharge horizontally (as pictured) or may project vertically. Unit heaters are primarily used in industrial or commercial applications where noise or air movement is not objectionable. Steam or hot water coils are the available hydronic choice. Electric resistance unit heaters are also available.	14.7 MBH to 520 MBH

Figure 10.9 (cont.)

Terminal Units Equipment Data Sheets

Terminal Units	**Grilles, Registers and Diffusers**	**Means No. 157-460**
Useful Life: 20-40 years	**Typical Use and Consideration**	**Size Range**
Floor Grille 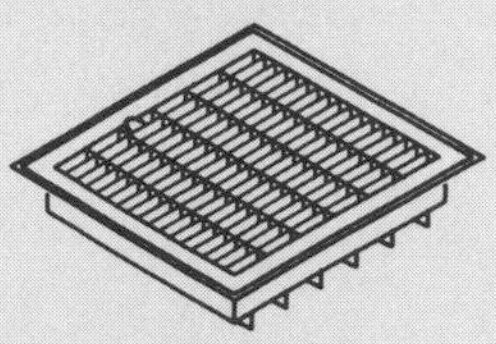	Grilles, registers and diffusers are the supply and return faces used in ductwork systems. Grilles are the least expensive and are generally used for exhaust and return air inlets requiring deflection but having no throttling capability. They are often used in non-critical supply applications as well.	6″ x 6″ to 48″ x 48″
Ceiling Register 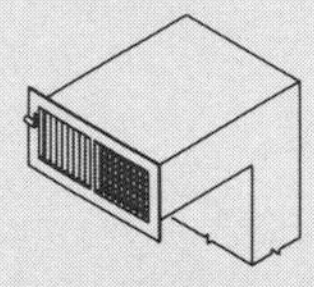	Registers are used on both supply and return duct openings. A register differs from a grille in that it has a volume control, whereas the grill does not have this throttling feature. Supply registers also have deflection capabilities.	**Means No. 157-470** 4″ x 6″ to 48″ x 24″
Supply Diffuser 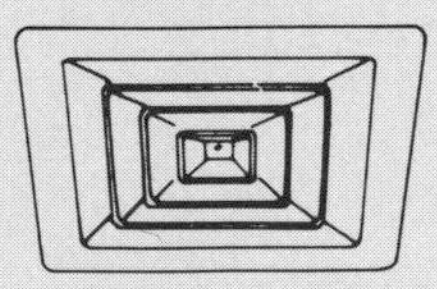	Supply diffusers distribute the conditioned air in accordance with the specified requirements for volume velocity and direction. A major concern with diffusers is noise criteria. Diffusers have the capability of various flow patterns. Diffusers are available as fixed or adjustable flow. Some diffusers include both supply and return air connections. Refer to manufacturer's data for the many shapes, materials, and finishes that are available for grilles, register, and diffusers.	**Means No. 157-450** 6″ x 6″ to 36″ x 36″
Useful Life: 20-40 years	**Air Mixing Boxes**	**Means No. 157-425, 480**
Air Mixing Box 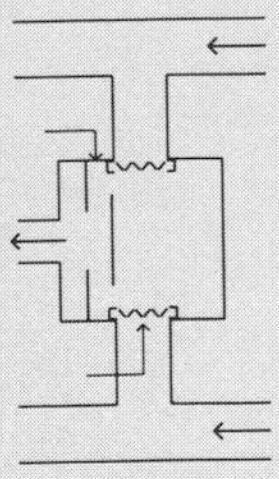	These mixing boxes are used to blend warm and cool air from dual duct supply systems to a final preset discharge temperature. This system has the capability of supplying heating or cooling simultaneously for special environmental requirements. Reheat coils and humidification capabilities can also be accommodated. Mixing boxes are used in constant volume systems as well as in variable air volume systems. Damper control is critical for this application.	150–1900 cfm
Useful Life: 20-40 years	**Variable Air Volume Boxes**	**Means No. 157-425, 480**
Variable Air Volume Box 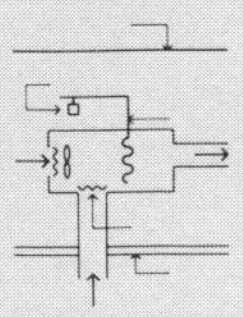	The VAV box detailed here incorporates a heating coil, a variable or modulating damper, plus has the capability of utilizing the warm air trapped in the ceiling plenum. This economizer feature is supplemented by the reheat coil when plenum temperatures are not sufficient. The control station air handling unit provides the capacity to vary the primary air flow.	150–1900 cfm

Figure 10.9 (cont.)

Terminal Units Equipment Data Sheets (continued)		
Terminal Units	**Electric Baseboard Heaters**	**Means No. 168-130**
Useful Life: 10-20 years	**Typical Use and Consideration**	**Capacity Range**
Electric Baseboard Heater	Electric baseboard heat, like most electric heat, is the least expensive type of mechanical heat to install. Operating costs of electric resistance heaters are high. Major features of this form of electric heat are built-in individual thermostatic controls, individual room thermostats, and the capability of individual metering. This type of heat is used in residential and light commercial applications.	375-1875 watts
Useful Life: 10-20 years	**Electric Unit Heaters**	**Means No. 168-130**
Electric Unit Heater	Electric unit heaters consisting of an electric resistance element and a fan within the same enclosure are for residential, commercial and industrial applications. Electric unit heaters can be floor or wall mounted and the heavy duty commercial and industrial models are commonly suspended from overhead. Not susceptible to freezing, these units can be located in extreme areas for localized or spot heating.	.75 to 50 kilowatts
Useful Life: 10-20 years	**Computer Room Units**	**Means No. 157-130**
Computer Room Unit	These dedicated-use heating and cooling units provide conditioned air with temperature and humidity control for the close tolerances required in computer rooms. They may be completely self-contained or may obtain heating or cooling from a remote source. Air distribution may be free blow or ducted. Air may be distributed from above a hung ceiling or from beneath a raised floor. Computer room units are built to much higher standards than conventional air conditioners to give trouble-free operation for their critical function.	

Figure 10.9 (cont.)

CHAPTER ELEVEN

CONTROLS

Controls are *manual* or *automatic* devices used to regulate HVAC systems. A bathroom light switch that also activates an exhaust fan is an example of a manual control; a fan connected to a time clock is automatically controlled. The simplest controls turn equipment on or off. Advanced controls may allow the equipment to respond based on demand.

Type of Operation

The two basic control systems most frequently used today are *pneumatic* and *electric* controls. (*Electronic* controls, also called direct digital controls, are becoming increasingly popular in medium- to large-sized buildings.) Electric controls are usually used in small residential and commercial buildings and run on 24, 48, or 120 volt systems. Pneumatic controls are used on larger buildings and operate on compressed air (between 0 and 15 psi) from a compressor generating 100 psi air, which is reduced. Electric systems use motors or solenoid valves and switches (magnetic) to turn devices on and off. Pneumatic systems use air pressure, often against a piston, to move the devices being activated.

The choice between electric or pneumatic systems often depends on the designer's preference. Electric systems are fast, accurate, and easy to install, but contain many more component parts (such as motors), which require upkeep. Pneumatic systems have simple components and are durable, but require a compressor and an air filter/dryer system.

There are also *self-powered* controls, such as thermostatic valves. These self-powered controls are powered by a gas which expands with temperature changes to move the device.

Detection/Activation

Controls can be divided into two categories; devices that detect (controllers) and devices that activate (controlled devices). A time clock detects the "on" or "off" time and a thermostat detects the temperature. They, in turn, activate a valve or a motor. Examples of other controllers and their relationship to controlled devices are shown in Figure 11.1.

Both types of devices may be either electrically or pneumatically controlled. The control system and control logic are usually developed according to the needs of the owner or facilities managers. Before designing a control system, it is necessary to determine when each piece of equipment in the generation and termination systems will be "on" or "off."

Internal/External Controls

Most pieces of HVAC equipment have internal controls. For example, chillers, oil burners, window air conditioners, and cooling towers all come from the factory with pre-wired controls, which activate the internal parts of the equipment in the proper sequence. Certain safety features are also provided with the equipment. Examples are a low water cut-off or stack temperature time delay. Both of these features are designed to ensure safety; the first monitors the water level in the boiler to make sure it is not too low, and the other detects when the fuel being injected has not ignited.

In addition to internal safety controls, manual or automatic external controls are added to start and stop the equipment or to regulate its output. A familiar example of an external control is a thermostat in a house. When it measures that the space temperature is below the set point, it automatically sends a

Simple Control Sequences

Controllers		Controlled Devices
Thermostat	⟶	VAV damper/motorized valve/zone pump
Pressure sensor	⟶	Fan motor, space heater
High temperature limit	⟶	Oil/gas burner
Duct smoke sensor	⟶	Equipment shutdown, damper motor
Outdoor air sensor	⟶	Boiler start-up
Time clock	⟶	Pump motor, package unit
Humidistat	⟶	Steam valves
Aquastat	⟶	Oil/gas burner, fan motor
Enthalpy meter	⟶	Return, exhaust, and intake dampers

Figure 11.1

signal to the circulating pump to turn on. Meanwhile, the internal controls of the boiler maintain the boiler water temperature.

Zoning

For greater design flexibility, an approach known as zoning is commonly used in HVAC systems. By subdividing a building into a series of zones, the set conditions can be maintained properly for a specific space and the equipment need only service the areas that require heating or cooling. Combining zoning with temperature set-back controls and a time clock is an effective and simple means of achieving building control. Although it is common to have many zones within one building, the number of stages is usually limited to three occupied zones and three unoccupied zones.

Figure 11.2 illustrates heating systems and operations common to residences, commercial buildings, and institutions. Combining the two-zone heating system with the two-stage heating system results in a two-zone, two-stage system.

Energy Management

In commercial and institutional buildings, *energy management systems* (EMS) are both elaborate and cost effective. These complex control systems are usually operated by computers to continually monitor energy consumption throughout a building. The detection and activation devices are controlled by the EMS and by the operator. In a large building, individual tenant spaces can be controlled from the EMS center. Variations in temperature can be made regarding the occupied and unoccupied times or heating and cooling may be shut down altogether, such as in hotel rooms where there is only partial occupancy. The outdoor temperature, read by an outdoor air sensor, can be used to automatically vary the hot water temperature for heating.

An EMS can also monitor electric loads. Lights can be turned on or off automatically depending on the time of day or outdoor brightness, as read by a photoelectric cell. Energy consumption for the building can be analyzed on a minute-by-minute basis to *peak shave* utility demand charges. Peak shaving involves reducing the energy demands of certain equipment to offset the drain at peak hours. For example, water coolers may be turned "down" (to a higher temperature) during periods of greatest energy demand.

Buildings with complete control and communication systems, such as infrared detectors to turn lights off when rooms are not occupied and cable, computer, satellite, and telecommunication hookups with a full automated scheme of operations, are referred to as "smart" buildings. Typical controls for automated buildings are outlined in Figure 11.3.

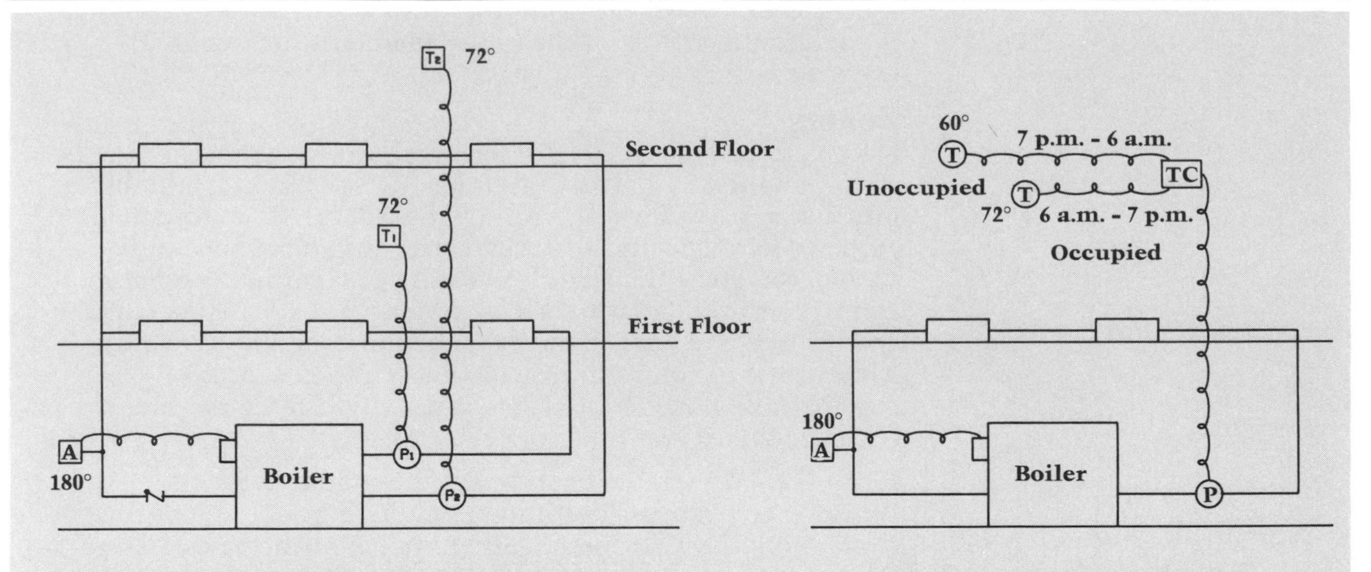

Two-Zone Heating System
T_1 activates Pump P_1 and maintains 72°F.
T_2 activates Pump P_2 and maintains 72°F.
A activates burner when return water falls below 180°F.

Two-Stage Heating System
72° F during the day (6 a.m.-7 p.m.).
60° F during the night (7 p.m.-6 a.m.).
A activates burner.

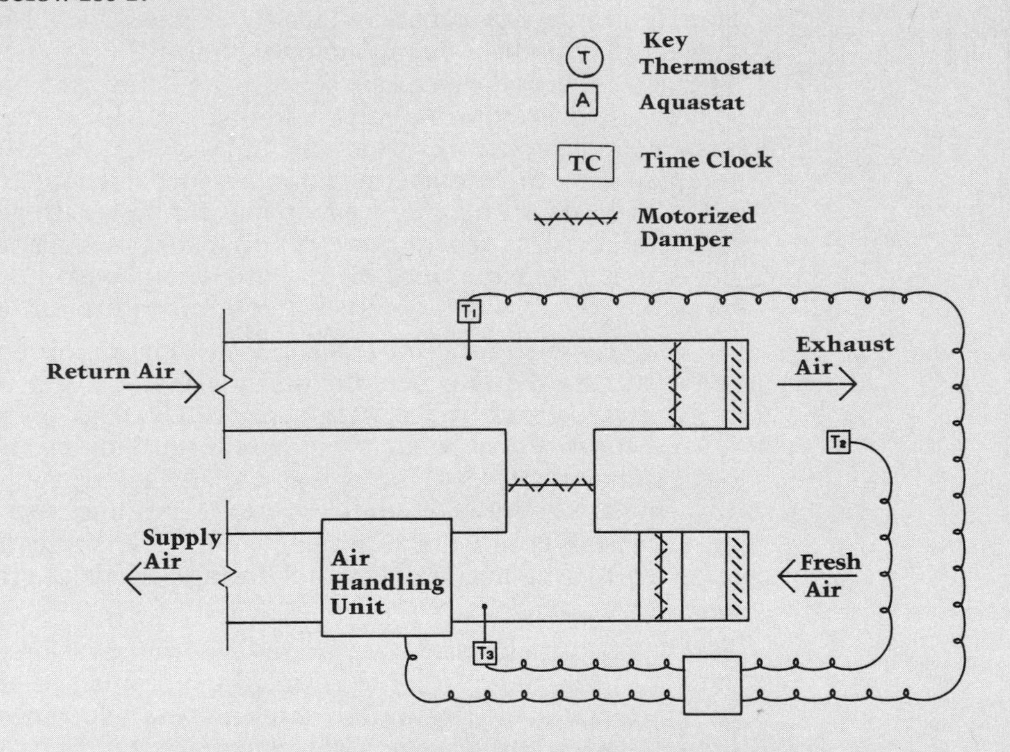

T_1 senses return air temperature and determines if heating or cooling is required.
T_2 senses outdoor temperature to determine optimum position of dampers.
T_3 senses mixed air temperature to determine if air handling unit heating or cooling is required.

Modulating Damper and Air Handling Unit

Basic Control Diagrams

Figure 11.2

Typical Controls in "Smart" Buildings

Control System

- Time shift load response (alternates equipment operation to avoid simultaneous peaks)
- Automatic temperature response for individual spaces
- Energy management control system for entire system

Heating, Ventilating, and Air Conditioning

- Zoning
- Heat reclamation
- Off-hour energy storage

Electrical

- Lighting
  - High voltage
  - Photocell actuators
  - Infrared switches
  - Delayed time ballast
- Power
  - Flexible grid for tenants
- Emergency systems
  - Fire alarm
  - Emergency lighting
  - Smoke evacuation
  - Security
  - Fire detection

Communications

- Telephone
  - Re-routing call
  - Intercom/conference
  - Modem/data bank/EMCS access
  - Automatic billing
- Computer access for tenants
  - Data processing
  - Word processing
  - Data base
- Fiber optics
- Satellite communications
- Pneumatic delivery
- Cable communications

Transportation

- Elevators
  - Voice synthesizers
  - Variable mode control

Building Services

- Trash removal
  - Vacuum system
  - Compactors
- Cleaning
  - Window system equipped for wash down

Figure 11.3

Equipment Data Sheets

The data sheets for controls (Figure 11.4) summarize the useful life, typical uses, capacity range, advantages and disadvantages, and special considerations for each device. These sheets can be used in the selection of each piece of equipment. A Means line number, where applicable, is provided. Costs can be determined by looking up the line number in the current edition of *Means Mechanical Cost Data.*

Typical Controls in "Smart" Buildings (continued)

Technical Support
- Computer programmers
- Secretarial pool
- Central library and information center
- Mail delivery
- Copy services
- Computer software security
- Banking services

Executive Support
- Dining room
- Meeting rooms
- Transportation services
- Messenger services
- Valet services
- On-call meal and office services

Building Organization
- Scheduled spaces
  - Meetings
  - Off-hours use
- Orientation
  - Sun
  - Traffic

Figure 11.3 (cont.)

Controls	Outdoor Reset	Means Number 157-420
Useful Life 15-25 years	**Typical Use and Consideration**	**Capacity Range** 0-70°F/210-100°F
Outdoor Sensor; Control Panel; Boiler	The outdoor sensor (bulb) for this control is mounted on the building wall out of direct sunlight or is provided with a sun shield. When there is a drop in outdoor temperature, this control will raise (reset) the boiler water operating temperature. Conversely, if there is a rise in outdoor temperature, the boiler operating temperature will fall.	
Controls	**Thermostatic Heating Control Valves (Non-Electric)**	**Means Number** 156-240
Useful Life 10-20 years	**Typical Use and Consideration**	**Size Range** 1/2″ to 1-1/4″
Heating Unit; Control Setting Dial	These non-electric, self-contained valves can be controlled by a thermostat via a capillary tube, or may have a built-in thermostat. A dial on the valve handle allows for temperature adjustment.	
Controls	**Controllers**	**Means Number** 157-420
Useful Life 10-20 years	**Typical Use and Consideration**	**Capacity Range** Various
	Compares measured quantity (temperature, humidity, pressure) with set point and sends or halts an activating signal. Controllers act like switches. When "on call" they make or break contact, depending on the controlled device requirements. Many controllers are integral with the sensing device. Most thermostats, for example, contain the controller as part of the thermostat unit. Coordinate controllers with control manufacturer and control system.	
Controls	**Gauges and Thermometers**	**Means Number** 157-240
Useful Life 10-20 years	**Typical Use and Consideration**	**Capacity Range** 40-240°F 30″—0-30 lbs.
	Thermometers are installed in equipment, piping, and ductwork to read water and air temperatures. Thermometers are provided as dial type, in diameters of 2″, 4″, or 6″, or as stem type, 5″, 7″, or 12″ long. Thermometers and gauges can be supplied as straight or angle configuration. For piping installation, thermometers are threaded into a separable socket or well so that the instrument may be removed without draining the pipe line.	

Figure 11.4

Controls	Gauges and Thermometers	Means Number 157-240
Useful Life **10-20 years**	**Typical Use and Consideration**	**Capacity Range** **40-240°F** **30″—0-30 lbs.**
	Gauges are installed in pipe lines and equipment to read pressure, vacuum, or both (compound gauge). A gauge cock is installed at the inlet to allow removal of the gauge without losing line pressure. Steam gauges are normally provided with a pigtail or siphon to protect the working parts from prolonged contact with the steam. Gauges in lines subject to pulsation or vibration should be installed with pressure snubbers.	
Controls	**Temperature Control Devices**	**Means Number 157-420**
T **Thermostat**	The most common temperature control instrument is the space thermostat. Thermostats sense temperature, and at a variation from their set point, transmit a signal or command to a piece of mechanical equipment to turn on or off. Thermostats operate burners (gas or oil), pumps, unit heaters, fan coils, valves, dampers, etc. Some of the options available when selecting a thermostat are: line or low voltage; pneumatic; indicating or non-indicating; occupied/unoccupied; night set-back locking enclosures; or built-in clock.	
A **Aquastat**	Aquastats may be surface-mounted (strap-on), remote bulb, or inserted into the piping or equipment. They sense water temperature and, at a variation from the set point, send a signal to the controlled equipment. Aquastats are used primarily as limit or regulating controls on boilers, condensate return lines, pumps, converters, etc.	
H **Humidistat**	Humidistats control the humidity in a conditioned space by sensing the moisture in the space or in the air supply duct. When the moisture content varies from the set point, a signal is sent to operate the humidifier or dehumidifier.	
	Time clocks control the hours of operation for designated periods. They are available in 24-hour or seven day arrangements. Holiday and weekend skips are an optional feature of time clocks, as are battery backups.	
	Other control devices such as sensors, which read certain conditions, and relays, which transfer a control signal from a controller of one voltage to an operator of another voltage, are used throughout temperature control systems. Switches are also available to convert electrical/pneumatic signals.	

Figure 11.4 (cont.)

Controls	Automatic Control Valves	Means Number 157-420
Useful Life 10-20 years	**Typical Use and Consideration**	**Capacity Range 1/2" - 6"**
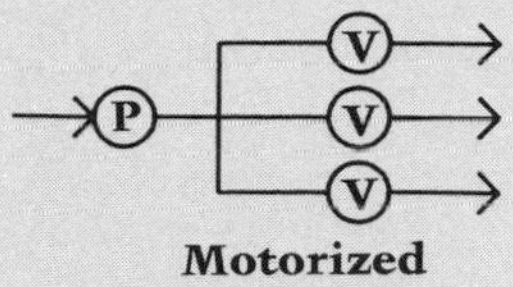 Motorized Zone Valve 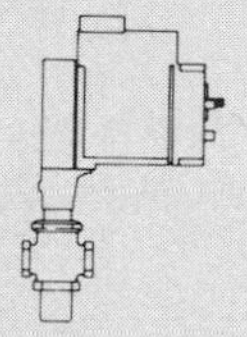 3 Way Valve 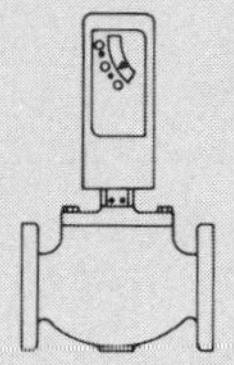 Globe Valve	Automatic control valves control the flow in hydronic circuits. These valves may have electric, pneumatic, or magnetic (solenoid) operators. Valve body determination is similar to manual valves, i.e., gate type for open and closed functions, and globe type for modulating or throttling. Motor-operated valves may be used to control individual components in a system or to separate various piping loops into zones.	

Figure 11.4 (cont.)

CHAPTER TWELVE

ACCESSORIES

Accessories are necessary adjuncts to HVAC equipment operation and performance. Flues, chimneys, motor starters, controllers, noise and vibration reducers, and air cleaning and filtration equipment are some of the important accessories used in HVAC systems.

Flues and Chimneys

All combustion systems produce hot gases, which are vented to the outside above the roof of a building through a system of flues and chimneys remotely located from air intake louvres and enclosed in fire-rated shafts, typically rated for two hours of fire resistance. The inside chamber that conducts the gases upward is called the flue. The entire structural assembly is called the chimney.

Chimneys are made from many materials — brick, clay tile, concrete, single- or double-wall steel or aluminum — and come in standardized sizes from 4" to 48". Local codes determine the materials permitted for a given use. Since combustion gases move at relatively low velocities (20 to 40 feet per second), the surface roughness of the chimney does not significantly affect the flow of gas. Flue diameters are relatively unaffected by the materials used for the flue.

The main principle of a chimney is that the hot combusted gases rise relative to colder outside air. The main factors that affect chimney size are the temperature, pressure, and velocity of the gas being conveyed.

Gas Temperature

As the temperature of the flue gas rises, its density relative to ambient air decreases. The gas is lighter and more buoyant, producing a higher pressure, or draft, to drive the gases up the chimney. Hotter flue gases produce better draft and require smaller chimneys.

Pressure

The following formula is used to calculate the pressure that acts to push air up a chimney:

$P_d = 0.255\ P_{(atm)}\ L\ (1/\{T_a - 1/T_f\})$

P_d = pressure in inches of water at the base of the chimney which is theoretically available to drive the flue gas

$P_{(atm)}$ = atmospheric pressure expressed in inches H_g (typically 29.92 inches H_g)

L = height of chimney (feet)

T_a = outside air temperature (boiler room temperature if conservative) (°$F_{(abs)}$)

T_f = average flue gas temperature (°$F_{(abs)}$)

$F_{(abs)}$ = temperature °F + 460

Common flue gas properties for chimneys are shown below.

Fuel	Average Temperature Rise (°F)	Approximate Value (T_f − °F $_{(abs)}$)	Mass (lbs. gas per 1,000 Btu)
Gas (no draft hood)	300–400	800	1.1 (1.6 w/hood)
Oil	450–550	950	1.2
Incinerators	1,300–1,400	1,800	1.6

Velocity

As the velocity of the flue gas increases, the required diameter of the flue decreases. This is expressed as follows:

$$V = \frac{H\ M\ T_f}{26\ P_{(atm)}\ D^2}$$

$$D^2 = \frac{H\ M\ T_f}{26\ P_{(atm)}\ V}$$

V = velocity of flue gas (ft./sec.)

H = heat input (MBH)

M = lbs. mass gas in flue per 1,000 Btu/hr.

T_f = average flue gas temperature °$F_{(abs)}$

$P_{(atm)}$ = atmospheric pressure (in H_g)

D = flue inside diameter (in.)

A velocity of 20 feet per second is a conservative estimation used for typical installations.

Always slightly increase chimney size to account for start-up flue temperatures, which are colder than during actual operation. Use the equivalent diameter charts in duct design to select a rectangular flue size of equivalent circular diameter.

Example: Find the diameter of a flue for a boiler with an input of 1,500,000 Btu/hour burning oil.

H = 1,500 MBH

M = 1.2

T_f = 950°F $_{(abs)}$

P = 29.92 in H_g

V = 20 ft./sec (minimum/conservative)

$$D^2 = \frac{1500 \times 1.2 \times 950}{26 \times 29.92 \times 20} = 109.9$$

D = 10.5″

Therefore, use a 12″ diameter flue.

Lining

Masonry chimneys should be lined with tile or metal to prevent deterioration of the mortar by acids; sulfuric acid, for example, forms with water vapor from combustion waste gases such as sulphur dioxide. For this reason, special care must be used in selecting chimneys for fuels with a high sulphur content.

Ventilation

Metal flues for gas systems typically require a type B vent, a double-wall metal vent with insulation. The metal should be corrosion resistant (e.g., stainless steel, aluminum, or galvanized steel), because gas exit temperatures are low, under 550°F, and an insulating wall is necessary to prevent the gas from cooling too much before it leaves the flue. Other flues include type BW, type L, and masonry chimneys. BW and L are also reserved for low-temperature flue gases. Local codes vary and should be checked for permitted flue types.

Examples of flues and chimneys are illustrated in Figure 12.1.

Motor Starters

Electric motors for HVAC equipment are turned on and off by motor starters. Motor starters are activated by either an *automatic controller*, such as a thermostat, or by a *manual switch*, which closes contacts in the starter. This, in turn, permits a current to flow in the motor, causing the motor to run.

The starter, a separate piece of equipment, may be mounted adjacent to the motor it serves or located remotely, as in a motor control center. Motor controllers typically are designed to:

- start and stop the motor as directed by control device;
- protect against short circuits or grounds;
- ignore the high initial motor inrush current, but protect the motor from excessive current draw, or *overloading*, which damages insulation and shortens motor life. The overload relays that protect the motor may be magnetic or thermal (melting alloy) or bi-metallic. Basically, they monitor the current or temperature of the motor and shut it down during overload.

Functions in operational controls, such as reversing, changing speed, and other rapid sequences, are more common in industrial plants.

Manual Starters

Manual starters (on/off switches) are often located so that they can be operated remotely to either draw the operator away from the machine or to allow the machine a few minutes to cool during, or to prevent, an overload.

Motors less than one horsepower (HP) single phase under 230 volts use *fractional horsepower* motor starters. Single-phase motors up to 5 HP and three-phase motors up to 10 HP commonly use *integral horsepower manual starters*. These operate similar to the fractional horsepower starters, but are rated for the heavier daily loads and multiple phases.

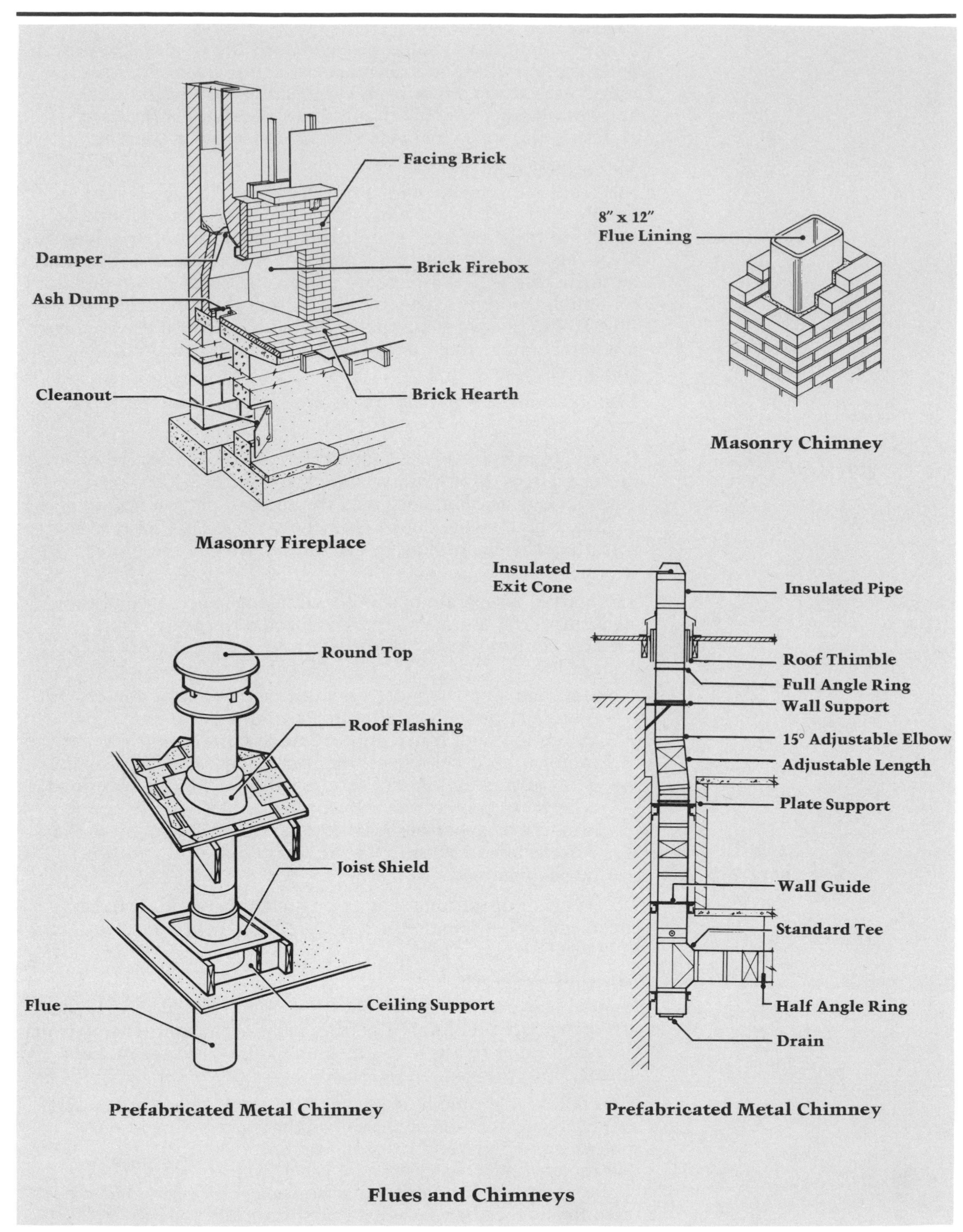
Facing Brick
Damper
Brick Firebox
Ash Dump
Cleanout
Brick Hearth
Masonry Fireplace
8" x 12"
Flue Lining
Masonry Chimney
Round Top
Roof Flashing
Joist Shield
Flue
Ceiling Support
Prefabricated Metal Chimney
Insulated
Exit Cone
Insulated Pipe
Roof Thimble
Full Angle Ring
Wall Support
15° Adjustable Elbow
Adjustable Length
Plate Support
Wall Guide
Standard Tee
Half Angle Ring
Drain
Prefabricated Metal Chimney
Flues and Chimneys

Figure 12.1

Magnetic (Automatic) Starters

When motors are to be activated automatically or by a remote device, magnetic starters are used. Magnetic starters are the most common type of starters used for HVAC equipment. Electromagnets, powered by the controllers, are used in magnetic starters to make contacts in relays, which remotely activate the motor. These starters must be selected according to the National Electric Manufacturers Association (NEMA) standards for voltage, horsepower, phase, current rating, and power.

Most major pieces of equipment have motor starters, usually tied to the automatic control system. They also have an automatic required reset on power failure to prevent any operator near a piece of equipment from being injured inadvertently when power is resumed.

Enclosures

Motor starters as well as other electrical equipment must be encased in properly specified enclosures. The following are the main NEMA-approved enclosure types:

NEMA 1	General purpose	Normal indoor use
NEMA 3	Dust tight, rain tight	Ordinary outdoor use
NEMA 3R	Rainproof, select resistant	Severe outdoor use
NEMA 4	Water tight	Shipyards, dairies, hose test stations, etc.

Other NEMA classifications include 4X, water tight, corrosion resistant enclosures; class I or class II, hazardous locations; 12, industrial use; or 13, oil tight and dust tight.

Magnetic starters are wired into the control circuit, usually incorporated into the control wiring sequence. Some examples of motor starters are shown in Figure 12.2.

Noise and Vibration

All mechanical equipment vibrates and makes unwanted noise, which must be reduced to minimize discomfort to those who must work near the equipment. It is not possible to eliminate all noise; some installations actually call for a small amount of background noise to mask conversations and incidental sounds. However, noise and vibrations must be modified when they are objectionable.

In rare situations it is possible to place equipment in remote locations, avoiding the problem of noise entirely. In other situations, the equipment is placed on isolated footings or encased to control sound.

Vibration Isolation

The most common solution for excessive noise or vibrations is to examine the mass and frequency of the equipment and then to select appropriate noise and vibration control devices. For example, vibration absorbers absorb, or dampen, the noise or vibrations. Another common technique is to absorb sound with insulation.

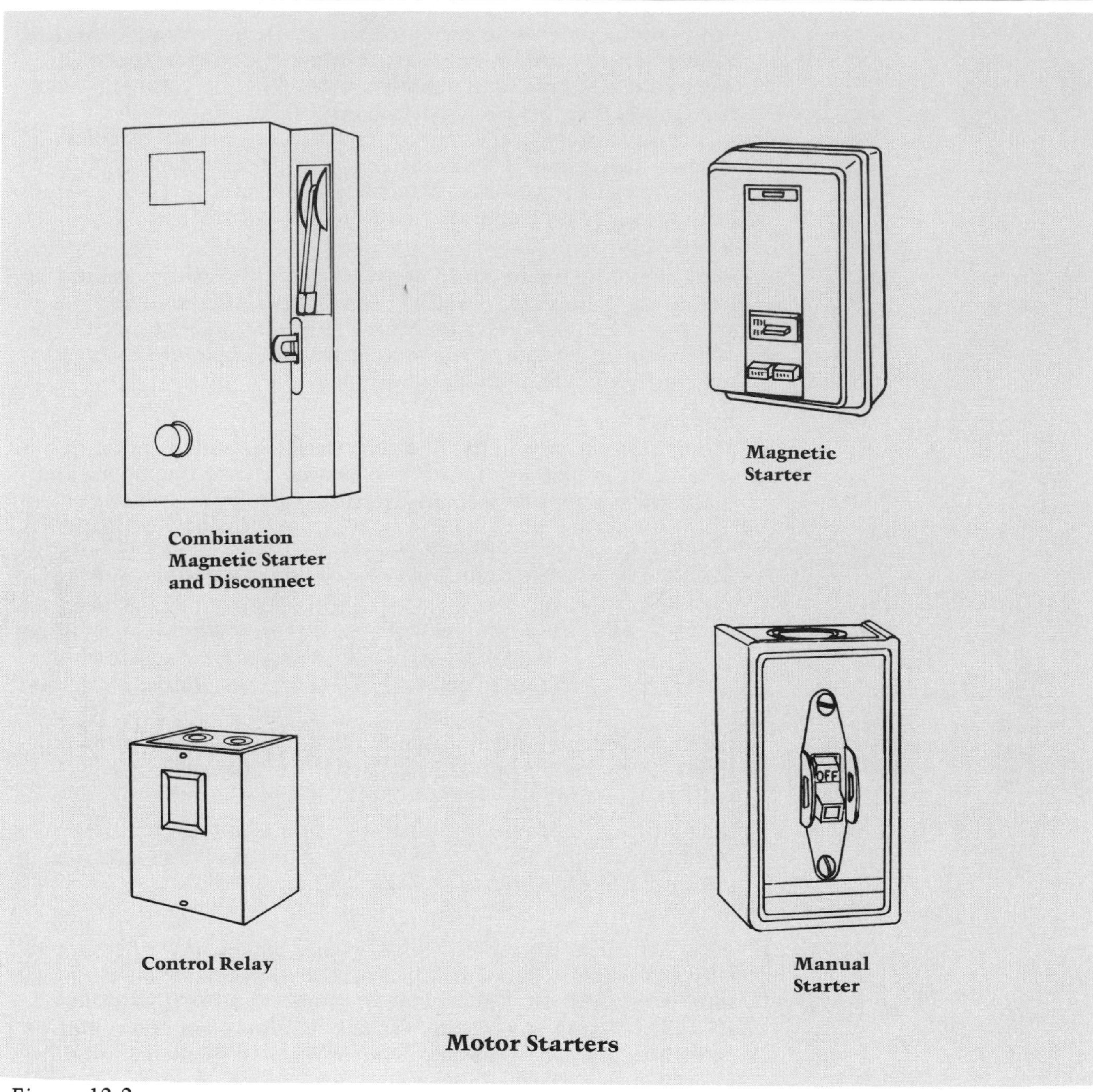
Combination
Magnetic Starter
and Disconnect
Magnetic
Starter
Control Relay
Manual
Starter
OFF
Motor Starters

Figure 12.2

A machine rigidly bolted to a structure will transmit all its vibration to the building. If a machine is mounted on springs or absorbent cushions (pads), its movements can be dampened considerably, effectively controlling sound.

A critical relationship exists between the frequency of the disturbing vibration (from a fan or pump) and the natural frequency of the mounting (vibration isolator). When properly selected, the natural frequency of the vibration isolator should be three to five times lower than that of the machine mounted. If it is less than three times lower than the force of the machine, very large forces may be transmitted to the structure; more than five has a negligible increase in overall effectiveness. Roughly ten percent of the forces on an isolator with one-third of the frequency of the machine and four percent of the forces on an isolator of one-fifth the frequency of the machine will be transmitted to the structure.

Example: Select a vibration isolator for a 1,000-pound piece of equipment operating at 600 rpm (the disturbing frequency) that develops a 50-pound force. 600 rpm = 600/60, or 10 hertz (Hz). The isolator that is selected should be at least 1/3 10 Hz, or 3.3 Hz. The isolator will transmit a maximum of 10 percent of the 50 pounds, or 5 pounds, to the structure, a negligible increase over 1,000 pounds. To determine the necessary vibration isolator for the selected 3.3 Hz, the spring constant for the isolator must be determined. The spring constant, k, is based on the mass of the equipment and is expressed in pounds per inch.

$$k = \frac{(\text{natural frequency of isolator (Hz)})^2 \text{ x equipment (lbs.)}}{9.8}$$

where the natural frequency of the isolator equals one-third to one-fifth of the disturbing frequency.

Therefore, for the above example,

$$k = \frac{(3.3)^2 \times (1{,}000)}{9.8} \text{ or 1,111 lbs./in. max.}$$

This is illustrated in Figure 12.3.

An important factor in vibration isolation is the need for *flexible connections*, because the equipment moves as it vibrates on the isolator pads or springs. The piping, ductwork, and even electrical connections must be isolated from these vibrations so that they will not be transmitted throughout the building or shake the pipes.

The natural frequency of the floor system should differ from the vibration isolator natural frequency in order to avoid catastrophic resonant reinforcement, which could initiate uncontrolled amplified vibrations. For this reason, the frequency should be verified with the structural designer.

Sound transmission in ductwork can be reduced by the use of insulation and elbows. An effective technique is to insulate the interior of ductwork and/or provide an elbow within ten feet of air handling equipment.

Vibration and noise control devices are shown in Figure 12.4.

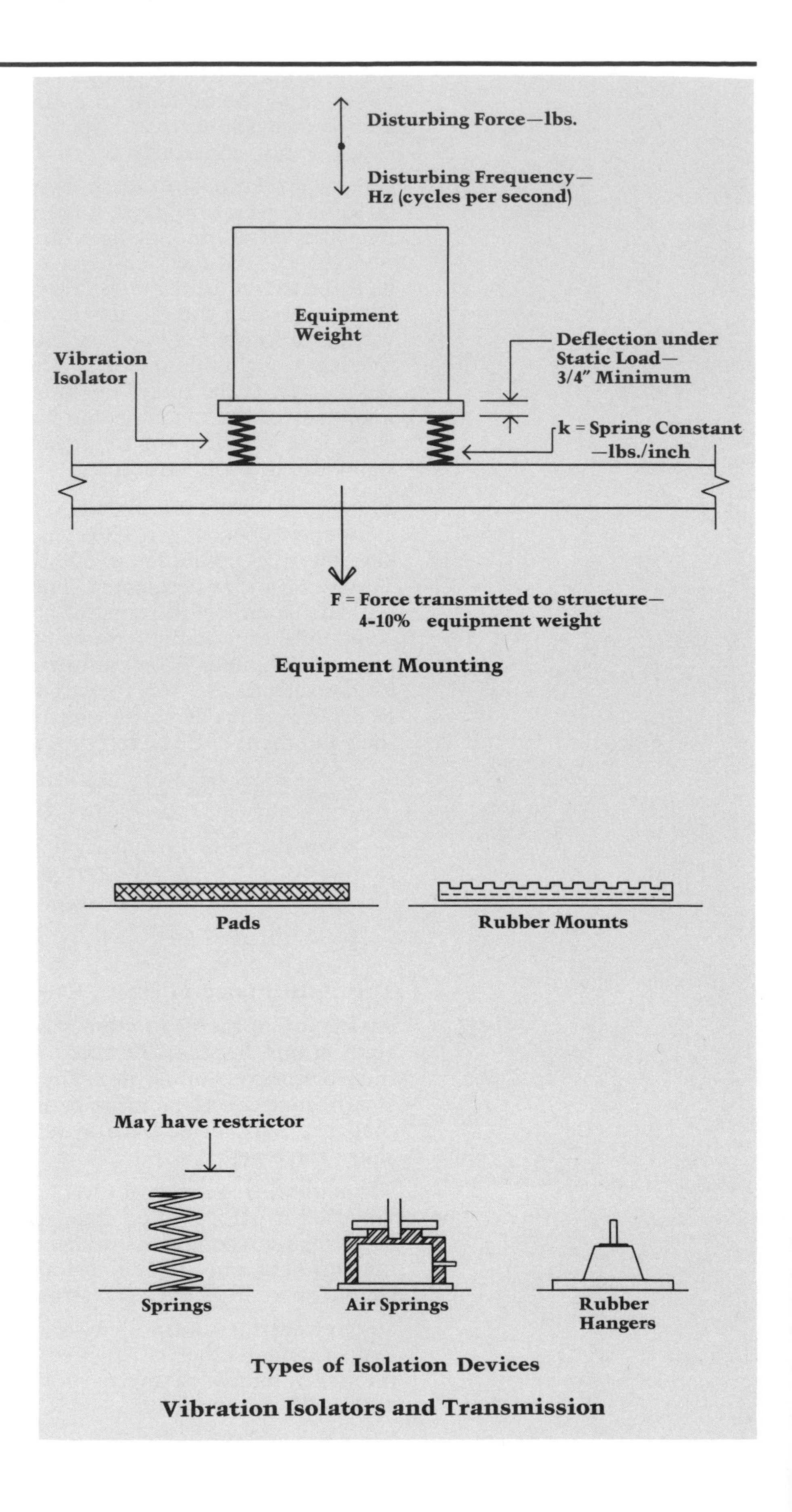
Disturbing Force—lbs.
Disturbing Frequency—
Hz (cycles per second)
Equipment
Weight
Vibration
Isolator
Deflection under
Static Load—
3/4" Minimum
k = Spring Constant
—lbs./inch
F = Force transmitted to structure—
4-10% equipment weight
Equipment Mounting
Pads
Rubber Mounts
May have restrictor
Springs
Air Springs
Rubber
Hangers
Types of Isolation Devices
Vibration Isolators and Transmission

Figure 12.3

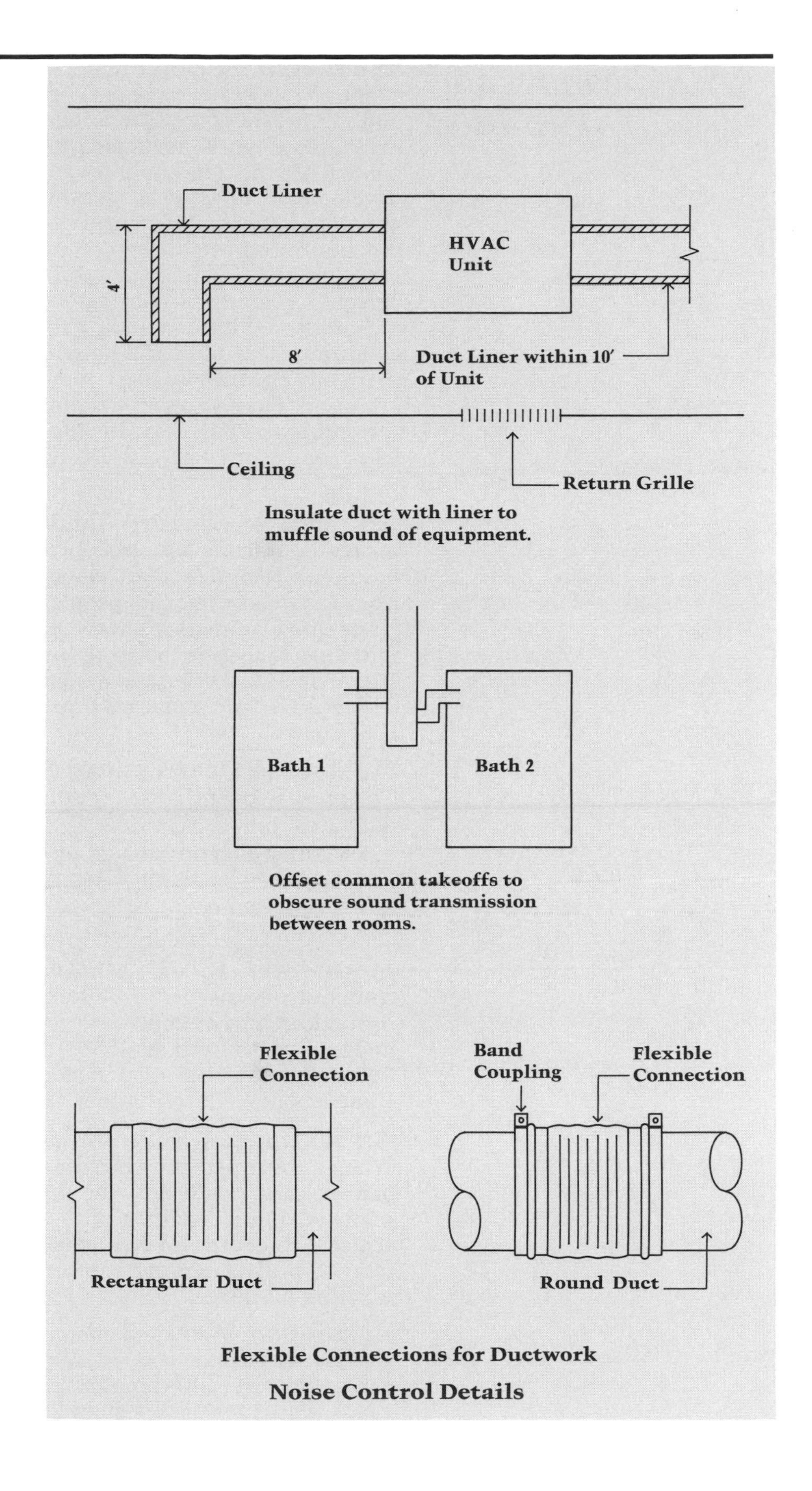
Duct Liner
HVAC Unit
4'
8'
Duct Liner within 10' of Unit
Ceiling
Return Grille
Insulate duct with liner to muffle sound of equipment.
Bath 1
Bath 2
Offset common takeoffs to obscure sound transmission between rooms.
Flexible Connection
Band Coupling
Flexible Connection
Rectangular Duct
Round Duct
Flexible Connections for Ductwork
Noise Control Details

Figure 12.4

Air Cleaning and Filtration

Besides being the proper temperature and humidity, the air supplied to a space must also be clean and free of odors and pollutants. There are several factors that have recently magnified the need for more attention to odor and pollution control. As modern buildings become tighter due to the development of better insulating and building techniques, odors are more likely to be trapped inside. Current energy codes and practice encourage lower amounts of outdoor air for ventilation. Five cfm per person is common today, whereas 15 cfm per person was used ten years ago, which means that there is less air introduced into a building to dilute any unpleasant condition. The outdoor air quality generally worsens each year; therefore, the introduction of outdoor air deserves careful attention. Finally, many buildings use 100 percent outdoor air in the spring and fall, dramatically increasing the need for air treatment, particularly in urban and industrial areas.

Odors

Most odors are generated within a building. These odors must be prevented from accumulating or being recirculated in the building's HVAC system. The odors controlled are primarily from tobacco smoke and people. Others include cooking fumes, paint spray, animals, bacteria, pollen, and cleaning chemicals. Buildings that have locker rooms, smoking rooms, bars, commercial kitchens, toilets, painting booths, animal shelters, dry cleaners, and laundries must have special provisions for odor control.

Methods of Odor Control

The most frequently used methods for controlling air quality include:

- Ventilation and exhaust (dilution or removal),
- Air washing (air cleaning, scrubbing),
- Adsorption (activated charcoal), and
- Odor masking (perfume).

Ventilation: For odor control, an exhaust system removes contaminated air from a building. Fresh outdoor air, when introduced into a supply air system, reduces odors by diluting inside air with fresh air. The American Society of Heating, Refrigeration, and Air Conditioning Engineers (ASHRAE) standards provides recommended air quantities for ventilation.

Air Washing: Water is the most common air washing fluid. When a fine mist spray is introduced into the air, odor-causing particles are absorbed by the liquid, collected in a pan, and removed. In process industries, chemicals other than water are selected to neutralize elements in the air. Another air washing method utilizes scrubbers, which are highly developed air washing machines.

Adsorption: Activated charcoal beds are the primary type of adsorption systems used to eliminate odors. The air (or gas) is passed over the adsorbing material. Because of the porosity and vast surface area of this granular material, odor-causing particles adhere to the surface. The beds of material, however, should be changed or regenerated periodically in order to work properly.

Odor Masking: The least desirable, but often the only available, cure for eliminating odors is by the use of "perfume." Introducing a more pleasant odor into the air to overcome an unpleasant one is often a manageable way to control odors.

Common methods of odor control are shown in Figure 12.5.

Pollutants

Automobile fumes, industrial pollutants, dust, and grime all exist to some extent in otherwise "fresh" outdoor air. These pollutants must be minimized, as they may contain sulfur oxide, carbon monoxide, hydrocarbons, nitrogen oxides, and exhausts from nearby buildings.

Methods of Pollution Control

Pollution control is often accomplished in conjunction with odor control. With pollution control, however, as the size of the particle to be removed decreases, the cost of the cleaning system increases. In addition to air washing, other common air cleaning devices are *filters* and *electronic cleaners*.

When selecting filters, it is necessary to know what is to be removed from the air. Bird screens filter birds and rodents; insect screens filter insects and smaller animals; and smaller screens control dust, grease, and specific particles. The smallest type of filter is an electronic filter, which removes molecules of an identified pollutant by ionization. Grease and tobacco smoke are usually removed by such filters. Examples of air filters are shown in Figure 12.6.

Filters are generally rated for their *efficiency* and *pressure drop*. The ratings are stated for a particular air quantity and velocity and must be converted for other design conditions. In addition, the performance of a filter varies dramatically over its life, decreasing as it becomes dirty. The efficiency rating is the percent of dust arrested by the filter. The pressure drop is the decrease in air static pressure that occurs as it passes through the filters.

Old filters should be periodically inspected and cleaned, as they may become clogged. If blockage impedes air flow, fans may become overloaded and a system failure could result.

Most filters are dry and fibrous. Some are available on continuous rolls, which continually feed new filter media to replace the old. *Dry filters* are commonly made from cellulose, glass fiber, or metal. Electronic filters (precipitators) are most effective on small particles, which are ionized and then collected on plates.

Hangers and Supports

Piping for mechanical installations must be supported by a wide variety of hangers and anchoring devices. Many different building components may be used for anchoring pipe hangers. The location of piping supports is also an important consideration. Some common locations for anchoring devices include the roof slab or floor slab, structural members, side walls, another pipe line, machinery, or building equipment (see Figure 12.7).

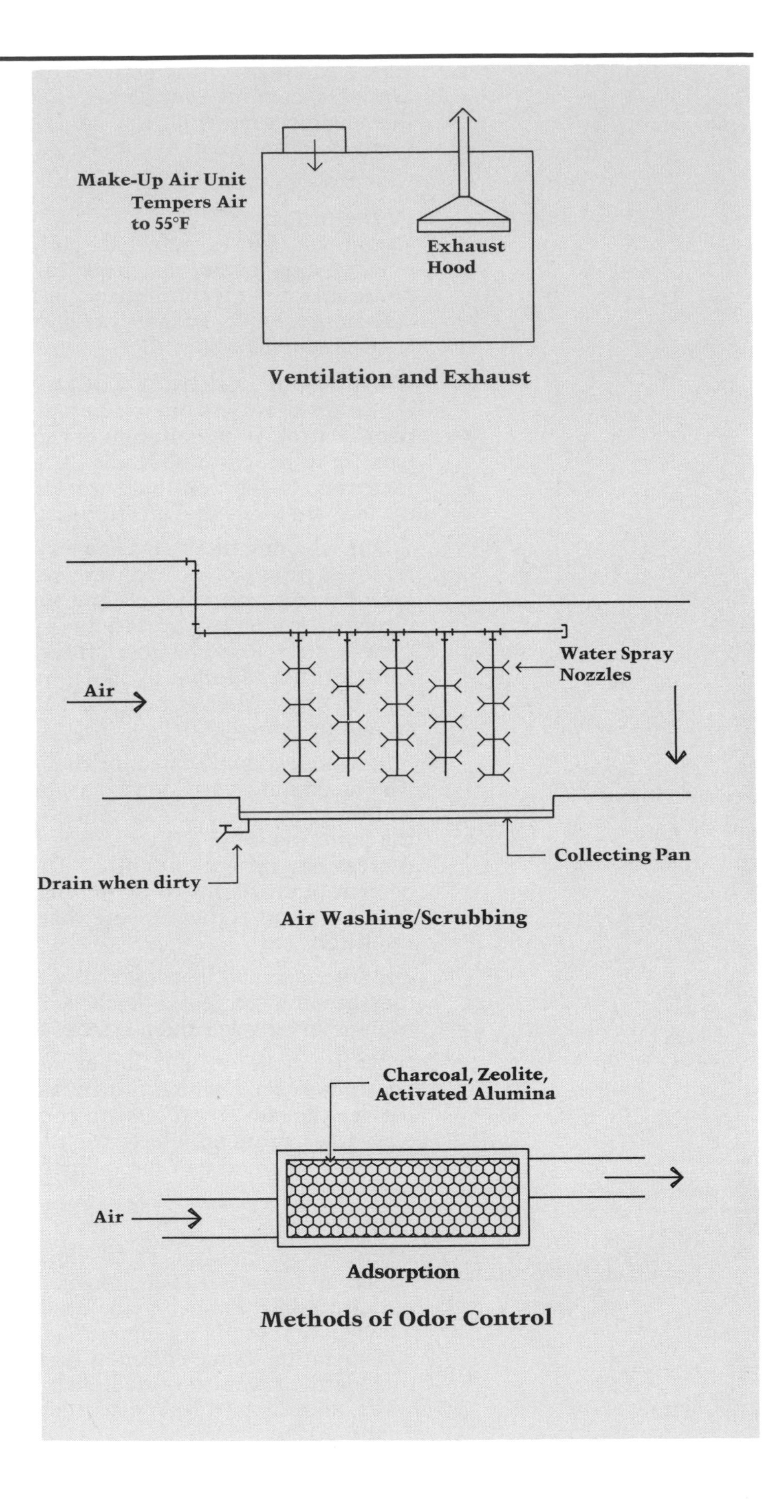
Make-Up Air Unit
Tempers Air
to 55°F
Exhaust
Hood
Ventilation and Exhaust
Water Spray
Nozzles
Air
Collecting Pan
Drain when dirty
Air Washing/Scrubbing
Charcoal, Zeolite,
Activated Alumina
Air
Adsorption
Methods of Odor Control

Figure 12.5

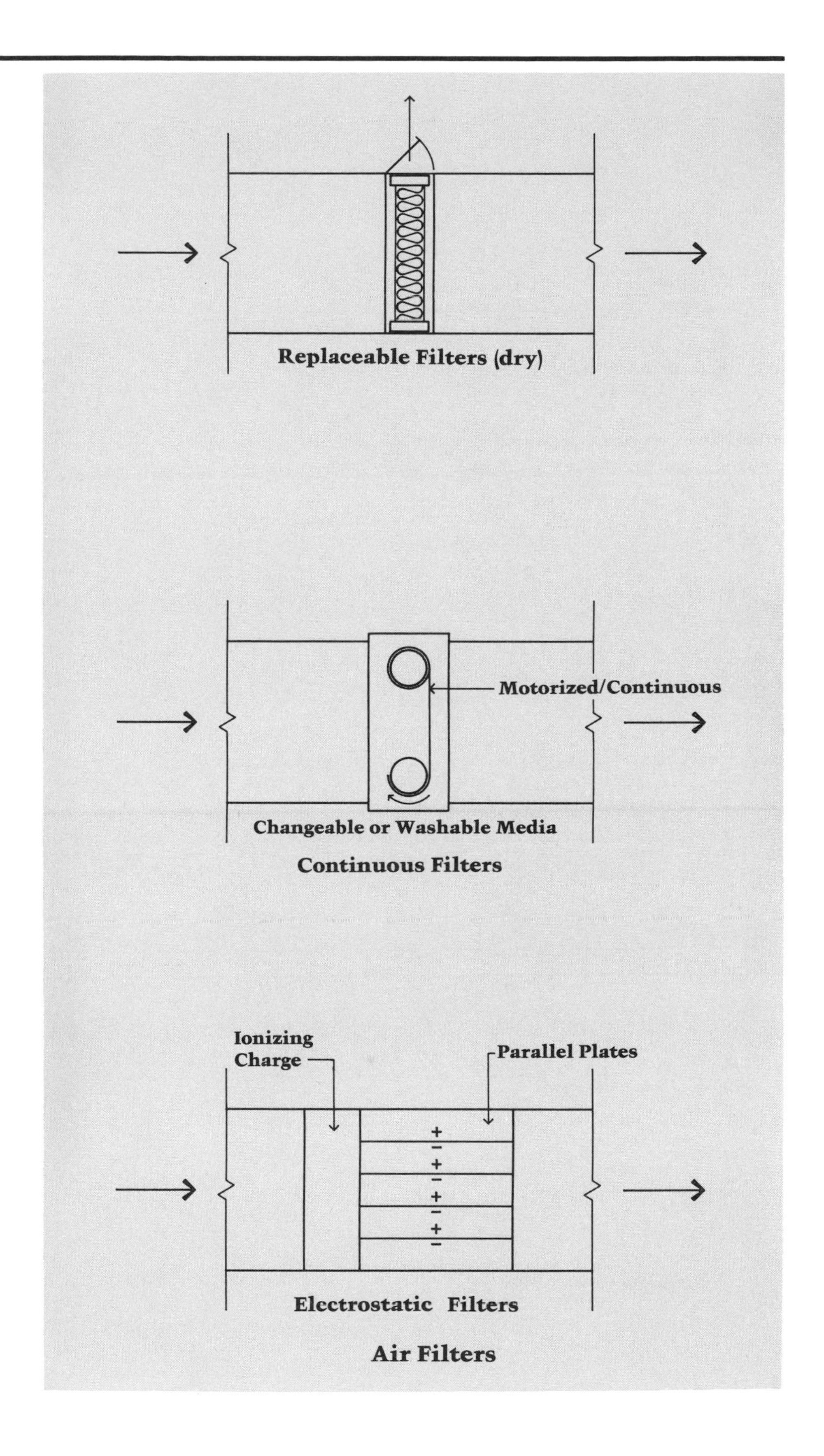
Replaceable Filters (dry)
Motorized/Continuous
Changeable or Washable Media
Continuous Filters
Ionizing Charge
Parallel Plates
+
−
+
−
+
−
+
−
Electrostatic Filters
Air Filters

Figure 12.6

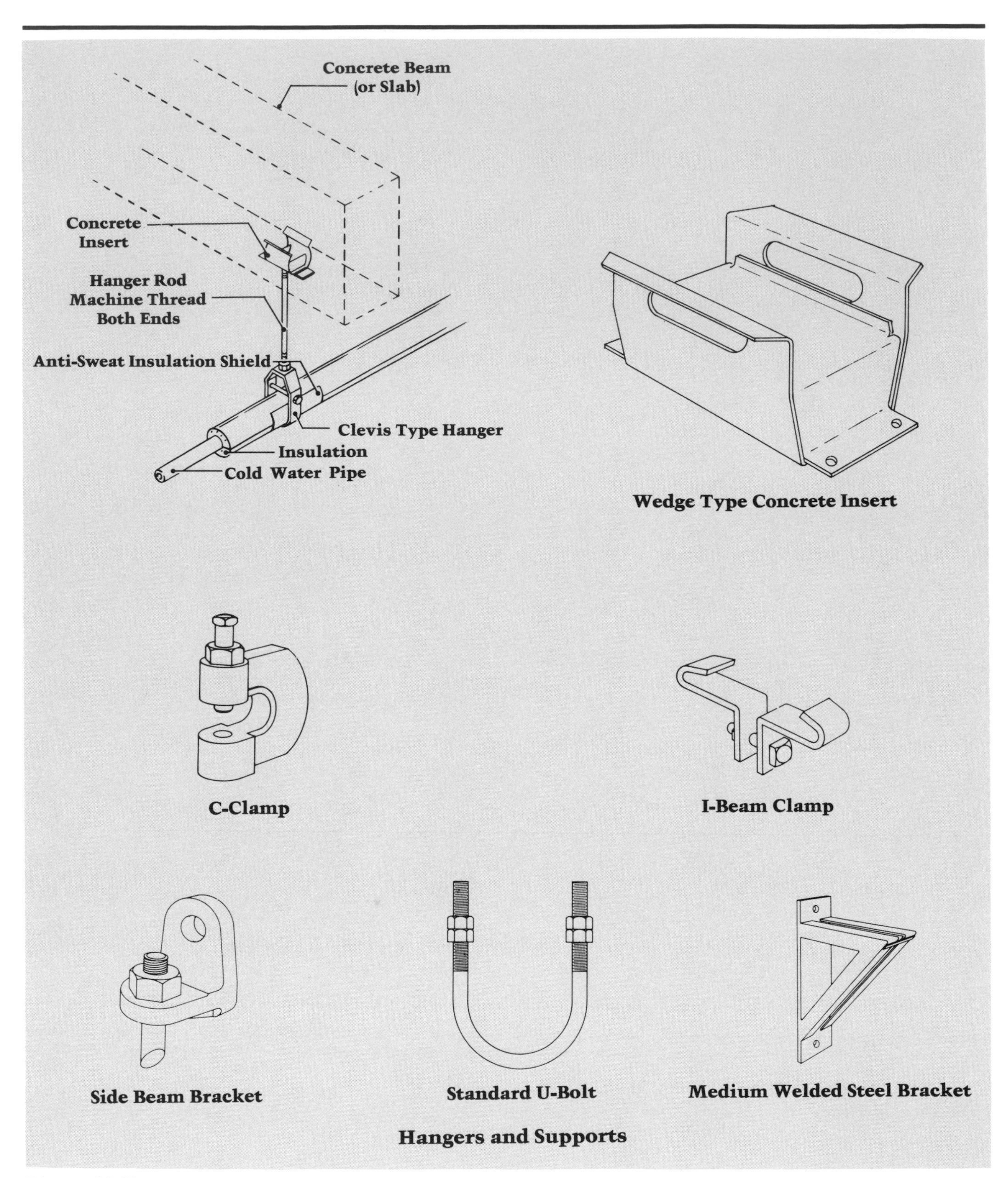
Concrete Beam
(or Slab)
Concrete
Insert
Hanger Rod
Machine Thread
Both Ends
Anti-Sweat Insulation Shield
Clevis Type Hanger
Insulation
Cold Water Pipe
Wedge Type Concrete Insert
C-Clamp
I-Beam Clamp
Side Beam Bracket
Standard U-Bolt
Medium Welded Steel Bracket
Hangers and Supports

Figure 12.7

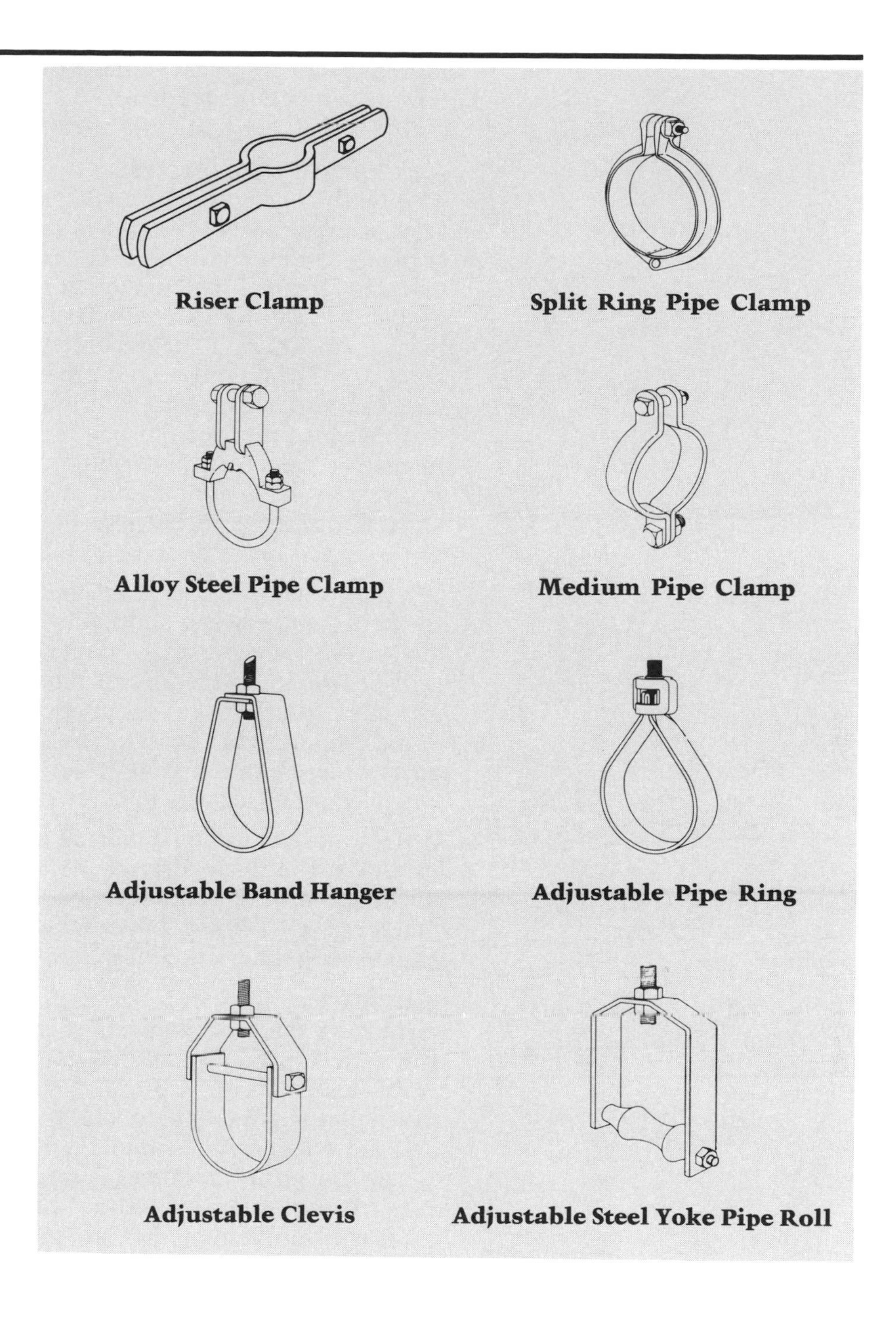

Figure 12.7 (cont.)

Pipe hangers are usually made out of black or galvanized steel. For appearance or in corrosive atmospheres, chrome or copper-plated steel, cast iron, or a variety of plastics may also be used.

Anchoring to Concrete

The method selected for anchoring the hanger depends on the type of material to which it is being secured. This is generally concrete, steel, or wood. If the roof or floor slab is constructed of concrete, formed, and placed at the site, concrete inserts may be nailed into the forms at the required locations, prior to the placing of concrete. These inserts may be manufactured from steel or malleable iron, and are either tapped to receive the hanger rod machine thread or contain a slot to receive an insert nut. Because the slotted type of insert requires separate insertable nuts for the various rod diameters, only one size of insert need be warehoused. For multiple side-by-side runs, long slotted insert channels in up to 10′ increments are available with several types of adjustable insert nuts.

When precast slabs are used, the inserts may be installed on site in the joints between slabs. They may be drilled or shot into the slab itself. Care must be taken to verify that no critical strands will be impaired. Electric or pneumatic drills and hammers are available to drill holes for anchors, shields, or expansion bolts. Another method of installing anchors on site utilizes a gunpowder actuated stud driver, which partially embeds a threaded stud into the concrete.

If the piping is to be supported from the sidewalls, similar methods of drilling, driving, or anchoring to those mentioned above are used for concrete walls. Where hollow-core masonry walls are to be fitted, holes may be drilled for toggle or expansion-type bolts or anchors.

Anchoring to Steel

Attaching the pipe supports to steel requires different methods from those used for concrete. When the piping is to be supported from the building's structural steel members, a wide variety of beam clamps, fish plates (rectangular steel washers), and welded attachments can be employed. If the piping is being run in areas where the structural steel is not located directly overhead, then intermediate steel is used to bridge the gap. This intermediate steel is usually erected at the piping contractor's expense. If the building is constructed of wood, then lag screws, drive screws, or nails are used to secure the support assembly.

From the anchoring device, a steel hanger rod, threaded on both ends, extends to receive the pipe hanger. One end of this rod is threaded into the anchoring device, and the other is fastened to the hanger itself by a washer and a nut. For cost effectiveness and convenience, continuous thread rod may be used. The pipe hanger itself may be a ring band, roll, or clamp, depending on the function and size of the piping being supported. Spring-type hangers are also used when it is necessary to cushion or isolate vibration.

Accommodating Expansion

Piping systems subject to thermal expansion must often absorb this expansion with the use of piping loops, bends, or manufactured expansion joints. The piping must be anchored to force the expansion movement back to the joint. In order to prevent distortion of the piping joint itself, alignment guides are installed. Figure 12.8 shows alignment guides and roll hangers used in these types of installations.

Specifically designed pipe hangers are used for fire protection piping with underwriters and factory mutual approvals.

Wood-Frame Construction

In wood-frame residential construction, holes drilled through joists or studs are often the sole support for certain pipes. These pipes may occasionally be reinforced with wedges cut from a two-by-four. A piece of wood, cut and nailed between two studs, is often used to support the plumbing stubouts to a fixture. Prepunched metal brackets are available to span studs for both 16″ and 24″ centers. These brackets can be used with both plastic or copper pipe and maintain supply stub spacing 4″, 6″, or 8″ on centers, giving perfect alignment and a secure time-saving support. Plastic support and alignment systems are also available for any plumbing fixture rough-in.

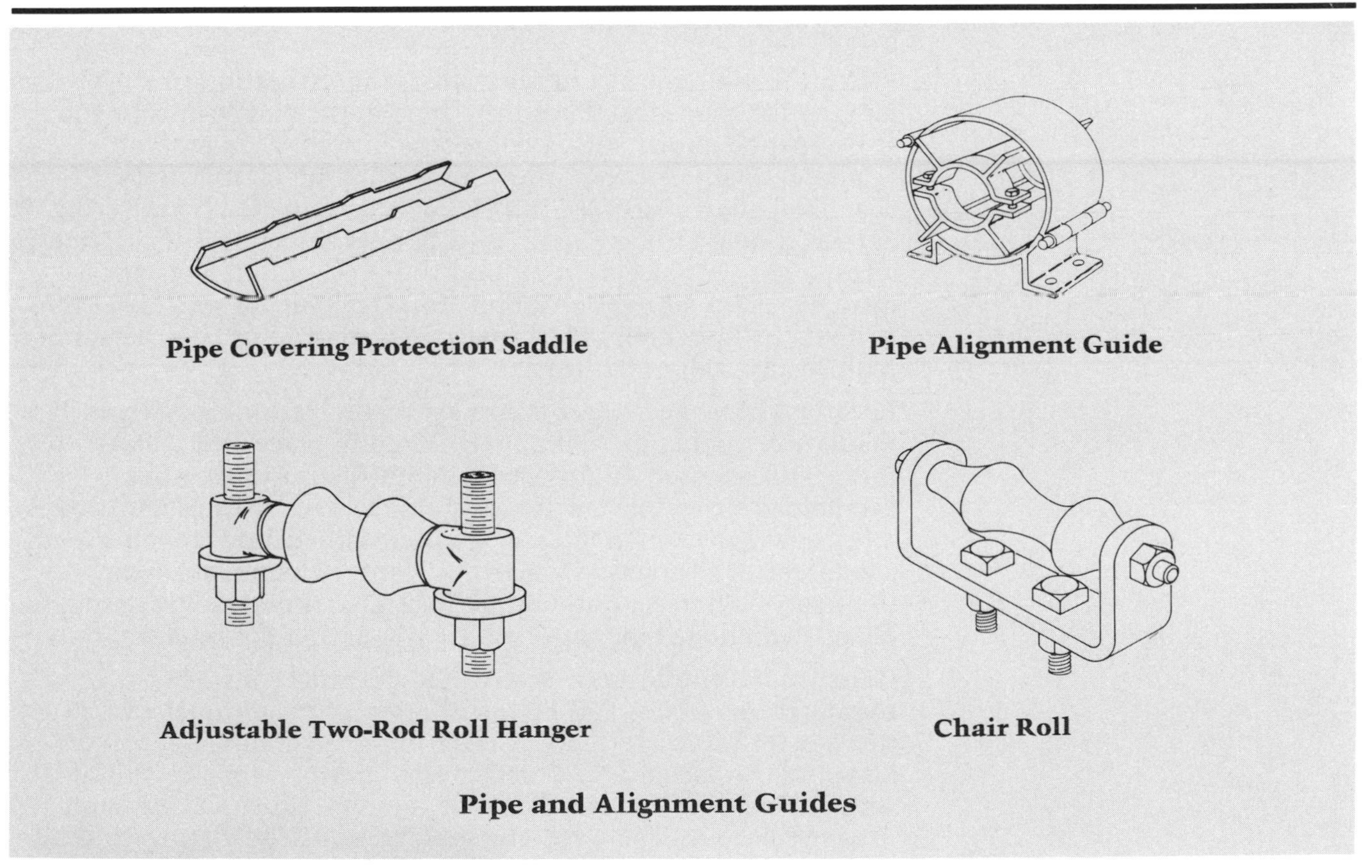

Figure 12.8

Insulation

Insulation is used in mechanical systems to prevent heat loss or gain and to provide a vapor barrier for piping and ductwork systems, boilers, tanks, chillers, heat exchangers, and air handling equipment casings. Most boilers, water heaters, chillers and air handling units are provided with insulated metal jackets and require little, if any, field insulation.

Insulation is manufactured from fiberglass, cellular glass, rock wool, polyurethane foam, closed cell polyurethane, flexible elastometric, rigid calcium silicate, phenolic foam, or rigid urethane, and is available in rigid or flexible form.

Insulation is produced in a variety of wall thicknesses and may be applied in layers for extreme temperatures. Rigid board or blocks, or flexible blanket insulation is used for ductwork and equipment. Perforated sections are available for use with pipe.

The standard length for pipe insulation has always been three feet. This is still true for fiberglass and calcium silicate, but foam insulations (e.g., polyurethane, polyethylene, and urethane) are produced in four-foot lengths, which means that less butt joints will have to be made in straight runs of pipe. Flexible elastometric insulation is shipped in six-foot lengths.

The simplest pipe covering installation method is slipping elastometric insulation over straight sections of pipe or tubing prior to pipe installation. This insulation can be formed around bends and elbows.

When the piping is already installed, the insulation must be slit lengthwise, placed, and both the butt joints and seams (at the slit) must be joined with a contact adhesive. This process requires more labor than the simple slipping-on method described above. Fittings, in this case, are covered by mitering or cutting a hole for a tee or a valve bonnet. Fiberglass insulation is already slit for placement around the pipe, and fittings are mitered. From straight sections, holes are cut for tees and valve bonnets using a knife. Elastometric insulation usually does not require any additional finish.

A variety of jackets and fittings are available for all types of insulation, including roofing felt wired in place and preformed metal jackets used as closures where exposed to weather. Premolded fitting covers are available to give fittings and valves a finished appearance. Factory-applied self-sealing jackets are an advantage of fiberglass insulation. Flanges, which are removed frequently, often require unique insulated metal jackets or boxes, which are fabricated by the insulation contractor.

Calcium silicate insulation is often specified for higher temperatures (above 850°F). Installation of calcium silicate is more labor intensive than installation of fiberglass or elastometric insulation. This rigid insulation is made in half sections for pipe and in three foot lengths. Multiple segments must be used in layers to achieve larger outside diameters. The segments are wired in place nine inches on center, normally requiring two workers. Fittings for calcium silicate insulation are made by mitering sections using a saw. Flanges and valves require oversized sections cut to fit. These valve and fitting covers are wired in place and finished with a troweled coat of insulating cement. This type of insulation has a high waste factor due to crumbling and breaking during cutting and ordinary use.

Special thickness of any insulation is obtained by adding multiple layers of oversized insulation, and sealing and securing using the same method used for the base layer. This additional work should be included in the labor estimate.

Hangers and supports require special treatment for insulated piping systems. The pipe hanger or support must be placed outside the insulation to allow for expansion and contraction in steam or hot water systems and to maintain the vapor barrier in cold water systems. As a result, oversized supports should be used to fit the outer diameter of the insulation, rather than the pipe.

For heated piping, preformed steel segments are welded to the bottom of the pipe at each point of support. These saddles are sized according to the insulation thickness in addition to the pipe's outer diameter. For cold water (anti-sweat) systems, where no metal-to-metal contact between pipe and support can be tolerated, the hanger is oversized to allow for placement of the insulation. The insulation is then protected by a sheet metal shield which matches the outer radius of the insulation.

For flexible insulation that cannot support the weight of the pipe, a rigid insert is substituted for the lower section of insulation at each point of support. The insert may be formed from calcium silicate insulation, cork, or even wood. The insulation jacket must enclose this insert within the vapor-proof envelope. The labor for insulation around hangers should be estimated carefully. The insert, for example, might be installed by the pipe coverer, while the metal shield or protector is furnished by the pipefitter or plumber.

Duct and equipment insulation may be a wrap-around blanket type insulation, wired in place and sealed with adhesive or tape. Rigid board is secured by the use of pins, secured to the duct exterior with mastic or spot welded into place. The insulation sheets are pressed onto these pins and secured with self-locking washers.

The butt joints and seams are sealed with adhesive and tape. Block and segment insulation, when installed on round or irregular shapes, is wired on and then covered with a chicken-wire mesh and coated with a troweled application of insulation cement.

Procedures for installing and finishing insulation will vary. Examples of insulation are illustrated in Figure 12.9.

Equipment Data Sheets

The equipment data sheets that follow can be used in the selection of accessories. A Means line number, where applicable, is provided for the component listed. Costs can be determined by looking up the line number in the current edition of *Means Mechanical Cost Data.*

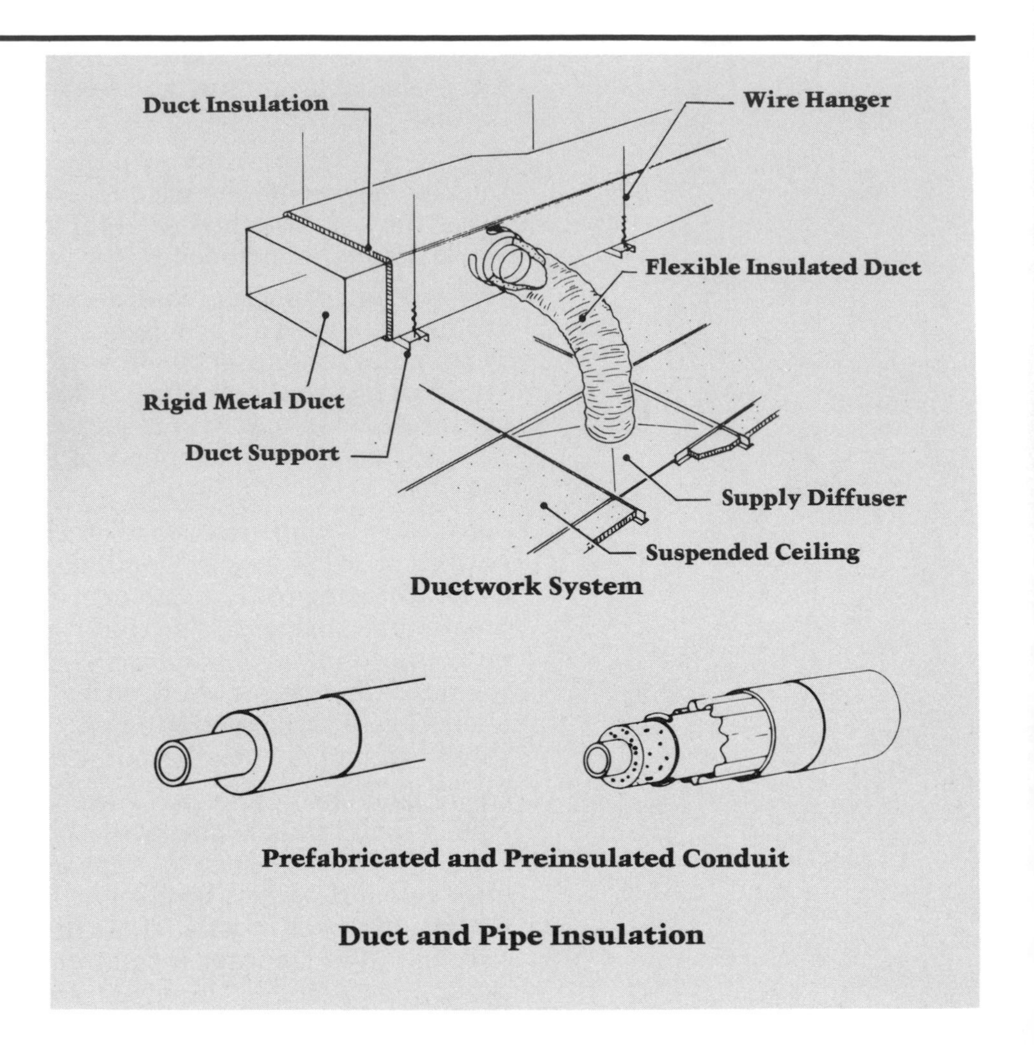
Duct Insulation
Wire Hanger
Flexible Insulated Duct
Rigid Metal Duct
Duct Support
Supply Diffuser
Suspended Ceiling
Ductwork System
Prefabricated and Preinsulated Conduit
Duct and Pipe Insulation

Figure 12.9

Accessory Equipment Data Sheets		
Accessory Equipment	**Chimneys**	**Means No. 155-680**
Useful Life: 10-20 years	**Typical Use and Consideration**	**Size Range (Diameter) 3"-48"**
Prefabricated Metal Chimney 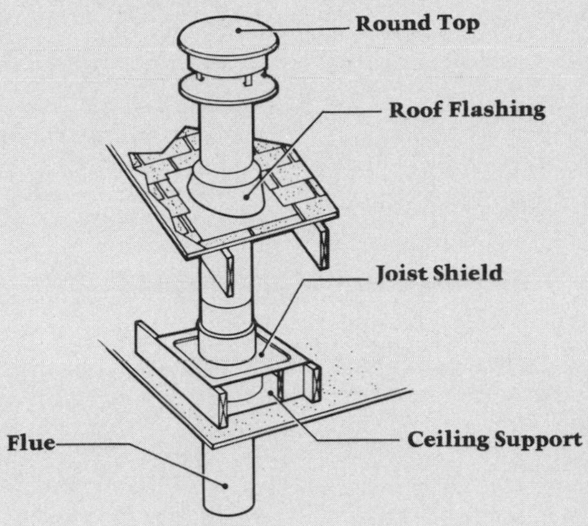 Round Top Roof Flashing Joist Shield Flue Ceiling Support	Metal vent chimneys are available in single wall and double wall type B construction and in galvanized steel, type 304 stainless steel liner with aluminized steel outer jacket (all fuels), and with 316 stainless steel liner (more durable) and 11 gauge galvanized jackets (2,000°F fuels—acid resisting with refractory lining).	
Masonry Chimney 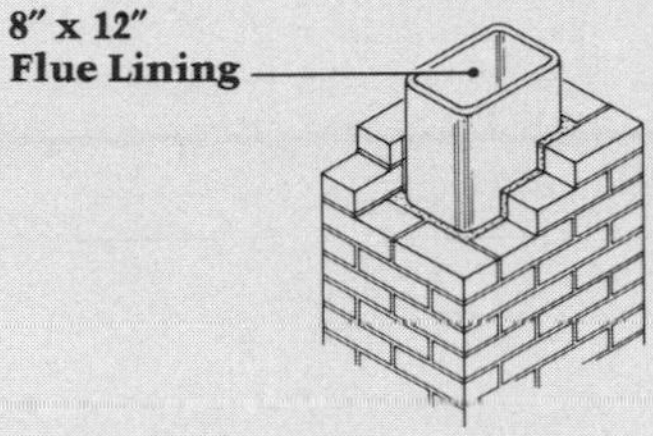 8" x 12" Flue Lining	Masonry flues and chimneys have the durability and adaptability for most fuels. The weight of such chimneys requires a substantial foundation.	
	Select proper flue material to match fuel.	

$$D^2 = \frac{HMT_f}{26(P_{atm})V}$$

D = Flue inside diameter - inches
H = Heat input (MBH)
M = lbs. mass gas in flue per 1,000 Btu/hr.
T_f = Average flue gas temperature (degree F_{abs})
$P_{(atm)}$ = Atmospheric pressure (in./H_s)—approx. 29.92
V = Velocity of flue gas (feet/sec.)—assume 20

Figure 12.10

Accessory Equipment Data Sheets (continued)

Accessory Equipment		Vibration Isolators	Means No. 157-485
Useful Life: 30+ Years	**Type**	**Typical Uses**	**Capacity Range: 10-30,000 lbs.**
Glass Fiber Pad	1	Advantages: Durable, provides damping under pads and slabs. Disadvantages: Concrete construction on top of pad.	1-4″ thick, 500 psi
Rubber Mounts and Hangers	2	Put under pipe supports or for suspended equipment. Disadvantages: Limited load capability.	0.5″ maximum deflection .25″ recommended
Spring Hangers and Isolators	3	Suspended equipment and pad or base mounted durable. Most common type. Disadvantages: Provide corrosion protection if outside.	Wide range capacities
Restrained Spring Isolator	4	Advantages: Hold down bolts or clips limit deflection when load of water or refrigerant is removed. Disadvantages: Most expensive. Horizontal restraint also important.	Use for large boilers, chillers, cooling towers.
Thrust Restraint	5	Used with HV duct systems to dampen vibrations and thrusts of air pressure, similar to spring hanger.	
Air Springs and Bellows		Large deflections and filtering of high frequency noise. Special applications. Disadvantages: Constant dry air supply required to maintain preset pressure.	6″ deflections

Consult manufacturer for recommendations of type of isolator and general guidelines. Check equipment weight and structural span for additional criteria. Recommended deflection varies from 1/4″ to 2.5″.

Type	Common Use
1	Under pads
2	Hung or mounted small equipment, pipes, terminal units
3	Compressors, pumps, fans, air handling equipment, package units
4	Chillers, condensers, cooling towers
5	Variable high velocity ductwork

Provide proper spacing and locations

$$\text{Isolator Frequency Hz} = \frac{\text{RPM of Machine}}{60 \times [3 \text{ to } 5]} \qquad \text{Spring Constant k (lb./in.)} = \frac{[\text{isolator frequency Hz}] \times \text{Equipment/(lbs.)}}{9.8}$$

Figure 12.10 (cont.)

Accessory Equipment Data Sheets (continued)			
Accessory Equipment		**Vibration Bases**	**Means No. 157-485**
Useful Life:	**Type**	**Typical Uses—Under Vibration Isolators Connecting to Structure**	**Common Use**
Direct Isolation Equipment Pad	A	Advantages: Rigid equipment with no additional support.	Chillers, tank compressors, cooling towers, air handlers, fans, package units
Structural Bases	B	Advantages: Transfers equipment concentrated loads to structural load points of building.	Fans greater than 300 rpm and 0-50 HP
Concrete Bases on Pad	C	Advantages: Provides needed mass to dampen impact on structure. Disadvantages: Expensive. Reinforced concrete.	Reciprocating compressors, fans greater than 300 rpm and greater than 50 HP
Curb	D	Advantages: Limits deflection to 1″ and provides narrow water-tight seal for equipment with curbs.	Special applications

Accessory Equipment	**Insulation**	**Means No. 155-651**
Useful Life: 20+ years	**Typical Uses—All Generation and Distribution Equipment**	**Size Range (Diameter) 1/2″-24″**
Equipment	Calcium silicate 1″-3″	
Ductwork—Blanket Type	Fiberglass liner 1/2″-2″ thick (1½-2 lb. density)	
Vinyl FRK wrap	1″-3″ thick	
Rigid board	1″-2″ thick—fire resistant or FRK barrier	
Pipe Covering		
Calcium silicate	1″-3″ thick—aluminum jacket available	
Rubber tubing	Closed cell foam 3/8″-3/4″ thick	
Urethane—ASJ cover	1″-1½″ thick—most common— widest temperature range (-60°-225°F)	

Special considerations:

Insulation is most effective if continuous. Make hangers large enough to accommodate insulation. Insulation protection saddles must be provided to prevent crushing the insulation.

Figure 12.10 (cont.)

Accessory Equipment Data Sheets (continued)			
Accessory Equipment	**Air Filters**		**Means No. 157-401**
Useful Life: 1-3 months	**Typical Uses—Odor and Dust Control**		**Capacity Range: .001-1,000 microns**
	Advantages	**Disadvantages**	**Remarks**
Activated Charcoal	Excellent for overall cleaning.	Expensive to install and recharge this system.	
Washers/Scrubbers	Most effective overall.	Grease and oil, requires additional filter.	Sludge must be removed.
Electronic Air Cleaners	Very effective for smaller particles. Low pressure drop.	Expensive. Additional filtration for large particles may be required.	Laboratories
Permanent Washable Air Filters.	System can be designed for specific filtration requirements.	Maintenance costs required. Reliability may be a problem.	
Renewable Roll Air Filters	Simple, maintains low pressure.	Maintenance costs required.	
Disposable Fiberglass Air Filters	Simple, light.	Air resistance increases with use.	Most common for small to medium sized systems.

Special Considerations:
Particle size to be removed should be known. Most systems should be able to handle particle sizes from 1 to 5 microns (1 micron = 0.001 millimeters) plus filter out larger particles such as smoke and dust.

The type of filter selected is heavily dependent on the maintenance regimen for the building. Sludge removal, cleaning of electronic filters, vacuuming of bag filters, and recharging charcoal beds requires an active, knowledgeable crew.

Accessory Equipment	**Motor Starters**	**Means No. 163-320**
Useful Life: 15-25 years	**Typical Uses—Control Electrical Devices**	**Capacity Range: Nema 1 through Nema 7**
Control Stations (START, STOP)	Used to remotely control operation of HVAC equipment from the work space.	
Motor Control Center (Pilot Lights, Starters, Push Buttons)	HVAC equipment that is controlled by automatic controls will commonly use automatic magnetic starters. Motor starters are specified by voltage, horsepower, phase, and current rating. Starters should ideally be in the view of the equipment operator.	**Means No. 163-130** **Capacity Range: 1/3-400 HP** **Size: 00 through 6**

Figure 12.10 (cont.)

Accessory Equipment Data Sheets (continued)		
Accessory Equipment	**Pipe Hangers and Supports**	**Means No. 151-900**
Useful Life: 25-40 years	**Typical Uses—Support Pipes**	**Size Range (Diameter) 1/4″-8″**
Brackets	For wall mounting	
Clamps	For attachment to steel beams	
Rings	Size to pipe and insulation. Provide plate for insulation	
Rods	1/4″-7/8″ diameter	
Rolls	Provide where expansion required—large temperature swings for pipe	
U hooks	Direct attachment—no rods	

Support Spacing for Metal Pipe or Tubing

Nominal Pipe Pipe Size (Inches)	Span Water (Feet)	Steam, Gas, Air (Feet)	Rod Size
1	7	9	↑
1½	9	12	⅜″
2	10	13	↓
2½	11	14	↑
3	12	15	½″
3½	13	16	↓
4	14	17	↑
5	16	19	⅝″ ↓
6	17	21	¾″
8	19	24	↑
10	20	26	⅞″
12	23	30	↓
14	25	32	↑
16	27	35	
18	28	37	1″
20	30	39	↓
24	32	42	↑
30	33	44	1¼″ ↓

Figure 12.10 (cont.)

PART THREE

SAMPLE ESTIMATES

Part III contains two sample projects, a residential building requiring heating only and a commercial building requiring some heating and year-round cooling, to illustrate HVAC system design and selection. These last two chapters of the book illustrate the design selection process from generation to distribution to termination, and produce complete projects from start to finish.

CHAPTER THIRTEEN

MULTI-FAMILY HOUSING MODEL

In this chapter, an HVAC system is designed for a low-rise multi-family building, using the principles discussed in the previous chapters. The sample building is a four-story, multi-family apartment building with eight units on each upper floor and six units on the first floor, in Boston, Massachusetts. It has a brick facade with wood casement windows. The windows have both operable and fixed glass sections. The plans for the building are illustrated in Figures 13.1 through 13.3.

Each of the steps for designing an HVAC system is discussed in detail in the sections that follow. References to design tables and equipment data sheets; examples of calculations for heating and cooling loads, boiler output, and tank and pipe sizing; and a system summary sheet guide the reader through the design process.

The four basic selections to make in order to establish the type of HVAC system for a building are listed below.

1. Make the initial selections (Chapters 4 and 5).
2. Determine the design criteria (Chapters 2 and 3).
3. Compute the loads (Chapter 2).
4. Select the equipment (Chapters 5 through 12).

Initial Selection

As described in Chapters 4 and 5, the initial selections to make are: (1) the type of system, (2) the type of fuel, (3) the type and make-up of the distribution system, and (4) the type of generation system.

Type of System

The types of heating and cooling systems for buildings are listed in Figure 4.2; a forced hot water system with separate window or through-the-wall air conditioners is recommended for multi-family residences. Combined systems utilizing water source heat pumps are also common. In this example, it is assumed that the building does not require cooling as part of the base system, and that the electrical designer and architect will provide for future through-the-wall air conditioners supplied by

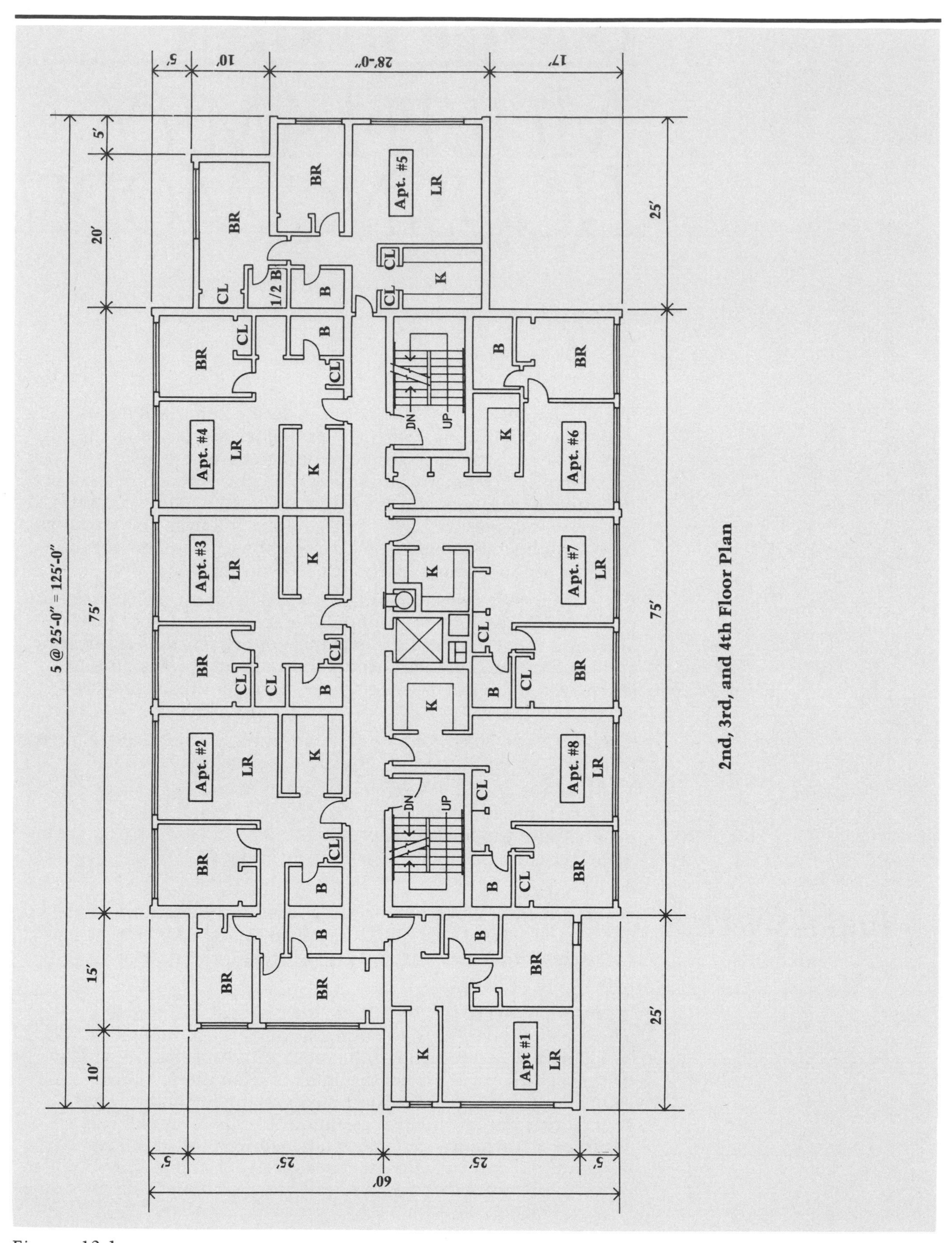
2nd, 3rd, and 4th Floor Plan
5 @ 25'-0" = 125'-0"
Apt #1
Apt. #2
Apt. #3
Apt. #4
Apt. #5
Apt. #6
Apt. #7
Apt. #8
LR
BR
K
B
1/2 B
CL
UP
DN
75'
25'
20'
15'
10'
5'
60'
28'-0"
17'

Figure 13.1

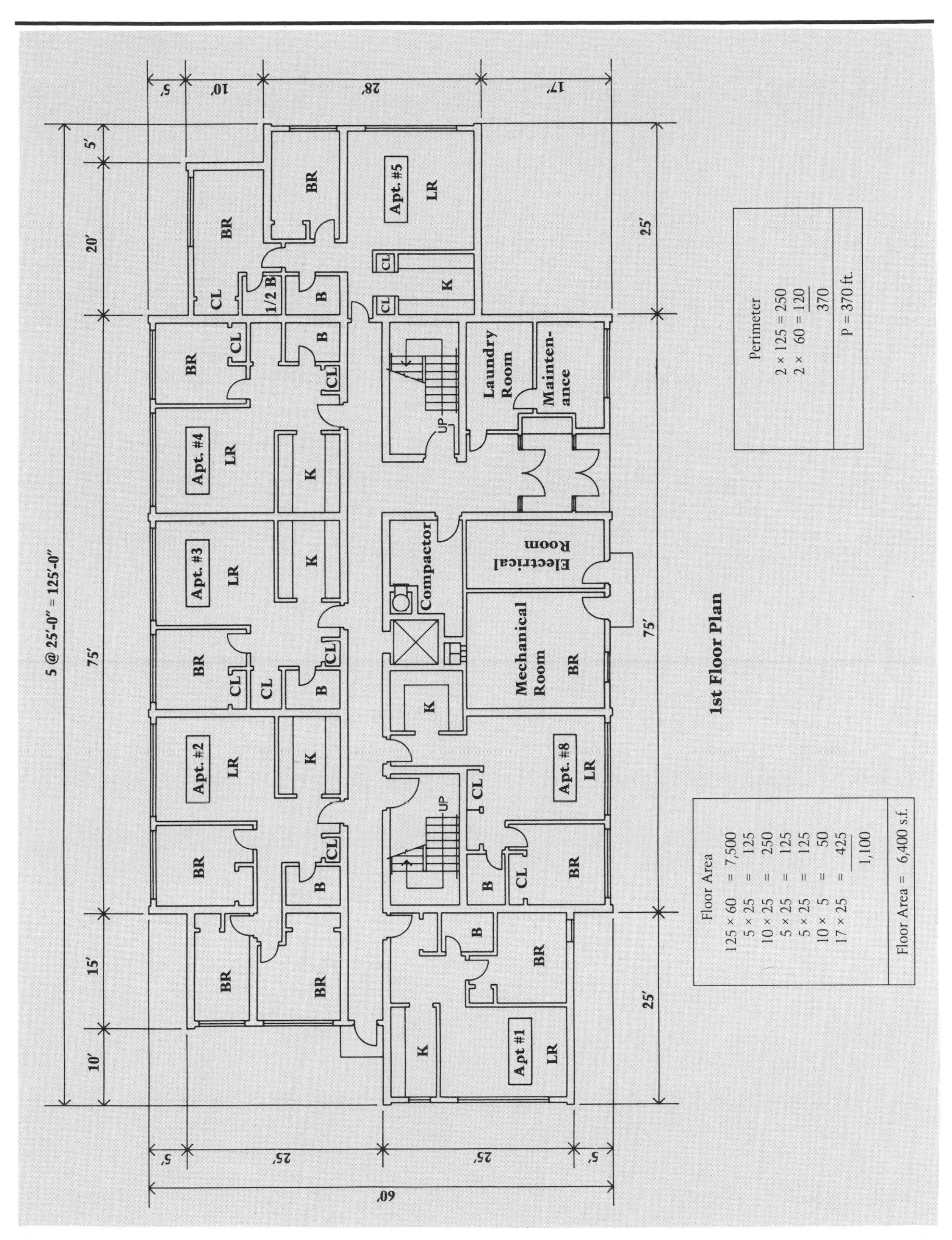

5 @ 25'-0" = 125'-0"
1st Floor Plan
Apt #1
Apt. #2
Apt. #3
Apt. #4
Apt. #5
Apt. #8
LR
BR
K
B
CL
1/2 B
Compactor
Electrical Room
Mechanical Room
Laundry Room
Mainten-ance
UP
Floor Area
125 × 60 = 7,500
5 × 25 = 125
10 × 25 = 250
5 × 25 = 125
5 × 25 = 125
10 × 5 = 50
17 × 25 = 425
1,100
Floor Area = 6,400 s.f.
Perimeter
2 × 125 = 250
2 × 60 = 120
370
P = 370 ft.

Figure 13.1a

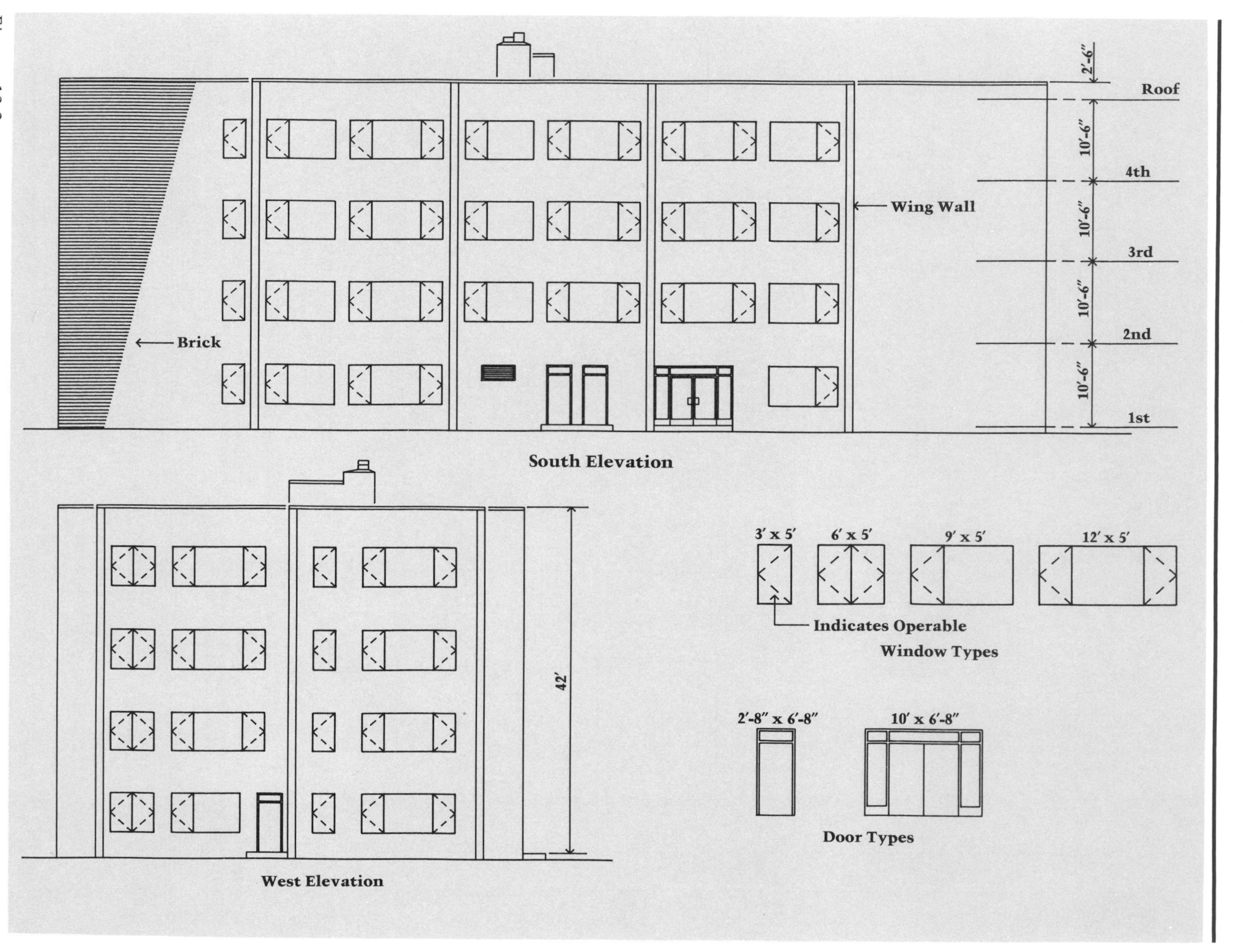
2'-6"
Roof
10'-6"
4th
Wing Wall
10'-6"
3rd
10'-6"
2nd
Brick
10'-6"
1st
South Elevation
42'
West Elevation
3' x 5'
6' x 5'
9' x 5'
12' x 5'
Indicates Operable
Window Types
2'-8" x 6'-8"
10' x 6'-8"
Door Types

Figure 13.2

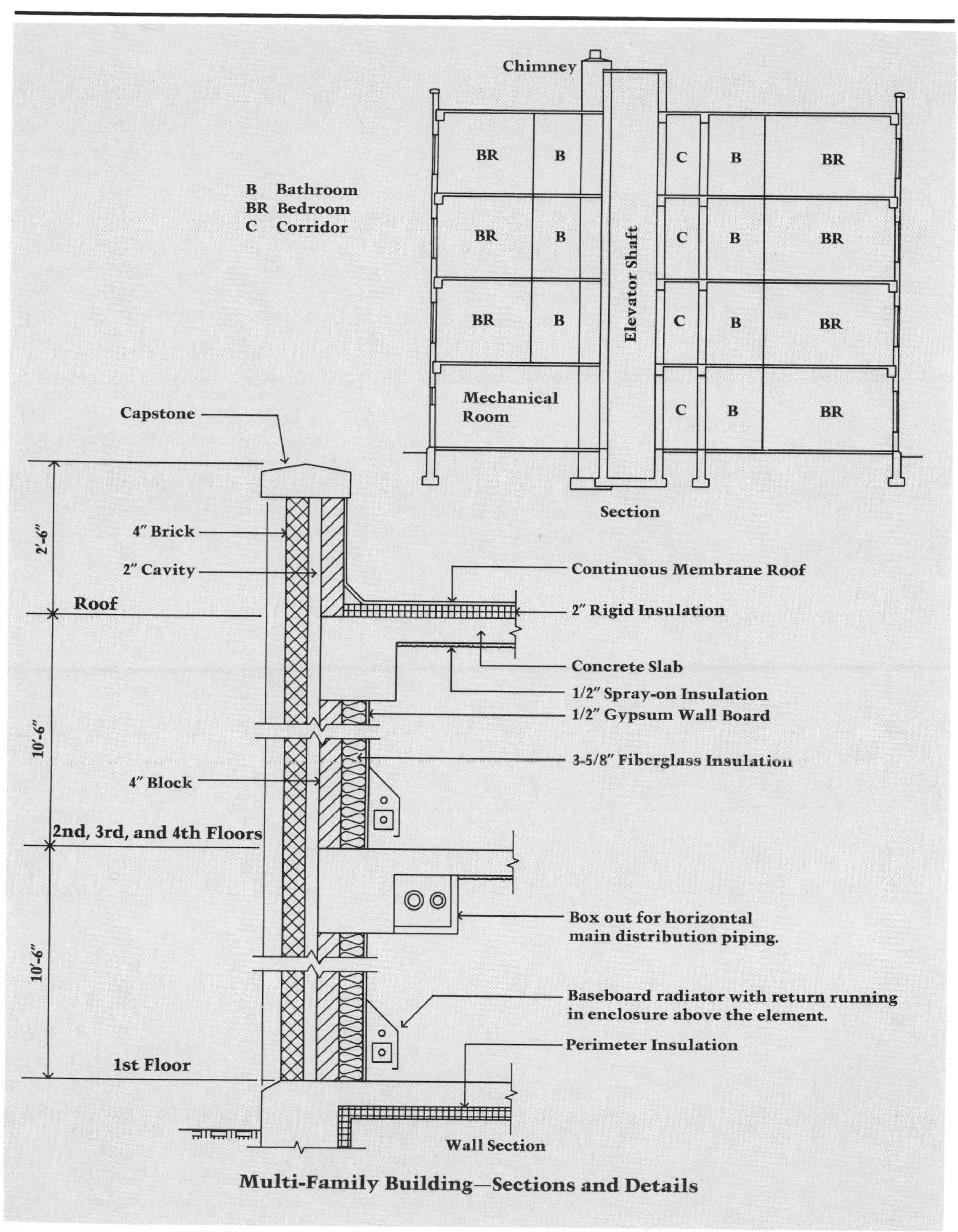

Multi-Family Building—Sections and Details

Figure 13.3

the tenants. Therefore, only heating by hot water will be provided. Thus, the type of system chosen is a separate hot water heating system.

Fuel

A new system should have high fuel efficiencies. Figure 4.3 lists basic properties and costs of fuel and indicates that #2 oil is the most economical. Therefore, #2 oil is selected as the choice of fuel. (Fuel efficiencies vary; a different fuel could be the best choice for a particular geographic location.)

Distribution System

In Figure 4.4, a hot water distribution system is recommended for continuously used buildings. Therefore, the distribution system selected for this building is a hot water distribution system.

Generation System

Hot water is generated by a steam boiler, a high-temperature water boiler, or a hot water boiler (see Figure 1.6). In this example, direct steam from a utility company is not available. Any system other than a hot water boiler also requires a heat exchanger, which adds to the overall cost of the system because of the relatively equal cost of steam and hot water boilers. Therefore, a hot water boiler is selected, as it is the most economical (see Figure 13.4).

Design Criteria

Once the initial selections have been made — a separate hot water system, #2 fuel oil, hot water distribution, and a hot water boiler — the design criteria must be determined. Design criteria are the indoor and outdoor conditions of temperature, humidity, and ventilation around which the building will be designed. The design criteria for this project are listed in Figure 13.5.

Design criteria are used to compute the heating and cooling loads. In all cases, check such criteria against local codes, manufacturers' literature, and overall architectural requirements.

Computation of Loads

The next step is to compute the total heating load (H_t). The total heating load for a building is the sum of all conduction and convection heat losses (see Figure 2.14).

$$H_t = H_c + H_e + H_v + H_s$$

$$H_c = UA\Delta T$$

$$H_e = FP\Delta T$$

$$H_v = 1.1\ \text{cfm}\Delta T$$

$$H_s = KA$$

Conduction Heat Losses

In order to compute the conduction heat losses (H_c, H_e, and H_s) through walls, floors, doors, windows, and slabs, first establish the U values and related coefficients. (Refer to Chapter 2, "Heating and Cooling Loads," for more information on the computation of U values.) The U values for this model are listed in Figure 13.6.

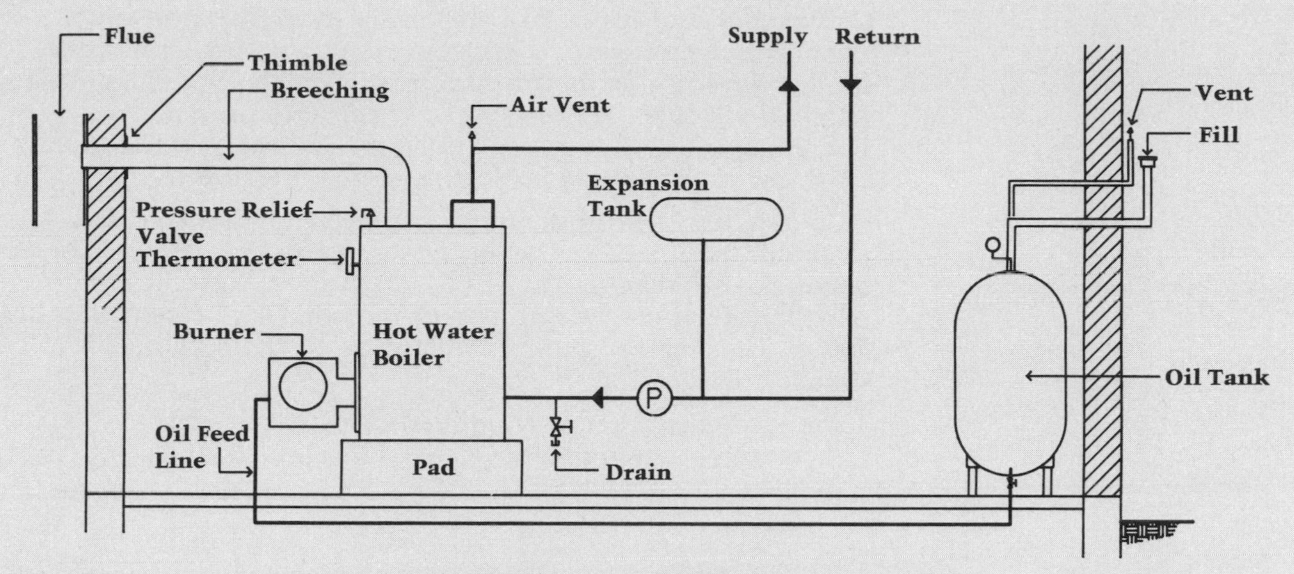

Basic Oil Fired Hot Water Heating System

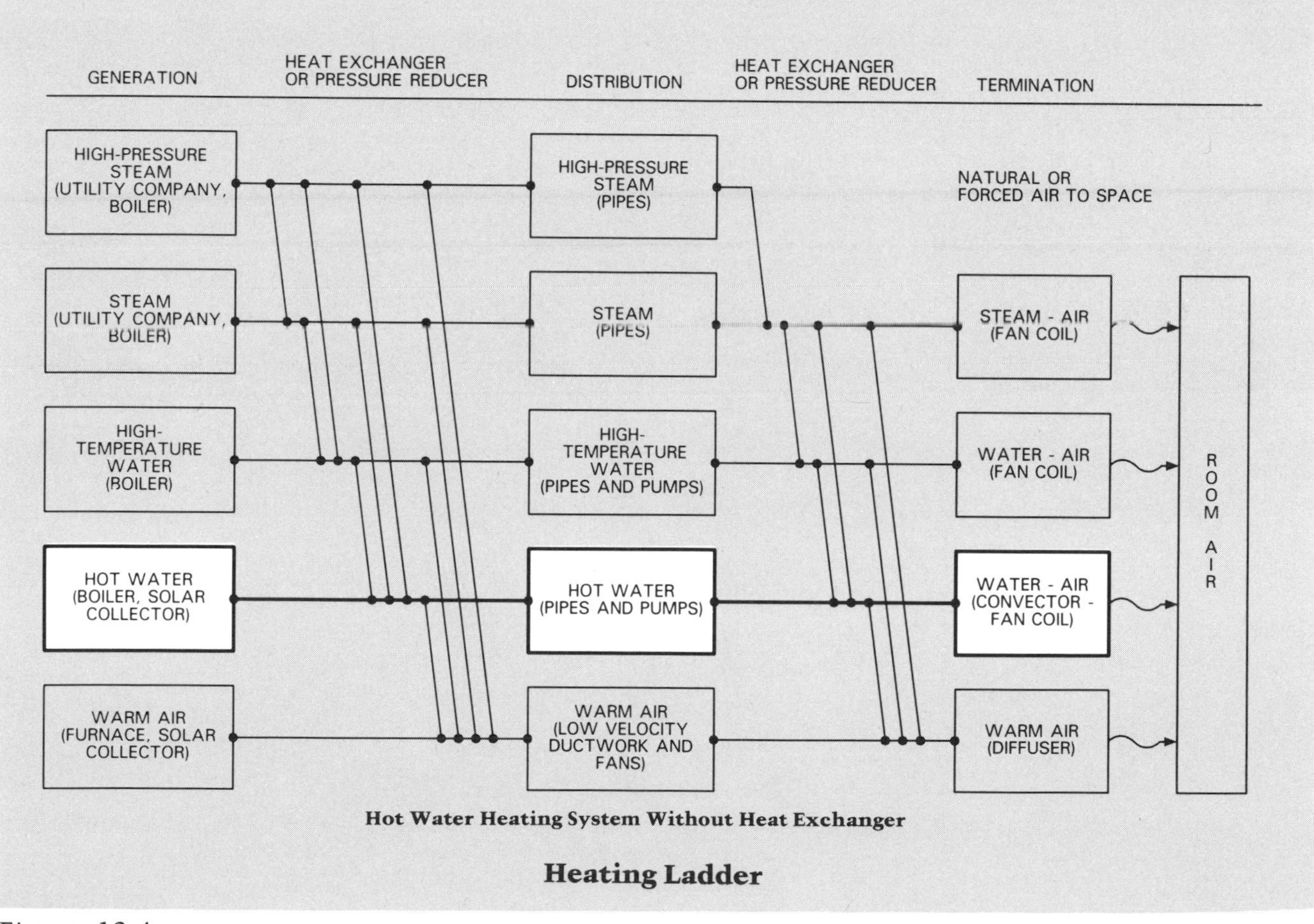

Hot Water Heating System Without Heat Exchanger

Heating Ladder

Figure 13.4

Convection Heat Losses

Convection heat losses (H_v) are caused by infiltration and kitchen and bathroom ventilation. Generally, the amount of convection loss due to infiltration is larger than the loss due to ventilation, since exhaust fans only run intermittently. Therefore, infiltration is used to compute heat loss and the crack coefficients must be determined (see Figure 13.6).

Calculate the areas of the roof, walls, windows, doors, and floors to obtain the linear footage of the cracks and perimeter. These quantities are then multiplied by the respective values of temperature or constant to obtain the building's convection heat losses. The area and crack calculations for the building are shown in Figure 13.7.

The heat losses are now computed using the design temperatures determined in Figure 13.5, the U values and

Design Criteria

Item	Reference Section	Value	Comments	Recommendation
Indoor Design Temp.	Fig. 2.8	68–72°F	Local code 68°F	68°F
Outdoor Design Temp.	Fig. 2.9	9°F	Local code 9°F	9°F
ΔT Adjoining Spaces	Fig. 2.10	T_I–T_{adj}	Lobby	Use fan coil unit
Ventilation Standards	Fig. 1.22	4% floor area	Windows meet criteria	
		Exhaust toilets and kitchens	Use as required	
U Values	Chap. 2	Btu/hr./s.f./°F		
	Walls	0.08	Verify	Compute
	Roof	0.06	Actual	Actual
	Windows	0.53	Actual	Actual

Figure 13.5

U Values and Constants			
Reference	**Material**	**U = 1/R_t**	**Remarks**
Walls			
Fig. 2.7	4″ brick	0.40	Face brick
Fig. 2.5	2″ cavity	0.97	Allow 1″–4″
Fig. 2.7	4″ block	0.71	Sand and gravel
Fig. 2.7	3′-5/8″ insulation	13.00	Blanket type
Fig. 2.7	1/2″ gypsum	0.45	Plaster
Fig. 2.5	Outside film	0.17	
Fig. 2.5	Inside film	0.68	
		R_t = 16.38	
		U = 1/R_t = 0.06 (0.08 req'd.)	
		U walls = 0.06 Btu/hr./s.f./°F	
Roof			
Fig. 2.7	Roof membrane	0.15	Assume roll roof
Fig. 2.7	2″ rigid insulation	14.40	Polyisocyanurate
Fig. 2.7	6″ concrete	0.67	Sand and gravel
Fig. 2.7	1/2″ spray-on insulation	1.00	Cement fiber-wood
Fig. 2.5	Outside film	0.17	
Fig. 2.7	Inside film (Table 34)	0.61 (horiz. surface)	
		R_t = 17.00	
		U = 0.059 (0.06 req'd.)	
		U roof = 0.059 Btu/hr./s.f./°F	
Windows			
Fig. 2.6	Double insulated glass	U = 0.55 (0.53 req'd.)	
		U windows = 0.55 Btu/hr./s.f./°F	
Door			
Fig. 2.6	Metal face 1″	U = 0.31 Btu/hr./s.f./°F	
	Front door	U = 1.10 Btu/hr./s.f./°F	
Edge Coefficient (F)			F = 0.49 Btu/hr./ft.
Fig. 2.11			4″ brick
			4″ block
Slab Constant (k)			
Fig. 2.10	Slab on grade		k = 2 Btu/s.f.
Crack Coefficient			
Windows			
Fig. 2.12	Residential casement/15 mph	Window coefficient = 0.87 cfm/ft. 1/32″ crack	
Doors			
Fig. 2.12	Weather-stop doors/15 mph		Door coefficient = 0.90 cfm/ft.

Figure 13.6

Areas and Crack Calculations

Facade	(1) Gross Wall Area (s.f.)	(2) Window Area (s.f.)	(3) Door Area (s.f.)	(4) Miscel-laneous	Net Area (1)-(2)-(3)-(4)	Crack Length (l.f.)
South	42′ x 125′ = 5,250 s.f.	4 x 3′ x 5′ + 11 x 9′ x 5′ + 10 x 12′ x 5′ 1,155 s.f.	2 x 2′ -8″ x 6′-8″ = 35.6 s.f.	Front door 10 x 6′-8″ = 66.7 s.f.	3,992.7 s.f.	Windows 35 x (2 x 3′ + 2 x 5′) = 560 l.f. Doors 4 x (2 x 2′-8″ + 2 x 6′-8″) = 74.7 l.f.
North	42′ x 125′ = 5,250 s.f.	16 x 9′ x 5′ + 12 x 12′ x 5′ 1,440 s.f.	—	—	3,810 s.f.	Windows 40 x (2 x 3′ + 2 x 5′) = 640 l.f. Doors —
East	42′ x 60′ = 2,520 s.f.	4 x 12′ x 5′ + 4 x 9′ x 5′ 420 s.f.	—	—	2,100 s.f.	Windows 12 x (2 x 3′ + 2 x 5′) = 192 l.f. Doors —
West	42′ x 60′ = 2,520 s.f.	9 x 9′ x 5′ + 7 x 12′ x 5′ 825 s.f.	2′-8″ x 6′-8″ = 17.8 s.f.	—	1,677.2 s.f.	Windows 23 x (2 x 3′ + 2 x 5′) = 368 l.f. Doors 2 x (2′-8″ + 2 x 6′-8″) = 18.7 l.f.
		3,840 s.f. Windows	53.4 s.f. Doors	66.7 s.f. Front door	11,579.9 s.f. Walls	1,760 l.f. Windows 93.4 l.f. Doors

Transom over doors assumed same as wall.
Intake louver at boiler room ignored.
Elevator penthouse ignored (not a heated space).

Figure 13.7

coefficients from Figure 13.6, and the areas and perimeters from Figure 13.7. The total heat loss calculation is shown in Figure 13.8.

The convection heat loss is from infiltration and has been computed for all windows simultaneously. As discussed in Chapter 2, only a portion of the air infiltrates simultaneously (see Figure 2.13). It is assumed that 50 percent of the 105 MBH convection load occurs at any given time. Therefore, the total heat load for the building is

$$H_t = H_c + H_e + H_s + H_v \text{ (MBH)}$$
$$= 193.2 + 10.7 + 12.8 + 104.8/2$$
$$= 268.7 \text{ MBH (load/s.f.} = 268{,}700/(4 \times 6{,}400) = 10 \text{ Btu/s.f.)}$$

The average load of 10 Btu/s.f. is a very low heat load compared to many existing buildings, because current strict energy codes require much better building insulation than they have in the past. Older buildings with large areas of glass, single-pane glass, and little roof insulation often have heat loads three times as large.

Equipment Selection

The final step in the design process is the selection of equipment, including generation, distribution, and termination equipment. The proposed layout is first drawn as a *system diagram*. The apartments of the building are characterized by a living room and bedroom adjacent to each other, with a considerable amount of glass in both rooms. Glass is the source of greatest heat loss. Therefore, it makes sense to place hot water radiators under the windows.

It also is possible to use baseboard radiators with the supply and return piping fed from the same end by running the return piping back under the enclosure and above the element. The layout of the radiation with a typical detail is illustrated in Figure 13.9.

Distribution

The object in laying out a distribution system is to supply from the generating system to each terminal unit the least linear footage of piping (or ductwork) in order to save on both the cost of installing the pipe and the operating cost of pumping through extra pipe.

As a general rule, the least amount of piping is needed when one horizontal main, located on the mechanical floor, distributes from the main vertically to each terminal unit. Rising vertically approximately nine feet to each radiator uses considerably less piping than running horizontally on each floor, where the average distance between radiators is approximately fifteen feet. When radiators are adjacent to each other (e.g., less than nine feet), they can be supplied from a common riser. Systems to consider are illustrated in Figure 13.10.

Two of the systems shown in Figure 13.10 combine direct and reverse piping systems (see Chapter 9, "Distribution and Driving Systems," for an explanation of direct and reverse return systems). In this example, the vertical risers are piped as direct return and the horizontal main is piped as a reverse return

Heat Loss Calculations

Conduction $H_c = U A \Delta T$	U (Btu/hr./s.f./°F) (Fig. 13.6)	A (s.f.) (Fig. 13.7)	ΔT (°F) (Fig. 13.5)	Heat Loss (Btu/hr.)
Walls	0.06	11,579.9	(68–9)	40,993
Windows	0.55	3,840	(68–9)	124,608
Doors	0.31	53.4	(68–9)	977
Front Door	1.1	66.7	(68–9)	4,329
Roof	0.059	6,400	(68–9)	22,278
				193,185 Btu/hr.

$H_e = FP \Delta T$	F	P	ΔT	Heat Loss (Btu/hr.)
Perimeter	0.49 Btu/hr./s.f.	370 s.f. (See plans for calculations)	(68–9)	10,696 Btu/hr.

$H_s = kA$	k	A	—	Heat Loss (Btu/hr.)
Slab on Grade	2 Btu/hr./s.f.	6,400 s.f. (See plans)	—	12,800 Btu/hr.

Total Conduction Loss = 216,681 Btu/hr.
rounded to 217 MBH

Convection $H_v = 1.1 \text{ cfm } \Delta T$	Crack Length	cfm	ΔT		Heat Loss (Btu/hr.)
Windows	1,760	x 0.87	x 59	x 1.1	= 99,375
Doors	93.4	x .90	x 59	x 1.1	= 5,455

Total Convection Loss = 104,830 Btu/hr.
rounded to 105 MBH

Figure 13.8

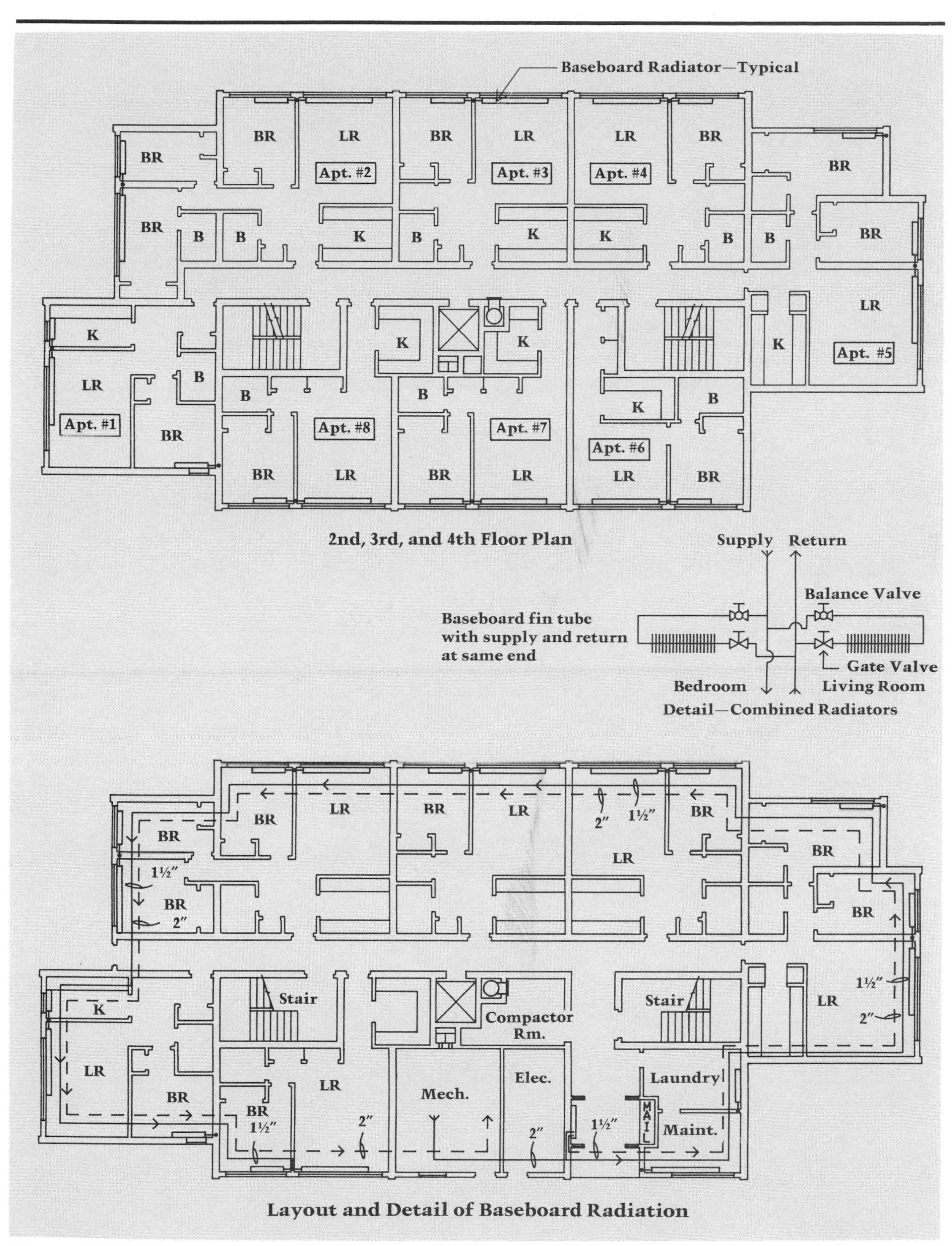
Baseboard Radiator—Typical
BR
LR
BR
LR
LR
BR
Apt. #2
Apt. #3
Apt. #4
BR
BR
B
B
K
B
K
K
B
B
BR
K
K
K
K
LR
LR
B
Apt. #5
B
B
K
B
Apt. #1
Apt. #8
Apt. #7
BR
Apt. #6
BR
LR
BR
LR
LR
BR
2nd, 3rd, and 4th Floor Plan
Supply
Return
Balance Valve
Baseboard fin tube
with supply and return
at same end
Gate Valve
Bedroom
Living Room
Detail—Combined Radiators
LR
BR
LR
2"
1½"
BR
BR
BR
LR
BR
1½"
BR
2"
BR
Stair
Stair
K
Compactor
Rm.
1½"
LR
2"
LR
Laundry
LR
Elec.
BR
Mech.
BR
1½"
2"
2"
1½"
MAIL
Maint.
Layout and Detail of Baseboard Radiation

Figure 13.9

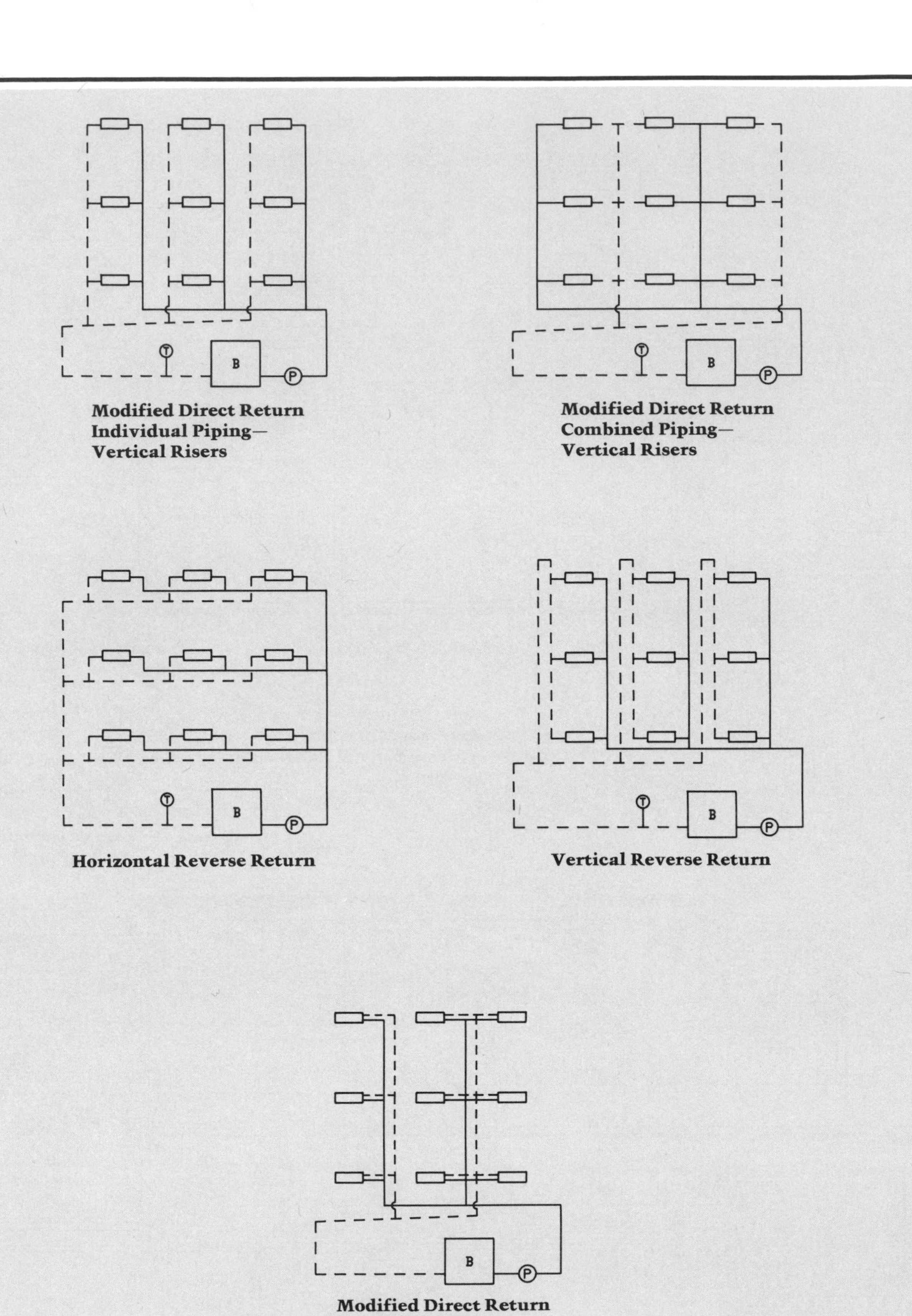
B
T
P
Modified Direct Return
Individual Piping—
Vertical Risers
Modified Direct Return
Combined Piping—
Vertical Risers
Horizontal Reverse Return
Vertical Reverse Return
Modified Direct Return
Combined Piping—
One End Connection
(Most Efficient)
Types of Systems to Consider

Figure 13.10

system. While a complete reverse return system is theoretically superior for overall distribution, a stack of two to six radiators can be effectively attached in a direct return stack and properly balanced with valves and still maintain similar flow characteristics of the reverse return system, but without the additional third riser. Figure 13.10 shows that the two complete reverse return systems have more piping than the modified systems, and that the combined piping system has the least.

The above analysis results in the following design objectives:

1. Place radiators under windows.
2. Distribute vertically.
3. Combine reverse return on the horizontal main with direct return on the risers.
4. Pipe radiators with supply and return fed at the same end (where possible).
5. Connect radiators to the same riser at living rooms and bedrooms.

The riser diagram shown in Figure 13.11 incorporates most of the above objectives and will be used to complete the design and selection of the heating system.

Generation Equipment

The following generation equipment is selected for this model (see Figure 13.4):

- Hot water boiler (oil-fired cast iron)
- Oil tank and accessories
- Expansion tank
- Circulating pump(s)

Each piece of generation equipment is discussed below and is summarized in Figure 13.15. A Means line number is provided in order to determine costs in the current edition of *Means Mechanical Cost Data.*

Hot Water Boiler—Oil (268.7 MBH Load): The capacity of the hot water boiler should be sized approximately 25 percent larger than the design load in order to allow for a reserve on excessively cold days and to account for a lowering of the boiler efficiency over its useful life (see Figures 5.1 and 5.3). The design output for the boiler is shown below:

Net building heat loss	268.7	MBH
Allowance for pipe loss (5%)	13	(insulated pipe)
Pickup allowance (10%)	26	(for pickup)
Total load	307.7	
Allow 25% oversizing	75	
Total design boiler output	382.7,	round to 380 MBH

For oil-fired boilers with capacities in the 380 MBH output range, a cast iron sectional boiler is selected. Cast iron is the customary boiler used in buildings of this type due to its life cycle, cost, and longevity.

The generation equipment data sheet on cast iron sectional boilers can be used to make the final selection (see Figure 5.11a). An oil-fired cast iron boiler comes in capacities ranging from

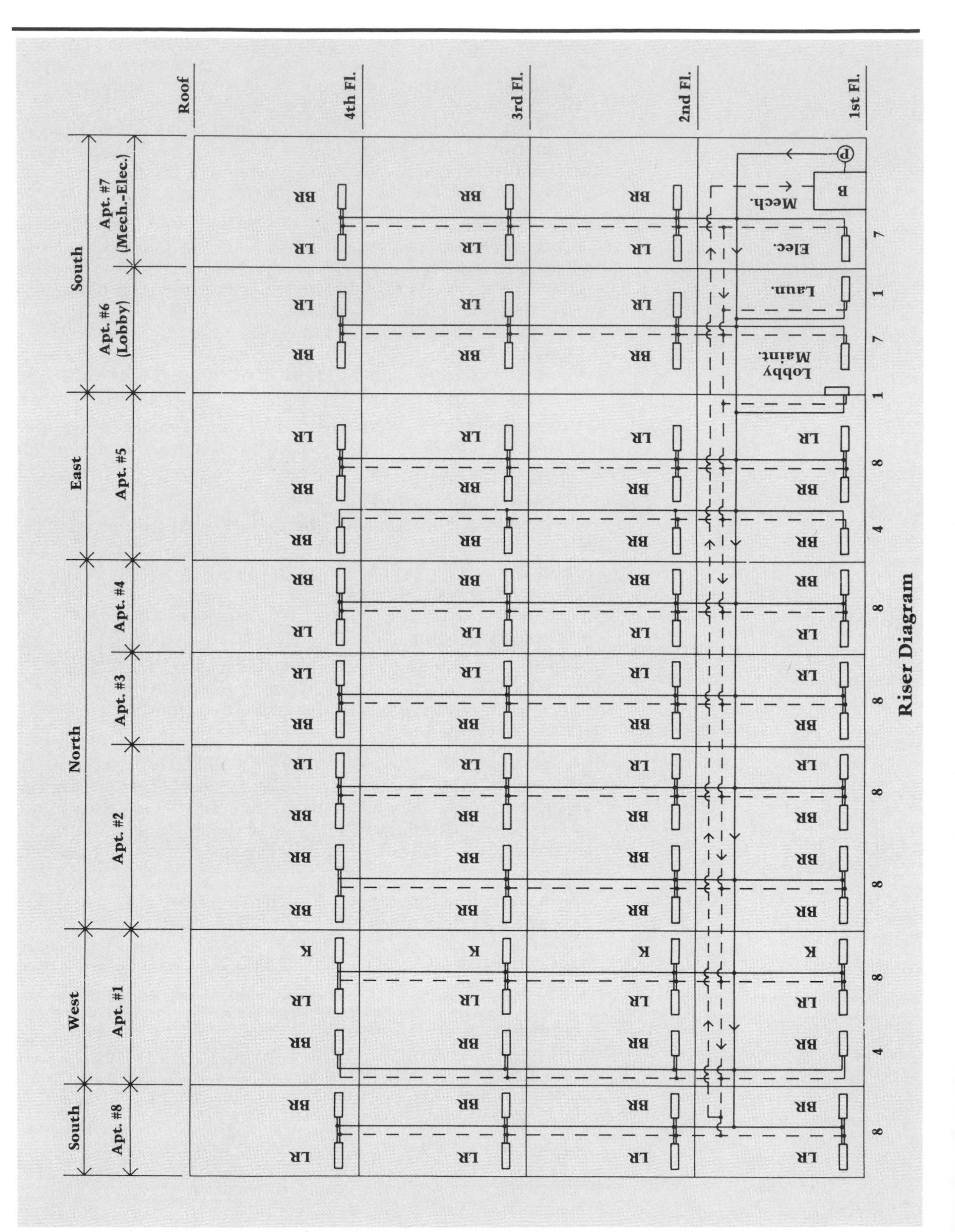

Riser Diagram
Roof
4th Fl.
3rd Fl.
2nd Fl.
1st Fl.
South
West
North
East
South
Apt. #8
Apt. #1
Apt. #2
Apt. #3
Apt. #4
Apt. #5
Apt. #6 (Lobby)
Apt. #7 (Mech.-Elec.)
LR
BR
K
Lobby
Maint.
Laun.
Elec.
Mech.
B
P
8
4
8
8
8
8
8
4
8
1
7
1
7

Figure 13.11

100 to 7,000 MBH. (See line number 155-120-1000/3000 from *Means Mechanical Cost Data*.)

Because this type of building may require a back-up system, current practice often uses two boilers, one sized for one-third and the second for two-thirds of the total load, and controlled to sequence as necessary to meet demand. If the boiler should fail for any reason during the heating season, great discomfort to the inhabitants, as well as considerable damage to the property from freezing, could result. Furthermore, the efficiencies noted for the equipment are based on continuous full load. On mild days, a single boiler would waste considerable heat warming itself and the chimney on each start-up. Using two pieces of equipment is effective because each boiler, when running, will be close to peak efficiency. Further, in the event of a boiler failure, the second boiler can be relied on for at least some heat. Thus, for this building, two boilers are chosen, one close to one-third and the other close to two-thirds of the required output load. By referring to *Means Mechanical Cost Data* and manufacturer's literature, two boilers, one boiler at 144 MBH and the other at 236 MBH, are selected. The total selected output now equals 380 MBH.

Oil Tank and Accessories: The size of an oil tank is determined from the total amount of fuel required during the heating season. The following formula is used for computing fuel consumption (see also Chapter 5, "Generation Equipment"):

$$F = \frac{24 \times DD \times H_t \times C}{E\ \Delta T \times V}$$

where

DD = degree days (5,600 Boston) (Figure 2.9)

H_t = design heat loss 258.6 MBH

e = boiler efficiency — assume 85% (0.85) (Figure 5.1)

ΔT = 68 – 9 = 59°F (Figure 13.5)

V = heating value of fuel 144 MBH/gal. (Figure 4.3)

C = coefficient (see Figure 5.3)

For a ΔT of 9 and $H_G/H_N = 370/258.6 = 1.43$, $C = 1.42$

Thus,

$$F = \frac{24 \times 5{,}600 \times 258.6 \times 1.42}{0.85 \times 59 \times 144} = 6{,}834 \text{ gal./yr.}$$

From Figure 5.11a, a 7,000-gallon steel tank is selected after comparing the advantages and disadvantages of fiberglass and steel tanks. (Accessories such as cathodic protection anchors, excavation, pad, pumps, and fuel lines are noted on the system summary sheet at the end of the chapter.)

In this example, because the tank is sized for the full year heating load, the fuel can be purchased off-season in bulk delivery, at a savings of from ten to thirty percent below peak season prices.

Expansion Tank: Expansion tanks are sized after the total volume of water necessary to fill the system is determined. For convenience, it is assumed that this was done after the piping was sized, and the total volume of water was determined to be 500 gallons.

Although the size of the expansion tank depends primarily on the volume of water in the system, other factors, such as system temperature and pressure, also affect the tank size, as shown in the equation below (see also Chapter 5, "Generation Equipment").

$$V_t = \frac{(0.00041t - 0.0466)V_s}{P_a/P_f - P_a/P_o}$$

where

t = 190°F (assumed average water temperature)

V_s = gallons of water in system (500 assumed)

P_a = atmospheric pressure 14.7 psi

P_o = maximum tank operating pressure 30 psi + P_a – 10% (i.e., 30 + 14.7 – 4) = 39 psi

P_f = initial tank pressure = $P_a + P_{pump} + P_{height} + P_{residual}$

= assume the pump is after the expansion tank and provides no head. Also assume that the tank is 6′ above the floor of a 42′ return riser and 1.5 psi residual pressure exists

= 14.7 + 0 + ((42 – 6) × 0.433) + 1.5 = 31.7 psi

$$V_t = \frac{(0.00041 \times 190 - 0.0466)\ 500}{14.7/31.7 - 14.7/39} = \frac{15.7}{0.46 - 0.38} = 196 \text{ gallons}$$

Use a 220-gallon expansion tank to allow for a 50 percent air cushion. Note that if the tank were placed at the top of the building, the 36-foot height would be eliminated from P_f, resulting in a smaller tank.

Circulating Pumps: Although most pumps are suitable for this application of hot water heating, a hot water circulation pump with high-temperature seals is selected, because it is recommended for pumps with a temperature over 150°F in order to prevent leaks. A frame-mounted type is chosen because the motor, which is separate, can be easily replaced. Multiple stages are not necessary for heating, since the boiler controls can be used to vary output temperature.

The pump capacity (in gpm) and the head (in feet) must be specified. The gpm is determined from the heat load:

1 gpm = 10 MBH at 20°F temperature drop

$$\frac{258.6 \text{ MBH}}{10 \text{ MBH/gpm}} = 25.9 \text{ gpm}$$

Based on this equation, a load of 258.6 MBH, 25.9 gpm is required.

The pump head is determined by examining the friction loss around the longest loop, as shown below.

Straight run of pipe	
Basement loop (Figure 13.9) 2 × perimeter = 2 × 370 ft. =	740.0 ft.
Boiler room piping, at 10% of basement	74.0
Vertical piping, one loop (Figure 13.11) 2 × 42 ft. =	84.0
One radiator 4 elbow equiv. × 2.7* =	10.8
One control valve at radiator 0.7 elbow equiv. × 2.7*	1.9
One balancing valve at radiator 17 elbow equiv. × 2.7*	45.9
Total straight run of pipe	956.6
Add 50% for fittings	478.3
Total equivalent length	1434.9 ft.

The total equivalent length is rounded to 1500 feet.

*See Figure 9.13 — use 3 fps on 1" pipe for elbow, copper pipe.

The 50 percent allowance for fittings is a convenient short cut, but should be verified against a thorough takeoff of all elbows, tees, strainers, valves, meters, and other components in the system that may cause friction losses.

Using 1,500 linear feet as the total equivalent straight run of pipe, the pump head required to drive water around the loop can be determined by selecting the maximum loss per foot of pipe that will be permitted. A design head loss of 2.5 feet of head per 100 feet is selected. This maintains a maximum flow of four feet per second in the pipes, which is considered acceptable. Therefore, the total head for the pump is:

$$H_p = \frac{2.5}{100} \times 1500 = 37.5 \text{ feet of head}$$

A pump with 26 gpm and 37.5 feet of head is required, and, therefore, a 60-16 type pump is selected (see Figure 7.3).

Accessories: In the mechanical room, additional fittings, thermometers, pressure gauges, and specialties are required (see Figure 13.12).

Distribution Equipment

The major component of the distribution system is the piping. The pipe sizes are determined by the flow (gpm) in each pipe. The flow for each line is determined by the heat load being supplied. Use Figure 13.11, the distribution system layout, to size the piping after the heat load for each room has been determined.

Room-by-Room Load Breakdown: A detailed room-by-room analysis is usually performed on a computer. However, because this building is fairly uniform, the loads are easily determined and then adjusted to the special conditions of the first floor and top floor. The room-by-room load breakdown is outlined in Figure 13.13.

As expected, the second and third floor have identical heat losses. The first-floor heat load is higher due to door, slab, and perimeter losses. The fourth floor heat load is also higher due to roof losses. For convenience, the floor loads are divided by the

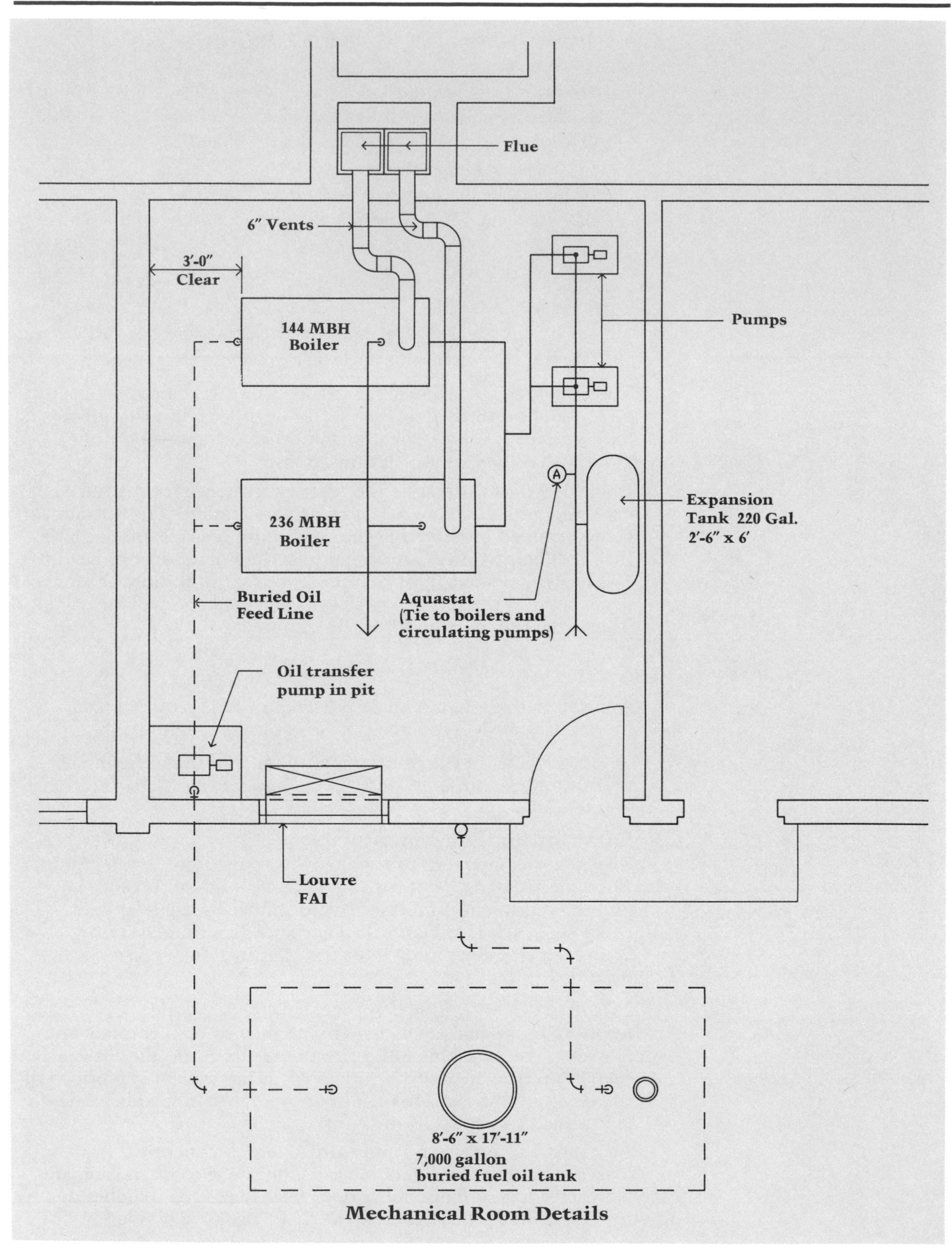
Flue
6" Vents
3'-0"
Clear
144 MBH
Boiler
Pumps
236 MBH
Boiler
A
Expansion
Tank 220 Gal.
2'-6" x 6'
Buried Oil
Feed Line
Aquastat
(Tie to boilers and
circulating pumps)
Oil transfer
pump in pit
Louvre
FAI
8'-6" x 17'-11"
7,000 gallon
buried fuel oil tank
Mechanical Room Details

Figure 13.12

perimeter to obtain the loads per foot that the active length of baseboard radiators will be designed to carry.

Now the loads must be determined for each room containing a radiator. The riser for apartment #8 (see Figure 13.11) will be used to illustrate the room-by-room loads, piping loads, and pipe sizes (see Figure 13.14).

Because the building loads are relatively low, a 3/4" diameter pipe will be sufficient for each of the risers. This size is selected because it exceeds the design requirement of 1/2" and it matches

Heating Load Breakdown by Floor

Conduction $H_c = UA\Delta T$	**Total Load (Btu/hr.)**	**1st Floor**	**2nd Floor**	**3rd Floor**	**4th Floor**	**Remarks**
Walls	40,993	10,250	10,250	10,250	10,250	Approx. equal
Windows	124,608	31,200	31,200	31,200	31,200	Approx. equal
Doors	977	1,000				1st floor only
Front Door	4,329	4,450				1st floor only
Roof	22,278				22,300	4th floor only
Total	193,185	46,900	41,450	41,450	63,750	193,550 Btu/hr. (rounded)
$H_e = FP\Delta T$ $H_c = kA$	**Total Load (Btu/hr.)**	**1st Floor**	**2nd Floor**	**3rd Floor**	**4th Floor**	**Remarks**
Perimeter	10,696	10,700				1st floor only
Slab on grade	12,800	12,800	—	—	—	1st floor only
Total	23,496	23,500				23,500 Btu/hr. (rounded)
Convection (50% at one time)	**Total Load (Btu/hr.)**	**1st Floor**	**2nd Floor**	**3rd Floor**	**4th Floor**	**Remarks**
Windows	99,375 x .5	12,500	12,500	12,500	12,500	Approx. equal
Doors	5,455 x .5	2,750				1st floor only
Total	52,415	15,250	12,500	12,500	12,500	52,750 Btu/hr. (rounded)
Floor-by-Floor Summary	—	85,600	53,950	53,950	76,250	259,300 Btu/hr. (rounded)
Loss per Linear Foot of Perimeter (P = 370 ft.)		231	146	146	206	

Figure 13.13

Loads and Pipe Sizes for Risers

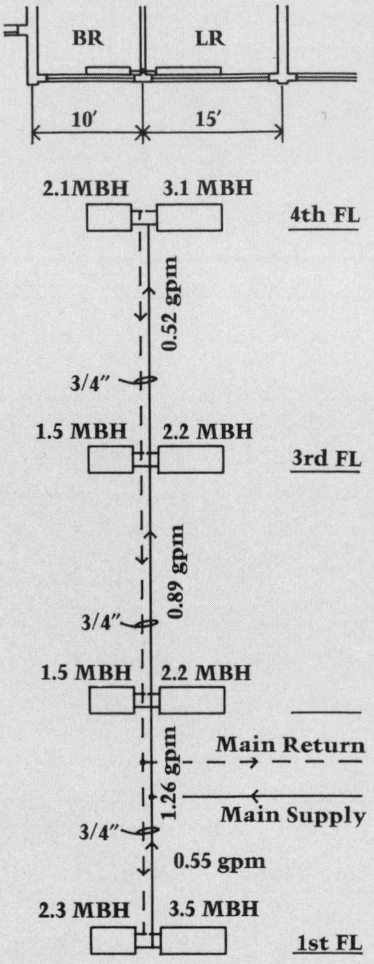

Riser Diagram Apt. #8[a]	Perimeter Load (Btu/l.f.)[b]	Room Load (Btu/hr.)		Flow Req'd (1 gpm = 10 MBH)	Riser Size Based on Copper Type L 2.5'/100 Loss[c]	Baseboard Length Based on 680 Btu/l.f.
		BR	LR			
BR LR 10' 15' 2.1MBH 3.1 MBH 4th FL	206 Btu/l.f.	206 x 10 = 2,060 rounded to 2.1 MBH	206 x 15 = 3,090 rounded to 3.1 MBH	BR = 0.21 gpm LR = 0.31 gpm 0.52 gpm Riser below = 0.52 gpm	0.52 gpm 1/2"	BR 3.00 use 4' LR 4.55 use 6'
0.52 gpm 3/4" 1.5 MBH 2.2 MBH 3rd FL	146 Btu/l.f.	146 x 10 = 1,460 rounded to 1.5 MBH	146 x 15 = 2,190 rounded to 2.2 MBH	BR = 0.15 gpm LR = 0.22 gpm 0.37 gpm 0.52 + 0.37 = 0.89 Riser below = 0.89 gpm	0.89 gpm 1/2" close, use 5/8"	BR 2.2' use 4' LR 3.2' use 4'
0.89 gpm 3/4" 1.5 MBH 2.2 MBH	146 Btu/l.f.	146 x 10 = 1,460 rounded to 1.5 MBH	146 x 15 = 2,190 rounded to 2.2 MBH	BR = 0.15 gpm LR = 0.22 gpm 0.37 gpm 0.89 + 0.37 = 1.26 Riser below = 1.26 gpm	1.26 gpm 5/8"	BR 2.2' use 4' LR 3.2' use 4'
1.26 gpm Main Return Main Supply 3/4" 0.55 gpm 2.3 MBH 3.5 MBH 1st FL	231 Btu/l.f.	231 x 10 = 2,310 rounded to 2.3 MBH	231 x 15 = 3,465 rounded to 3.5 MBH	BR = 0.23 gpm LR = 0.35 gpm 0.58 gpm Riser below = 0.58 gpm	0.52 gpm 1/2"	BR 3.4 use 4' LR 5.1 use 6'

The flow rates in each length of pipe are indicated on the diagram. Since the standard radiator uses 3/4" diameter tube, all sizes are increased to 3/4" minimum for compatibility and to eliminate the cost of furnishing and installing reducers.

[a]Reference Figure 16.11
[b]Reference Figure 16.13
[c]Reference Figure 9.11

Figure 13.14

the diameter of the typical radiator connections. Copper tubing is selected for its ease of installation and adaptability to the copper element in the radiation.

The main horizontal loop in the first-floor ceiling must also be sized. Steel pipe is used for the main and for all piping in the boiler room. The total load of 256.8 MBH requires 26 gpm, which at 2.5′/100′ of friction loss results in a 2″ diameter pipe for the horizontal supply and return. This could be reduced to 1-1/2″ approximately halfway around the loop, where the required flow drops below 15 gpm. Note that as the supply pipe size decreases, the return pipe size, increases (see Figure 13.9).

The results of the sizing of the distribution system are outlined below.

3/4″ pipe	11 risers × 2 pipes each riser × 42 ft.	= 924 l.f.
3/4″ pipe	80 radiators × 10 l.f. average each	= 800 l.f.
1-1/2″ pipe (steel)	horizontal main × 2 pipes × 185 ft.	= 370 l.f.
2″ pipe (steel)	horizontal main × 2 pipes × 185 ft.	= 370 l.f.

Terminal Units

The terminal units for this model are supplied with forced hot water. Fin tube baseboard radiation is selected because it is economical, will fit the room dimensions, and can be supplied and returned from the same end. Cast iron baseboard could have been selected at a slight cost premium. Fan coil units are used for public spaces (e.g., lobby).

By referring to Figure 10.3, the length of the radiator elements can be determined. For the unit selected at 190°F average operating temperature (180–200°F), each foot of this type of radiator yields 680 Btu/l.f. Figure 13.14 shows the selection of fin tube baseboard radiators based on design loads and an output of 680 Btu/l.f. All radiators selected are four feet of active fin length except the first and fourth floor living rooms, which require six foot radiators. Of the 80 terminal units, 14 will be 6 feet long. Fan coil units for the lobby are selected in a similar manner.

Controls

The controls for a heating system are straightforward. In the mechanical room, the controls turn the pumps and boilers on or off and, if desired, may regulate the water temperature according to the outdoor temperature. In each apartment there may also be room or unit thermostats to regulate room temperature. Each of the controls selected is discussed in the following sections.

System On/Off: The heating system operates during the colder weather (heating season) and is off during the rest of the year. A master control switch is used to deactivate both boilers during the off season.

Outdoor Reset: During the heating season, the boiler water temperature is set at 200°F for an outdoor design temperature of 9°F. On less cold days, the boiler water temperature can be automatically reset to a lower temperature, down to 140°F, for example, when the outdoor temperature is 60°F.

Room or Unit Thermostats: Depending on the requirements of the building as established by the client, there may or may not be thermostats to control heat in each apartment. Individual nonelectric thermostatically controlled valves are usually provided at each fin tube radiator. These operate by sensing the room air and, depending on the type of valve, use a refrigerant gas to operate the position of the valve. By using the self-contained thermostatic valves on each radiator, not only is each room individually controlled, but elaborate control wiring and extensive use of motorized valves is eliminated.

Boiler Sequencing: Two boilers have been selected to complement each other and to stage through the heating sequence. A simple temperature control first activates the small boiler. When the small boiler can no longer handle the load, the large boiler cycles on and the small boiler is shut down. In extreme conditions, when the large boiler alone cannot handle the load, both boilers are called online. With these controls in place, the heating system can be expected to function with good overall efficiency.

Controllers: Whenever the water temperature falls below the preset design temperature, the boilers are activated by an aquastat mounted on the boiler (see Figure 13.12). When the boiler water temperature is at the design temperature, a signal is generated to activate the circulating pumps. The controller for the oil transfer pump is integral with the transfer pump unit.

Night Set-Back/Time Clock: Because this is a residential building occupied 24 hours a day, 7 days a week, it is not possible to predict when the entire system could be set back for heating. Some building codes require that public hallways be ventilated, in which case supply or exhaust fans would be controlled to shut down at night. However, for this building, it is assumed that they are not required.

Accessories

The heating system for this building has the accessory devices and equipment described below.

- Breeching
- Motor starters
- Vibration isolators
- Hangers and supports
- Insulation

Breeching: The breeching from the boilers to the vertical flue should not be less than the diameter of the boiler breeching connection. It is assumed that the breeching size for the boilers selected is six inches. The boilers are individually connected to the flue and each run uses three elbows. The 6″ diameter can be determined using the following formula:

$$D^2 = \frac{HMT_f}{26\ P(atm)V}$$

H = 370 MBH (Boiler sizes selected)
M = 1.2
T_f = 950°F
P(atm) = 29.92 inches Hg
V = 20 ft./sec. (assumed—conservative)

therefore,

$$D^2 = \frac{370 \times 1.2 \times 950}{26 \times 29.92 \times 20} = 27$$

D = 5.2″ diameter

Figure 13.12 shows approximately 27 linear feet of 6 inch flue piping in addition to six elbows.

Motor Starters: In this system, the electrically driven pieces of equipment which require motor starters are the two burners, the two circulating pumps, and the oil transfer pump. Each of these starters is automatic (magnetic) and activated by a controller. The starters are in NEMA I (normal indoor) enclosures.

Vibration Isolators: The circulating pumps should be mounted so as to minimize transmission of vibrations. The pumps selected run at 1,750 rpm and weigh approximately 300 pounds, or 75 pounds per spring.

$$1{,}750 \text{ rpm} = \frac{1{,}750}{60} \text{ or } 29 \text{ Hz}$$

The isolator should have a frequency of one-third to one-fifth 29 Hz, or 8 Hz. From this figure, the recommended spring constant (k) is determined using the following formula:

$$k = \frac{(\text{nat. freq. of isolator (Hz)})^2 \times (\text{equipment (lbs.)})}{9.8}$$

$$k = \frac{8^2 \times 75}{9.8} = 489, \text{ round to 500 lbs./in. maximum}$$

The piping from the pump should be isolated by flexible connectors and the entire assembly placed on a pad similar to that shown in Figure 12.5.

Hangers and Supports: Pipe hangers and supports are selected for the diameter of the pipe and noise control. In the boiler room and on the horizontal mains (steel pipe), the pipes are supported on clevis-type hangers. Note that the cost of hanger assemblies 10 feet on center are included in the cost of piping. Riser clamps are also required.

22 vertical pipes × 3 supports per riser = 66 clamps

Insulation: All piping is insulated with fiberglass insulation and ASJ (all service jacket) cover. The 3/4″, 1-1/2″, and 2″ lengths of pipe are entered on the system summary sheet.

Summary

All components of the system — generation equipment, distribution equipment, terminal units, controls, and accessories — have now been determined. The basic system is shown in Figures 13.4 and 13.11, and the layout of the mechanical room is shown in Figure 13.12. Figure 13.15 outlines each major component of the system, including a Means line number to determine costs in the current annual edition of *Means Mechanical Cost Data*.

System Summary Sheet

Item	Size	Quantity	Type	Accessories	R.S. Means Line No.	Remarks
Generation						
Oil-Fired Boiler	144 MBH	1	Cast iron	Insulated jacket	155-120	
Oil-Fired Boiler	236 MBH	1	Cast iron	Insulated jacket	155-120	
Oil Tank	5,000 gal.	1	Steel	Underground ST1-P3	155-671	Excavate and pour pad
Oil Pump and Piping	30 l.f.	1	Copper	Remote gauge	155-250	
Expansion Tank	220 gal.	1	Steel	ASME	155-671	
Circulating Pumps	26 gpm	2	Frame mounted	37.5 ft. head	152-410	High-temperature seals
Distribution						
Copper Tubing	3/4"	1,740 l.f.	Type L	Supports and anchors	151-401	
Steel Pipe	1-1/2"	380 l.f.	Schedule 40	Supports and anchors	151-701	Dielectric fittings
Steel Pipe	2"	380 l.f.	Schedule 40	Supports and anchors	151-701	
Valves and Fittings	Various	380 l.f.	Various			
Terminal Units						
Fin Tube Baseboard	3/4"	348 l.f.	7" high	66-4' lengths, 14-6' lengths	155-630	
Lobby, Fan Coil Unit		300 cfm	Wall hung	Heating only	157-150	
Controls						
On/off Switch	Unit	1	Winter "on"	Limit controls	157-420	Tie to boiler and pumps
Outdoor Reset	Unit	1	Automatic	Adjust boiler temperature	157-420	Adjust for electronic
Thermostatic Valves	Unit	81	Individual	Self-contained	156-240	Non-electric
Controllers	Unit		Aquastat	Set at 180°F	157-420	Tie to boilers and pumps
Accessories						
Breeching	6"	27 l.f.	Single wall	6 elbows	155-680	
Motor Starters	Unit	5	NEMA O, 5 HP		163-130	
Vibration Isolation	Unit	8	k—500 lbs./in.		157-485	4 each pump
Flexible Hose	Unit	4	2" pipe/pump		156-235	2 each pump
Riser Clamps	Unit	66	3/4"		151-901	
Pipe Insulation	3/4"	1,725 l.f.	Fiberglass		155-651	With ASJ Cover
Pipe Insulation	1-1/2"	372 l.f.	Fiberglass		155-651	With ASJ Cover
Pipe Insulation	2"	372 l.f.	Fiberglass		155-651	With ASJ Cover

Figure 13.15

CHAPTER FOURTEEN

COMMERCIAL BUILDING MODEL

In this chapter, an HVAC system is designed for a low-rise commercial office building using the principles discussed in the previous chapters. The sample building, a three-story office building with a lower-level garage, is located in Albuquerque, New Mexico, and thus requires cooling. The building is an exposed concrete frame with fixed windows, hung ceilings, and a penthouse, provided to house the HVAC equipment. Each floor has an open plan. The entire building will be occupied by one tenant. The plans for the building are illustrated in Figures 14.1 through 14.3.

The four basic selections to make in order to establish the appropriate type of HVAC system for a building are listed below.

1. Make the initial selections (Chapters 4 and 5).
2. Determine the design criteria (Chapters 2 and 3).
3. Compute the loads (Chapter 2).
4. Select the equipment (Chapters 5 through 12).

Initial Selection

As described in Chapters 4 and 5, the initial selections to make are: (1) the type of system, (2) the type of fuel, (3) the type and make-up of the distribution system, and (4) the type of generation system.

Type of System

The types of heating and cooling systems for buildings are listed in Figure 4.2; for a multi-story business occupied by one tenant, one central system is recommended. Because the building has fixed glass, it is necessary to supply fresh outdoor air for ventilation. A hung ceiling plenum is provided for air ducts; and an air distribution system will heat, cool, and ventilate, which are required, as well as filter and humidify/dehumidify, which are desirable. Thus, an air conditioning system is provided.

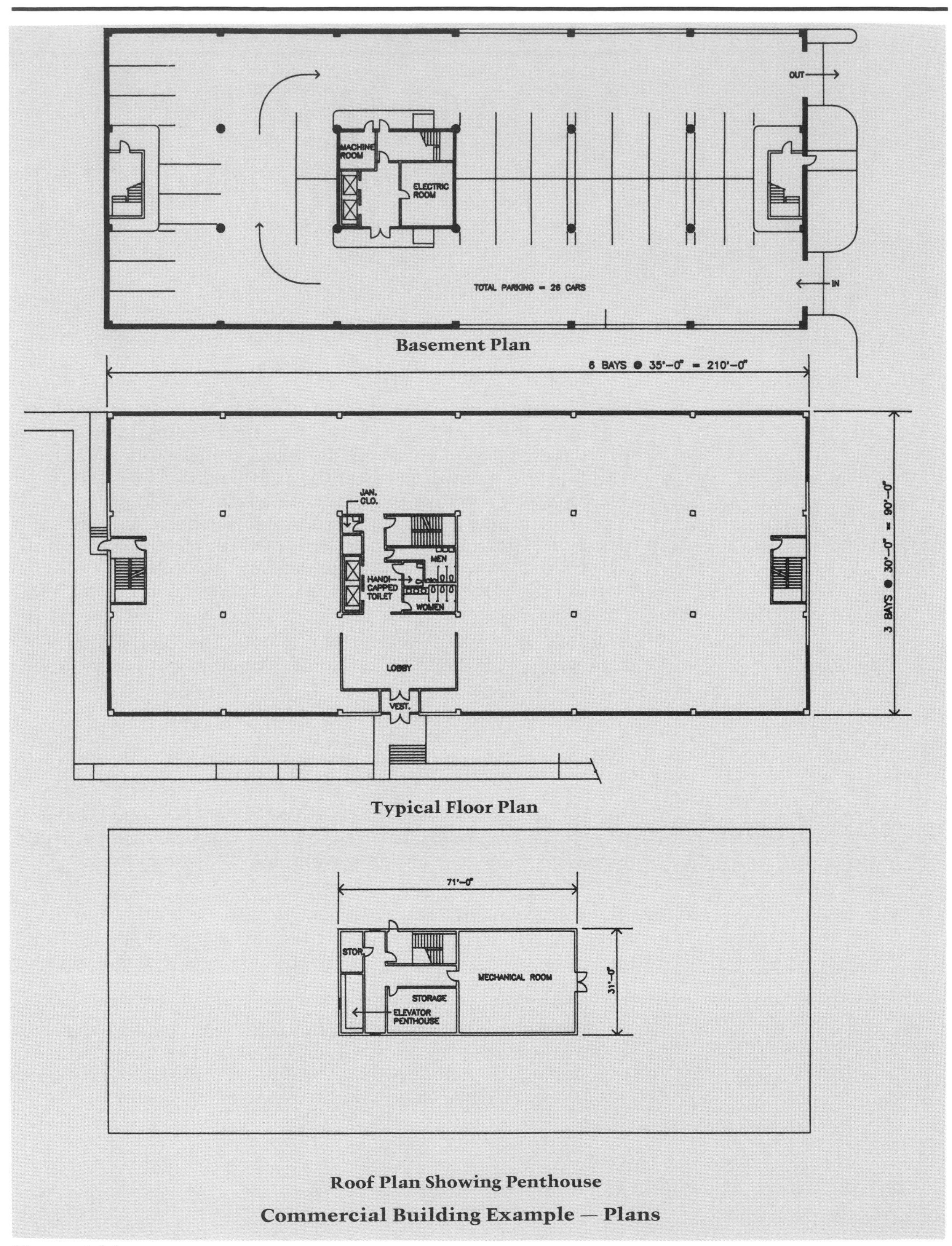

Commercial Building Example — Plans

Figure 14.1

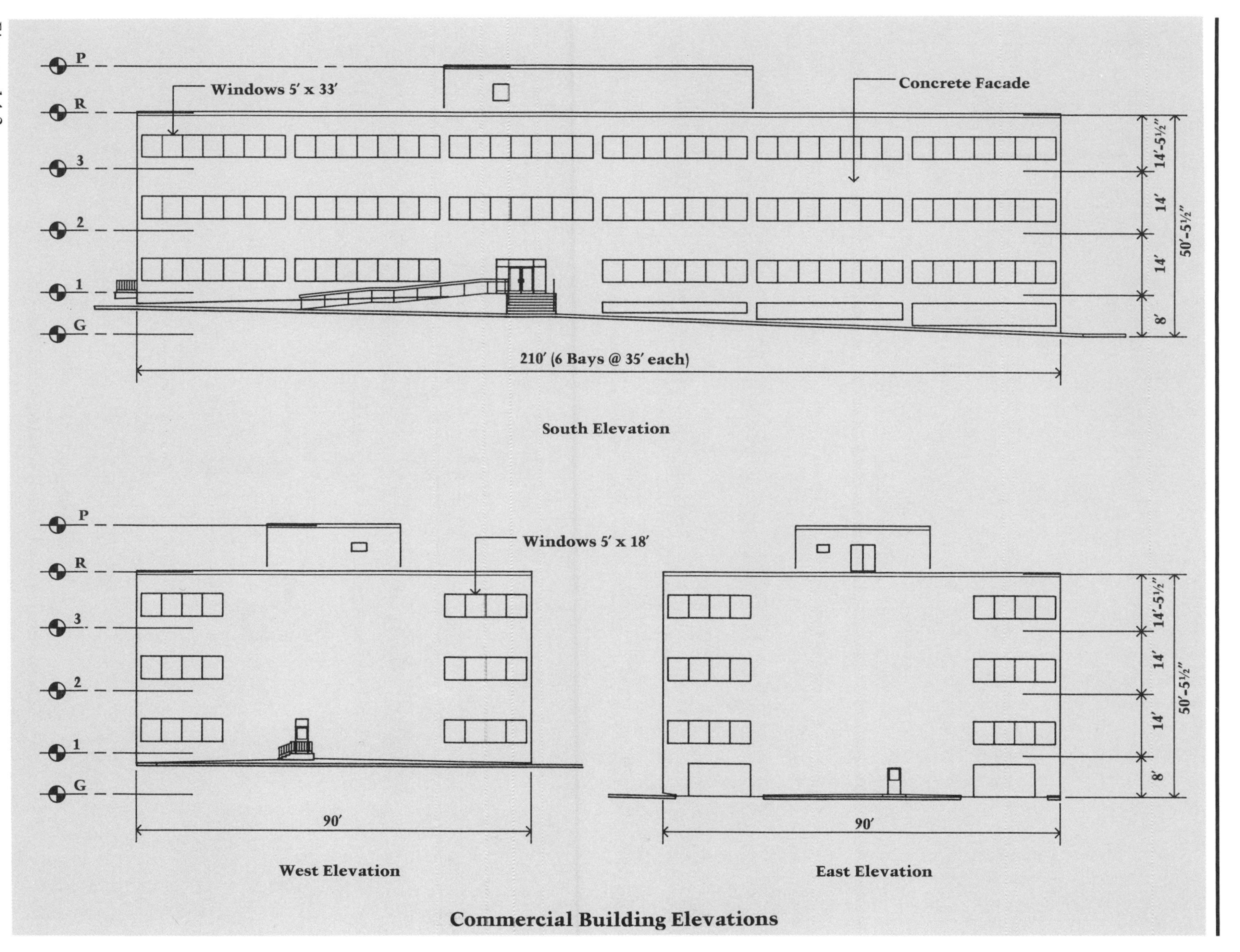

Windows 5′ x 33′
Concrete Facade
P
R
3
2
1
G
14′-5½″
14′
14′
8′
50′-5½″
210′ (6 Bays @ 35′ each)
South Elevation
Windows 5′ x 18′
P
R
3
2
1
G
14′-5½″
14′
14′
8′
50′-5½″
90′
90′
West Elevation
East Elevation
Commercial Building Elevations

Figure 14.2

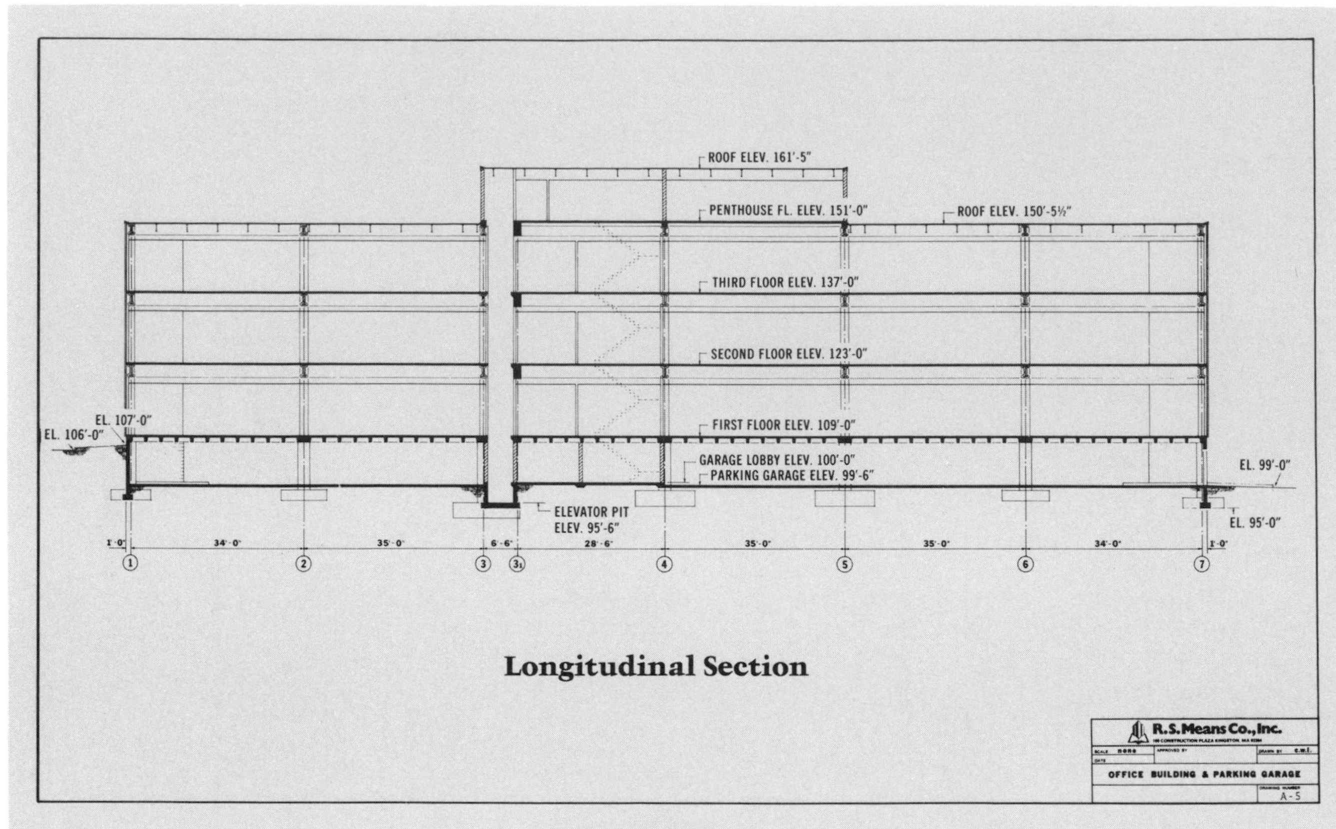

Longitudinal Section

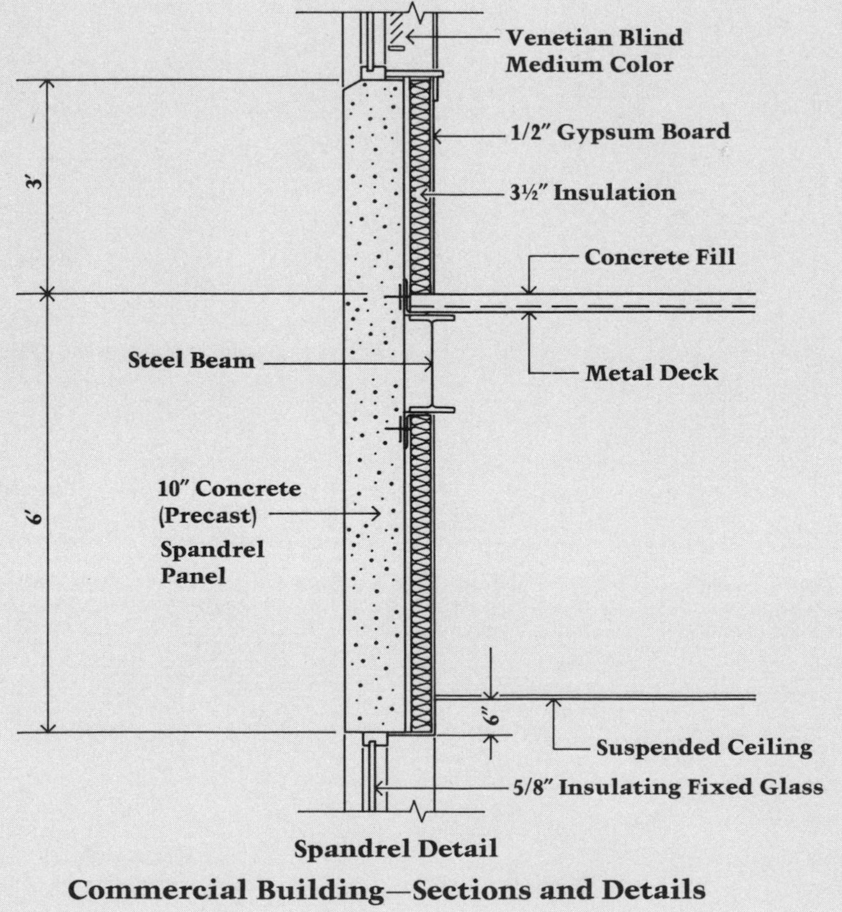

Spandrel Detail

Commercial Building—Sections and Details

Figure 14.3

Fuel

Figure 4.3 lists the basic properties and cost of fuels. Even though electricity is listed as the most expensive fuel, it is the likely choice for cooling for this building, because electricity must be purchased in sizeable quantities for the light and power of the building and is an essential fuel for cooling.

In winter, the actual heat load of the office building may in fact be low, as people, lights, and equipment add warmth to the space, thereby reducing the heat load. However, it is also assumed that gas is available; and since gas is less expensive than electricity, it will be used for heating.

Distribution System

An air distribution system is selected for this building because it can heat and cool, as well as provide the required ventilation. Moving the heating and cooling medium from the mechanical room to the terminal units can be done with ductwork as an all air system or by piping from the mechanical room to air handling units. A piping distribution system from the mechanical room to the terminal units is chosen for the following reasons:

- To save floor space that would be consumed by larger vertical air duct risers;
- To allow for future flexibility, because additional piping can be run above the ceiling to new or relocated units with minimal disturbance to the floor layout; and
- To independently operate each floor during off-hours with minimum fan power.

Therefore, the distribution system consists of both hot and chilled water piping to distribute hot/chilled water from the mechanical room to air handling units on each floor and ductwork from each unit to distribute the conditioned air to the ceiling diffusers.

Generation System

For this model, the generation equipment selected is a gas-fired hot water boiler; it is the most direct choice that does not involve heat exchangers (see Figure 1.23). Chilled water will be made by a chiller, requiring a piece of equipment to reject heat, such as a cooling tower or air cooled condensing unit.

Figure 14.4 shows the path selected for this system on the air conditioning ladder and also the schematic diagram for the mechanical system.

Design Criteria

With the overall basic parameters now established, the design criteria must be determined. The design criteria for this sample project are listed in Figure 14.5.

These criteria are used in the computation of the heating and cooling loads. Local codes, manufacturer's literature, and architectural requirements must also be checked to verify building design criteria.

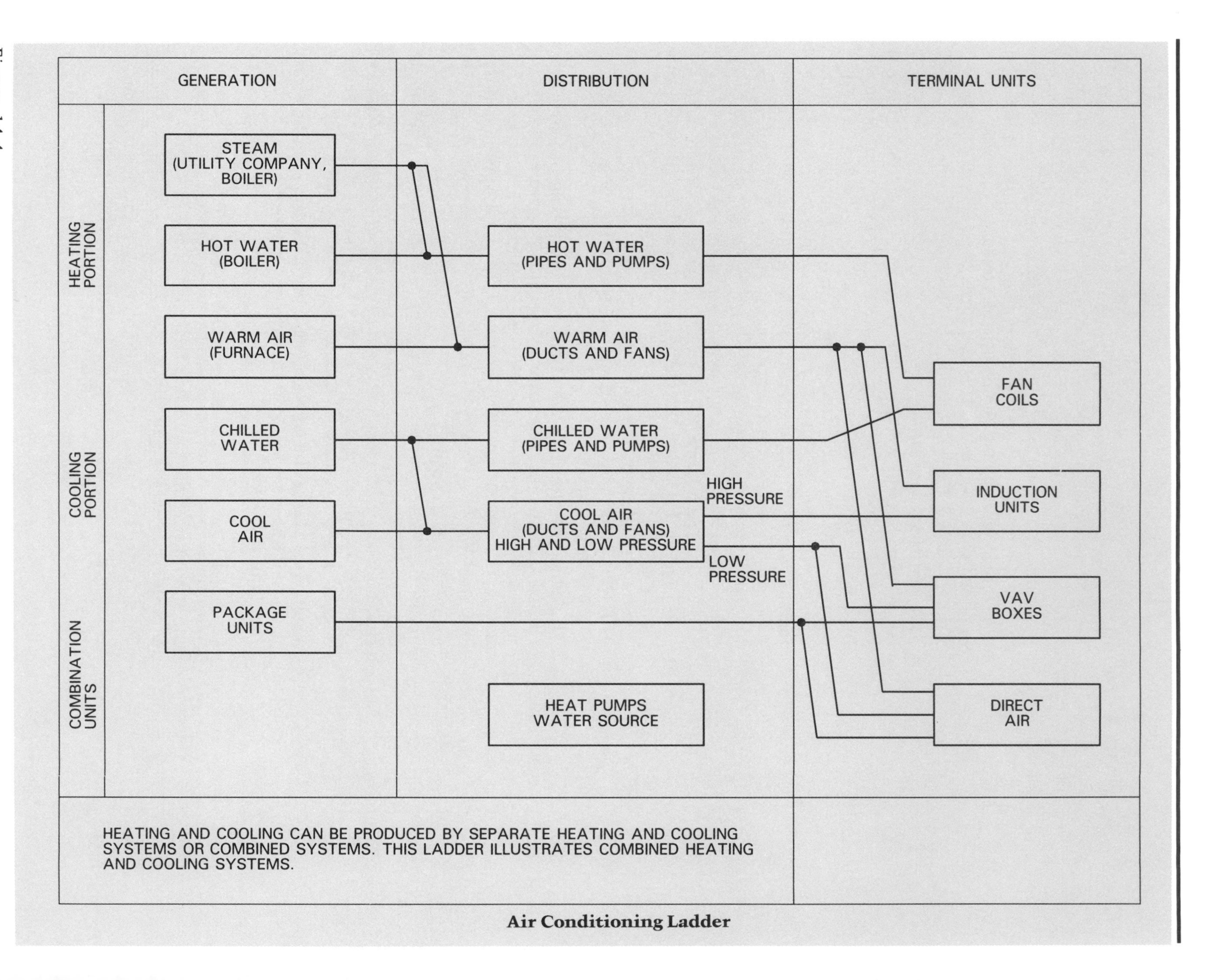

Air Conditioning Ladder

Figure 14.4

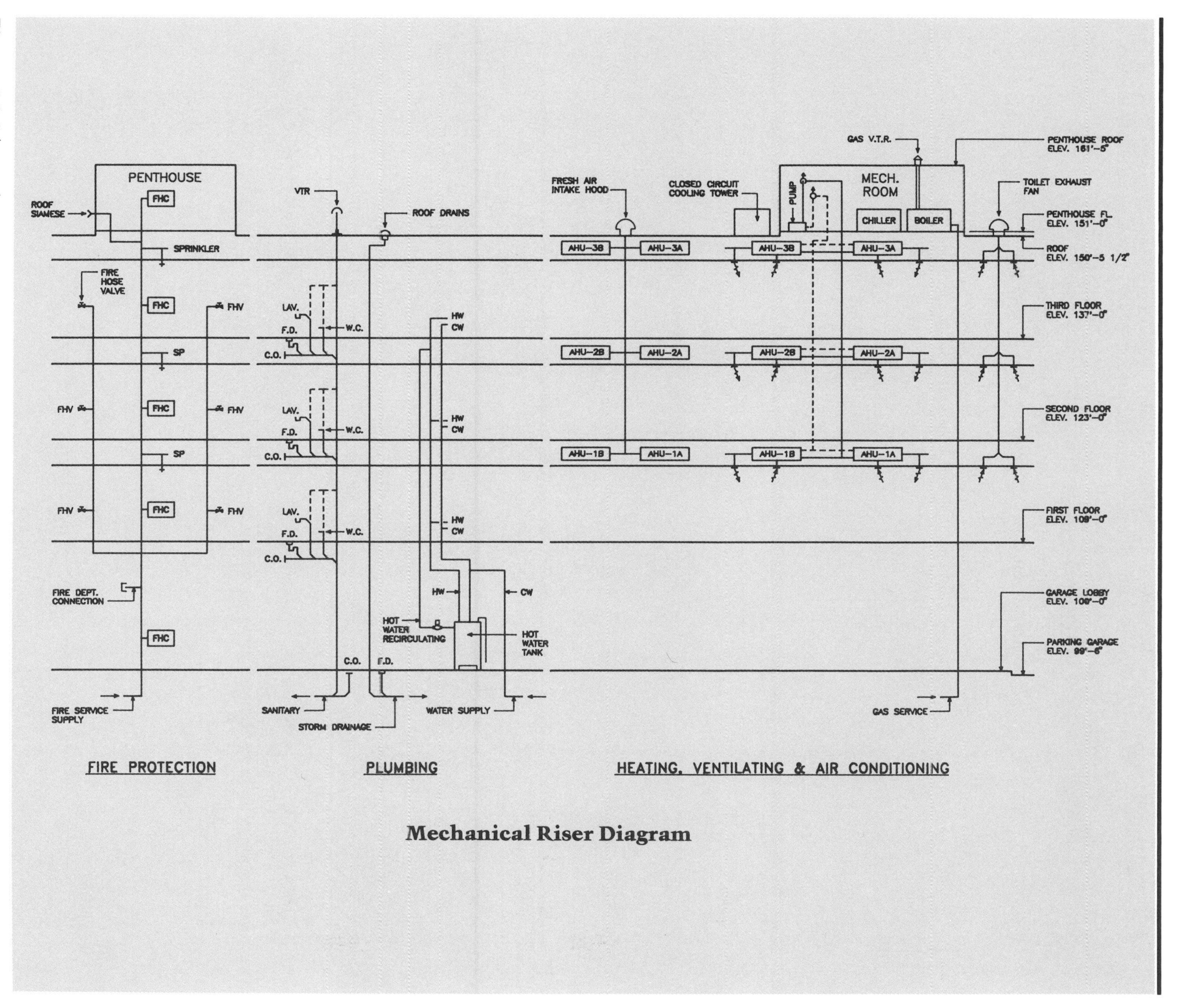

Figure 14.4 (cont.)

Computation of Loads

The next step is to compute the total heating and cooling loads for the building. (For more information, refer to Chapter 2, "Heating and Cooling Loads.")

Heating Loads

The total heating load, H_t, for a building is the sum of all conduction and convection heat losses (see Figure 2.14).

Design Criteria

Item	Reference	Value	Comments	Recommendations
Indoor Design Temp.	Fig. 2.8			
Business	Chap. 3			
Summer		74–78°F/30–55% RH	78°F	78°F/40% RH
Winter		70–74°F	68°F typical code	68°F
Garage		N/A		Not heated or cooled
Outdoor Design Temperature	Fig. 2.9			
Summer (2½%)		94°F/61% RH		94°F/61% RH
Winter (97½%)		16°F		16°F
ΔT Adjoining Spaces	Fig. 2.10			
Stairs			Assume 10°F outdoor	10°F
Garage				98°F Summer 16°F Winter
Ventilation Standards	Fig. 1.22	5–25 cfm/person	Toilet exhaust serves to relieve pressure	7 cfm/person
U values	Chap. 3			
Walls		0.08 Btu/hr./s.f./°F		0.08 Btu/hr./s.f./°F
Roof		0.06 Btu/hr./s.f./°F		0.06 Btu/hr./s.f./°F
Windows		0.53 Btu/hr./s.f./°F		0.53 Btu/hr./s.f./°F
Population	Chap. 2	100 s.f./person Floors 3 x 90′ x 210′ = 56,700 s.f. –Stairs 3 x 2 x 10′ x 20′ = 1,200 –Core 3 x 30′ x 35′ = 3,150 52,350 s.f.		524 People

Figure 14.5

$$H_t = H_c + H_e + H_v + H_s$$
$$H_c = UA\Delta T$$
$$H_e = FP\Delta T$$
$$H_v = 1.1 \text{ cfm } \Delta T$$
$$H_s = KA$$

Conduction Heat Losses: The conduction heat losses through all surfaces — walls, floors, doors, windows, and slabs — are determined by first calculating the area of the surfaces and then multiplying by the temperature differences and U values (see Figure 14.6).

Convection Heat Losses: Because the windows are fixed and the building has a forced air system, the infiltration losses are zero, except at the front door. The ventilation losses, however, are substantial, because the building has an occupancy of 524 persons, with 7 cfm of fresh outdoor air recommended (see Figure 14.5). The crack lengths (in linear feet) for this project are shown in Figure 14.7.

The total heat loss is outlined in Figure 14.8.

$$H_t = H_c + H_e + H_v + H_s = 620 + 0 + 212 + 0 = 832 \text{ MBH}$$

Cooling Loads

The cooling loads are now determined for the building and are summarized in Figure 14.9.

Several times of day and several days of the year could be selected to establish the design temperature. For this example, 3 p.m. (1500 hours solar time) in July is selected. For a more thorough analysis, several other times and days would be considered to determine the overall worst condition. The ability to analyze several dozen trials is an obvious benefit of using a computer. Because the cooling load temperature difference (CLTD), solar heat gain factor (SHGF), shading coefficient (SC), and cooling load factors (CLF) depend on orientation, the heat gain calculations must be done for each orientation separately and then added together to obtain the total cooling load. The total cooling load is the sum of

$$H_t = H_c + H_v + H_s + H_i + H_p$$
$$H_c = \text{conduction heat gain} = UA \text{ (CLTD) (roof, walls, glass and floors)}$$
$$H_v = \text{convection heat gain} = 1.1 \text{ cfm } \Delta T + 4840 \text{ cfm } \Delta W \text{ (ventilation and infiltration)}$$
$$H_s = \text{solar heat gain} = A \times \text{(SHGF)(SC)(CLF)}$$
$$H_i = \text{internal heat gain} = \text{(sensible load)} \times \text{(CLF)} + \text{(latent load)}$$
$$H_p = \text{people heat gain} = N_o \times P_s \times \text{(CLF)} + N_o \times P_i$$

Conduction Heat Gain: The conduction heat gain (UA × CLTD), or the heat transmitted to the building through the enclosure walls, is computed for the roof, walls, glass, doors, and floor. The U and A values have already been determined in Figures 14.6 and 14.7.

U Values and Coefficients			
Reference	**Material**	$U = \frac{1}{R_t}$	**Remarks**
Walls			
Fig. 2.7	10″ concrete	1.10	Sand and gravel
Fig. 2.7	3-5/8″ insulation	13.00	Blanket type
Fig. 2.7	1/2″ gypsum	0.32	(plaster)
Fig. 2.5	Outside air film	0.17	
Fig. 2.5	Inside air film	0.68	
		R_t = 15.17	
		U = 0.066 (0.08 req'd) OK	
		U walls = 0.066 Btu/hr./s.f./°F	
Stair Wall	2.6 Double partition	U stair wall = 0.34 Btu/hr./s.f./°F	Interior wall - 3/8″ gypsum board
Roof			
Fig. 2.7	Roof membrane	0.15	Assume roll roof
Fig. 2.7	2″ rigid insulation	14.40	Polyisocyanurate 2″
Fig. 2.7	6″ concrete	0.67	Sand and gravel
Fig. 2.7	1/2″ spray-on insulation	1.00	Cement fiber-wood
Fig. 2.5	Outside film	0.17	
Fig. 2.5	Inside film	0.61 for horizontal surface	
		R_t = 17.00	
		U = 0.059 (0.06 req'd) OK	
		U roof = 0.059 Btu/hr./s.f.	
Floor (First)			
Fig. 2.7	3″ concrete and 3″ average for joists	0.67	Equiv. to waffle slab
Fig. 2.7	Carpet and fibrous pad	2.08	Add insulation recommended
Fig. 2.5	Inside film	0.61 (horizontal surface)	
Fig. 2.5	Outside film	0.17	Conserves little wind
		R_t = 3.53	
		U floor = 0.28 Btu/hr./s.f./°F	
Windows Fig. 2.6	Double insulated glass	U windows = 0.55 Btu/hr./s.f./°F	
Doors 2.6	Glass door—single	U = 1.1 Btu/hr./s.f./°F	
Edge Coeff. (F)	Not applicable—building above grade		F = N/A
Slab Const. (k)	Not applicable		k = N/A
Crack Coeff. Fig. 2.12 Fig. 2.12	Windows—fixed —no leakage Doors—weatherstop 15 mph (winter) 7½ mph (summer)		Window coeff. = 0 Door coeff. = 0.90 Door coeff. = (0.45 + 0.60)/2 = 0/53 cfm/ft.

Figure 14.6

Figure 2.17 is used to determine the cooling load temperature difference. The roof has a suspended ceiling with a metal deck and concrete slab approximately 4-1/2" thick and weighing 55 pounds per square foot; roof No. 9, close enough at 53 pounds per square foot, is selected. The cooling load temperature difference can be corrected as noted in the table.

$CLTD_{corr}$ = [(CLTD + LM) × K + (78 − T_r) + (T_o − 85)] × f
CLTD = 32°F (Figure 2.17)
LM = latitude month correction (Figure 2.17)
latitude of Albuquerque = 35 (Figure 2.9).
LM at 32, July, Horizontal Surface is 1 (32 ≈ 35) (Figure 2.17).
K = 1.0, assume dark roof
T_r = indoor room design temperature = 78°F (Figure 14.5)
T_o = outdoor design temp = 94°F (Figure 14.5)
f = 1.0 (no attic ducts or fans)

Therefore,

$CLTD_{corr}$ = [(32 + 1) × 1.0 + (78 − 78) + (94 − 85)] 1.0
$CLTD_{corr}$ = 42

For the heat gained by the walls, the cooling load temperature difference depends on the type of wall, time of day, and orientation. As noted in Figure 14.3, the wall weighs 160 pounds per square foot—a heavy wall. According to Figure 2.17, a type B wall is selected (h.w. concrete wall = 156 lbs./s.f.). The time selected is 1500 hours and the cooling load temperature difference is listed for each orientation in Figure 14.9. The correction factor for the walls is similar to that for the roof, except that no fan factor is applied. The correction factor is listed at the bottom of the CLTD tables for walls and computed below for each orientation:

$CLTD_{corr}$ = (CLTD + LM) × K + (78 − T_r) + (T_o − 85)
CLTD = north 9°F, south 14°F, east 24°F, west 14°F (Figure 2.17)
LM = 32° latitude July, north 1°F, south -3°F, east 0°F, west 0°F (Figure 2.17)
K = 1.0 dark room
T_r = 78°F (Figure 14.5)
T_o = 94°F (Figure 14.5)

Thus, for the north wall

$CLTD_{corr}$ = (9 + 1) × 1 + (78 − 78) + (94 − 85) = 19°F

The $CLTD_{corr}$ for south, east, and west walls are computed similarly and are listed in Figure 14.9.

The floor over the garage is a special case. The cooling load temperature difference for the north wall is used to compute the heat gain for the underside of the floor, because it acts as a shaded surface in much the same way as the north wall does.

Windows are thin and transparent compared to building walls, because they do not store heat during the day regardless of their orientation. Using the detailed (ASHRAE) method, the temperature may be varied during the day. However, many designers prefer to take the full load through the glass. This method is used here. Thus, the conduction through all windows is determined simultaneously, and the temperature difference equals

$$T_o - T_r$$

$$94 - 78 = 16°F$$

Areas and Crack Calculations						
Facade	**(1) Gross Wall Area (s.f.)**	**(2) Window Area (s.f.)**	**(3) Door Area (s.f.)**	**(4) Misc. Area (s.f.)**	**Net Area (1)-(2)-(3)-(4)**	**Crack Length (l.f.)**
North	50.46′ x 210′ 10,596 s.f.	18 x 5′ x 33′ 2,970 s.f.	— —	— —	7,626 s.f.	Windows - 0 Doors - 0
South	50.46′ x 210′ 10,596 s.f.	17 x 5′ x 33′ 10′ x 10′ - 42 2,863 s.f.	2 x 3′ x 7′ 42 s.f.	— —	7,691 s.f.	Windows - 0 Doors 2 x 6 + 3 x 7 33 l.f.
East	50.46′ x 90′ 4,541 s.f.	6 x 5′ x 18′ 540 s.f.	— —	— —	4,001 s.f.	Windows - 0 Doors - 0
West	50.46′ x 90′ 4,541 s.f.	6 x 5′ x 18′ 540 s.f.	—	—	4,001 s.f.	—
Total		Windows 6,913 s.f.	Doors 42 s.f.		23,319 s.f.	Doors 33 l.f.

Penthouse and stair doors ignored.
Roof Area = 90′ x 210′ = 18,900 s.f.
1st Floor Area = 90′ x 210′ = 18,900 s.f.
Stair Towers - Wall Area = 2 (towers) x 50.46′ (high) x [10′ + 20′ + 10′] (interior wall) = 4,036 s.f.

Figure 14.7

Note: At 1500 hours, this method produces results nearly equal to the results of the ASHRAE method. At other times of the day, it produces a slightly larger (conservative) load when compared to the ASHRAE method.

Heat Loads

Conduction
$H_c = UA\Delta T$

	U Btu/hr./s.f./°F (Fig. 14.6)	A (s.f.) (Fig. 14.7)	ΔT (°F) (68-16) = 52°F (Fig. 14.5)	Heat Loss (Btu/hr.)
Walls (net)	0.06	23,319	52	72,755
Windows	0.55	6,913	52	197,712
Doors	1.1	42	52	2,402
Roof	0.059	18,900	52	57,985
First Floor	0.28	18,900	52	275,184
Stairs	0.34	4,036	10	13,722

H_c = 619,760 Btu/hr.

H_e = FP = 0 (F = 0, no perimeter losses—building above grade over garage)

H_e = 0 Btu/hr.

H_s = kA = 0 (k = 0, no slab on grade in heated space) H_s = 0 Btu/hr.

Total Conduction = 619,760 Btu/hr.
H_c = 0 Btu/hr.
= 620 MBH

Convection
$H_v = 1.1$ cfm ΔT

	Crack Length (Fig. 14.7)		cfm/ft. (Fig. 14.6)		ΔT (Fig. 14.5)		Constant	Heat Loss Btu/hr.
Windows			0					0
Doors	33′	x	0.90	x	52	x	1.1=	1,699 Btu/hr.
Ventilation (Fig. 14.5)	524	x	7.22	x	52	x	1.1=	209,810

Total Convection = 211,509 Btu/hr.
= 212 MBH

Note that all of the 212 MBH from ventilation infiltrates simultaneously; no reduction is taken, as was done for the windows in the multi-family housing model in Chapter 13.

Figure 14.8

Cooling Loads

Conduction Heat Gain
H_c = U x A x (CLTD)[1]

	U (Btu/hr./s.f./°F) (Fig. 14.65)	A (s.f.) (Fig. 14.7)	CLTD (°F) (Fig. 2.17)	$CLTD_{corr}$ (°F)	H_t	Remarks Solar Time 1,500 hrs.
Roof	0.059	18,900	32	42	46,834	Suspended ceiling Roof # 9, Type B
Walls						
North	0.065	7,626	9	19	9,418	K = 1.0, T_r = 78, T_o = 94
South	0.065	7,691	14	20	9,998	
East	0.065	4,001	24	33	8,582	
West	0.065	4,001	14	23	5,981	
Floor	0.28	18,900		19	100,548	Use north CLTD T_o for under floor
Glass						
Doors	1.1	42		16	739	
Windows	0.55	6,913		16	60,834	94–78 = 16
Partitions	0.34	4,036		10	13,722	Fig. 2.6—double partition-stair

Total = 256,656 Btu/hr. H_c = 260 MBH

Convection Heat Gain
H_v = 1.1 cfmΔT = 4,840 cfmΔW

	Sensible (Btu/hr.)			+	Latent (Btu/hr.)		
Windows	0			+	0	=	0
Doors	[1.1 x (0.53 x 33) x 20]	=	385	+	[4,840 x (0.53 x 33) x (-0.0131)]	=	-1,108
Ventilation	[1.1 x (524 x 7) x 20]	=	80,696	+	[4,840 x (524 x 7) x (-0.0131)]	=	-232,565
	H_v (sensible)	=	81,081		H_v (latent)	=	-233,673
		=	82 MBH			=	234 MBH[3] (117)

Computation of Moisture Content
(ΔW)

Outdoor (94°F/61%) = 0.0214 lb./lb. air[2]
Indoor (78°F/40%) = 0.0083 lb./lb. air[2]
ΔW= -0.0131 lb. moisture/lb. dry air

[1]CLTD as corrected—see discussion in text.
[2]Refer to Figures 2.18 and 14.5.
[3]Use of 50% of 234 to achieve more conservative design due to unusual dry air condition.

Figure 14.9

Cooling Loads (continued)

Solar Heat Gain
H_s = A (SHGF) (SC) (CLF)

	A (s.f.) (Fig. 14.7)	SHG (Fig. 2.19)	SC (Fig. 2.20)	CLF (Fig. 2.17)	H_s (Btu/hr.)
Windows (July–36° lat.)					
North	2,970	39	0.56	0.82	53,189
South	2,863	90	0.56	0.35	50,503
East	540	216	0.56	0.20	13,064
West	540	216	0.56	0.72	47,029
Doors (south)	42	90	1.0	0.50	1,890

Total H_s = 165,675
= 166 MBH

Internal Heat Gain
H_i + (Sensible Load) x CLF + (Latent Load)

		H_i (Sensible) (Btu/hr.)	CLF (Fig. 2.19)	H_i (Latent) (Btu/hr.)
Lights	2.5 watts/s.f./ x 3.41 Btu/watt x 3 floors (18,900 s.f./floor)	483,368	1.0	0
Computers	262 x 450	117,900	1.0	0
Office Machines	3 x 10,000	30,000	1.0	0
Coffee	3 x 6,500	19,500	1.0	3 x 2,000 = 6,000
Power	2.0 watts/s.f. x 3.41 Btu/watt x 3 floors (18,900 s.f./floor)	386,694		

H_i (Sensible) = 1,037,462
= 1,038 MBH

H_i (Latent) = 6,000
= 6 MBH

People Heat Gain +
$H_p = N_oP_s$ (CLF) + N_o P_L

Activity	N_o	P_s (Fig. 2.21)	CLF	H_p (Sensible)	P_L (Fig. 2.21)	H_p (Latent)
Office Work	524	255	1.0	133,620	255	133,620

H_p(Sensible) =134 MBH
H_p(Latent) =134 MBH

Figure 14.9 (cont.)

Convection Heat Gain: The quantities for infiltration (in cfm) in the summer are lower than they are in the winter, because the design velocity of the air in the summer is 7.5 mph; in the winter it is 15 mph. In this example, only the doors infiltrate air and the majority of the convection load comes from the ventilation air requirement of 7 cfm per person. The convection sensible heat load (1.1 cfm ΔT) is computed separately from the convection latent heat load (4,840 cfm ΔW), and then they are added together. From Figure 2.18, the moisture content of the design room condition (78°F/40 percent relative humidity) is 0.0083 pounds of moisture per pound of dry air. This is larger than the moisture content of the outdoor air (94°F/61 wet bulb), which is 0.0214 pounds of moisture per pound of dry air. Therefore, the dryer outdoor air will minimize some of the latent heat load, which is indicated by the negative values for ΔW in Figure 14.9. Normally, outdoor air adds dramatically to the latent and overall load, but Albuquerque is unusual in this regard.

Humidification would generally be indicated; however, in the summer, dry air is preferable, so the unusual choice of dehumidification is used to offset internal latent heat gains. The toilet exhaust will remove some moisture and dry outdoor air will be introduced to replace it.

Solar Heat Gain: Solar heat gain, the heat gain due to radiant solar energy through glass, varies with orientation. Therefore, the loads for each of the four sides (and skylights, if any) are computed individually. The solar heat gain factor (SHGF) is obtained from Figure 2.19. For each orientation in July, the selected month of the design temperature, 36° latitude is chosen as close enough to Albuquerque.

The shading coefficient is obtained from Figure 2.20. For single-pane glass, the transmissibility is 86 percent. One pane of glass with medium-color venetian blinds has a shading coefficient of 0.65. Double-insulated glass, in this example, reduces the shading coefficient by 86 percent, or:

$$SC = 0.65 \times 0.86 = 0.56$$

For the door, the shading coefficient equals 1.0. Although the shading coefficient could be reduced further because the windows are recessed and a portion of the head and one jamb are in shadow, this would account for only a small overall savings and therefore is ignored for this model.

The cooling load factors are obtained from Figure 2.21. The window cooling load factor is obtained from the chart with interior shading. The door cooling load factor is selected from the charts with no interior shading. Because the building walls and floors are heavy (over 75 lbs./s.f.), there is good reason to be comfortable with the cooling load factor reduction, which accounts for the thermal mass of the building and its effect on drawing down the heat gain. If this were a metal curtain wall building, it would be examined more carefully to determine if the cooling load factor values listed would be used or if a value closer to 1.0 should be used to be conservative.

Internal Heat Gain: The main internal heat gains in offices are from lights, computers, office equipment, and small appliances. The lighting load should be obtained from the architect, who often consults a lighting designer or the electrical engineer. A typical layout for offices uses 2′ × 4′ fluorescent fixtures arranged in a suspended ceiling on an 8′ x 12′ spacing, or 96 square feet for each fixture. If each fixture has four 40-watt fluorescent lamps, the energy density would be (4 × 40 watts)/96 square feet or 1.67 watts per square foot. Increasing this by 25 percent for the ballast energy yields 2.08 watts per square foot for general lighting. This should be adjusted for task lights and special lighting to about 2.5 watts per square foot. The lights are one of the most significant heat gain loads to be considered when calculating the overall cooling load.

Computers and other office machines add heat to a building. For this example, personal computers are assumed to be dispersed throughout the building for office tasks. The building does not have a specific computer room. Each personal computer produces 450 Btu/hour sensible heat. One personal computer for each 2 occupants, or 262 computers, are assumed to be operating at any one time.

An allowance of 10,000 Btu/hour on each floor will be made for copy machines, typewriters, and other office machines. It is also assumed that each floor will be allocated an electric coffee warmer. From Figure 2.22, the sensible load for this appliance is 6,500 Btu/hour and the latent load is 2,000 Btu/hour.

Power allowances must also be made for receptacles and machine power, such as fans and portions of pump energy that will wind up in the occupied spaces and will add to the heat load. To account for this, 1-1/2 watts per square foot is assumed for general receptacles and 1/2 watt per square foot for machine loads, or a total of two watts per square foot for power.

People Heat Gain: For office work, the latent and sensible heat gains are each equal to 255 Btu/hour per person. The cooling load factor equals 1.0, a figure which is slightly conservative and generally accepted.

Thus, the total heat gain for the building is:

	Gain	**Sensible Load (MBH)**	**Latent Load (MBH)**
Conduction	H_c	260	
Convection	H_v	82	117
Solar	H_s	166	
Internal	H_i	1,038	6
People	H_p	134	134
	H_t	1,680 MBH +	23 MBH

H_t = 1,703 MBH = 142 tons (round to 145 tons)

Equipment Selection

The basic system shown in Figure 14.4 indicates a mechanical room on the roof with a boiler, chiller, and air cooled condenser. The overall simplicity of the open plan office layout lends itself to minimizing the number of pieces of equipment. All ductwork and air handling units can be easily accommodated above the suspended ceiling. The ductwork above the ceiling provides for heating or cooling and allows for a degree of flexibility in future layouts. By placing the air handling units near the central core, the piping is minimized and accessible, and fresh outdoor air can be easily supplied to each unit. The supply air ducts distribute air out to the perimeter, as well as to some interior zones, and the return air ducts are located at the center of the interior zones. In this way, supply air is delivered to the perimeter, where most of the heat loss and gain originates, and also supplies cooling to the general office area. The central return air removes warmed air for cooling. Toilet exhaust fans are sized for 80 percent of the fresh air supply requirements to keep the building slightly pressurized. A layout of the first floor is shown in Figure 14.10.

Because the piping to a cooling tower or air cooled condenser is located outside the building, consideration must be given to freeze protection. The choices are either to drain the condenser water in the winter (which is not a possibility if water operation is necessary) or to use a glycol/water solution in the exposed piping network. This office building requires cooling in cold weather, and therefore glycol is used in the outside loop. This is somewhat conservative, but it will avoid a severe repair cost should a power failure in winter cause the pipes to freeze. The building size is considered small to intermediate, and would probably not have a full-time attendant on duty; and so the extra glycol protection is good overall insurance. The use of glycol means that an air cooled condenser (closed circuit) must be used to cool the condenser glycol/water solution from the chiller. The glycol/water solution will be in a closed loop. If plain water were used, a cooling tower would be selected, which causes the condenser water to spill over the tower fill in an open loop. The objection to using a cooling tower with glycol is that it mandates the use of the closed loop, such as an air cooled condenser. The selection is made later in this chapter.

The previous analysis results in the following design objectives:

1. Provide supply ducts in ceiling near perimeter and exterior zones.
2. Provide return air at center of interior zones.
3. Distribute hydronic piping from mechanical room to air handling units.
4. Provide fresh air for ventilation from roof to air handling units — exhaust air from toilet rooms.
5. Use a glycol/water solution in the condenser loop.
6. Use an air cooled condenser.

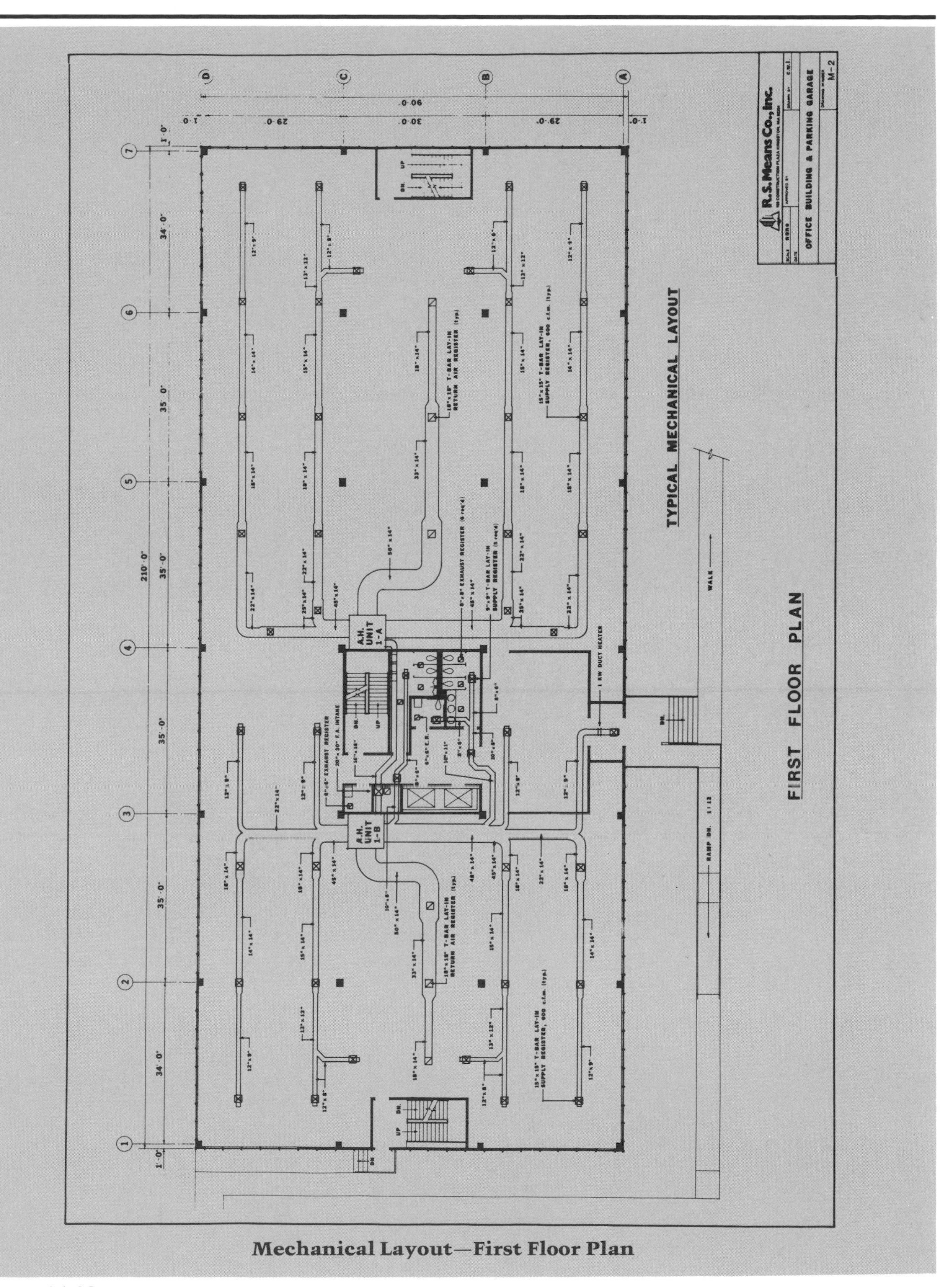

Mechanical Layout—First Floor Plan

Figure 14.10

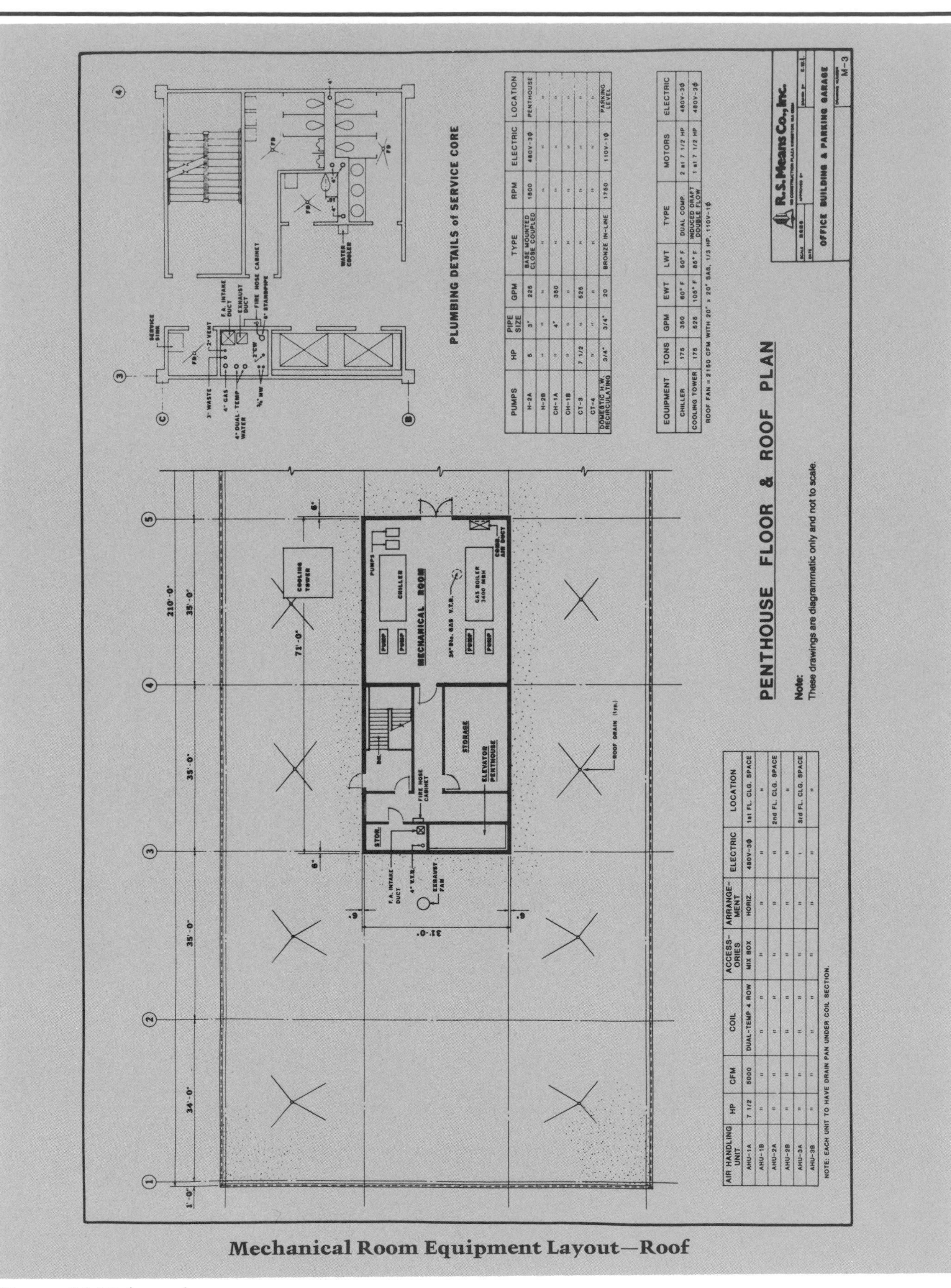

Mechanical Room Equipment Layout—Roof

Figure 14.10 (cont.)

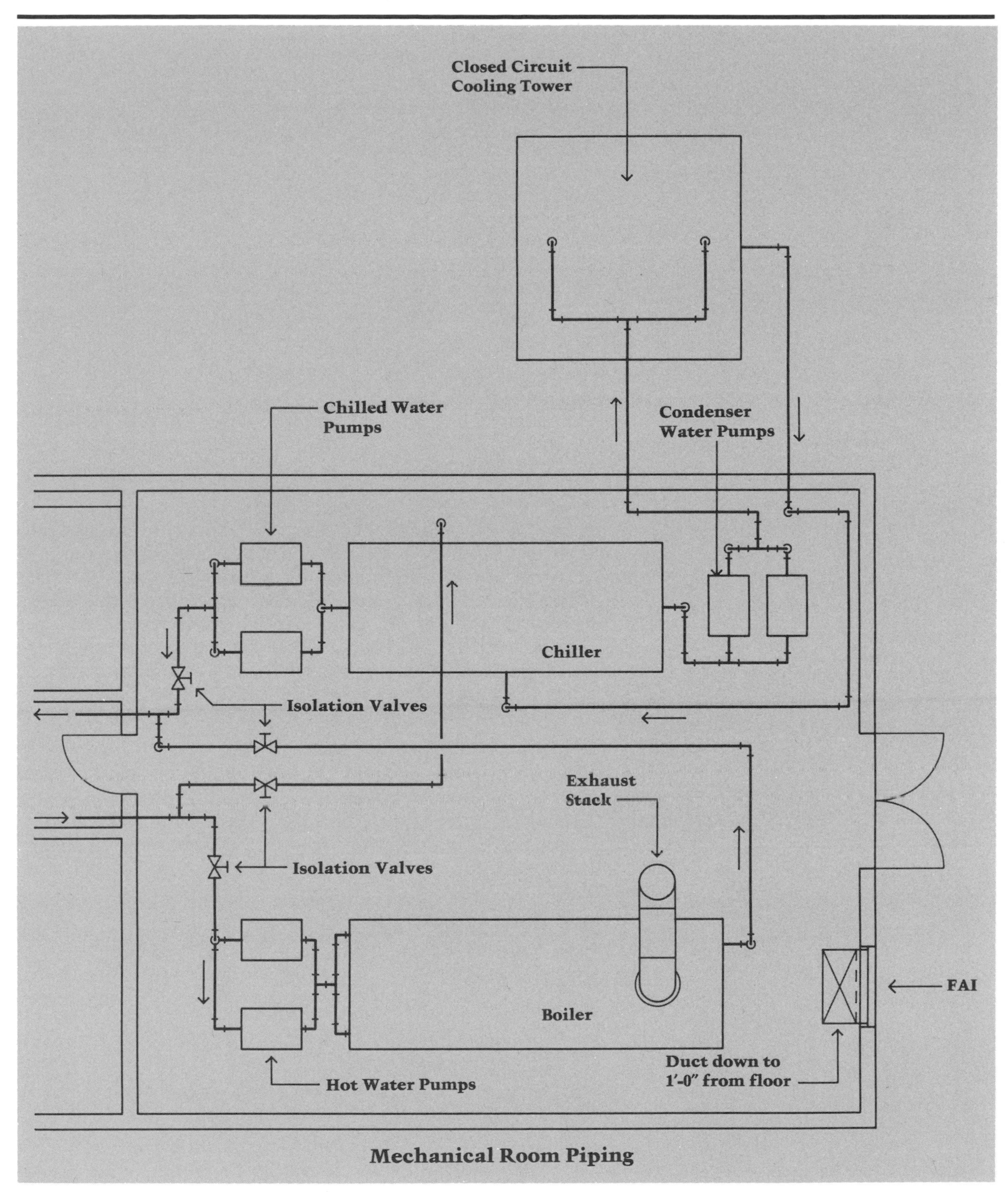
Closed Circuit
Cooling Tower
Chilled Water
Pumps
Condenser
Water Pumps
Chiller
Isolation Valves
Exhaust
Stack
Isolation Valves
FAI
Boiler
Hot Water Pumps
Duct down to
1'-0" from floor
Mechanical Room Piping

Figure 14.11

Generation Equipment

The following generation equipment is selected (see Figure 14.4):

- Hot water boiler — gas-fired
- Hot water expansion tank
- Chiller
- Chilled water expansion tank
- Air cooled condenser
- Hot water circulating pumps
- Chilled water pumps
- Condenser glycol/water pumps
- Fresh air supply fan
- Toilet exhaust fan

Hot Water Boiler — Gas-Fired: The boiler output size is determined as follows:

Net building heat loss	832	MBH (H_t)
Allowance for pipe losses (5%)	41	(insulated pipe)
Pickup allowance (5%)	43	(small startup short piping)
Total load	916	
Allow 25% oversizing	229	
Total design boiler output	**1,145**	**MBH**

Using the boiler selection chart (Figure 5.1), cast iron boilers are chosen for this example because they are available in the range needed. Gas-fired boilers are commonly available with a low-fire and high-fire operation as an option, and this feature is selected. This option gives a boiler the same general efficiency as two boilers, but at a substantial cost savings. Since the building has substantial heat gains from lights, people, and computers, the risk of a heat loss problem due to boiler failure is minimized. A gas-fired hot water boiler with an output of 1,275 MBH is selected.

Hot Water Expansion Tank: The hot water system is assumed to contain 1,000 gallons. This should be double-checked during a final review of all piping totals. The appropriate tank in terms of volume is obtained from the expansion tank data sheet. The following equation is used to calculate the required volume of the tank:

$$V_t = \frac{(0.00041t - 0.0466)}{P_a/P_f - P_a/P_o} \times V_s \text{ (for water between 160°F and 280°F)}$$

V_t = tank volume

t = 190°F — average hot water temperature

V_s = 1,000 gallons of water in system

P_a = 14.7 — atmospheric pressure in psia

P_f = $P_a + P_{height} + P_{residual}$ — initial tank pressure. With the tank at top of building, $P_{height} = 0$.
14.7 + 0 + 1.5 = 16.2 psia

P_o = maximum tank pressure — 30 psi (relief valve) + 14.7 (P_a) – 10% = 39 psig

$$V_t = \frac{(0.00041 \times 190 - 0.0466) \times 1{,}000}{14.7/16.2 - 14.7/39} = \frac{31.3}{0.91 - 0.38} = 59.1 \text{ gal.}$$

An 80-gallon ASME-approved steel tank is selected.

Chilled Water Expansion Tank: The size of the chilled water expansion tank is determined similar to a hot water expansion tank, but using the adjusted formula. The volume of water is assumed to be the same (1,000 gallons) and the average system temperature, t, is assumed to be 50°F.

$$V_t = \frac{(0.006)V_s}{14.7/16.2 - 14.7/39} = \frac{6}{0.53} = 11.3 \text{ gal.}$$

A 30-gallon ASME steel expansion tank (the smallest practical size) is selected.

Chiller: The chiller is sized with a 20 percent reserve capacity.

Heat gain	145 tons – 1,741 MBH
20% reserve	29 tons – 348 MBH
Chiller size	174 tons – 2,089 MBH

Therefore, use a 175 ton chiller (2,100 MBH).

The chiller produces water at 45–50°F, which is sent to the coils in the air handling units. The warmed return water (52–60°F) is sent back to the chiller for cooling. The chiller cools the water using a refrigeration cycle, which takes the heat from the water in the air handling unit loop and sends it to the air cooled condenser. This is then sent out to be cooled at approximately 88–90°F and returns to the chiller at approximately 80–85°F. As previously discussed, a glycol/water mixture will be used instead of water to cool the system to prevent freezing.

Using the selection criteria for chillers (Figure 1.14), either a reciprocating or centrifugal type could be selected. The small size indicates that an absorption cycle chiller is too impractical to use and the required steam is not available. Therefore, a reciprocating water cooled chiller at 175 tons with dual compressors and direct drive is chosen. The chiller will be glycol cooled with adjustments made for glycol in sizing the pumps. The dual compressors allow for some load modulation, and the direct drive is available from the manufacturer. The manufacturer's literature notes that the chiller handles 600 gpm of water, requires 28 feet of head to drive the water through the chiller at design conditions, and that the head loss for the condenser water through the chiller is 18 feet.

Air Cooled Condenser: The air cooled condenser is sized to match the requirements of the chiller, which, in this case, is 175 tons. The condenser water from the chiller could be cooled by a cooling tower or by an air cooled condenser. In a cooling tower, the water spills over a series of baffles (fill), and the droplets are cooled by circulating air, which also causes some evaporation, adding to the cooling effect. A cooling tower is an open loop, because the water leaves the piping to spill over the fill (see Figure 9.1). A cooling tower is not effective in this case, however, because the glycol does not operate well in an open cooling tower and requires a closed loop to contain the glycol/water mixture. Therefore, a 175-ton closed loop air cooled condenser is selected to cool the glycol/water mixture as it passes through coils. Manufacturer's literature indicates that approximately 800 gpm of water against a 57 foot head is the necessary flow through the condenser.

Hot Water Circulating Pump(s): Two hot water circulating pumps will be provided, one as a standby. The hot water pumps should have high-temperature seals. Using Figure 7.2, a frame-mounted pump is selected. To specify the pump, the capacity in gpm and the head in feet must be determined.

Capacity: Using the relationship that 1 gpm = 10 MBH at a 20°F temperature drop, the capacity to move the design heat load of 832 MBH is 83.2 gpm, rounded to 84 gpm.

Head: The head is calculated by determining the friction loss around the longest loops:

Straight run of pipe		
Mechanical room supply and return 2 × 30′	=	60.0 l.f.
Vertical supply and return to AHU (156′–114′)	=	42.0
Horizontal supply and return at AHU 2 × 6′	=	12.0
straight run	=	114.0 l.f.

Fittings

1 elbow equivalent = 2.7 feet (see Figure 9.12) – 3 fps on 1″ pipe

Boiler Room		
1 boiler (Fig. 9.12)	3 × 1	3.0 l.f.
10 elbows	10 × 2.7′ × 1	27.0
5 gate valves	5 × 2.7′ × 0.7	9.5
2 globe valves	2 × 2.7′ × 17	91.8
AHU Fittings		
4 elbows	4 × 2.7′ × 1	10.8
2 gate valves	2 × 2.7′ × 0.7	3.8
2 globe valves	2 × 2.7′ × 17	91.8
		237.7
Add 50% for misc. fittings, strainers		118.9
Straight run of pipe		114.0
Total equivalent pipe length =		470.6 l.f.

The 50 percent allowance for fittings is a short cut. A detailed take-off of the final loop would be required in an actual job to double-check this figure. The air handling unit coil also has a head loss. For heating, it will be taken at 3 feet for the anticipated flow — check with specific manufacturers for other specific recommendations.

The head loss on 471 linear feet of pipe at the general design loss of 2.5′/100′ equals:

$$H_p = 471' \times 2.5/100 + 3'(\text{AHU}) = 14.8, \text{ or } 15 \text{ feet of head}$$

This is very conservative, since the chilled water demand is much higher and, if the piping is sized for the chilled water capacity, the heating loop will be oversized, resulting in lower head for the hot water pump. A pump with 84 gpm and 15 feet of head is therefore required. From Figure 7.3, a 60-14 type pump is selected.

Chilled Water Pump(s): The chilled water system piping loop is identical to the hot water loop, except that the chilled water will flow through the chiller instead of the boiler. This produces a different friction loss through the system, which circulates at approximately 50°F instead of 200°F.

Capacity: The required flow through the system is determined from the heat gain plus the reserve, which totals 175 tons (2,100 MBH). In cooling, the supply and return water temperatures typically differ by approximately 7°F, instead of the 20°F drop associated with heating. Thus, the standard flow equation, 1 gpm = 10 MBH at 20°F drop, must be modified by the ratio of 20 to 7 to correct for this difference. The seven degree differential and required flow must be verified with the particular chiller selected.

$$\text{gpm req'd} = \frac{2{,}100 \text{ MBH}}{10 \text{ MBH/gal.}} \times \frac{20}{7} = 600 \text{ gpm}$$

The 600 gpm agrees with the chiller data previously stated.

Head: The head from the heating loop can be adjusted to account for the minor differences of the chiller and the temperature. From the chiller selection, the chiller has a head loss of 28 feet, the boiler 0.08 feet (3 × 2.5/100). The head loss for the air handling unit coil in cooling is assumed to be nine feet for cooling, which is greater than the three feet needed for the heating coils.

Since the friction loss flow charts, Figure 9.11, are based on 60°F water, no temperature correction is necessary for the ± 50°F chilled water. Therefore, the friction head required for the chilled water piping is:

Head loss piping loop (same as hot water pumps)	= 15.0 feet
Correction for chiller (+28′ – 0.08′)	= 28.0
Correction for AHU (+9′ – 3′)	= 6.0
Design chilled water pump	= 49.0 feet
Friction head required	= 50.0 feet

The chilled water pump is designed for 600 gpm flow and 50 feet of head. Beyond the range of those shown in Figure 7.3, manufacturers' literature should be consulted. The requirements of the pumps are listed on the system summary sheet at the end of this chapter.

Condenser Glycol/Water Pumps: The condenser glycol/water pumps circulate the glycol/water mixture between the chiller and the air condenser. The capacity and head are now determined.

Capacity: In this example, the condenser water enters at 88°F and leaves at 83°F — a 5°F drop in temperature, which is about the average temperature drop. In some cases, particularly Albuquerque, evaporative coolers perform well because of the relatively dry air. However, an air cooled condenser has been selected for this example; therefore, the 5°F temperature drop is used.

$$\text{gpm req'd} = \frac{2{,}100 \text{ MBH}}{10 \text{ MBH/gal.}} \times \frac{20}{5} = 840 \text{ gpm}$$

Because glycol is used, the flow must also be corrected to account for the specific heat of the mixture, using Figure 9.12. For a 100°F fluid temperature, the flow increases by a factor of 1.16. Hence, the capacity of the pumps required is:

840 gpm x 1.16 = 975 gpm

Head: The manufacturer's literature for this particular air cooled condenser indicates a head loss of 57 feet, and the manufacturer's literature for the chiller indicates a head loss of 18 feet. In addition to these losses, the following piping losses must be accounted for:

Straight run of pipe

Supply and return chiller to condenser			80.0 l.f.
Fittings (1 elbow equivalent = 2.7 feet)			
6 gate valves	6 × 2.7′ × 0.7		11.3
2 globe valves	2 × 2.7′ × 17		91.8
10 elbows	10 × 2.7′ × 1		27.0
	Equivalent pipe length	=	210.0
	Add 50% for strainers	=	105.0
	Use		315.0 l.f.

Pump head (315 l.f. × 2.5′/100′)		7.9 ft.
+ chiller		57.0
+ air cooled condenser		18.0
Head for water system	**=**	**82.9 ft.**

This figure must be adjusted for glycol using the formula provided in Figure 9.12. For 100°F water, the pressure drop correction factor is 1.49. Thus, the design head loss is:

82.9 × 1.49 = 123.5 ft.

Glycol may dissolve certain types of gaskets, and therefore, glycol-resistant seals are specified.

A 975 gpm, 124 foot head pump with glycol-resistant seals is entered on the system summary sheet.

Fresh Air Supply Fan: The purpose of the fresh air supply fan is to supply ventilation air to the building. The design requirement is 7 cfm of air per person for the design capacity of 524 persons, which has already been accounted for as part of the convection loss for heating and cooling (Figures 14.8 and 14.9). The selection guide for fan types (Figure 7.7) indicates that a centrifugal fan with forward curved or radial blades would be appropriate for this low-volume application. The fan requires curbs and flashing accessories for roof mounting. In order to avoid heating coil freeze-up in winter, an electric preheat coil set at 55°F minimum must be provided. The capacity and head of the supply fan are now determined.

Capacity: 524 persons × 7 cfm each = 3,668 cfm, or 3,700 cfm.

Head: The fan will direct air vertically down to the six air handling units. The length of the ductwork is:

Straight run of duct	
Roof to first floor (151–120)	31 l.f.
Fittings	
roughly 3,700 cfm @ 1,000 fpm = 3.7 s.f.	
duct = 24″ × 24″	
3 elbows (smooth elbow (Figure 10.22))	
L/D = 9 D = 2′ (24″) L = 9 × 2′	18
	49
Add 50% fittings	25
Total duct length	75 l.f.

Minimum fan pressure head

75 l.f. × 0.1″/100′ = 0.075″ (pressure including the duct heater)

This is low when compared to the capability of such fans. Therefore, a 3,800 cfm at 0.5″ pressure supply fan is selected.

Toilet Exhaust: The exhaust requirements are now sized for 80 percent of the supply air:

0.8 × 3,800 = 3,040 cfm

The pressure requirements are similar to the supply air and, therefore, a roof exhaust fan is selected. Louvres should be provided in the doors or transoms to the toilet rooms to permit the air to move from the space to the exhaust system.

Distribution Equipment

There are two major components that make up a distribution system — the piping and the ductwork. For both, the cooling loads govern the design of the distribution system.

The cooling load is broken down by floor using the data from Figure 14.9, which shows the computation of the cooling load for this commercial building model.

Generally, the loads on the three floors are equal except for adjustments for the roof, first floor slab, and door loads. The load breakdown for the building is shown in Figure 14.12.

Using these loads, the loads for each air handling unit are established. (The loads and pipe sizes are detailed in Figure 14.13.)

AHU 1A and 1B = 760 MBH/2 = 380 MBH
AHU 2A and 2B = 638 MBH/2 = 319 MBH
AHU 3A and 3B = 693 MBH/2 = 347 MBH

The flow to each air handling unit is based on a 7°F temperature drop. Therefore:

$$\text{AHU 1A and 1B} = \frac{380\text{ MBH}}{10\text{ MBH/gal.}} \times \frac{20}{7} = 109\text{ gpm}$$

$$\text{AHU 2A and 2B} = \frac{319\text{ MBH}}{10\text{ MBH/gal.}} \times \frac{20}{7} = 91\text{ gpm}$$

$$\text{AHU 3A and 3B} = \frac{347\text{ MBH}}{10\text{ MBH/gal.}} \times \frac{20}{7} = 99\text{ gpm}$$

The total flow to all air handling units matches the 600 gpm supplied by the chiller.

The pipe sizes are indicated in Figure 14.13. Steel pipe is used throughout.

Ductwork: The ductwork consists of the following (see Figure 14.10):

- supply air from each air handling unit
- return air to each air handling unit
- fresh air supply from the roof
- toilet exhaust

The constant pressure drop method is used to size the ductwork, which, for architectural consideration, has been limited to 14 inches deep in this building. Because all six air handling units are relatively close in size (319 to 380 MBH), the ductwork is designed for only one air handling unit, 1A, to illustrate the procedure. Figure 14.14 illustrates the schematic layout of the ductwork and sizing calculations.

The supply registers are laid out approximately 30 feet on center around the building perimeter, and the return air is taken from the center of each area. This draws the warmer inner core air to the air handling unit for cooling directly, distributes the cool air to the perimeter, and draws it across the floor to the inner core.

Load Breakdown by Floor

Adjustments	
Roof Loads (third floor)	
Conduction	46.8 MBH
First Floor Slab and Door Loads	
Conduction	100.5 MBH
Doors	.7 MBH
Convection—Doors	.4 MBH
Solar	
Doors	.5 MBH
	102.1 MBH

	MBH		MBH		MBH
Roof	46.8	Slab & Door	102.1		
Third Floor	+531.0	First Floor	+531.0	Second Floor	531.0
Roof	577.8	First Floor	633.1	Second Floor	531.0
Increase by 20%	+115.0		+127.0		+106.0

Roof	693 MBH
First Floor	760 MBH
Second Floor	637 MBH
Total	2,090 MBH

Figure 14.12

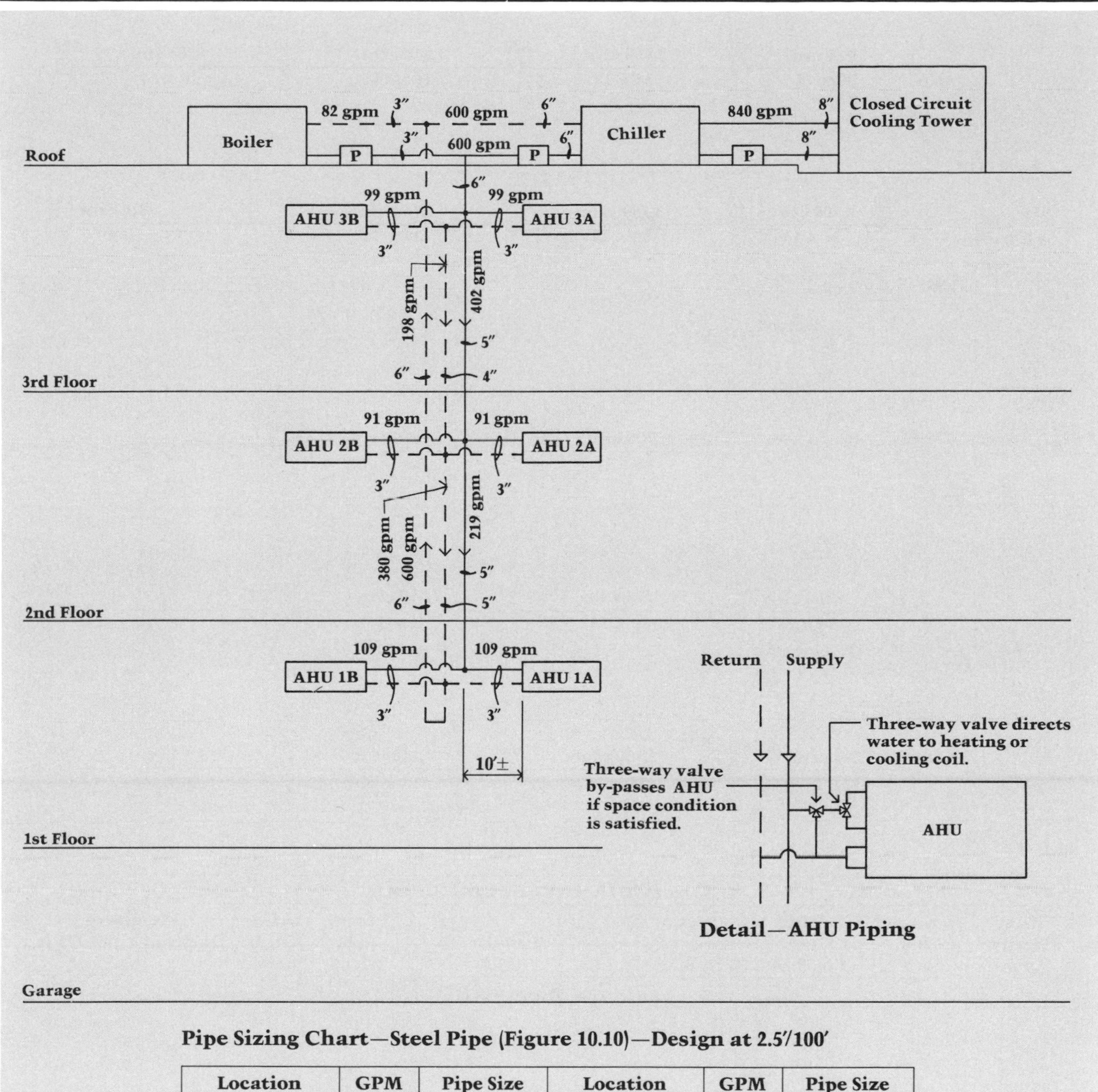

Pipe Sizing Chart—Steel Pipe (Figure 10.10)—Design at 2.5′/100′

Location	GPM	Pipe Size	Location	GPM	Pipe Size
Boiler	82	3″	Return	398	5″
Chiller	600	6″	Return	600	6″
Condenser	840	8″	AHU 1A and 1B	109	3″
Supply	402	5″	AHU 2A and 2B	91	3″
Supply	219	5″	AHU 3A and 3B	99	3″
Return	198	4″			

Loads and Pipe Sizes for Distribution Systems

Figure 14.13

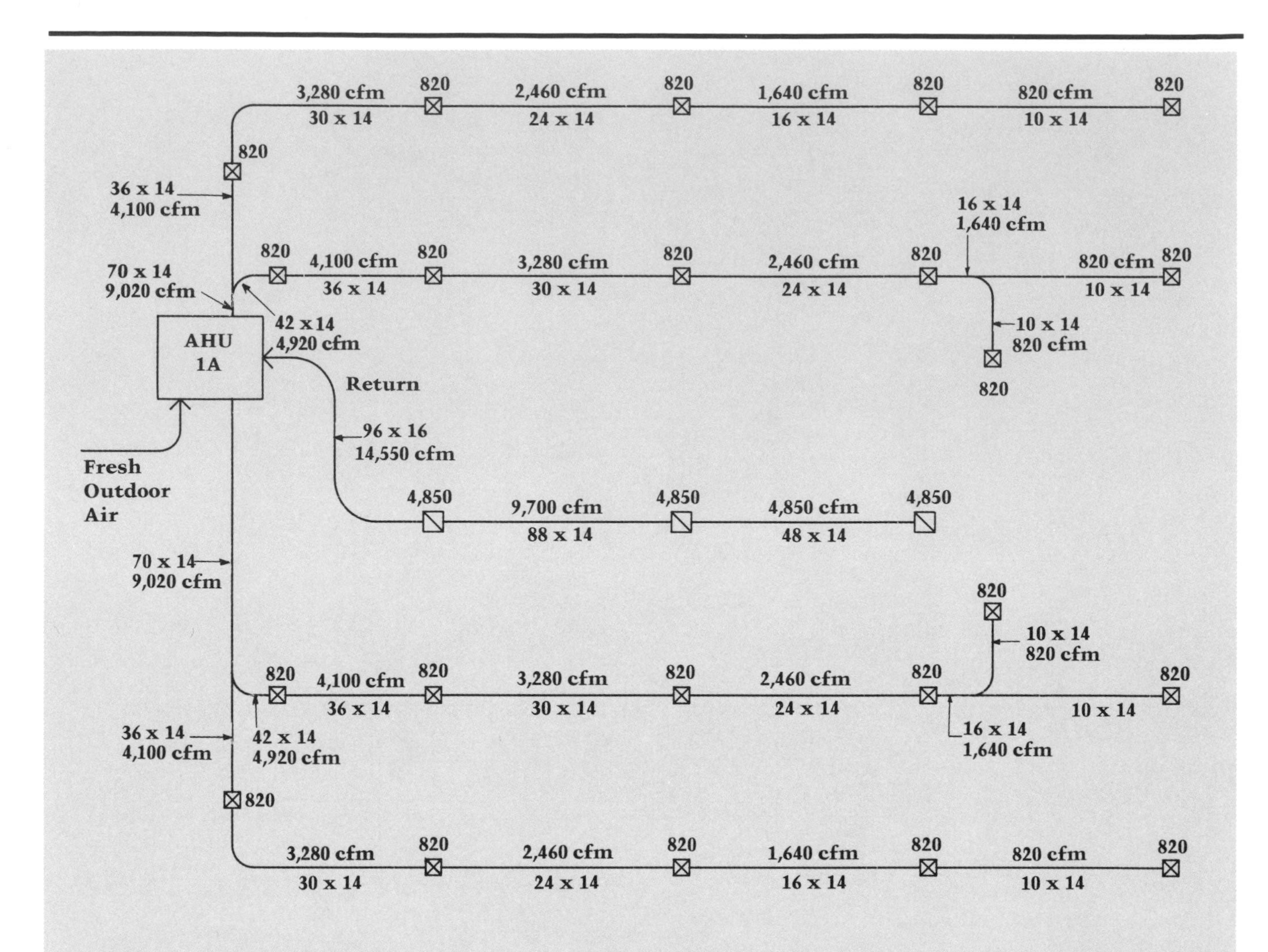

Ductwork Layout and Sizes

Duct Sizing (cfm)	Loss per 100′ (Fig. 9.22)	Round Duct (Fig. 9.20)	Rect. Equiv. (Fig. 9.21)	Length (l.F.)	Perim-eter (in.)	14 Ga. Alum. (lbs./l.f.)	Weight of Ductwork per Unit (lbs.)
				Supply Air			
820	.1	13	10 x 14	2 x 30 + 2 x 25 + 2 x 10 = 130	48	4	520
1,640	.1	16.5	16 x 14	2 x 30 + 2 x 5 = 70	60	5	350
2,460	.1	19.5	24 x 14	4 x 30 = 120	76	7	840
3,280	.1	22	30 x 14	2 x 30 + 2 x 35 = 130	88	7.5	975
4,100	.1	24	36 x 14	2 x 30 + 2 x 25 = 110	100	8	880
4,920	.1	25.5	42 x 14	2 x 15 = 30	112	9	270
9,020	.1	32	70 x 14	8 + 30 = 38	168	13	494
				Return Air			
4,850	.07	27	48 x 14	30	124	9.5	285
9,700	.07	35	88 x 14	30	204	15	450
14,550	.07	40	96 x 16	30	224	16	480
							5,544
							10% Fittings 554
							6098 lbs.
							Rounded to 6100

Figure 14.14

Fresh air from the roof is injected into the air handling unit along with the return air.

Air handling unit 1A has a heat load of 380 MBH. The amount of air to be distributed is determined using the following equation:

$H_v = 1.1 \text{ cfm } \Delta T$

H_v = sensible heat gain

cfm = air quantity

$\Delta T = T_{room} - T_{supply}$

The sensible load for air handling unit 1A equals the total load (380,000 Btu/hour) minus the latent load. The latent load for the building equals 134,000 Btu/hour for people, plus 6,000 Btu/hour for equipment (see Figure 14.9). Air handling unit 1A will take 1/6 of the 140,000, or 23,300 Btu/hour. Therefore,

$H_s = 380{,}000 - 23{,}300 = 356{,}700$ Btu/hr.

Because of the relatively dry air in Albuquerque, the latent load is small. Generally, approximately 25 percent of the cooling load is sensible heat. Thus,

$356{,}700 = 1.1 \text{ cfm } (78\text{–}60)$

$$\text{cfm} = \frac{356{,}700}{1.1 \times 18} = 18{,}015 \text{ cfm, rounded to } 18{,}100 \text{ cfm}$$

From the layout, there are 22 supply air diffusers, each handling approximately equal areas and loads. Thus, each diffuser must supply 18,100/22, or 820 cfm each. The cfm for each run of duct is shown in Figure 14.14.

The recommended friction loss is 0.1″/100′ for supply ducts and 0.07″/100′ for return ducts (see Figure 9.22). For the duct carrying 820 cfm, Figure 9.20 indicates a 13″ duct at 0.1″/100′. This is shown in Figure 14.14, along with other circular duct sizes. After each round duct is determined, Figure 9.21 is used to convert it to an equivalent rectangular duct. The 13″ round duct is equivalent to a 10″ x 14″ rectangular duct.

The return duct system is designed to accommodate 80 percent of the supply air, or

$18{,}100 \times 0.80 = 14{,}480$ cfm

Each of the three return air registers takes 14,480/3 = 4,826, rounded to 4,850 cfm. Return air ducts are sized for 0.07″/100′.

The ductwork for air handling unit 1A weighs approximately 6,100 pounds, which is typical for all six units. Thus, the total ductwork equals:

$6 \times 6{,}100 = 36{,}600$ lbs.

The fresh air and toilet exhaust ductwork is sized in a similar manner. The total fresh air requirement is 3,800 cfm, 1,267 cfm to each floor. The exhaust air equals 3,040 cfm, 1,013 cfm from each floor. They are sized as follows:

Location		cfm	Loss per 100′	Round Duct	Use	Length (l.f.)
Fresh air	1-2	1,267	0.1	15″	16″	14′
	2-3	2,534	0.1	20″	20″	14′
	3-roof	3,800	0.1	23″	24″	17′
Exhaust air	1-2	1,013	0.07	15″	16″	14′
	2-	2,026	0.07	20″	20″	14′
	3-roof	3,040	0.07	23″	24″	17′

Spiral preformed aluminum is selected for all ductwork.

Terminal Units

The terminal units for this building consist of the following:

- Air handling units (6)
- Supply air diffusers (11 per air handling unit)
- Return air grilles (3 per air handling unit)
- Toilet exhaust grilles (6)

Sizing for each of these components is described in the following sections.

Air Handling Units: The six air handling units are sized for 30 tons each to make them uniform and, therefore, easier to maintain. Each has a heating and a cooling coil, two-speed fan, and two connections, one for the mixture of fresh and return air, and one for supply air. The low fan speed is used for heating.

Supply Air Diffusers: Each of the 11 supply air diffusers has been designed for 820 cfm. The actual diffuser characteristics vary with each manufacturer. The design considerations for this system are to keep the exit velocity low for noise control (under 800 fpm), to select the diffuser for the proper throw (3/4 × 30/2 or ± 12′), and to keep the pressure losses through the diffuser between 0.05″ and 0.15″, if possible. A review of manufacturers' catalogs for available sizes of diffusers indicates 21″ × 21″ four-way diffusers with volume control meet these criteria. For the 6 air handling units, 66 diffusers are required.

Return Air Grilles: Eighteen return air grilles are required, three per air handling unit. Each grille is designed for 4,850 cfm. The return air is assumed to travel at 1,200 fpm (see Figure 9.18). This results in a free area of:

$$Q = VA$$
$$50 = 1{,}200 \times A$$
$$A = 4.04 \text{ s.f.}$$

The free area must be increased by fifty percent to account for the blade area of the grilles:

$$A = 4.04 \times 1.5 = 6.06 \text{ s.f.}$$

A 30″ × 30″ grille meets the area requirement. Note that a velocity of 1,200 is high and often lower values are used.

Toilet Exhaust Grilles: The total toilet room exhaust of 3,040 cfm will be removed by the exhaust grilles with capacities of 506 cfm each. The free area required is:

$$Q = VA$$
$$506 = 1{,}200 \times A$$
$$A = 0.42 \text{ s.f.}$$

The free area must be increased by fifty percent to account for the blade area of the grilles:

$$A = 0.42 \times 1.5 = 0.63 \text{ s.f.}$$

A 10″ × 10″ grille is adequate. However, use 12″ × 12″ to reduce velocity noise.

Controls

The control system for this model is more sophisticated than the controls for the apartment model (Chapter 13), because both the heating and cooling systems must be controlled; the building has occupied and unoccupied hours; there is opportunity for "free cooling"; and both piping and ductwork must be controlled.

Time Clock: A 24-hour, 7-day time clock is provided to switch equipment from occupied to unoccupied settings. During occupied hours, the equipment operates at an indoor temperature of 68°F in the winter and 78°F in the summer. During unoccupied hours (6 p.m. to 6 a.m. week nights and all weekend), the fresh air supply fan, the toilet exhaust fan, and all cooling equipment are turned off. In the winter, the night/weekend setback temperature is 55°F. The time clock has a 16-hour spring-wound reserve so that it does not need to be reset in the event of a power failure.

Summer/Winter/Off Switch: The piping to the air handling units is used for both heating or cooling. The boiler and chiller cannot run simultaneously in this layout, which is designed to save energy. The summer/winter/off switch activates the boiler and its pumps in the winter and simultaneously deactivates the chiller, air cooled condenser, and pumps. The reverse happens in the summer. The summer/winter/off switch also sets the proper temperature, and sets the fan speed of the air handling units to low in winter.

Outdoor Reset: The outdoor reset control measures the outdoor temperature and modulates the boiler water temperature accordingly to achieve maximum efficiency.

Automatic Control Valves: The hot and chilled water coil to each air handling unit has an automatic three-way valve. On demand for heating or cooling, the valve diverts water into the respective coil; otherwise, it bypasses the unit (see Figure 14.13). Twelve valves, two per unit, are required.

Return Air Sensors: The room temperature is measured by sensors in the return air ductwork. Each air handling unit has one temperature sensor tied to the control system to activate the three-way valves.

Controllers: Aquastats are used for the boiler, chiller, and air cooled condenser to modulate the internal controls. It is assumed that these come with the equipment selected.

Accessories

The HVAC system for this building has accessory devices and equipment as described below. (See Chapter 12 for more information on accessories.)

- Breeching
- Motor starters
- Vibration isolators
- Hangers and supports
- Insulation

Breeching: The breeching from the boiler to the chimney should equal at least the diameter of the breeching outlet. It is assumed that the breeching size for the boiler is ten inches. The boiler is connected to the flue and vented directly through the roof. The ten-inch diameter is verified using the following formula:

$$D^2 = \frac{H \times M \times T_f}{26\ P(atm)\ V}$$

H = 1,275 MBH (boiler size selected)
M = 1.1
T_f = 800°F
P(atm) = 29.92 inches H_g
V = 20 feet/second (assumed-conservative)

therefore,

$$D^2 = \frac{1{,}275 \times 1.1 \times 800}{26 \times 29.92 \times 20} = 72$$

$$D = 8.5'' \text{ diameter}$$

Use 10″ diameter as the next manufactured size larger than 8″ diameter. Figure 14.4 shows approximately 10 linear feet of 10″ flue.

Motor Starters: In this system, the equipment that requires motor starters are the boiler, supply fan, toilet fan, chiller, air cooled condenser, the six circulating pumps, and the air handling units. Each of these starters are automatic (magnetic) and activated by a controller. The starters are in NEMA I (normal indoor) enclosures.

Vibration Isolators: The circulating pumps are mounted so as to minimize transmission of vibrations. The pumps selected run at 1,750 rpm and weigh approximately 500 pounds, or 125 pounds per spring.

$$1{,}750 \text{ rpm} = \frac{1{,}750}{60} \text{ or } 29 \text{ Hz}$$

The isolators should have a frequency of one-third to one-fifth 29 Hz, or 8 Hz. From this, the recommended spring constant, k, can be found using the following formula:

$$k = \frac{(\text{rat. freq. of isolator (Hz)})^2 \times (\text{equipment (lbs.)})}{9.8}$$

$$k = \frac{8^2 \times 125}{9.8} = 816, \text{ rounded to } 1{,}000 \text{ lbs./in.}$$

The piping connections at the pump are isolated by flexible connecters and the entire assembly placed on a pad (see Figure 12.5). The chiller and air cooled condenser also require vibration isolation. Details vary with the specific equipment selected.

Hangers and Supports: Hangers and supports are selected for the ducts and pipes and for noise control. In the mechanical room and on all horizontal mains, the supports used are clevis-type hangers, with rods into concrete inserts. Note that the cost of hangers 10 feet on center is included in the cost of piping. Riser supports are also required.

Insulation: All piping is insulated with fiberglass insulation and ASJ jacket. This is entered on the system summary sheet (Figure 14.15). Ductwork is sound-insulated within ten feet of each air handling unit.

Summary

All components of the system — generation equipment, distribution equipment, terminal units, controls, and accessories — have now been determined. The basic system is illustrated in Figures 14.4 and 14.10. Figure 14.15 lists the major components, sizes, quantities, models, and accessories, as well as line numbers from *Means Mechanical Cost Data* in order to determine the cost of the overall system.

System Summary Sheet

Item	Size	Quantity	Type/Model	Remarks	R.S. Means Line No.
Generation					
Boiler—gas	1,275 MBH	1	Gas, hot water	Two-stage	155-115
Expansion tank	80 gal.	1	Steel, ASME	Galvanized Hot water	151-671
Expansion tank	30 gal.	1	Steel, ASME	Galvanized Chilled water	155-671
Chiller	175 tons	1	Glycol-cooled	Dual compressors Dir. drive/reciprocating	157-190
Air-cooled condenser	175 tons	1	Air-cooled	Closed circuit cooling tower	157-225
Pumps—hot water	84 gpm	2	Frame-mounted	15 ft. head High-temp. seals	152-410
Pumps—chilled water	600 gpm	2	Frame-mounted	50 ft. head Casing drain	152-410
Pumps—condenser	975 gpm	2	Frame-mounted	124 ft. head Glycol-resistant seals	152-410
Fresh air fan	3,800 cfm	1	Centrifugal	Roof curb Elec. preheat coil	157-290
Toilet exhaust fan	3,040 cfm	1	Roof exhaust	Roof curb Self-flashing curb	157-290
Distribution					
Steel pipe	3″	200 l.f.	Sched. 40	Supports and Anchors	151-701
	4″	12 l.f.	Sched. 40	″ ″	151-701
	5″	40 l.f.	Sched. 40	″ ″	151-701
	6″	70 l.f.	Sched. 40	″ ″	151-701
	8″	40 l.f.	Sched. 40	″ ″	151-701
Ductwork	Varies	36K lbs.	14 ga. alum.	Aluminum	157-250
	16″	28 l.f.	Spiral 18 ga.	Aluminum	157-250
	20″	28 l.f.	Spiral 18 ga.	Aluminum	157-250
	24″	34 l.f.	Spiral 18 ga.	Aluminum	157-250
Valves and fittings	Varies				
AHU	30 ton	6	Ceiling mount	Hot water coil	157-150
Diffusers	21″ x 21″	66	4-way	Volume damper	157-450
Grilles, office	30″ x 30″	18	Grille	Aluminum	157-460
Grilles, toilet	12″ x 12″	6	Grille	Aluminum	157-460

Item	Size	Quantity	Remarks	R.S. Means Line No.
Controls				
Time clock	6 contacts	1	24 hr.-7 day Spring wound	
Switch	Main	1	Summer/winter Manual	
Outdoor Reset	Unit	1	Automatic Adj. temp.	157-420
3-way motor-operated valves	3″	12	Automatic	157-420
Thermostat	Unit	6	Return air duct Tie to control valves	157-425
Accessories				
Breeching	10″	10 l.f.	Double wall	155-680
Motor starters	Unit	17 l.f.	NEMA 1	163-130
Vibration isolation	Unit	24 l.f.	k = 1000 lbs./in., 4 ea. pump	157-485
Flexible hose conn.	Unit	12 l.f.	Two 3″, two 6″, two 8″	156-235
Riser clamps	Unit	—	Varies, alternate floors	151-901
Pipe insulation	Varies	—	Fiberglass w/ASJ	155-651
Duct liner	1/2″	—	Liner 2 lb. density	155-651

Figure 14.15

APPENDICES

Glossary

ANSI
American National Standards Institute.

API
American Petroleum Institute.

ASHRAE
American Society of Heating, Refrigerating, and Air Conditioning.

ASME
American Society of Mechanical Engineers.

ASTM
American Society for Testing Materials.

Backing Ring
A metal ring used during the welding process. Its purpose is to prevent melted metal from entering a pipe when making a butt-weld joint, and to provide a uniform gap for welding. Often referred to as "chill rings."

BE
Bevelled end. (The end of a pipe or fitting prepared for welding.)

Bell and Spigot Joint
The most commonly used joint in cast iron soil pipe. Each length is made with an enlarged bell at one end into which the spigot end of another piece is inserted. The joint is then made tight by lead and oakum or a rubber ring caulked into the bell around the spigot.

Black Steel Pipe
Steel pipe that has not been galvanized.

Blank Flange
A flange in which the bolt holes are not drilled.

Blind Flange
A flange used to close off the end of a pipe.

Branch
The outlet or inlet of a fitting that is not in line with the run, and takes off at an angle to the run (e.g., tees, wyes, crosses, laterals, etc.).

British Thermal Unit
The heat required to heat 1 pound of water 1°F.

Building Sewer
The pipe running from the outside wall of the building drain to the public sewer.

Building Storm Drain
A drain for carrying rain, surface water, condensate, etc., to the building drain or sewer.

Bull Head Tee
A tee with the branch larger than the run.

Butt Weld Joint
A welded pipe joint made with the ends of the two pipes butting each other.

Butt Weld Pipe
Pipe welded along the seam and not lapped.

Carbon Steel Pipe
Steel pipe which owes its properties chiefly to the carbon which it contains.

Companion Flange
A pipe flange which connects with another flange or with a flanged valve or fitting. This flange differs with a flange that is an integral part of a fitting or valve.

Cooling Tower
A device for cooling water by evaporation. A natural draft cooling tower is one where the air flow through the tower is due to its natural chimney effect. A mechanical draft tower employs fans to force or induce a draft.

Couplings
Fittings for joining two pieces of pipe.

Drainage System
The piping system in a building up to where it discharges to the sewer system.

Elbow
A fitting that makes an angle in a pipe run. The angle is 90 degrees unless another angle is specified.

Equivalent Length
The resistance of a duct or pipe elbow, valve, damper, orifice, bend, fitting, or other obstruction to flow expressed in the number of feet of straight duct or pipe of the same diameter which would have the same resistance.

ERW
Electric resistance weld. (A method of welding pipe in the manufacturing process.)

Expansion Joint
A joint whose primary purpose is to absorb the longitudinal expansion and contraction in the line due to temperature changes.

Expansion Loop
A large radius loop in a pipe line which absorbs the longitudinal expansion and contraction in the line due to temperature changes.

Flange
A ring-shaped plate at right angles to the end of the pipe. It is provided with holes for bolts to allow fastening of the pipe to a similar flange.

Flat Face
Pipe flanges which have the entire face of the flange faced straight across and which use a full face gasket. These are commonly employed for pressures less than 125 pounds.

Galvanizing
Coating iron or steel surfaces with a protective layer of zinc.

Gate Valve
A valve utilizing a gate, usually wedge-shaped, which allows fluid flow when the gate is lifted from the seat. Gate valves have less resistance to flow than globe valves, and should always be used fully open or fully closed.

Globe Valve
A valve with a rounded body utilizing a manually raised or lowered disc which, when closed, seats so as to prevent fluid flow. Globe valves are ideal for throttling in a semi-closed position.

Header
A large pipe or drum into which each of a group of boilers, chillers, or pumps are connected. (see Manifold)

Horizontal Branch
In plumbing, the horizontal line from the fixture drain to the waste stack.

Hot Water Heating System
One in which hot water is the heating medium. Flow is either gravity or forced circulation.

ID
Inside diameter.

IDHA
International District Heating Association.

Insulation
Thermal insulation is a material used for covering pipes, duct, vessels, etc., to effect a reduction of heat loss or gain.

Lapped Joint
A lapped joint, like a van stone joint, is a type of pipe joint made using loose flanges on lengths of pipe. The ends of this pipe are lapped over to give a bearing surface for a gasket or metal to metal joint.

Lap Weld Pipe
Pipe made by welding along a scarfed longitudinal seam in which one part is overlapped by another.

LCL
Less carload lot. (Pipe is usually ordered from the mill by the carload, or LCL.)

Lead Joint
A joint made by pouring molten lead into the space between a bell and spigot, and making the joint tight by caulking.

Malleable Iron
Cast iron which has been heat treated to reduce its brittleness.

Manifold
A fitting with several branch outlets.

Mill Length
Also known as "random length". The usual run-of-the-mill pipe is 16 to 20 feet in length. Line pipe for power plant or oil field use is often made in double random lengths of 30 to 35 feet.

Nipple
A piece of pipe less than 12 inches long and threaded on both ends. Pipe over 12 inches long is regarded as a cut measure.

Nominal
Name given to standard pipe size designations through 12 inches nominal O.D. For example, 2" nominal is 2-3/8" O.D.

OD
Outside diameter.

O.S. & Y.
Outside screw and yoke. A valve configuration where the valve stem, having exposed external threads supported by a yoke, indicates the open or closed position of the valve.

P.E.
Plain end. (Used to describe the ends of pipe which are shipped from the mill with unfinished ends. These ends may eventually be threaded, beveled, or grooved in the field.)

Pipe
A hollow cylinder or tube for conveyance of a fluid.

Plug Valve
A valve containing a tapered plug through which a hole is drilled so that fluid can flow through when the holes line up with the inlet and the outlet, but when the plug is rotated 90 degrees, the flow is stopped.

Plumbing Fixtures
Devices which receive water, liquid, or water-borne wastes,and discharge the wastes into a drainage system.

Plumbing System
Arrangements of pipes, fixtures, fittings, valves, and traps, in a building which supply water and remove liquid borne wastes, including storm water.

Potable Water
Water suitable for human consumption.

PSI
Pounds per square inch.

Reducer
A pipe coupling with a larger size at one end than the other. The larger size is designated first. Reducers are threaded, flanged, welded, etc. Reducing couplings are available in either eccentric or concentric configurations.

Riser
A vertical pipe extending one or more floors.

Roof Drain
A fitting which collects water on the roof surface and discharges it into the leader.

Schedule Number
Schedule numbers are American Standards Association designations for classifying the strength of pipe. Schedule 40 is the most common form of steel pipe used in the mechanical trades.

Screwed Joint
A pipe joint consisting of threaded male and female parts joined together.

Seamless Pipe
Pipe or tube formed by piercing a billet of steel and then rolling, rather than having welded seams.

Service Pipe
A pipe connecting water or gas mains into a building from the street.

Slip-on Flange
A flange slipped onto the end of a pipe and then welded in place.

Socket Weld
A pipe joint made by use of a socket weld fitting which has a female end or socket for insertion of the pipe to be welded.

Stainless Steel
An alloy steel having unusual corrosion-resisting properties, usually imparted by nickel and chromium.

Storm Sewer
A sewer carrying surface or storm water from roofs or exterior surfaces of a building.

Street Elbow
An elbow with a male thread on one end, and a female thread on the other end.

Swing Joint
An arrangement of screwed fittings to allow for movement in a pipe line.

Swivel Joint
A special pipe fitting designed to be pressure tight under continuous or intermittent movement of the equipment to which it is connected.

TBE
Thread both ends. Term used when specifying or ordering cut measures of pipe.

T & C
Threaded and coupled; an ordering designation for threaded pipe.

Tee
A pipe fitting that has a side port at right angles to the run.

TOE
Thread one end. Term used when specifying or ordering cut measures of pipe.

Union
A fitting used to join pipes. It commonly consists of three pieces. Unions are extensively used, because they allow dismantling and reassembling of piping assemblies with ease and without distorting the assembly.

Van Stone
A type of joint made by using loose flanges on lengths of pipe whose ends are lapped over to give a bearing surface for the flange.

Vents
Vents are used to permit air to escape from hydronic systems, condensate receivers, fuel oil storage tanks, as a breather line for gas regulators, etc.

Vent Stack
A vertical vent pipe which provides air circulation to and from the drainage system.

Vent System
Piping which provides a flow of air to or from a drainage system to protect trap seals from siphonage or back pressure.

Welding Fittings
Wrought steel elbows, tees, reducers, saddles, and the like, beveled for butt welding to pipe. Forged fittings with hubs or with ends counter-bored for fillet welding to pipe are used for small pipe sizes and high pressures.

Welding Neck Flange
A flange with a long neck beveled for butt welding to pipe.

Wye
A pipe fitting with a side outlet that is any angle other than 90 degrees to the main run or axis.

HVAC

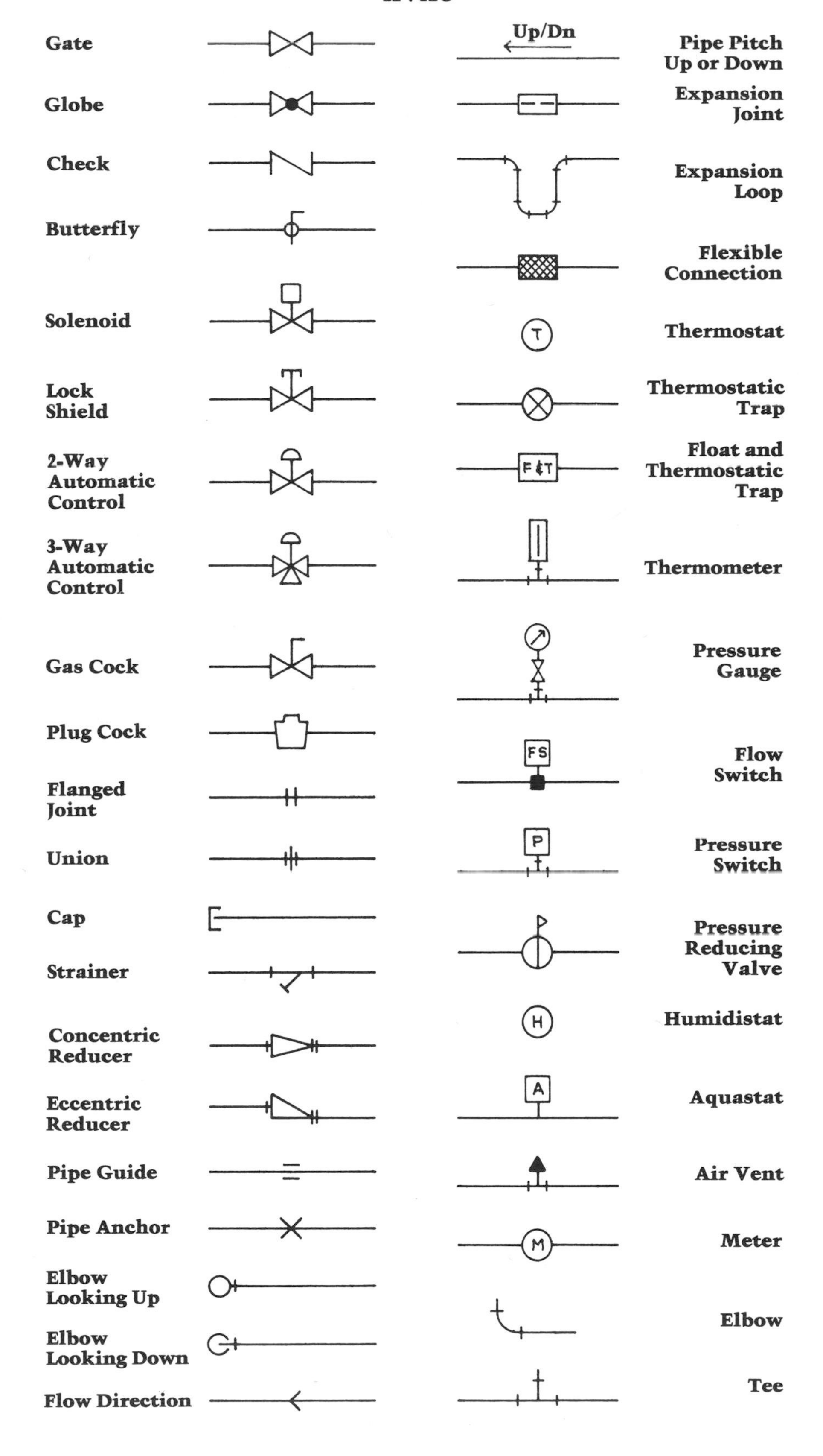
Valves, Fittings & Specialties
Gate
Globe
Check
Butterfly
Solenoid
Lock Shield
2-Way Automatic Control
3-Way Automatic Control
Gas Cock
Plug Cock
Flanged Joint
Union
Cap
Strainer
Concentric Reducer
Eccentric Reducer
Pipe Guide
Pipe Anchor
Elbow Looking Up
Elbow Looking Down
Flow Direction
Up/Dn
Pipe Pitch Up or Down
Expansion Joint
Expansion Loop
Flexible Connection
T
Thermostat
Thermostatic Trap
F&T
Float and Thermostatic Trap
Thermometer
Pressure Gauge
FS
Flow Switch
P
Pressure Switch
Pressure Reducing Valve
H
Humidistat
A
Aquastat
Air Vent
M
Meter
Elbow
Tee

Plumbing

Item	Symbol
Floor Drain	
Indirect Waste	W
Storm Drain	SD
Combination Waste & Vent	CWV
Acid Waste	AW
Acid Vent	AV
Cold Water	CW
Hot Water	HW
Drinking Water Supply	DWS
Drinking Water Return	DWR
Gas-Low Pressure	G
Gas-Medium Pressure	MG
Compressed Air	A
Vacuum	V
Vacuum Cleaning	VC
Oxygen	O
Liquid Oxygen	LOX
Liquid Petroleum Gas	LPG

HVAC

Item	Symbol
Hot Water Heating Supply	HWS
Hot Water Heating Return	HWR
Chilled Water Supply	CHWS
Chilled Water Return	CHWR
Drain Line	D
City Water	CW
Fuel Oil Supply	FOS
Fuel Oil Return	FOR
Fuel Oil Vent	FOV

HVAC (Cont.)

Symbol	Item
FOG	Fuel Oil Gauge Line
PD	Pump Discharge
	Low Pressure Condensate Return
LPS	Low Pressure Steam
MPS	Medium Pressure Steam
HPS	High Pressure Steam
BD	Boiler Blow-Down

Fire Protection

Symbol	Item
F	Fire Protection Water Supply
WSP	Wet Standpipe
DSP	Dry Standpipe
CSP	Combination Standpipe
SP	Automatic Fire Sprinkler
	Upright Fire Sprinkler Heads
	Pendent Fire Sprinkler Heads
	Fire Hydrant
	Wall Fire Dept. Connection
	Sidewalk Fire Dept. Connection
FHR	Fire Hose Rack
FHC	Surface Mounted Fire Hose Cabinet
FHC	Recessed Fire Hose Cabinet

HVAC Ductwork Symbols

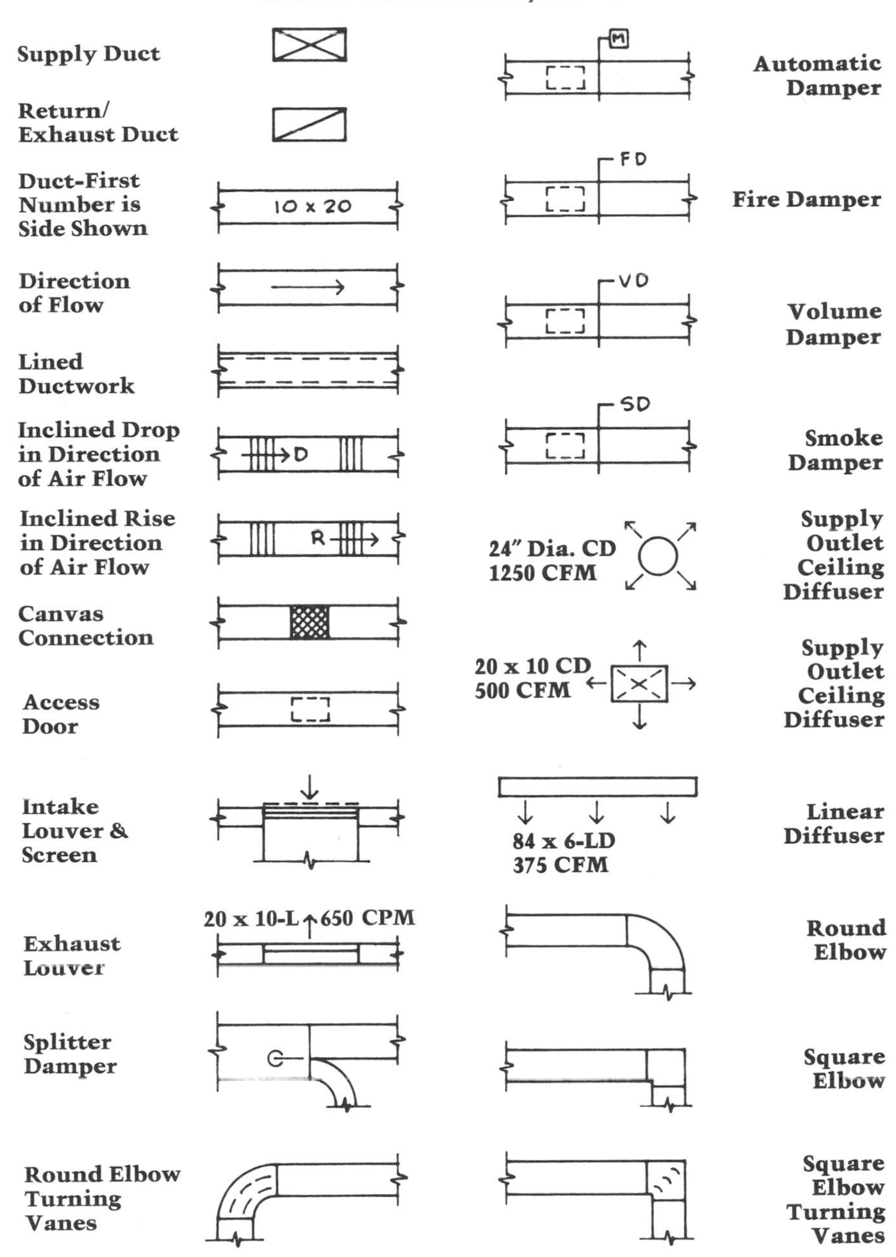

Double Duct Air System

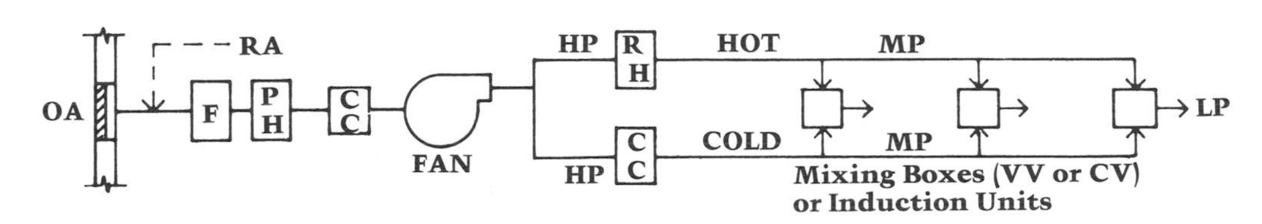

OA = Outside Air
RA = Return Air
F = Filter
PH = Preheat Coil

CC = Cooling Coil
RH = Reheat Coil
HP = High Pressure Duct
MP = Medium Pressure Duct

LP = Low Pressure Duct
VV = Variable Volume
CV = Constant Volume

Appendix C

A	Area Square Feet; Ampere
ABS	Acrylonitrile Butadiene Styrene; Asbestos Bonded Steel
A.C.	Alternating Current ; Air Conditioning; Asbestos Cement
A.C.I.	American Concrete Institute
Addit.	Additional
Adj.	Adjustable
af	Audio-frenquency
A.G.A.	American Gas Association
Agg.	Aggregate
A.H.	Ampere Hours
A hr	Ampere-hour
A.I.A.	American Institute of Architects
AIC	Ampere Interrupting Capacity
Allow.	Allowance
alt.	Altitude
Alum.	Aluminum
a.m.	ante meridiem
Amp.	Ampere
Approx.	Approximate
Apt.	Apartment
Asb.	Asbestos
A.S.B.C.	American Standard Bld. Code
Asbe.	Asbestos Worker
A.S.H.R.A.E.	American Society of Heating, Refrig. & AC Engineers
A.S.M.E.	American Society of Mechanical Engineers
A.S.T.M.	American Society for Testing and Materials
Attchmt.	Attachment
Avg.	Average
Bbl.	Barrel
B.&B.	Grade B and Better; Balled & Burlapped
B.&S.	Bell and Spigot
B.&W.	Black and White
b.c.c.	Body-centered Cubic
B.F.	Board Feet
Bg. Cem.	Bag of Cement
BHP	Brake Horse Power
B.I.	Black Iron
Bit.; Bitum.	Bituminous
Bk.	Backed
Bkrs.	Breakers
Bldg.	Building
Blk.	Block
Bm.	Beam
Boil.	Boilermaker
B.P.M.	Blows per Minute
BR	Bedroom
Brg.	Bearing
Brhe.	Bricklayer Helper
Bric.	Bricklayer
Brk.	Brick
Brng.	Bearing
Brs.	Brass
Brz.	Bronze
Bsn.	Basin
Btr.	Better
BTU	British Thermal Unit
BTUH	BTU per Hour
BX	Interlocked Armored Cable
c	Conductivity
C	Hundred; Centigrade
C/C	Center to Center
Cab.	Cabinet
Cair.	Air Tool Laborer
Calc	Calculated
Cap.	Capacity
Carp.	Carpenter
C.B.	Circuit Breaker
C.C.F.	Hundred Cubic Feet
cd	Candela
cd/sf	Candela per Square Foot
CD	Grade of Plywood Face & Back
CDX	Plywood, grade C&D, ext. glue
Cefi.	Cement Finisher
Cem.	Cement
CF	Hundred Feet
C.F.	Cubic Feet
CFM	Cubic Feet per Minute
c.g.	Center of Gravity
CHW	Commercial Hot Water
C.I.	Cast Iron
C.I.P.	Cast in Place
Circ.	Circuit
C.L.	Carload Lot
Clab.	Common Laborer
C.L.F.	Hundred Linear Feet
CLF	Current Limiting Fuse
CLP	Cross Linked Polyethylene
cm	Centimeter
CMP	Corr. Metal Pipe
C.M.U.	Concrete Masonry Unit
Col.	Column
CO_2	Carbon Dioxide
Comb.	Combination
Compr.	Compressor
Conc.	Concrete
Cont.	Continuous; Continued
Corr.	Corrugated
Cos	Cosine
Cot	Cotangent
Cov.	Cover
CPA	Control Point Adjustment
Cplg.	Coupling
C.P.M.	Critical Path Method
CPVC	Chlorinated Polyvinyl Chloride
C. Pr.	Hundred Pair
CRC	Cold Rolled Channel
Creos.	Creosote
Crpt.	Carpet & Linoleum Layer
CRT	Cathode-ray Tube
CS	Carbon Steel
Csc	Cosecant
C.S.F.	Hundred Square Feet
C.S.I.	Construction Specification Institute
C.T.	Current Transformer
CTS	Copper Tube Size
Cu	Cubic
Cu. Ft.	Cubic Foot
cw	Continuous Wave
C.W.	Cool White
Cwt.	100 Pounds
C.W.X.	Cool White Deluxe
C.Y.	Cubic Yard (27 cubic feet)
C.Y./Hr.	Cubic Yard per Hour
Cyl.	Cylinder
d	Penny (nail size)
D	Deep; Depth; Discharge
Dis.; Disch.	Discharge
Db.	Decibel
Dbl.	Double
DC	Direct Current
Demob.	Demobilization
d.f.u.	Drainage Fixture Units
D.H.	Double Hung
DHW	Domestic Hot Water
Diag.	Diagonal
Diam.	Diameter
Distrib.	Distribution
Dk.	Deck
D.L.	Dead Load; Diesel
Do.	Ditto
Dp.	Depth
D.P.S.T.	Double Pole, Single Throw
Dr.	Driver
Drink.	Drinking
D.S.	Double Strength
D.S.A.	Double Strength A Grade
D.S.B.	Double Strength B Grade
Dty.	Duty
DWV	Drain Waste Vent
DX	Deluxe White, Direct Expansion
dyn	Dyne
e	Eccentricity
E	Equipment Only; East
Ea.	Each
Econ.	Economy
EDP	Electronic Data Processing
E.D.R.	Equiv. Direct Radiation
Eq.	Equation
Elec.	Electrician; Electrical
Elev.	Elevator; Elevating
EMT	Electrical Metallic Conduit; Thin Wall Conduit
Eng.	Engine
EPDM	Ethylene Propylene Diene Monomer
Eqhv.	Equip. Oper., heavy
Eqlt.	Equip. Oper., light
Eqmd.	Equip. Oper., medium
Eqmm.	Equip. Oper., Master Mechanic
Eqol.	Equip. Oper., oilers
Equip.	Equipment
ERW	Electric Resistance Welded
Est.	Estimated
esu	Electrostatic Units
E.W.	Each Way
EWT	Entering Water Temperature
Excav.	Excavation
Exp.	Expansion
Ext.	Exterior
Extru.	Extrusion
f.	Fiber stress
F	Fahrenheit; Female; Fill
Fab.	Fabricated
FBGS	Fiberglass
F.C.	Footcandles
f.c.c.	Face-centered Cubic
f'c.	Compressive Stress in Concrete; Extreme Compressive Stress
F.E.	Front End
FEP	Fluorinated Ethylene Propylene (Teflon)
F.G.	Flat Grain
F.H.A.	Federal Housing Administration
Fig.	Figure
Fin.	Finished
Fixt.	Fixture
Fl. Oz.	Fluid Ounces
Flr.	Floor
F.M.	Frequency Modulation; Factory Mutual
Fmg.	Framing
Fndtn.	Foundation
Fori.	Foreman, inside
Foro.	Foreman, outside

Fount.	Fountain
FPM	Feet per Minute
FPT	Female Pipe Thread
Fr.	Frame
F.R.	Fire Rating
FRK	Foil Reinforced Kraft
FRP	Fiberglass Reinforced Plastic
FS	Forged Steel
FSC	Cast Body; Cast Switch Box
Ft.	Foot; Feet
Ftng.	Fitting
Ftg.	Footing
Ft. Lb.	Foot Pound
Furn.	Furniture
FVNR	Full Voltage Non Reversing
FXM	Female by Male
Fy.	Minimum Yield Stress of Steel
g	Gram
G	Gauss
Ga.	Gauge
Gal.	Gallon
Gal./Min.	Gallon Per Minute
Galv.	Galvanized
Gen.	General
Glaz.	Glazier
GPD	Gallons per Day
GPH	Gallons per Hour
GPM	Gallons per Minute
GR	Grade
Gran.	Granular
Grnd.	Ground
H	High; High Strength Bar Joist; Henry
H.C.	High Capacity
H.D.	Heavy Duty; High Density
H.D.O.	High Density Overlaid
Hdr.	Header
Hdwe.	Hardware
Help.	Helper average
HEPA	High Efficiency Particulate Air Filter
Hg	Mercury
H.O.	High Output
Horiz.	Horizontal
H.P.	Horsepower; High Pressure
H.P.F.	High Power Factor
Hr.	Hour
Hrs./Day	Hours Per Day
HSC	High Short Circuit
Ht.	Height
Htg.	Heating
Htrs.	Heaters
HVAC	Heating, Ventilating & Air Conditioning
Hvy.	Heavy
HW	Hot Water
Hyd.; Hydr.	Hydraulic
Hz.	Hertz (cycles)
I.	Moment of Inertia
I.C.	Interrupting Capacity
ID	Inside Diameter
I.D.	Inside Dimension; Identification
I.F.	Inside Frosted
I.M.C.	Intermediate Metal Conduit
In.	Inch
Incan.	Incandescent
Incl.	Included; Including
Int.	Interior
Inst.	Installation
Insul.	Insulation
I.P.	Iron Pipe
I.P.S.	Iron Pipe Size
I.P.T.	Iron Pipe Threaded
J	Joule
J.I.C.	Joint Industrial Council
K.	Thousand; Thousand Pounds
K.D.A.T.	Kiln Dried After Treatment
kg	Kilogram
kG	Kilogauss
kgf	Kilogram force
kHz	Kilohertz
Kip.	1000 Pounds
KJ	Kiljoule
K.L.	Effective Length Factor
Km	Kilometer
K.L.F.	Kips per Linear Foot
K.S.F.	Kips per Square Foot
K.S.I.	Kips per Square Inch
K.V.	Kilo Volt
K.V.A.	Kilo Volt Ampere
K.V.A.R.	Kilovar (Reactance)
KW	Kilo Watt
KWh	Kilowatt-hour
L	Labor Only; Length; Long
Lab.	Labor
lat	Latitude
Lath.	Lather
Lav.	Lavatory
lb.; #	Pound
L.B.	Load Bearing; L Conduit Body
L. & E.	Labor & Equipment
lb./hr.	Pounds per Hour
lb./L.F.	Pounds per Linear Foot
lbf/sq in.	Pound-force per Square Inch
L.C.L.	Less than Carload Lot
Ld.	Load
L.F.	Linear Foot
Lg.	Long; Length; Large
L. & H.	Light and Heat
L.H.	Long Span High Strength Bar Joist
L.J.	Long Span Standard Strength Bar Joist
L.L.	Live Load
L.L.D.	Lamp Lumen Depreciation
lm	Lumen
lm/sf	Lumen per Square Foot
lm/W	Lumen Per Watt
L.O.A.	Length Over All
log	Logarithm
L.P.	Liquefied Petroleum; Low Pressure
L.P.F.	Low Power Factor
Lt.	Light
Lt. Ga.	Light Gauge
L.T.L.	Less than Truckload Lot
Lt. Wt.	Lightweight
L.V.	Low Voltage
M	Thousand; Material; Male; Light Wall Copper
m/hr	Manhour
mA	Milliampere
Mach.	Machine
Mag. Str.	Magnetic Starter
Maint.	Maintenance
Marb.	Marble Setter
Mat.	Material
Mat'l.	Material
Max.	Maximum
MBF	Thousand Board Feet
MBH	Thousand BTU's per hr.
M.C.F.	Thousand Cubic Feet
M.C.F.M.	Thousand Cubic Feet per Minute
M.C.M.	Thousand Circular Mils
M.C.P.	Motor Circuit Protector
MD	Medium Duty
M.D.O.	Medium Density Overlaid
Med.	Medium
MF	Thousand Feet
M.F.B.M.	Thousand Feet Board Measure
Mfg.	Manufacturing
Mfrs.	Manufacturers
mg	Milligram
MGD	Million Gallons per Day
MGPH	Thousand Gallons per Hour
MH	Manhole; Metal Halide; Man Hour
MHz	Megahertz
Mi.	Mile
MI	Malleable Iron; Mineral Insulated
mm	Millimeter
Mill.	Millwright
Min.	Minimum
Misc.	Miscellaneous
ml	Milliliter
M.L.F.	Thousand Linear Feet
Mo.	Month
Mobil.	Mobilization
Mog.	Mogul Base
MPH	Miles per Hour
MPT	Male Pipe Thread
MRT	Mile Round Trip
ms	millisecond
M.S.F.	Thousand Square Feet
Mstz.	Mosaic & Terrazzo Worker
M.S.Y.	Thousand Square Yards
Mtd.	Mounted
Mthe.	Mosaic & Terrazzo Helper
Mtng.	Mounting
Mult.	Multi; Multiply
MVAR	Million Volt Amp Reactance
MV	Megavolt
MW	Megawatt
MXM	Male by Male
MYD	Thousand yards
N	Natural; North
nA	nanoampere
NA	Not Available; Not Applicable
N.B.C.	National Building Code
NC	Normally Closed
N.E.M.A.	National Electrical Manufacturers Association
NEHB	Bolted Circuit Breaker to 600V.
N.L.B.	Non-Load-Bearing
nm	nanometer
No.	Number
NO	Normally Open
N.O.C.	Not Otherwise Classified
Nose.	Nosing
N.P.T.	National Pipe Thread
NQOB	Bolted Circuit Breaker to 240V.
N.R.C.	Noise Reduction Coefficient
N.R.S.	Non Rising Stem
ns	nanosecond
nW	nanowatt
OB	Opposing Blade
OC	On Center
OD	Outside Diameter
O.D.	Outside Dimension
ODS	Overhead Distribution System
O & P	Overhead and Profit
Oper.	Operator
Opng.	Opening
Orna.	Ornamental
O.S.&Y.	Outside Screw and Yoke
Ovhd	Overhead

Oz.	Ounce
P.	Pole; Applied Load; Projection
p.	Page
Pape.	Paperhanger
PAR	Weatherproof Reflector
Pc.	Piece
P.C.	Portland Cement; Power Connector
P.C.F.	Pounds per Cubic Foot
P.E.	Professional Engineer; Porcelain Enamel; Polyethylene; Plain End
Perf.	Perforated
Ph.	Phase
P.I.	Pressure Injected
Pile.	Pile Driver
Pkg.	Package
Pl.	Plate
Plah.	Plasterer Helper
Plas.	Plasterer
Pluh.	Plumbers Helper
Plum.	Plumber
Ply.	Plywood
p.m.	Post Meridiem
Pord.	Painter, Ordinary
pp	Pages
PP; PPL	Polypropylene
P.P.M.	Parts per Million
Pr.	Pair
Prefab.	Prefabricated
Prefin.	Prefinished
Prop.	Propelled
PSF; psf	Pounds per Square Foot
PSI; psi	Pounds per Square Inch
PSIG	Pounds per Square Inch Gauge
PSP	Plastic Sewer Pipe
Pspr.	Painter, Spray
Psst.	Painter, Structural Steel
P.T.	Potential Transformer
P. & T.	Pressure & Temperature
Ptd.	Painted
Ptns.	Partitions
Pu	Ultimate Load
PVC	Polyvinyl Chloride
Pvmt.	Pavement
Pwr.	Power
Q	Quantity Heat Flow
Quan.; Qty.	Quantity
Q.C.	Quick Coupling
r	Radius of Gyration
R	Resistance
R.C.P.	Reinforced Concrete Pipe
Rect.	Rectangle
Reg.	Regular
Reinf.	Reinforced
Req'd.	Required
Resi	Residential
Rgh.	Rough
R.H.W.	Rubber, Heat & Water Resistant Residential Hot Water
rms	Root Mean Square
Rnd.	Round
Rodm.	Rodman
Rofc.	Roofer, Composition
Rofp.	Roofer, Precast
Rohe.	Roofer Helpers (Composition)
Rots.	Roofer, Tile & Slate
R.O.W.	Right of Way
RPM	Revolutions per Minute
R.R.	Direct Burial Feeder Conduit
R.S.	Rapid Start
RT	Round Trip
S.	Suction; Single Entrance; South
Scaf.	Scaffold
Sch.; Sched.	Schedule
S.C.R.	Modular Brick
S.D.R.	Standard Dimension Ratio
S.E.	Surfaced Edge
S.E.R.; S.E.U.	Service Entrance Cable
S.F.	Square Foot
S.F.C.A.	Square Foot Contact Area
S.F.G.	Square Foot of Ground
S.F. Hor.	Square Foot Horizontal
S.F.R.	Square Feet of Radiation
S.F.Shlf.	Square Foot of Shelf
S4S	Surface 4 Sides
Shee.	Sheet Metal Worker
Sin.	Sine
Skwk.	Skilled Worker
SL	Saran Lined
S.L.	Slimline
Sldr.	Solder
S.N.	Solid Neutral
S.P.	Static Pressure; Single Pole; Self Propelled
Spri.	Sprinkler Installer
Sq.	Square; 100 square feet
S.P.D.T.	Single Pole, Double Throw
S.P.S.T.	Single Pole, Single Throw
SPT	Standard Pipe Thread
Sq. Hd.	Square Head
S.S.	Single Strength; Stainless Steel
S.S.B.	Single Strength B Grade
Sswk.	Structural Steel Worker
Sswl.	Structural Steel Welder
St.; Stl.	Steel
S.T.C.	Sound Transmission Coefficient
Std.	Standard
STP	Standard Temp. & Pressure
Stpi.	Steamfitter, Pipefitter
Str.	Strength; Starter; Straight
Strd.	Stranded
Struct.	Structural
Sty.	Story
Subj.	Subject
Subs.	Subcontractors
Surf.	Surface
Sw.	Switch
Swbd.	Switchboard
S.Y.	Square Yard
Syn.	Synthetic
Sys.	System
t.	Thickness
T	Temperature; Ton
Tan	Tangent
T.C.	Terra Cotta
T.D.	Temperature Difference
TFE	Tetrafluoroethylene (Teflon)
T. & G.	Tongue & Groove; Tar & Gravel
Th.; Thk.	Thick
Thn.	Thin
Thrded	Threaded
Tilf.	Tile Layer Floor
Tilh.	Tile Layer Helper
THW.	Insulated Strand Wire
THWN; THHN	Nylon Jacketed Wire
T.L.	Truckload
Tot.	Total
T.S.	Trigger Start
Tr.	Trade
Transf.	Transformer
Trhv.	Truck Driver, Heavy
Trir.	Trailer
Trlt.	Truck Driver, Light
TV	Television
T.W.	Thermoplastic Water Resistant Wire
UCI	Uniform Construction Index
UF	Underground Feeder
U.H.F.	Ultra High Frequency
U.L.	Underwriters Laboratory
Unfin.	Unfinished
URD	Underground Residential Distribution
V	Volt
VA	Volt/amp
V.A.T.	Vinyl Asbestos Tile
VAV	Variable Air Volume
Vent.	Ventilating
Vert.	Vertical
V.G.	Vertical Grain
V.H.F.	Very High Frequency
VHO	Very High Output
Vib.	Vibrating
V.L.F.	Vertical Linear Foot
Vol.	Volume
W	Wire; Watt; Wide; West
w/	With
W.C.	Water Column; Water Closet
W.F.	Wide Flange
W.G.	Water Gauge
Wldg.	Welding
Wrck.	Wrecker
W.S.P.	Water, Steam, Petroleum
WT, Wt.	Weight
WWF	Welded Wire Fabric
XFMR	Transformer
XHD	Extra Heavy Duty
Y	Wye
yd	Yard
yr	Year
Δ	Delta
%	Percent
~	Approximately
Ø	Phase
@	At
#	Pound; Number
<	Less Than
>	Greater Than

INDEX

D

E

I

L

M

N

W

Z